Topics in Applied Physics   Volume 53

Topics in Applied Physics Founded by Helmut K. V. Lotsch

Glassy Metals II

Atomic Structure and Dynamics, Electronic Structure, Magnetic Properties

Edited by H. Beck and H.-J. Güntherodt

With Contributions by   M. von Allmen   H. Beck
P. Chaudhari   J. Durand   P. H. Gaskell   U. Gonser
H.-J. Güntherodt   R. Harris   H.-U. Künzi
P. Oelhafen   R. Preston   H. Rudin   J. F. Sadoc
F. Spaepen   P. J. Steinhardt   J. O. Strom-Olsen
J.-B. Suck   C. N. J. Wagner

With 186 Figures

Springer-Verlag
Berlin Heidelberg GmbH 1983

Professor Dr. *Hans Beck*

Institut de Physique, Université de Neuchâtel, 1, Rue A.-L. Brequet,
CH-2000 Neuchâtel, Switzerland

Professor Dr. *Hans-Joachim Güntherodt*

Institut für Physik, Universität Basel, Klingelbergstraße 82,
CH-4056 Basel, Switzerland

ISBN 978-3-662-31171-4   ISBN 978-3-540-38733-6 (eBook)
DOI 10.1007/978-3-540-38733-6

2153/3130-543210

Preface

The quantitative treatment of the physical properties of systems which do not have long-range periodicity is an outstanding challenge in modern fundamental research in solid-state physics. Metallic glasses offer the unique possibility to study the disordered state of matter with metallic properties down to the very lowest temperature. The experimental results obtained for various physical quantities have greatly stimulated theoretical work in this domain.

On the other hand, some unusual properties of glassy metals, such as high mechanical ductility, high magnetic permeability, low coercive forces, temperature-independent resistivities and others, make them suitable for industrial applications. Let us mention foils for magnetic shielding, which are already available, and applications for transformer materials and tape-recorder heads, which will soon be on the market. Thus not only physicists, but also metallurgists, surface chemists, metal engineers and representatives of other branches of science and development have a strong interest in these materials.

In this volume various experimentalists and theoreticians active in the field present the state of the art in the following domains of the physics of metallic glasses:

— Structure determination (J. F. Sadoc and C. N. J. Wagner) and modelling (P. H. Gaskell)
— Mössbauer spectroscopy (U. Gonser and R. Preston)
— Mechanical properties (H.-U. Künzi), defects (P. Chaudhari, F. Spaepen, and P. J. Steinhardt), and vibrational dynamics (J.-B. Suck and H. Rudin)
— Laser quenching (M. von Allmen)
— Electron spectroscopy (P. Oelhafen) and electronic transport at low temperature (R. Harris and J. O. Strom-Olsen)
— Magnetic properties (J. Durand).

A more detailed characterisation of the various contributions is given in the introductory chapter.

It is a pleasure to thank the authors of these contributions for their collaboration and their effort to review their domain of research as thoroughly and concisely as possible.

Neuchâtel, Basel, April 1983 *H. Beck · H.-J. Güntherodt*

Contents

Contributors

Allmen, Martin von
  Institut für Angewandte Physik, Universität Bern, Sidlerstraße 5,
  CH-3012 Bern, Switzerland

Beck, Hans
  Institut de Physique, Université de Neuchâtel, 1, Rue A.-L. Brequet,
  CH-2000 Neuchâtel, Switzerland

Chaudhari, Praveen
  IBM Thomas J. Watson Research Center, P.O. Box 218,
  Yorktown Heights, NY 10598, USA

Durand, Jacques
  Université de Nancy I, Laboratoire de Physique du Solide, B.P. No. 239,
  F-54506 Vandoevre-Les-Nancy-Cedex, France

Gaskell, Philip H.
  Cavendish Laboratory, University of Cambridge, Madingley Road,
  Cambridge CB3 OHE, U.K.

Gonser, Ulrich
  Fachbereich Angewandte Physik, Universität des Saarlandes,
  D-6600 Saarbrücken, Fed. Rep. of Germany

Güntherodt, Hans-Joachim
  Institut für Physik, Universität Basel, Klingelbergstraße 82,
  CH-4056 Basel, Switzerland

Harris, Richard
  Department of Physics, Ernest Rutherford Physics Bld., McGill University,
  3600 University Street, Montreal PQ, H3A 2T8, Canada

Künzi, Hans-Ulrich
  Département des Matériaux, Ecole Polytechnique Fédérale de Lausanne,
  34, ch. de Bellerive, CH-1007 Lausanne, Switzerland

Oelhafen, Peter

Institut für Physik, Universität Basel, Klingelbergstraße 82,
CH-4056 Basel, Switzerland

Preston, Richard

Department of Physics, Northern Illinois University,
DeKalb, IL 60115, USA

Rudin, Hermann

Institut für Physik, Universität Basel, Klingelbergstraße 82,
CH-4056 Basel, Switzerland

Sadoc, Jean François

Université de Paris-Sud, Centre d'Orsay, Lab. de Physique des Solides,
F-91405 Orsay, France

Spaepen, Frans

Division of Applied Sciences, Harvard University,
Cambridge, MA 02138, USA

Steinhardt, Paul J.

University of Pennsylvania, Department of Physics E1,
Philadelphia, PA 19104, USA

Strom-Olsen, John O.

Department of Physics, Ernest Rutherford Physics Bld., McGill University,
3600 University Street, Montreal PQ, H3A 2T8, Canada

Suck, Jens-Boie

Inst. Max von Laue–Paul Langevin, 156 Avenue des Martyrs,
F-38042 Grenoble Cedex, France

Wagner, Christian N. J.

Materials Science and Engineering Department, University of California,
Los Angeles, CA 90024, USA

1. Introduction

H. Beck and H.-J. Güntherodt

Glassy Metals I, the first volume on metallic glasses in this series [1.1], has appeared in spring 1981. In the mean time the fourth international conference on rapidly quenched metals, RQ 4, has taken place in Japan in summer 1981. Over 500 participants from industrial and university laboratories participated at this meeting. This respectable number and the continuing flow of publications on various aspects of metallic glasses and amorphous metals in general demonstrate the importance of this domain, both in fundamental and applied research.

The general framework of the physics of glassy metals has been presented in the introduction of [1.1], where the main preparation methods and various experimental and theoretical approaches to the characterization and understanding of their physical properties have been sketched. The contributions to this second volume can also be placed in this general scheme and they complete the detailed description of the general program put forward in the introduction of [1.1].

The main emphasis of this book is on various aspects of atomic structure and dynamics and on electronic structure. There is some overlap between the topics of both volumes. Electronic transport, for example, has been discussed in detail in [1.1] and is taken up again here. Moreover, [1.1] contained already some contributions concerning structural aspects, whereas structural models and further techniques for determining ionic structures are analyzed in detail in the present book. These overlaps could hardly be avoided – and we hope our readers will not find it too inconvenient – since it is sign that the field of metallic glasses is still rapidly developing. It is easy to present a well-ordered, systematic overview about a domain of research which is well established and where the development is essentially restricted to refine some details. Here it seemed to us to be more important to present the different topics on as up-to-date a level as possible.

Actually the field of metallic glasses is still expanding. This incited us to plan a third volume which will include such subjects as corrosion, theory of electronic structure and applications which were already scheduled for the present volume. Moreover there are other exciting new developments which will be covered.

In the following we give a short overview over the various contributions, grouped together under three main aspects:

1.1 Atomic Structure and Dynamics

Information about the arrangement of the ions in an amorphous material can be obtained by various scattering techniques – measuring structure factors in Fourier space – or by "local" methods – probing the environment of a given ion in real space. None of these experiments really yields a full three-dimensional picture of the way the ions are arranged (say, the coordinates of all ions in a reasonably large domain). Therefore, it is important to complement such experimental structure information by developing model structures (Chap. 2 by P. H. Gaskell), which are built by the computer on the basis of some structural and topological restrictions. The validity of such a model structure can then be assessed by comparing its predictions for various correlation functions etc. with experimental data.

J. F. Sadoc and C. N. J. Wagner (Chap. 3) summarize experimental and theoretical problems arising in x-ray and neutron diffraction experiments and present results on the static (equilibrium) structure of various glassy metals obtained this way, including partial structure factors, information on local anisotropy and the order of magnetic alloys.

In contrast to this, Mössbauer spectroscopy (Chap. 4 by U. Gonser and R. S. Preston) is an important "local" tool in order to elucidate the atomic arrangments and magnetic structures of condensed matter. Fortunately, ^{57}Fe – the most suitable nucleus for this technique – is an isotope of iron, which is a constituent of the Fe–B family and is important for applications.

The fifth chapter by P. Chaudhari, F. Spaepen and P. J. Steinhardt – devoted to defects and atomic transport – deals will static and dynamical structural aspects. It is obvious that the very definition and classification of defects and their motion, giving rise to atomic transport, in a disordered system is difficult, since there is no regular ideal structure to which defects could be contrasted. In this context the possible identification of dislocations in an amorphous medium appears specially interesting.

Mechanical properties (Chap. 6 by H. U. Künzi) are an immediate consequence of the microscopic interactions and the defect structure on a macroscopic level. The emphasis of this chapter is on magnetoelastic behaviour, which is also of interest for applications, and on the "anelasticity" of metallic glasses. The temperature and frequency dependence of the elastic response can reveal insight into various microscopic relaxation processes.

Recent inelastic neutron scattering experiments (Chap. 7 by J. B. Suck and H. Rudin) have revealed interesting information on the vibrational motion of the ions in a metallic glass. Obviously, such spectra, taken on systems without long-range order, are not easy to analyze and to interpret. Nevertheless, some special features of the dispersion of collective modes and of the vibrational density of states have been identified successfully.

As explained in [1.1] glassy metals are prepared by rapid cooling of the corresponding melt. Besides the traditional "splat-cooling" and "melt-spinning" techniques, more recently laser pulses have been used to heat a metallic surface locally which is then cooled by heat conduction into the bulk of the sample. In Chap. 8 M. Von Allmen shows the application of this technique to the production of metallic glasses. Since quench rates of up to 10^{10} K/s can be achieved, a wider class of metallic material is amenable to be produced in the amorphous state like this.

1.2 Electronic Structure and Transport

Many physical properties, such as electronic specific heat, ferromagnetism, superconductivity etc., depend in one way or another on the electronic density of states. Experimental results on this quantity obtained by electron spectroscopy for various families of glassy metal alloys were reviewed by P. Oelhafen in Chap. 9. The implication of electronic structure on questions like bonding and stability are also investigated.

Then the subject of low temperature electronic transport, which is in a rather controversial state, is taken up again (Chap. 10 by R. Harris and J. O. Strom-Olsen). The most striking feature seems to be the negative $\ln T$ dependence of the resistivity found in many metallic glasses at very low T. It is still not settled wether this Kondo-like behaviour is of structural or magnetic origin.

1.3 Magnetic Properties

This is a very broad subject, equally important from the point of view of basic research as from the point of view of various technological applications of glassy metals, owing to the novel and exceptional magnetic behaviours of some such alloys. Chapter 11 by J. Durand is mainly concerned with basic magnetism, more precisely with the use of magnetic investigations as a tool to developing a better understanding of the atomic structure of amorphous metals. Moreover a summary of the most salient magnetic features of transition metal base and of rare-earth base amorphous alloys is given. Some of the intriguing problems of crystalline magnetism such as spin glass behaviour or the question of "localized" versus "itinerant" behaviour, are of course of equal importance in amorphous magnetism.

Reference

1.1 H.-J. Güntherodt, H. Beck (eds.): *Glassy Metals I*, Topics Appl. Phys. Vol. 46 (Springer, Berlin, Heidelberg, New York 1981)

2. Models for the Structure of Amorphous Metals

P. H. Gaskell

With 21 Figures

The art of modelling the structure of glassy metals is central to any understanding of their microscopic properties. It seems fairly safe to assume that no single experimental structural technique, nor perhaps any combination, will in the near future provide a sufficiently large bank of information that we can "solve" the structure in the sense that the structures of crystals have been solved. The reason is that, since glasses have aperiodic, isotropic structures, experimental data are averaged (scrambled might be a better description). The experimentalist and theoretician alike are therefore presented with a set of figures which can, and often does, satisfy several alternative structural descriptions.

In order to proceed at all it is necessary to be prepared to construct atomic models – either physically or in a computer – to calculate the structural properties which can be obtained experimentally [structure factor, $S(Q)$, radial distribution function, RDF] and then to compare experimental and calculated results. Moreover, since data are so limited, it is vital that postulated models and experimental facts are compared in as much detail as possible. It will generally be necessary to compare the predictions of a given model with the results of at least two experimental techniques – x-ray, neutron scattering, EXAFS, for example – in order to obtain even a partial answer to questions concerning the *local* arrangement of atoms, that is, distances to nearest neighbours, coordination numbers, etc. In order to proceed further, it may be important to establish the *symmetry* of the environment around a given atom, for which NMR or Mössbauer measurements provide a clue. Information on more distant correlations – in the range 1–2 nm – will generally require a further set of techniques – small angle x-ray or neutron scattering, high-resolution electron microscopy, for example. Validation of a model or even the task of establishing just what structural information exists for a particular glass entails a search for, and analysis of, data from several techniques. Moreover, it is unusual for sufficient information to exist for a single alloy and it is important that information for a particular alloy is compared with measurements obtained on chemically or compositionally related alloys. For example, studies of binary alloys over a wide composition range or of a series of alloys of transition metals with several, different, metalloids will often prove necessary.

Any correlation between experimental and theoretical (for want of a better word) structural data must be established in some detail. Often the *general* features of the RDFs for models of binary alloys reproduce those found by experiment. Such functions are, however, relatively insensitive to variations in structural details beyond nearest neighbours and, if further information is to be obtained, simulated and experimental data should agree to the precision of measurement or of computation. This often places a constraint on the extent to which conclusions can be drawn, either because of the known errors in experimental measurements or the known inadequacies of the modelling technique.

An account of models for amorphous metals must include, therefore, a critical *analysis* of the results of a particular technique, and *synthesis* of such data with information from other techniques or from compositionally-related alloys. This part of the study, in addition to building up a body of results to be interpreted, will suggest likely explanations and impose constraints on the range of possible solutions.

The strategy for this chapter is to begin by describing the various conceptual notions which have guided modelling studies of amorphous metals over the brief history of the subject. We begin with the structures which represent the extremes of the spectrum – the so-called dense random-packed hard sphere (DRPHS) and microcrystallite models. Then, recognising that these *are* extremes and do not necessarily represent reality, we introduce structures in which the order extends to the nearest-neighbour shell – and perhaps beyond – the so-called *designed* or *stereochemically-defined* models.

In Sect. 2.2 we describe the evidence provided by experiment. Emphasis is placed on the results themselves rather than the techniques by which they are obtained (other chapters in this series deal with such points adequately). However, it is not possible to neglect the details of the experimental procedure, particularly the effect of errors, convolution with thermal and termination broadening functions, etc.

Section 2.3 deals with the techniques used in constructing and evaluating models. Comparison of the properties of models with experimental data occupies Sect. 2.4 and finally, an attempt is made to present a view of the state of structural knowledge – what is known and what lies undiscovered – together with some suggestions for new avenues of research.

Attention will be confined to a particular family of amorphous metals, the alloys of certain transition metals with the so-called metalloid elements. These transition metal-metalloid (T–M) glasses are currently the most important series of metallic glasses, both scientifically and technologically. Moreover, since they have been more thoroughly investigated than others, experimental data are more complete and informative. Occasional discussion of alloys of two or more transition metals is appropriate but the scientifically fascinating alloys of simple metals, e.g., Mg–Zn, Ca–Al will not be discussed as they have already featured in an earlier article in this series by *Hafner* [2.1].

2.1 Conceptual Models for Amorphous Metals

2.1.1 Microcrystallite Models

Since the measured structure factors $S(Q)$ for many amorphous metals show some similarities with data for corresponding crystalline alloys, it is natural to ask whether the properties of the disordered phase can be represented by an assembly of microcrystals. Differences resulting from an absence of long-range periodicity can be the result of the small size of the (well-defined) crystals, random orientation, strain introduced by mismatch of lattices at grain boundaries, presence of a disordered interfacial region and defects such as dislocations, disclinations, or stacking faults within the microcrystallites. *Dixmier* et al. [2.2] suggested that amorphous Ni–P alloys could be represented by a model in which Ni atoms (which make the major contribution to X-ray scattering) lie in close-packed hexagonal planes arranged irregularly in packets of dimensions less than 2 nm. In this way, they derived an x-ray intensity curve which qualitatively agrees with experiment. *Cargill* [2.3] re-examined this and several other microcrystalline models and concluded that simple random-stacking models were unable to reproduce all the experimental evidence known at that time. He calculated the effect on the structure factor of Ni, of crystallite size, stacking fault density, uniform dilatation and lattice symmetry and was unable to fit calculated values of $S(Q)$ with his experimental measurements. Furthermore, it was not possible to reconcile the strain associated with grain boundaries with the small density difference between amorphous and crystalline materials nor the elastic strain necessary to broaden $S(Q)$ with the heat of crystallisation of a-Ni_3P. More promising but still inadequate results were obtained using a microcrystallite model based on the Ni_3P structure.

Subsequent investigations of simple microcrystallite models have fared little better. The problem centres on the nature of the interface between crystallites. An upper limit for microcrystallites, set by the character of diffraction data, is 1.5–2 nm. In a cube of this size, about two thirds of the atoms lie within one atom diameter of the surface. Unless the interfacial structure can be specified, therefore, radial distribution functions beyond first neighbours are quite meaningless. Moreover, the excess enthalpy of typical glasses is relatively small – only a few percent of the cohesive energy and strains would therefore be of this magnitude. The value of the strain obtained from the breadth of the distribution of nearest-neighbour distances is about 5%. Consequently, the energy associated with interfaces cannot be large: broken, or even extensively distorted bonds would incur a large energy penalty. High energies associated with an interface between crystallites would also lead to instability with respect to crystal growth by Ostwald ripening. We must therefore assume that the range of interfacial structures is limited to boundaries such as stacking faults, twin planes, etc., where the energy penalty is tolerably small.

With such constraints the model cannot be described as a *simple* microcrystalline assembly. Where aperiodic packing involving twin boundaries has been studied in polytetrahedral models for amorphous semiconductors, for example [2.4], the resulting structure is better described as an amorphous cluster (Sect. 2.1.3).

2.1.2 Dense Random-Packed Hard Sphere Models

Pioneering experiments by *Bernal* [2.5] on the structure of monoatomic liquids have been immensely influential in directing subsequent investigations of the structure of binary and ternary alloys. Bernal's early experiments involved ball bearings packed into rubber bladders, kneaded, set in black paint with positions painstakingly analysed by hand and eye. These experiments have been replaced by several generations of computer-simulated models without, until recently at least, any step forward comparable to that taken by the first investigators. Bernal and his coworkers were able to show that it is possible to produce a randomly-packed model which is sufficiently dense to reproduce experiment and contains no crystalline regions, provided constraints introduced by a regular surface are avoided. The coordination of each atom is found to be variable, with the number of neighbours ranging from 5 to 11.

Bernal analysed the shape of the cavities defined by nearest neighbours and showed that there are five, and only five, polyhedra with equal triangular faces (deltahedral) which are small enough not to admit another sphere of the same size. Bernal's "canonical holes" are shown in Fig. 2.1a.

Bernal showed that an indefinite number of structures, with almost equal nearest-neighbour distances, could be formed by packing the five deltahedra together. Studies of packed (foam) models of these polyhedra indicated that it is possible to fit them together in at least 197 different ways. The relative proportion of each polyhedron was established and this showed a strong preference for tetrahedra and half octahedra (48 % and 27 % by volume, respectively). The larger cavities (trigonal prisms 8 %, Archimedean antiprisms 2 % and tetragonal dodecahedra 15 %) appear to be wrapped around by a network of the smaller polyhedra.

Work by *Scott* [2.6] using similar techniques (ball-bearings coated with wax rather than paint and an optical method of measurement) produced a model which proved to have remarkably similar properties to *Bernal*'s. Further developments followed the construction by *Finney* [2.7] of a large (7994-atom) model with a packing density, $\eta = 0.6366 \pm 0.0004$. With this model it was possible to calculate an accurate RDF (Fig. 2.1b).

The connection between structural models for liquids and experimental data for amorphous metallic solids [2.8] became more evident when *Cargill* [2.9] compared his RDF for a-$Ni_{76}P_{24}$ with the predictions of Finney's model, Fig. 2.1. Treating the hard sphere radius as the only adjustable parameter, the fit between the two was considered to be encouragingly good. Peaks agree in position and the splitting of the second-neighbour distribution is reproduced.

It is difficult to overestimate the effects these pioneering investigations of sphere packing have had on the subsequent development of structural studies

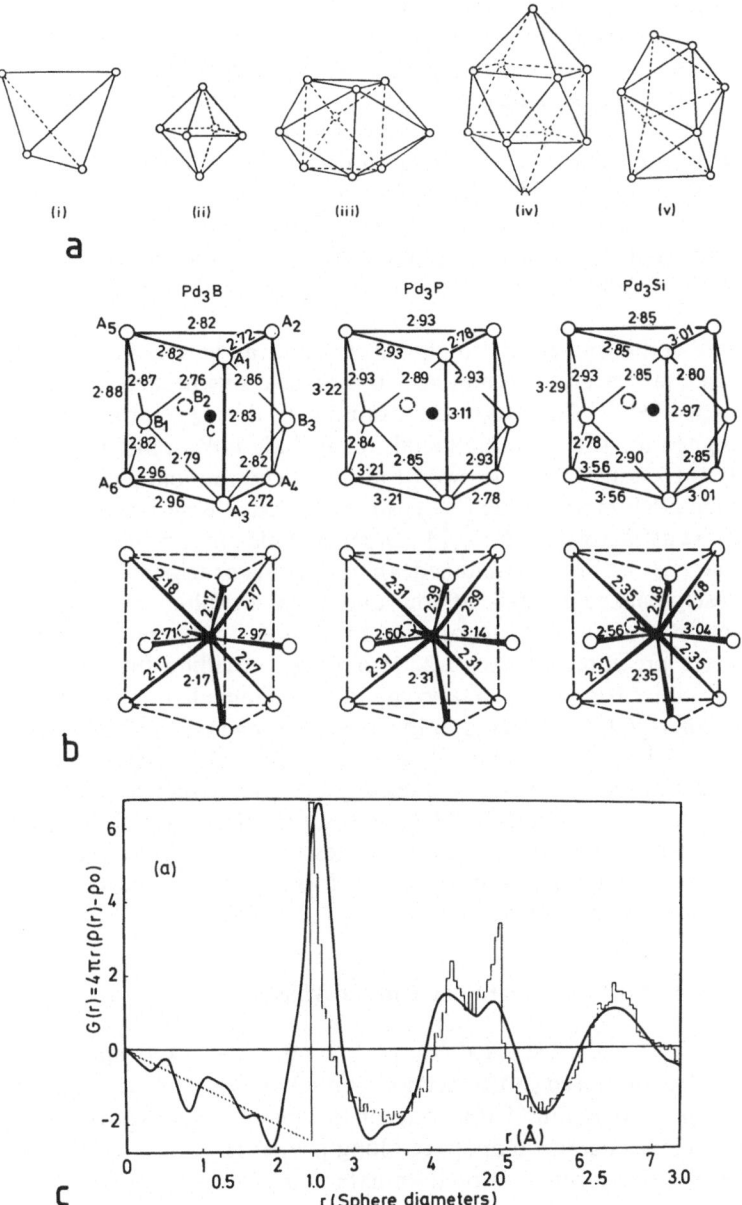

Fig. 2.1. (a) Polyhedra formed by packing equal spheres; after *Bernal* [2.5]; (i) tetrahedron, (ii) octahedron, (iii) trigonal prisms "capped" with three half octahedra, (iv) Archimedean antiprisms capped with two half-octahedra, (v) tetragonal dodecahedron. (b) Distorted trigonal prisms found in three crystalline phases related to the cementite (Fe_3C) structure with interatomic distances in Å (after *Aronsson* and *Rundqvist* [2.73]). (c) Comparison of the reduced radial distribution functions $G(r)$ for Finney's 7994-atom DRPHS model [2.7] and experimental data for a-$Ni_{76}P_{24}$ [2.2]

of polyatomic metallic glasses. *Finney* et al. [2.10] note that the presentation of the RDF of the hard sphere model, together with Cargill's Ni–P data, at the same session of the 8th Congress of the International Union of Crystallography, marked a critical turning point in the development of the subject. The framework of thought, initially dominated by crystallographic concepts, was transformed and "dense random-packing models became not only respectable but fashionable". (It would not be unreasonable to suggest that, by majority consent if for no other reason, dense random packings have come to be regarded as the idealised structure for amorphous metals, bearing the same relation to the real structure as idealised crystal structures do to real crystals).

The influence of this model on subsequent thought can be seen in the extent to which its essential features are retained in modifications needed to accommodate new knowledge. An important example is the extension of DRP structures, designed for monoatomic metals, to include polyatomic materials. One possibility is to assign atoms of two or more species randomly to the sites of a monoatomic DRP model. *Polk* [2.12] suggested an alternative, which has proved to be more attractive for the T–M glasses, in which the smaller M atoms occupy the larger Bernal holes (Archimedean antiprisms or trigonal prisms) in densely or loosely-packed models at low and high metalloid contents, respectively. Each M atom is thus surrounded by T atoms only and this agrees with the evidence presented in Sect. 2.2. Moreover, providing the metalloid preferentially occupies trigonal prismatic cavities, the immediate environment of the atom is similar to that observed in the crystalline phase (Fig. 2.1b). Using Bernal's estimate of the number of large holes in a *dense* random-packed model, *Polk* suggested that if metalloids were to occupy the three larger cavities, then this would provide a composition $T_{80}M_{20}$, that is, near the centre of the glass-forming range in these alloys. This latter suggestion has been criticised by several workers; re-examination of hole sizes in *dense* random models suggests that Polk's estimate was over-optimistic.

2.1.3 Amorphous or Noncrystallographic Cluster Models

In contrast to the models discussed above – randomly-packed spheres or crystallites – models discussed in this section contain an element of structural selection or *design*. If it is found that one of the many groupings of atoms possesses advantageous energetic or space-filling properties, it may be reasonable to assume that this group will predominate in a given structure.

a) Noncrystallographic Clusters Based on Tetrahedral Close Packing

Clusters containing a symmetry element inconsistent with translational periodicity and therefore with a crystal lattice are produced by repeated local tetrahedral close packing. Figure 2.2 shows examples of the generation of structures with pentagonal symmetry introduced by sharing triangular faces.

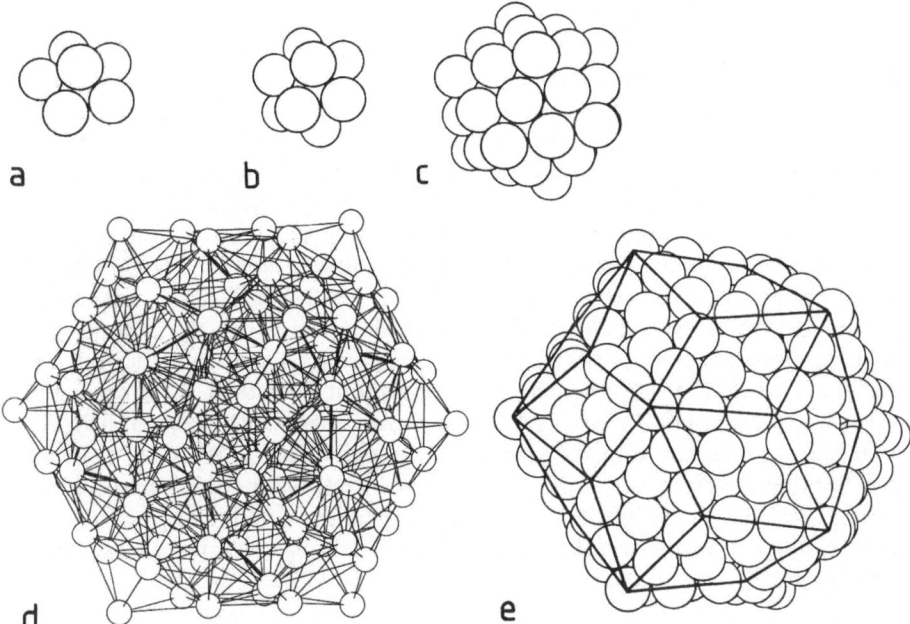

Fig. 2.2. (**a, b**) Noncrystallographic clusters produced by tetrahedral close packing; (**c**) The 55-atom "Mackay" icosahedron [2.14]; (**d**) Thirteen, 13-atom icosahedra (2b) bridged by octahedra producing an icosahedron of icosahedra; (**e**) 471-atom triacontrahedron produced from thirteen interpenetrating 55-atom Mackay icosahedra (after *Barker* [2.13])

Hoare and co-workers [2.13] have enumerated several families of noncrystallographic clusters which can be shown to be more stable than crystallographic clusters containing the same number of atoms. Clearly, such "polytetrahedral" structures include the icosahedron consisting of twelve atoms surrounding a central thirteenth atom (Fig. 2.2b). This configuration is particularly stable and is likely to represent an absolute minimum energy configuration for thirteen atoms.

Tetrahedral close packing is not space filling. The angular mismatch in Fig. 2.2a is small (7.5°) but the effect is cumulative so that large structures contain considerable strain and, in fact, a soft potential is required in order to preserve their symmetry. Clusters of this type ultimately become self-limiting, the elastic strain eventually becoming so large that it is then energetically favourable to nucleate a new structure.

Noncrystallographic structures containing tetrahedra and octahedra can be grown without any obvious size limit [2.14]. Figure 2.2d shows an example: thirteen 13-atom icosahedra bridged by octahedra form a 127-atom cluster – an icosahedron of icosahedra. Figure 2.2e shows a 471-atom cluster which may be regarded as thirteen interpenetrating 55-atom "Mackay" icosahedra [2.15], each unit being shown in Fig. 2.2c.

Evidence for the presence of noncrystallographic clusters in amorphous monoatomic solids is fragmentary. *Farges* et al. [2.16] have shown that small particles of argon, homogeneously nucleated in the absence of a substrate, exhibit electron diffraction patterns (Fig. 2.3a) which can be reproduced by scattering from 22 to 143-atom clusters (Fig. 2.3b), formed as minimum energy structures in molecular dynamics simulations with Lennard-Jones interactions. The final structures clearly show evidence for polyicosahedral packing and Mackay icosahedra appear in the larger cluster (Fig. 2.3c). *Briant* and *Burton* [2.17] presented similar evidence and proposed a general model for glasses based on packed 13-atom icosahedra. *Barker* et al. [2.18] have relaxed the central core of Finney's DRPHS model and observe that the second peak splitting becomes more realistic and the proportion of distorted icosahedral units and related structures increases from about 29% to 33%. More recent work by *Finney* and *Wallace* [2.19] has shown that while polytetrahedrality increases on relaxation, there is no evidence for polyicosahedral (or spherical polytetrahedral) clusters in either the original or relaxed models. Finally, *Sadoc* et al. [2.20], using an algorithm which tends to force a high degree of icosahedrality, found that the resulting interference functions show the characteristic splitting in the second peak.

b) Stereochemically Defined Models Based on Trigonal Prismatic Packing

While not discounting the possibility of some form of tetrahedral close packing to describe the arrangement of the metallic atoms in an amorphous binary alloy, there are arguments in favour of extending the "design" concept to produce trigonal prismatic coordination for the metalloid. Several authors note that there are close parallels between the properties of amorphous and crystalline alloys, suggesting some similar structural elements. Perhaps the most convincing evidence for trigonal prismatic packing (TPP) around the metalloid is also purely circumstantial. This is the useful working rule that the nearest-neighbour coordination of certain atoms in a glass is almost identical to that observed in the corresponding crystal.

The environment of silicon in amorphous silica is an obvious example. X-ray and neutron scattering data [2.21] show a silicon-oxygen bond length distribution with standard deviation $\sigma_s = 0.036$ nm (this figure includes contributions from static disorder and thermal vibrations but excludes termination broadening). The silicon bond angle is also well defined ($\sigma = 0.6°$ [2.22]). The environment of silicon can thus be described with some accuracy in terms of an almost undistorted tetrahedron. However, the nearest-neighbour arrangement around oxygen is less ordered, illustrated by the breadth of the oxygen angle distribution ($\sigma = 15°$). Strict tetrahedral coordination of silicon and more variable oxygen bond angles are characteristics of most of the crystalline forms of SiO_2.

Crystalline borides, phosphides, carbides, silicides, etc., of those transition metals which readily form glasses have lattices (near glass-forming com-

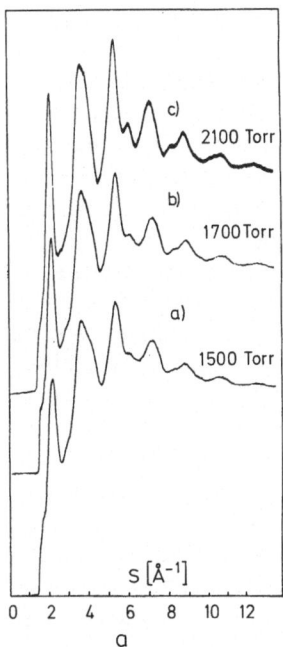

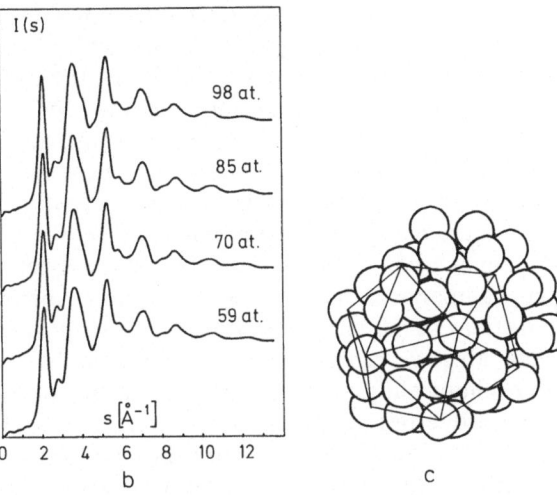

Fig. 2.3. (a) Electron diffraction from small argon particles homogeneously nucleated at pressures indicated. (b) Diffracted intensities from 59 to 98-atom clusters simulated by molecular dynamics techniques. (c) Projection of the 70-atom cluster revealing the presence of a 55-atom icosahedron

positions) based on trigonal prismatic coordination of the metalloid (Fig. 2.1b). This structure persists over a wide range of compositions and appears to be stable over a range of radius ratios from about 0.6–0.9, whereas the ideal radius ratio – for a metalloid within an undistorted trigonal prism – is 0.53. It thus appears that this type of packing may be especially stable.

There are, therefore, reasons (to be expanded in Sect. 2.2) for suggesting that the specific *stereochemistry* of the local environment of the metalloid *defined* by the corresponding crystal should be one of the elements in the *design* of random packings. It does not seem adequate to assume that for all amorphous alloys of this type, geometrical considerations alone are important and that the non-metallic alloys fit into holes whose shape and size are determined solely by the packing of the metallic atoms. In view of the increasing evidence [2.23, 24] that the metalloid-metal bond is strong (and may be directional), it is *likely* that the metalloid elements play a dominant role in determining not only the shape and size of their own coordination polyhedra but those of the metallic atom also.

2.1.4 Random and Designed Models: Similarities and Differences

Summarising this section, there are two types of model for amorphous metals which need to be taken seriously: dense random-packed assemblies and cluster models in which a structural element has been defined by chemical intuition or some similarity to crystalline structures.

Given an adequate knowledge of the potential energy function, it would be sufficient to choose a *random* assembly of atoms and allow the structural problem to solve itself – via molecular dynamics or Monte Carlo simulation. Even for monatomic metals, however, this procedure requires care since the resulting structures are sensitive to the choice of the potential energy function. For diatomic systems, potential energy functions are rarely better than guesses, except for alloys of simple metals for which pseudopotential calculations are possible and indeed successful [2.1]. In the specific case of the T–M alloys, while the T–T potential function may be obtainable by experiment or calculation, the M–M and, more important, the T–M interactions are not even *qualitatively* understood. Modelling such structures with a single type of potential energy function is thus fraught with danger. Computer simulation techniques using energy minimisation, Monte Carlo algorithms or molecular dynamics techniques can introduce the polytetrahedrality which seems to be the hallmark of monatomic amorphous metals. With T–M and T–T interactions of the same type, it would be surprising if this type of random model were able to reproduce the more complex behaviour of crystalline and amorphous alloys – particularly the details of the M–T coordination – unless these properties are functions of the relative size of the atoms and *strength* of the interaction alone. These are the only parameters, apart from the starting structure, which are at the disposal of the modeller.

In a *stereochemically defined* model [2.6, 58], other constraints can be added which, however inadequate, emphasise the particular aspect of the structure: polytetrahedrality, polyicosahedrality for the local structure around the metal, or trigonal prismatic local coordination for the metalloid.

a) Characteristics of Dense Random-Packed and Stereochemically Defined Models

It is difficult to implement the constraints mentioned above (even randomness is achieved only with care); consequently, the two types of model may be less distinct than appears at first sight. It is therefore appropriate to set down what appear to be salient features of each type of model – in their idealised forms – before looking at the experimental evidence for T–M glasses.

Characteristics of DRP *Models*

i) In general, alloys of two elements A and B consist of a random mixture of the two species. The probabilities of AA, AB, and BB contacts are essentially independent of the chemical nature of A and B. [For the *particular* case of the T–M alloys, this element of the random model is relaxed and replaced by the requirement that M atoms are predominantly surrounded by T atoms. The evidence, to be presented below, appears overwhelming so that perfect chemical short-range order (which clearly is a more appropriate characteristic of SD models) can be considered the common feature of both types of model for T–M alloys, although it seems unlikely to be a general characteristic.]

ii) The spatial arrangement of atoms in local coordination polyhedra is essentially random, subject only to space-filling constraints.

iii) There is thus no preference, deriving from consideration of energy or space filling, for a particular local symmetry or coordination number. The average local geometry thus becomes a function of the radius ratio and concentration of each species. This fact is commonly overlooked: it is often assumed that local structure similar to that of a crystalline phase will develop automatically in DRP structures, whatever the radius ratio. The simulations of *Blétry* [2.25], *Jansen* et al. [2.26] and *Boudreaux* and *Frost* [2.27] clearly demonstrate the contrary.

iv) The structure is essentially homogeneous. Subject only to statistical fluctuations, each atom in the structure is equivalent to any other of that type. Longer range spatial variations in density are also subject to random fluctuations alone.

v) Randomness and homogeneity at the *local* level of structure imply similar characteristics in medium and longer-range structures.

Characteristics of Stereochemically Defined Models

i) One of the possible local configurations found in a DRP model is preferred on energetic grounds or because the packing conserves space.

ii) This arrangement may represent the dominant coordination polyhedron over a wide range of concentration and radius ratio. The nearest-neighbour coordination number is thus relatively independent of these quantities.

iii) Exceptional stability of one type of cluster may be suggested by a preference for this structure in crystals also, leading to potential *equivalence* of local structures of the crystalline and amorphous phases.

iv) Differences in structure with concentration and the chemical nature of the two types of atoms, the distinction between amorphous and crystalline phases, are described by variations in the way local structural units are connected.

v) The structure may be essentially *heterogeneous*. For example, compact units may be interspersed in a less dense matrix, or the topology at local or medium-range level may be essentially inequivalent for two or more subsets of atoms of the same type.

vi) Structural definition need not be confined to nearest and next-nearest neighbours. Larger crystallographic or noncrystallographic structures may exist.

2.2 Experimental Investigations of the Local Structure of Transition Metal-Metalloid Glasses

In this section we review the experimental information which atomic models must seek to reproduce, explain and extend. Much of the detail has been

Table 2.1. Experimental mean $\bar{r}_1$, standard deviation $\sigma_s(r_1)$, and coordination numbers for nearest-heighbour $M-T$ and $T-T$ distributions in several amorphous alloys. The first four measurements are perhaps the most complete as all three partials have been extracted. The standard deviation $\sigma_s(r_1)$ represents an approximation to the disorder-induced broadening since termination broadening has been substracted. Thermal broadening is included, however. In the column labelled Technique, n, x stand for neutron and x-ray scattering, respectively, i for isotope substitution, and [ ] for concentration

Alloy	Atom Pair	$\bar{r}_1$ [nm]	$\sigma_s(r_1)$ [nm]	N	Q_{max} [nm]	Tech-nique	Comments	Ref.
$Co_{81}P_{19}$	P–Co	0.232	0.0105	8.9 ± 0.5	150	n, x		[2.31]
	Co–Co	0.254	0.0155	$10\ \pm0.4$	150			
$Fe_{80}B_{20}$	B–Fe	0.214	0.0096	8.6	118	n, i, x		[2.32]
	Fe–Fe	0.257	0.0165	12.4	113			
$Ni_{81}B_{19}$	B–Ni	0.211	0.0142	8.9	130	n, i		[2.32]
	Ni–Ni	0.252	0.0147	10.5	130			
$Fe_{75}P_{25}$	P–Fe	0.238	0.018	8.1	70	x	Anomalous scattering	[2.78]
	Fe–Fe	0.261	–ve	10.7	70			
$Pd_{84}Si_{16}$	Si–Pd	0.240	0.0106	9.0 ± 0.9	150	n, x	Only 2 Measurements	[2.34]
	Pd–Pd	0.276	0.0154	11.0 ± 0.7	150			
$Pd_{80}Si_{20}$	Si–Pd	0.242	0.0081	6.56	250	n	1 measurement approx. Gaussian fit	[2.35]
	Pd–Pd	0.280	0.0167	10.6	250			
$Pd_{80}Ge_{20}$	Ge–Pd	0.253	0.0078	5.6	250	n	1 measurement approx. Gaussian fit	[2.35]
	Pd–Pd	0.281	0.019	10.3	250			
$Co_{81.5}B_{18.5}$	B–Co	0.207	–	6.6		n, x	1 measurement approx. Gaussian fit	[2.74]
	Co–Co	0.250	–	12.7				
$Co_{80}P_{20}$	P–Co	0.22	~0.01	9	–	EXAFS	Exponential	[2.75]
$Pd_{78}Ge_{22}$	Ge–Pd	0.249	<0.01	8.6 ± 0.5		EXAFS		[2.33]
$Fe_{83}B_{17}$	B–Fe	0.227	–	6.9		n, x	Only 2 measurements	[2.37]
	Fe–Fe	0.256	–	10.7				
$Fe_{80}B_{20}$	B–Fe			9.5		Möss-bauer		[2.76]

covered in Chap. 3 and is not repeated here. Table 2.1 contains a summary of the more important structural parameters of amorphous T–M alloys.

2.2.1 Chemical Short-Range Order Around the Metalloid: M–M Avoidance

Information on the immediate environment of the metalloid atom is contained in partial pair distribution functions (PPDFs)

$$G_{\alpha\beta}(r) = 4\pi r[\varrho_{\alpha\beta}(r) - \varrho_0]/c_\beta \qquad (2.1)$$

derived from the partial structure factors $S_{\alpha\beta}(Q)$.

Partial pair functions $G_{\alpha\beta}(r)$ are components of the total distribution function

$$G(r) = 4\pi r \sum_{\alpha} \sum_{\beta} c_{\alpha} c_{\beta} f_{\alpha}(Q) f_{\beta}(Q) \varrho_{\alpha\beta}(r) / |\langle f(Q) \rangle| , \qquad (2.2)$$

where c_{α}, $f_{\alpha}(Q)$ are the concentration and atomic scattering factors for atom α, $\varrho_{\alpha\beta}(r)$ is the atomic density of β atoms at a distance r from an average α atom, ϱ_0 is the average density and $\langle f(Q) \rangle = c_{\alpha} f_{\alpha} + c_{\beta} f_{\beta}$.

There are also advantages in dissecting $S(Q)$ into "Bhatia-Thornton" partials, $S_{NN}(Q)$, $S_{NC}(Q)$ and $S_{CC}(Q)$. These are defined elsewhere in this volume [2.2] (see also *Blétry* [2.25]). If the three B–T partials can be separated, and this may be easier for elements with approximately the same radius (since $S_{NC} \simeq 1$), $S_{CC}(Q)$ gives directly the degree of *chemical* order (M–M avoidance, for example) and is particularly informative when Fourier transformed to give a "radial concentration correlation function".

Accurate "partials" are difficult to extract except in particularly favourable cases (see, for example, *Sadoc* and *Wagner* [2.30] and *Blétry* [2.25]). The investigations of *Sadoc* and *Dixmier* [2.31] on a-$Co_{81}P_{19}$ showed a nearest-neighbour coordination number of 9 cobalt atoms with no P–P nearest neighbours at the expected distance of 0.21 nm. The first P–P peak is found at 0.35 nm beyond the first-neighbour shell of cobalt atoms. Moreover, *Blétry* has shown that the chemical short-range order parameter derived from $S_{CC}(Q)$ is almost equal to the value predicted for perfect ordering [2.31].

Partial pair distribution functions have recently been obtained for $Fe_{80}B_{20}$ and $Ni_{81}B_{19}$ [2.32]. These data are reproduced in Figs. 2.4, 2.5 and salient distances, coordination numbers, etc., are collected in Table 2.1. Again there is no indication of B–B correlation below about 0.36 nm, although the diameter of the B atom is less than half this value, 0.17 nm.

The Ge K-edge EXAFS measurements of *Hayes* et al. [2.33] indicate a coordination number of 8.6 ± 0.5 Pd atoms surrounding Ge in a-$Pd_{78}Ge_{22}$. Calculations show that replacement of even one palladium atom by germanium in the first-neighbour shell significantly reduces the quality of fit to experiment.

2.2.2 The Metalloid Coordination Number

Metalloid coordination numbers for $Co_{81}P_{19}$, $Fe_{80}B_{20}$, and $Ni_{81}B_{19}$ are 9.0 ± 0.9, 8.6, and 8.9, respectively: see Table 2.1, where other measurements are also enumerated. With four exceptions – certain measurements for Pd–Si, Pd–Ge, Fe–B, and Co–B – the metalloid coordination number has been found to be *constant* in the range 8.5–9.0.

The exceptions warrant some comment. (a) Data for Pd–Si has been obtained by analysis of x-ray and neutron diffraction data [2.34] giving a coordination number of 9.0 ± 0.9, and by decomposition of the first peak in the neutron RDF into symmetrical Gaussian components corresponding to Si–Pd

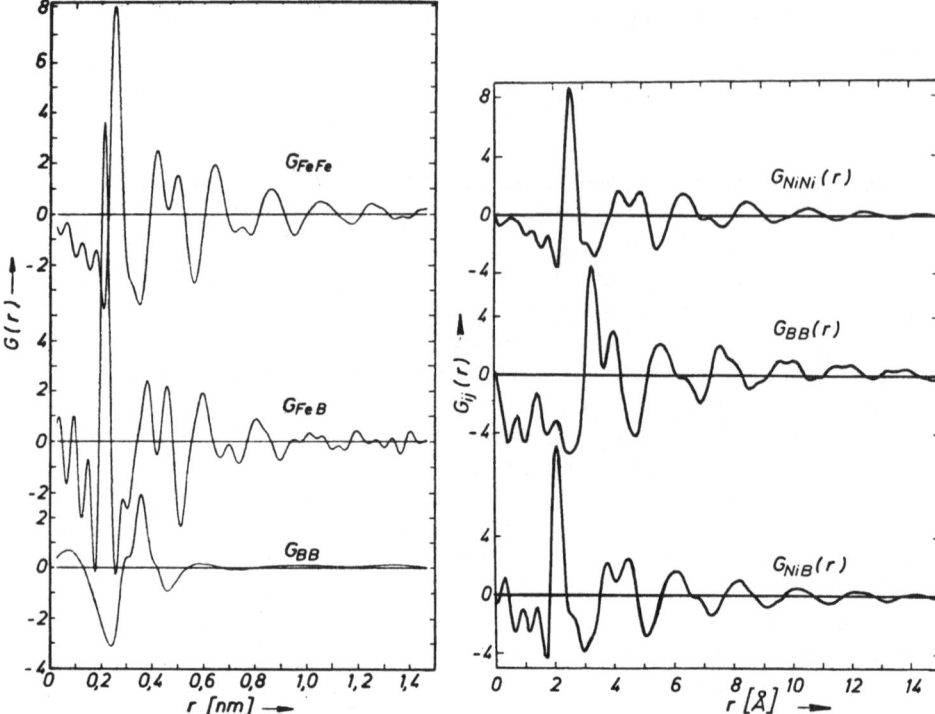

Fig. 2.4. Partial reduced pair distribution function for amorphous $Fe_{80}B_{20}$ (*Nold* et al. [2.32])

Fig. 2.5. Partial pair distribution functions for a-$Ni_{81}B_{19}$ (*Lamparter* et al. [2.32]). $G_{BB}(r)$ was obtained by measuring the neutron scattering intensity from an alloy with a mixture of Ni isotopes, giving zero coherent scattering from Ni. Note the split first-neighbour distance and the presence of oscillations beyond 10 Å

and Pd–Pd leading to a value of 6 [2.35]. *Sadoc* and *Dixmier*'s [2.34] partial RDF clearly shows that the first-neighbour distribution of $G_{Si-Pd}(r)$ is not symmetrical, with a shoulder at high r values (this behaviour would be anticipated by comparison with crystalline Pd_3Si). It seems likely, therefore, that the value of 9 is more accurate. A similar argument applies for $Co_{81.5}B_{18.5}$ [2.74]. (b) This argument is more difficult to sustain for Pd–Ge alloys [2.35], however. Firstly, there is no independent measurement of $G_{Ge-Pd}(r)$, but comparison with crystalline palladium-germanium alloys suggests a *symmetrical* Ge–Pd distribution. Further evidence is needed to establish the coordination number of this alloy, although it should be said that EXAFS which gives a value of 8.6 is the more direct technique. Furthermore, a recent preliminary report of EXAFS data for a- and c-$Pd_{80}Ge_{20}$ suggests that the environment of Ge is almost identical in both forms of the alloy [2.36]. (c) The coordination number for $Fe_{84}B_{16}$ given by *Waseda* and *Chen* [2.37] (6.9 ± 0.8) was obtained using x-ray anomalous scattering. It is not clear whether the errors inherent in this technique are responsible for the differences with other measurements. *Dini* et al. [2.38] have shown that the conflict between various measurements of the x-ray and neutron scattering from Fe–B alloys may be due to differences in specimen quenching rate or to the data analysis technique (see Sect. 2.4.1). It seems likely that the lower value obtained by Waseda and Chen is unreliable in the light of the consensus reached by several other groups using different techniques.

2.2.3 The Breadth and Shape of the M–T Distribution

Values for the breadth of the first peak in the M–T distribution are given in Table 2.1. A value for termination broadening has been subtracted (see Sect. 2.3.5) so that the resultant σ_s represents the standard deviation of the M–T bond length distribution, containing contributions from instrumental and thermal broadening in addition to static (disorder) broadening. In all cases, *narrow distributions*, suggesting a *well-defined* environment, are observed for M–T distances, with somewhat larger values of σ_s for M–M distributions.

There are important differences in the symmetry of M–T distributions in glasses, for which the corresponding crystalline phases are isotypic with Fe_3C or Fe_3P (or Ti_3P). *Gaskell* [2.39] has pointed out that the asymmetric distribution of Si–Pd distances in $Pd_{84}Si_{16}$ [2.34] is not observed in the P–Co distribution for $Co_{81}P_{19}$ [2.29]. There is a marked difference also in the symmetry of the corresponding distributions in the crystalline lattices (Fig. 2.6). Note, however, that a recent study of the phosphorus K-edge EXAFS in a-CoP suggested an asymmetry in the P–Co distribution [2.75]. Similar, but less marked, asymmetry is observed in the B–Ni distribution in $Ni_{81}B_{19}$ [in contrast to the symmetrical B–Fe peak in a-$Fe_{80}B_{20}$ (Ti_3P-type)] [2.32].

Differences in the total and partial RDFs of Fe- and Ni–B alloys (Figs. 2.4, 5) reveal the effects of local "bonding" interactions most clearly. The radii of Fe and Ni atoms are almost equal (to 2%) so that a purely geometrical model would predict almost identical properties. Changes in the RDFs are not restricted to first peaks, as discussed below. *Aur* et al. [2.40] have also pointed out that the structures of Fe and (Fe, Ni)-B alloys are not identical. They show that the total RDFs are *qualitatively* similar to Gaussian-broadened, near-neighbour bonding distributions for c-Fe_3B and a weighted sum of crystalline Fe_3P and Ni_3P.

The symmetry of the metal atom shell around boron can be obtained from spin-echo NMR spectra of ^{11}B in $Mo_{70}B_{30}$, $Mo_{48}Ru_{32}B_{20}$, and $Ni_{78}P_{14}B_8$ [2.41]. The electric field gradient, EFG, at the boron atom is characterised by values of the quadrupolar coupling constant and the asymmetry η of the EFG tensor. Values for η and the shape of the NMR spectra were different for the Mo- and Ni-based alloys, reflecting differences observed in the respective crystalline borides. The authors note that large values of η would be incom-

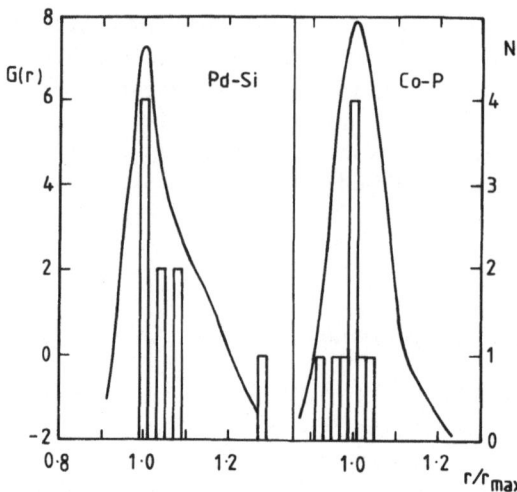

Fig. 2.6. Comparison of the profiles of the first-neighbour M–T distributions in crystalline and amorphous alloys. Bar charts represent the distribution of first neighbours (right-hand ordinate) in crystalline $Pd_3Si(Fe_3C)$ and Fe_3P lattices, as a function of reduced interatomic distance r/r_{max}, where r_{max} is the value corresponding to the maximum of the distribution. Curves represent Si–Pd and P–Co partial pair distribution functions (left-hand ordinate) for glassy $Pd_{84}Si_{16}$ and $Co_{81}P_{19}$ alloys [2.31]

patible with a random (Polk-type) model. This conclusion has been questioned by *Czjek* [2.42] who has calculated a preference for large values of η in random charge models and suggests that it is the *small* values of η observed in amorphous molybdenum borides which are difficult to reconcile with random structures.

2.2.4 Macroscopic Physical Properties

Fe-(P, B) alloys crystallise to the Fe_3P structure for B : P ratios less than about 0.5 and to the Ti_3P structure otherwise. The transition is marked by changes in several physical properties. *Durand* and *Yung* [2.43] have shown that the composition-dependence of the Curie temperature and saturation magnetic moment of *amorphous* alloys shows breaks at compositions corresponding to the transition from Fe_3P to Ti_3P in the *crystalline* alloys. These data suggest that the local and (probably) the medium-range structures of amorphous and crystalline alloys are equivalent.

The density of amorphous and crystalline phases are generally similar, T–M glasses being less dense than the crystalline phase by about 2%. *Gaskell* [2.44] has shown that the composition-dependence of the densities of a wide range of crystalline and amorphous binary T–M alloys can be reduced to a simple linear relationship between the *metal atom* packing fraction η_M and the radius ratio p of the M and T atoms:

$$\eta_M(x) = \eta^0[1 - x\gamma(p - p_c)].$$

Here, x is the atom fraction of the metalloid, η^0 the packing fraction of the hypothetical pure amorphous T alloy, γ a constant and $p_c (= 0.53)$ is the radius ratio for an atom within an undistorted trigonal prism. Values for the parameters in this empirical equation are almost identical for crystalline and amorphous alloys, indicating similar local (and medium-range) structures.

2.2.5 Summary

Experimental data from several different techniques have converged to suggest two important features which characterise the local structure of T–M glasses.

(a) The first coordination shell around the metalloid is well defined, with a narrow distribution of bond lengths. Moreover, there is little evidence for a coordination number which departs significantly from 8 or 9.

(b) The environment around the metalloid is similar to that observed in the corresponding crystalline phase. In certain cases this correlation extends in some detail: the distinctive properties of two (similar) crystalline structures are observable in glasses also.

2.3 Modelling the Structure of T–M Glasses Using Random-Packed and Stereochemically Defined Models

Development of a suitable model for amorphous materials can be considered in four stages:

(a) construction of an initial "starting" structure;
(b) refinement of this model using energy minimisation, Monte Carlo or molecular dynamics techniques;
(c) characterisation of the topology of the refined model in terms of polyhedron statistics, etc.;
(d) calculation of microscopic and macroscopic properties and comparison with experimental data.

The relative importance of the first two stages depends on the nature of the modelling technique. In Monte Carlo or molecular dynamics simulations, for example, the starting structure may be a set of arbitrarily chosen coordinates (with obvious precautions to avoid overlap). The refined structure is reached in Monte Carlo algorithms by a sequence of random displacements of randomly chosen atoms, with, optionally, a test to ensure convergence towards a target structure, as defined by an energy criterion or agreement with an experimental RDF. Molecular dynamics simulations proceed by repeated solutions of Newton's equations over time periods corresponding to a fraction of a vibration. In both approaches, the key to success lies more in the selection of a representative potential energy function than in the choice of a starting structure – although the importance of the latter in Monte Carlo processes has been demonstrated by simulation of As_2Se_3 [2.45].

Where energy minimisation is employed as the refinement technique (and of course in investigations where this stage is omitted entirely), the choice of starting structure will be crucial. Energy minimisation algorithms of this type attempt to choose a minimum potential energy configuration for a N-atom cluster in 3N dimensional configurational space by, for example, displacing atoms sequentially in the direction of maximum decrease in energy, by an amount proportional to the gradient of the energy at that point. Normally, stability of the calculation requires that displacements shall be small fractions of a nearest-neighbour distance. Consequently, there is no guarantee that the minimum energy structure obtained by iterative processes of this type represent anything other than a *local* rather than a *global* minimum for an amorphous structure. This can be turned to advantage, however. In each of the refinement techniques, the modeller loses some degree of control of the character of the model: a random model may start to crystallise or, more likely, a model containing elements of symmetry becomes more random. (It is this loss of control which may make it more difficult to recognise the salient features of the resulting structure unless the characterisation stage is included.) It is often useful to employ a refinement technique in which the defects of the initial model are removed (voids, compressed areas, distorted bonds) without introducing such extensive displacements that local groups of atoms are destroyed.

In the following paragraphs we outline the various stages of typical simulations. Recent review articles [2.30, 46–48] present more detailed accounts.

2.3.1 Construction of the Initial Model

a) Random Monatomic Models

These have been constructed by packing spheres as described earlier, but computer-modelling is clearly more convenient. A useful technique consists of serial deposition of atoms onto a seed, normally a four-atom tetrahedral cluster, such that the added atom is placed without overlap in contact, or within a specified distance of three existing spheres. Some criterion such as the closest available site to the centre, for example, is introduced to govern the overall shape of the cluster [2.30, 49–52]. *Ichikawa* [2.53] introduced the additional constraint that for the three surface spheres of diameter D which comprise the possible site, the internuclear distance should be less than kD. The constant k defines the degree of tetrahedral "perfection". Other methods include the search for suitable sites at random. This leads to the growth of irregular clusters containing so many voids [2.48] that the process can be considered to be a more realistic representation of deposition from the vapour with only limited surface atomic mobility. Alternatively initial coordinates can be chosen completely at random and the sphere radius increased from zero until overlap occurs when either an overlapping atom is removed or each atom is displaced to remove overlap. This method has been used recently by *Maeda* and *Takeuchi* [2.52] instead of serial deposition of atoms in an attempt to produce somewhat denser models with no suggestion of any radial inhomogeneity.

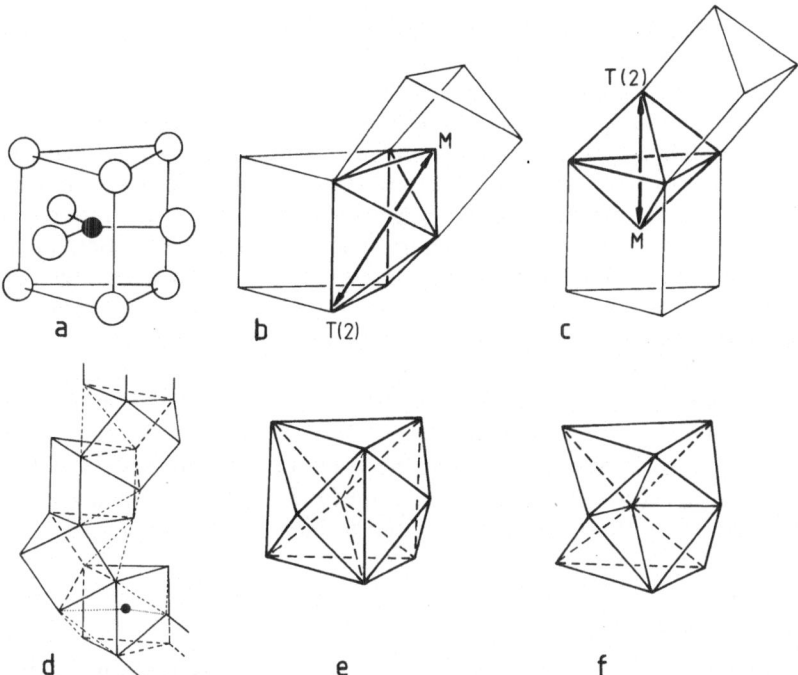

Fig. 2.7. (a) Trigonal prism with capping atoms. **(b)** Two trigonal prisms with a common edge linked as in the cementite structure. **(c)** TPs with the edge-sharing arrangement observed in Fe_3P and Ti_3P lattices. **(d)** Disordered chain of TPs produced by random edge-sharing. **(e, f)** Dissection of capped trigonal prisms into tetrahedra and octahedra. Slightly distorted TPs appropriate for alloys with small metalloid radius ratio are shown in **(e)**, and **(f)** corresponds to the local arrangement around large metalloids. These are observed in Fe_3C and Fe_3P lattices, respectively

b) Diatomic Random Models

These have been constructed using similar principles. Although physical models using spheres of two sizes have been constructed [2.55], experimental observation of short-range ordering in T–M alloys requires the inclusion of an additional constraint, M–M avoidance. An extensive series of investigations by *Blétry* [2.25] and *Boudreaux* and *Gregor* [2.56] has resulted in a modified serial deposition process for producing large random models of binary alloys over a wide range of composition and radius ratio.

An alternative procedure is to start from a monatomic model and label sites as one of two or more types of atom. Despite its apparent simplicity, this approach has not been investigated in detail.

c) Stereochemically-Defined Models

While the physical process of randomly packing spheres can be reproduced by computer simulation, packing of polyatomic structural units presents a more formidable problem. In constructing a model of packed trigonal prisms, *Gaskell* [2.57, 58] found that in order to achieve an adequately dense starting structure, it was necessary to first build a physical model. This has the advantage that the builder can see several moves ahead and, to some extent, prevent voids. The topology of the physical model was then used as input to a computer program which generated an accurate set of atom coordinates. The algorithm is illustrated in Fig. 2.7b. An undistorted prism forms the seed and subsequent units are generated by adding a new atom to a vacant square face to complete a half octahedron. One of the resulting triangles now forms the base for a further trigonal prism. The new prism is related to the original by a clockwise rotation of 215° around the common axis and reproduces the edge-sharing arrangement in Fe_3C lattices. A random arrangement of units is achieved by choosing the common axis at random, constrained only by the need to avoid overlap and produce an adequately dense structure (Fig. 2.7d). The final model contained only 542 atoms; at this point discrepancies between the physical model and the computer-generated version had become so large (due to difficulties in positioning spheres accurately) that the former ceased to be a reliable guide.

d) Dislocation Models

Ninomiya and co-workers [2.59] have constructed models for amorphous semiconductors by inserting screw dislocations into a diamond-cubic lattice. By removing atoms corresponding to one of the sublattices, a model for amorphous monoatomic metals is created. Morris has also discussed the formation of a model containing a three-dimensional network of mixed disclinations and glide dislocations [2.59].

2.3.2 Refinement of the Zero-Order Model

a) Energy Minimisation

This has been used extensively to produce more realistic structures. Popular algorithms are the methods of steepest descents or conjugate gradients [2.60].

Relaxation changes the characteristics of hard sphere models in several important respects and differences in the resulting structures depend to some extent on the type of potential energy function. These two points have been demonstrated by *Barker* et al. [2.18] and *von Heimendahl* [2.62] who have relaxed the central core (999 atoms) of *Finney*'s 7934-atom model [2.6] and a 888-atom core of *Bennett*'s [2.49] 3999-atom DRPHS model respectively. Both investigations demonstrated the important point that the RDFs for the relaxed structures are considerably more realistic than those for the zero-order models. In particular, the relative heights of the second split peak near two atom diameters (Fig. 2.8) are reversed so that the relaxed model provides a more reasonable representation of the RDF for monoatomic amorphous metals.

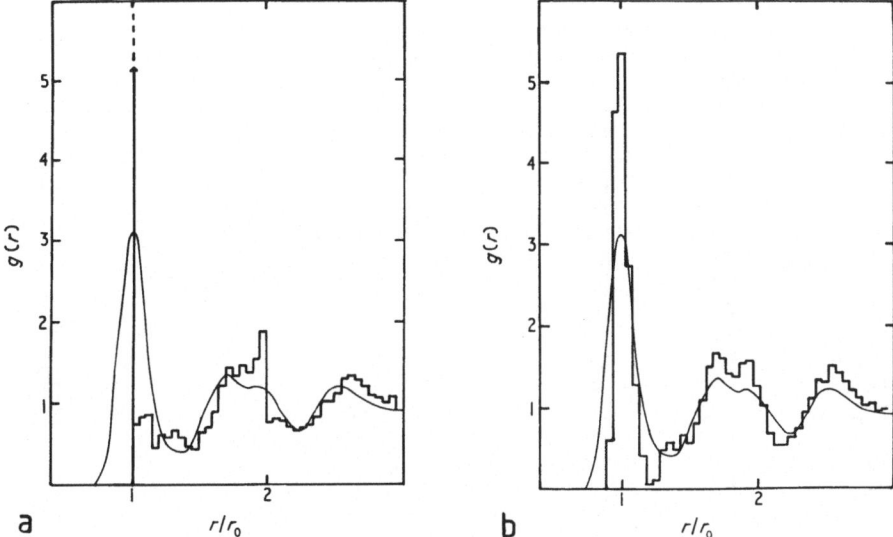

Fig. 2.8. Pair distribution functions for an 888-atom central core of the 3999-atom model of *Bennett* [2.49]: (**a**) before and (**b**) after energy minimisation [2.62]. Histograms show the distribution functions for the model and the full curves are for a-Co

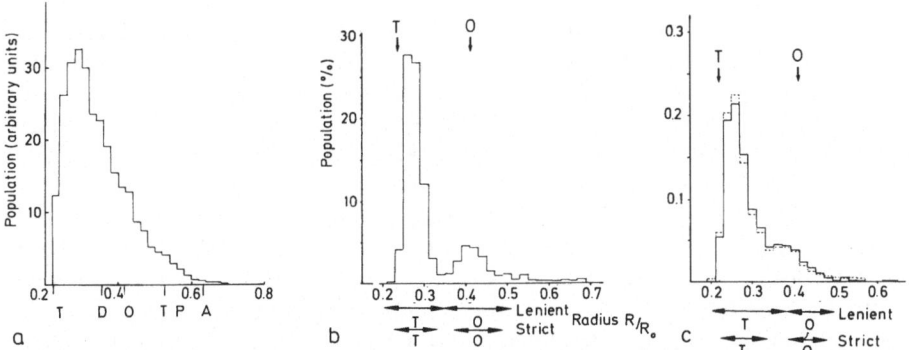

Fig. 2.9. (**a**) Distribution of interstice radii in the central 1825-atom core of Finney's 3994-atom model. (**b**) and (**c**) are the distribution after relaxation with Morse and Lennard-Jones potentials, respectively. Note that energy minimisation causes the larger cavities to be "squeezed out" leaving a population dominated by tetrahedra and octahedra [2.19]

A recent re-examination of the changes in structure which accompany relaxation has been reported by *Finney* and *Wallace* [2.19] who calculated the distribution of interstice radius for Finney's hard sphere model relaxed with Lennard-Jones 6–12 and Morse potentials. Figure 2.9 shows the fundamental change in character from an unrelaxed structure with a distribution of interstices comprising tetrahedra, octahedra, trigonal prisms and Archimedean

Table 2.2. Parameters used in several energy minimisation calculations. The potential function is a Lennard-Jones 6–12 function unless otherwise stated (ε: cohesive energy, p: radius ratio of metalloid). The second value quoted for the non central force-field refers to T atoms at the vertices of a trigonal prism

Alloy	p	$\varepsilon_{TM}/\varepsilon_{TT}$	$\varepsilon_{MM}/\varepsilon_{TT}$	Comments	Ref.
$Fe_{75}P_{25}$	0.84	1.43	0.29		[2.55]
$Fe_{100-x}P_x$	0.72	1.42	0.29	$10 < x < 30$	[2.64]
$Fe_{100-x}P_x$	0.86	1.66	0.38	Morse potential	[2.65]
$Fe_{75}P_{25}$	0.9)	1.5)		*Non central force fields	
	2.1)*	0.5)*	1.0		
$Ni_{75}P_{25}$	0.80	1.46	0.27		[2.55]
$Co_{75}P_{25}$	0.82	1.38	0.27		[2.55]
$Co_{80}P_{20}$	0.89	0.69	0.38		[2.39]
$Pd_{80}Si_{20}$	0.72	1.42	0.29		[2.64]
$Pd_{80}Si_{20}$	0.70*)			*TP vertex atoms	
	0.94**)	0.72	0.29	**Capping atoms (see text)	[2.59]
$Pd_{80}Ge_{20}$	0.8	2.0(?)	0.05(?)		[2.64]
$Fe_{100-x}B_x$	0.52	2	0.05	$15 < x < 25$	[2.64]
$Fe_{85}B_{15}$	0.58	1.07	0.05		[2.37]
$Nb_{75}Ge_{75}$	0.8	1.0	1.0		[2.77]

antiprisms to one which, to a good approximation, can be described as an assembly of distorted (dilated) tetrahedra and octahedra only. An obvious parallel is suggested with close-packed crystalline arrays, which are *regular* arrangements of *regular* tetrahedra and octahedra in a prescribed ratio. A secondary point, demonstrated by Fig. 2.9b, c, is that relaxation with a Lennard-Jones and with a softer Morse potential leads to similar but not identical results.

Potential energy functions used in these calculations are rarely realistic, as discussed earlier, although potentials with some experimental basis have been derived for T–T interactions, for example, the Johnson and Pak-Doyama potentials for Fe [2.61]. A systematic examination of the effect of changes in potential energy function for binary alloys has not been reported. Lennard-Jones potentials are often used:

$$V_i = 4\varepsilon_{ij} \sum_j (R_{ij}^{12} - R_{ij}^6), \tag{1.3}$$

where

$$R_{ij} = 2^{-1/6} R_{ij}^0 / r_{ij}.$$

r_{ij} is the distance between atoms i and j and R_{ij}^0 the equilibrium distance. The quantity ε_{ij} is approximately the cohesive energy, values for elements being tabulated by *Gschneider* [2.63]. Unless absolute energies are required, only the ratios of the various ε_{ij} values are needed. *Ching* and *Lin* [2.55] and *Fujiwara* and *Ishii* [2.65] suggest a prescription for the calculation of ε_{M-M} and ε_{T-M} but, as stated earlier, these can only be regarded as zero-order estimates. Relative magnitudes of the cohesive energies for a number of investigations are given in Table 2.2.

2.3.3 Characterisation of the Topology

a) Analysis of the Interstice Distribution

This provides a convenient graphic description of a model, especially if the representation is simplified by restricting the classification of holes to the canonical polyhedra of the model by accepting some degree of distortion. Cavities can be represented by listing the set of numbers, $n_i (i > 3)$, where n_i is the number of vertices of the polyhedron at which i edges cross. The tetrahedron, octahedron and capped trigonal prism are represented in this notation as $(4, 0, 0)$, $(0, 6, 0)$ and $(0, 3, 6)$, respectively [2.67].

Finney and *Wallace* [2.19] have extended the method by representing the spatial distribution of specified cavity types as radial distribution functions, for example, the probability of finding a tetrahedron at a distance r from a given tetrahedral origin. It then becomes possible to search random soft-packed models, say, for particular arrangements of tetrahedra and octahedra – icosahedral units (relatively infrequent), linear or branched tetrahedral chains (more likely) or close-packed crystal-like units.

b) Voronoi Polyhedra

Voronoi polyhedra can also be defined for atoms in an irregular lattice [2.6]. The model may then be represented in terms of the fraction of polyhedra with n faces or the number of edges per face, etc., or by the indices mentioned above. Since the Voronoi polyhedron and the dual of the cavity are isomorphous, the indices r_i now represent the number of faces with i edges. *Gellatly* and *Finney* [2.68] have examined the Voronoi construction for random packings of two sizes of spheres and conclude that the method can produce misleading results. An alternative, the radical plane construction, is preferred.

c) Local Structural Parameters

Coordination polyhedra in amorphous structures are extensively distorted and *Egami* et al. [2.54] have shown that the environment may be inadequately characterised without some reference to local levels of stress and strain. The rotational invariants of the local stress tensor, the hydrostatic stress, p^* and the average stress tensor τ, are defined as follows:

$$p^* = \tfrac{2}{3}(\sigma_1 + \sigma_2 + \sigma_3)/3$$

$$\tau^2 = [(\sigma_1 - \sigma_2)^2 + (\sigma_2 - \sigma_3)^2 + (\sigma_3 - \sigma_1)^2]/6,$$

where σ_1, etc. are the principal stresses.

In addition, local atomic site symmetry coefficients are defined: the spherically averaged force constant α_0 and a coefficient β, which measures the departure from spherical symmetry around the origin atom.

The quantities p^*, τ, α_0, and β can be evaluated at each site for a given force field. Mean and standard deviations of the distribution functions for these parameters can be used to follow, for example, the approach of a structure to thermal equilibrium, thus simulating "relaxation" in the region of the glass transition. A more detailed investigation of the translational motion involved is obtained by plotting values of p^*, say, at each site, or by obtaining RDFs for subsets of atoms comprising the high and low extremes of the distribution of p^*. The technique has been shown to be a powerful method of characterising monoatomic structures and modelling their behaviour.

2.3.4 Anisotropy

Many of the early DRPHS models were unrealistic in that the atom density decreased with distance from the centre of the cluster. *Boudreaux* and *Gregor*'s unrelaxed diatomic models [2.56] suffered from the same defect; moreover, they observed that their models were orientationally anisotropic. RDFs were distinctly different for correlations measured towards the centre, outwards and tangentially. The almost complete lack of correlation in the tangential direction was taken to imply an orientational anisotropy introduced by the Bennett packing algorithm [2.49]. The density deficit and orientational anisotropy were largely removed by energy minimisation.

A problem with a less obvious solution surrounds the anisotropy observed in models for vitreous SiO_2, amorphous tetrahedral semiconductors and monoatomic metals [2.69]. Anisotropy is revealed by the presence of distorted planes of atoms with a mean spacing near that of the corresponding crystal. Consequently, a marked anisotropy in $S(Q)$ is observed for Q corresponding to Bragg orientation of the "lattice planes". The presence of "warped sheets" of atoms have been seen in a physical model of randomly packed spheres [2.70] and subsequently *Barker* [2.13] observed structural anisotropy in drawings of the central cluster of Finney's model. This anisotropy does not disappear on relaxation, however.

2.3.5 Validation of the Model:
Calculation of Microscopic and Macroscopic Properties

Given a set of atomic properties, it is then possible to calculate several microscopic properties, RDF, $S(Q)$, $S(Q)$ and, with further assumptions, the electronic and vibrational densities of states [2.23, 65]. Macroscopic physical properties, i.e., heat of crystallisation, density, elastic moduli, gas solubility and diffusivity, are also obtainable. In principle, each property can be measured experimentally and used to test the validity of the model. For the limited purpose of assessing the viability, the structure factor and RDF are the most useful properties. *Total* structure factors are obtainable with some precision, by time of flight neutron scattering techniques, for example, and, in certain cases, partial pair functions are measurable, as discussed in Sects. 2.2.1, 2. Even radically different models *appear* to be similar when characterised only by $S(Q)$ or $G(r)$, so it is only by examining the *detail* of these functions that the differences can be discriminated.

a) Bond Length Distributions: Termination Broadening

A successful model must, of course, reproduce the positions of the peaks of an experimental RDF and their relative heights. Early unrelaxed monoatomic models were only partially successful in this regard. Energy-minimised structures are more adequate (see Sect. 2.4.1). However, as stressed in Sect. 2.2.3, the breadth and asymmetry of near-neighbour distribution functions are important structural characteristics and represent valuable diagnostic information, and any successful model must, therefore, reproduce them. It is vital that a model RDF is compared with experimental data with allowances for the differences between the two. Experimental RDFs are derived from $S(Q)$ by a Fourier transform over a finite range in Q and are distorted versions of the "true" RDF. Specifically, termination of the transform at $Q = Q_{max}$ leads to convolution of $G(r) = 4\pi r[\varrho(r) - \varrho_0]$ with a *peak shape function*. A δ-function in $G(r)$ is transformed into a broadened peak with side lobes – the so-called termination broadening and termination ripple [2.71].

The effect is to replace the weighting factor $c_\alpha c_\beta f_\alpha(Q) f_\beta(Q)/|\langle f(Q)\rangle|^2$ in (2.2) (Sect. 2.2.1) with a convolution of terms in $\varrho_{\alpha\beta}(r)$ by a peak shape function $P_{\alpha\beta}(r - r')$:

$$G(r) = 4\pi r \sum_\alpha \sum_\beta \int_0^\infty P_{\alpha\beta}(r - r')\varrho_{\alpha\beta}(r')dr' , \qquad (2.4)$$

where

$$P_{\alpha\beta}(x) = \frac{1}{\pi} \int_0^{Q_{max}} \frac{c_\alpha c_\beta f_\alpha(Q) f_\beta(Q) M(Q) \cos(Qx)dQ}{|\langle f\rangle|^2} . \qquad (2.5)$$

The peak shape function is plotted for three reciprocal space window functions $M(Q)$ in Fig. 2.10. If the ratio of x-ray scattering factors $f_\alpha(Q) f_\beta(Q)/|\langle f\rangle|^2$ is independent of Q or for neutron scattering where the scattering factors are Q-independent,

$$P_{\alpha\beta}(x) = \frac{1}{\pi} \frac{c_\alpha c_\beta f_\alpha f_\beta}{|\langle f\rangle|^2} \int_0^{Q_{max}} M(Q) \cos(Qx)dQ . \qquad (2.6)$$

Further simplification results if a smooth window function is used, since termination ripple is almost eliminated, and it may be sufficient to represent the

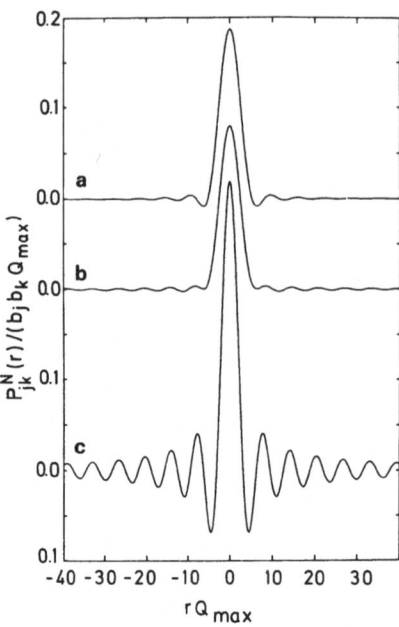

Fig. 2.10. Reduced neutron peak shape function [2.71] for three reciprocal space window functions $M(Q)$.

(Curve a) $M(Q) = \dfrac{\sin(\pi Q/Q_{max})}{(\pi Q/Q_{max})}$.

(Curve b) $M(Q) = e^{-BQ^2}$; $B = \ln(10/Q_{max}^2)$

(Curve c) $M(Q) = \begin{cases} 1 & Q < Q_{max} \\ 0 & Q > Q_{max} \end{cases}$

peak shape function by a Gaussian of half-width $\simeq 2.7/Q_{max}$ or $\sigma_t \simeq 2.3/Q_{max}$ (or, less accurately, $\sigma_t \simeq 1.6/Q_{max}$ if the structure is truncated at Q_{max}).

An alternative is to compare calculated and experimental measurements for $S(Q)$ directly: in this case an unbroadened experimental $S(Q)$ is compared with a computed $S(Q)$ – broadened due to finite model size. Consequently, the method only represents an advantage for large models where $G(r)$ may be trusted to values of r greater than, say, 1.5 nm.

It is good practice to compare values in real and reciprocal space anyway. $G(r)$ and $S(Q)$ emphasise different aspects of the structure. For example, the so-called "pre-peaks" which appear at values of Q lower than the first strong peak may be related to local fluctuations in chemical short-range order or to longer-range fluctuations in density [2.30].

b) Thermal Broadening

A second contribution to the broadening of experimental RDFs, σ_{th}, results from the effects of thermal atomic motion. An average value of the thermal root mean square displacements $(\overline{u^2})^{1/2}$ observed in crystalline solids gives some guide to the value of σ_{th}. Due to correlation of the motion of near-neighbour atoms, $(\overline{u^2})^{1/2}$ overestimates thermal broadening of the first two or three peaks in the RDF: near neighbours tend to vibrate with a small phase difference so that their separation is normally less than $(\overline{u^2})^{1/2}$. A more satisfactory treatment is to extract broadening terms from measurements of the RDF of an amorphous and crystalline alloy – the measurements being made under identical experimental conditions – particularly Q_{max} and temperature. Assuming that the vibrations of atoms in a glass are equivalent to those in a crystal – that is, the vibrational densities of states are similar [2.27] – this method gives an estimate of all broadening terms other than static (disorder) broadening. The latter can thus be obtained by subtraction in quadrature. The method has been used with success in several examinations of models for amorphous germanium since *Temkin* et al. [2.29] have provided equivalent RDF data for crystalline and amorphous germanium. Unfortunately, this practice has not been repeated for amorphous metallic alloys and it seems *highly desirable* that it should in future.

An alternative, which is also rarely practised, is to measure $S(Q)$ at low temperatures where σ_{th} may be neglected.

For most of the model calculations considered here, values of σ_{th} are not available. The only remaining approach is to add an arbitrary broadening term or terms which, together with σ_t, reproduces the breadth of first, second and higher-order peaks. Since these values are now *integral* components of the model, they should be quoted and, of course, be reasonable.

2.4 Models for Transition Metal-Metalloid Glasses

In this section, we review the status of different structures for TM–M glasses by comparing the predictions of models with experimental data. The word prediction is not unreasonable since development of most of the models described here *preceded* many of the most accurate experimental measurements presented in Sect. 2.2. Attention will be focused on those studies which allow comparison of simulated functions with partial distribution functions. Consequently, the much larger body of information on DRP models for monoatomic amorphous metals and which, it has been claimed, gives some indication of the local structure around metal atom sites in TM–M glasses, has been given lower priority.

2.4.1 Dense Random-Packed Models for Binary Alloys

a) Correlation Functions

Sadoc et al. [2.72] examined the properties of a number of models for binary alloys, $T_{1-x}M_x$ for x in the range 0.05 to 0.3 and radius radios p from 0.7 to 1.0. Computer-generated, unrelaxed models were studied and the building algorithm included M–M avoidance and a coordination number of 9 to simulate "chemical" interactions observed in the crystal. Like the monoatomic clusters also examined by these authors, the algorithm appears to generate incomplete icosahedral local structures rather than trigonal prisms.

RDFs were calculated for Ni–P alloys using a parabolic broadening function, corresponding to a Gaussian broadening parameter, $\sigma \simeq 0.005$ nm. (For Ni–P, $Q_{max} = 170$ nm^{-1} and thus $\sigma_t \simeq 0.014$ nm.) A typical example is given

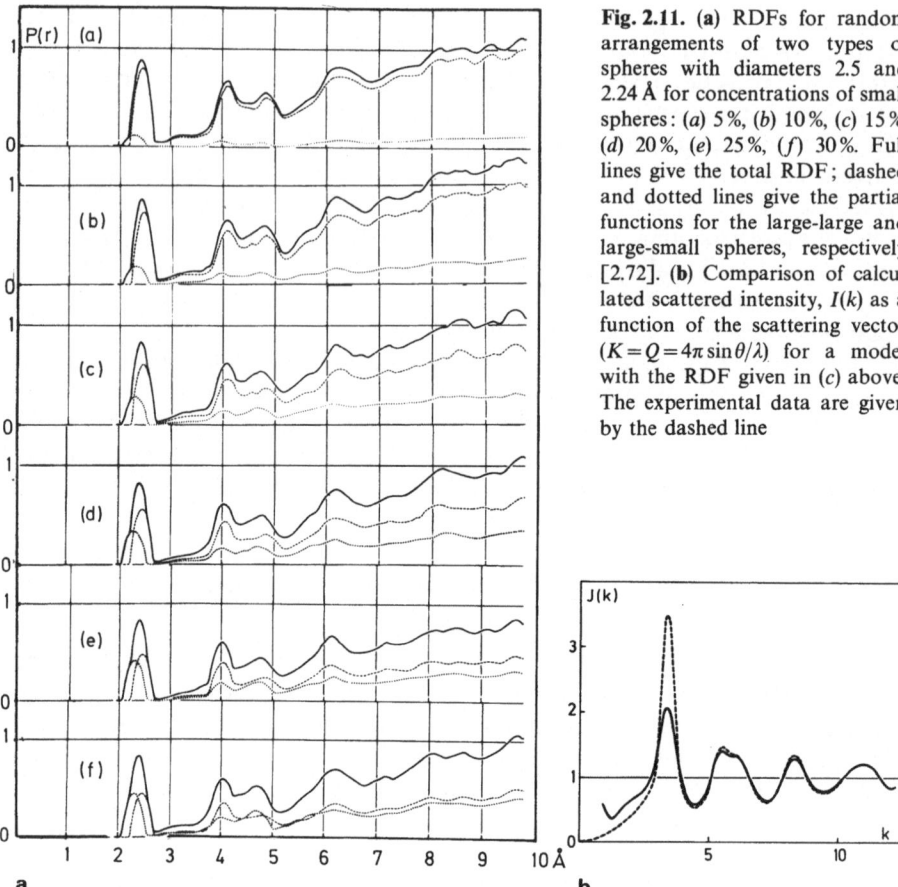

Fig. 2.11. (a) RDFs for random arrangements of two types of spheres with diameters 2.5 and 2.24 Å for concentrations of small spheres: (a) 5%, (b) 10%, (c) 15%, (d) 20%, (e) 25%, (f) 30%. Full lines give the total RDF; dashed and dotted lines give the partial functions for the large-large and large-small spheres, respectively [2.72]. (b) Comparison of calculated scattered intensity, $I(k)$ as a function of the scattering vector $(K = Q = 4\pi \sin\theta/\lambda)$ for a model with the RDF given in (c) above. The experimental data are given by the dashed line

in Fig. 2.11a: the major features – the positions of peaks and their relative intensities – are represented more accurately by this model than by monatomic models. Comparison with experimental $I(Q)$ data is shown in Fig. 2.11b.

Finney [2.11] has argued that although this model is *relatively* successful in reproducing x-ray scattering data, the density is 18% lower than random close packing, suggesting considerable voidage between the distorted icosahedral units.

Ching and *Lin* [2.55] constructed a small (500-atom) physical model, followed by energy minimisation using a Lennard-Jones potential function, to simulate the structure of amorphous Fe_3P, Ni_3P, and Co_3P. RDFs were limited to 0.6 nm by the small size of the model; nevertheless, the major features are reproduced with reasonable accuracy.

An extensive series of investigations into the construction and properties of large, computer-generated and energy-minimised models for binary TM–M alloys has been conducted by Boudreaux and co-workers. *Boudreaux* and

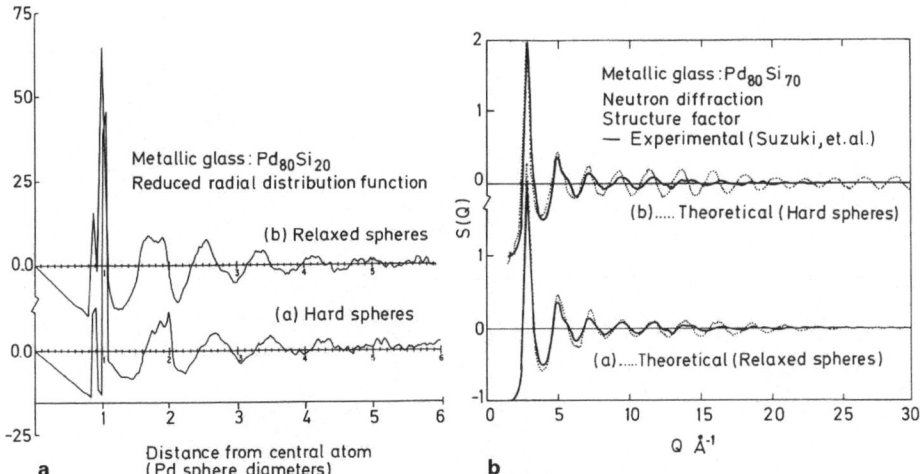

Fig. 2.12a, b. Reduced RDF for $Pd_{80}Si_{20}$ before and after energy minimisation using a Lennard-Jones potential. **(b)** Comparison of the calculated and experimental neutron structure factor [2.35] for relaxed and unrelaxed models [2.56]

Gregor [2.56] examined the properties of a 2000-atom model, scaled to simulate $Pd_{80}Si_{20}$ or $Fe_{80}P_{20}$. The computed "reduced radial distribution function" is shown in Fig. 2.12a and calculated and observed values of $S(Q)$ – derived by neutron scattering [2.35] – are given for comparison in Fig. 2.12b. The calculated curves reproduce the principal features of experimental data, but good agreement with experiment cannot be claimed since the second peak shoulder at $Q \sim 55 \ nm^{-1}$ is not reproduced.

Investigations also include some calculations for Fe–B and Fe–P alloys and a single model for $Pd_{80}Ge_{20}$ [2.64]. Boudreaux's reduced partial distribution functions for Fe–P and Fe–B are shown in Figs. 2.13, 14.

It is difficult to assess the accuracy of data for Fe–P alloys since accurate, high resolution partials have not been obtained. There is, however, adequate information on Fe–B, although conflict exists over important detail. The total x-ray RDF for $Fe_{80}B_{20}$ [2.32] shows no indication of the sharp peak near 1.5 atom diameters (Fig. 2.14a) and while the positions of both components of the second peak near 2 atom diameters are correctly predicted, their relative intensities are reversed. The same comments apply to the second peak in the partial RDFs although the intensity ratio for Fe–B correlations is the more correct.

The breadth of the first peaks of the partial RDFs are $\sigma \sim 0.008$–$0.018 \ nm$ for $G_{Fe-B}(r_1)$ and $G_{Fe-Fe}(r_1)$, respectively. Experimental values for σ_s are 0.01 and 0.017 nm. Since thermal broadening is not included in the calculated curve, the breadth of the first Fe–B peak is likely to be in excellent agreement with experiment. When comparing the calculated and experimental RDFs directly, note that the computed curve should be broadened with a Gaussian of width $\sigma > 2.3/115 = 0.02 \ nm$. This would lead to (further) smoothing of the features of

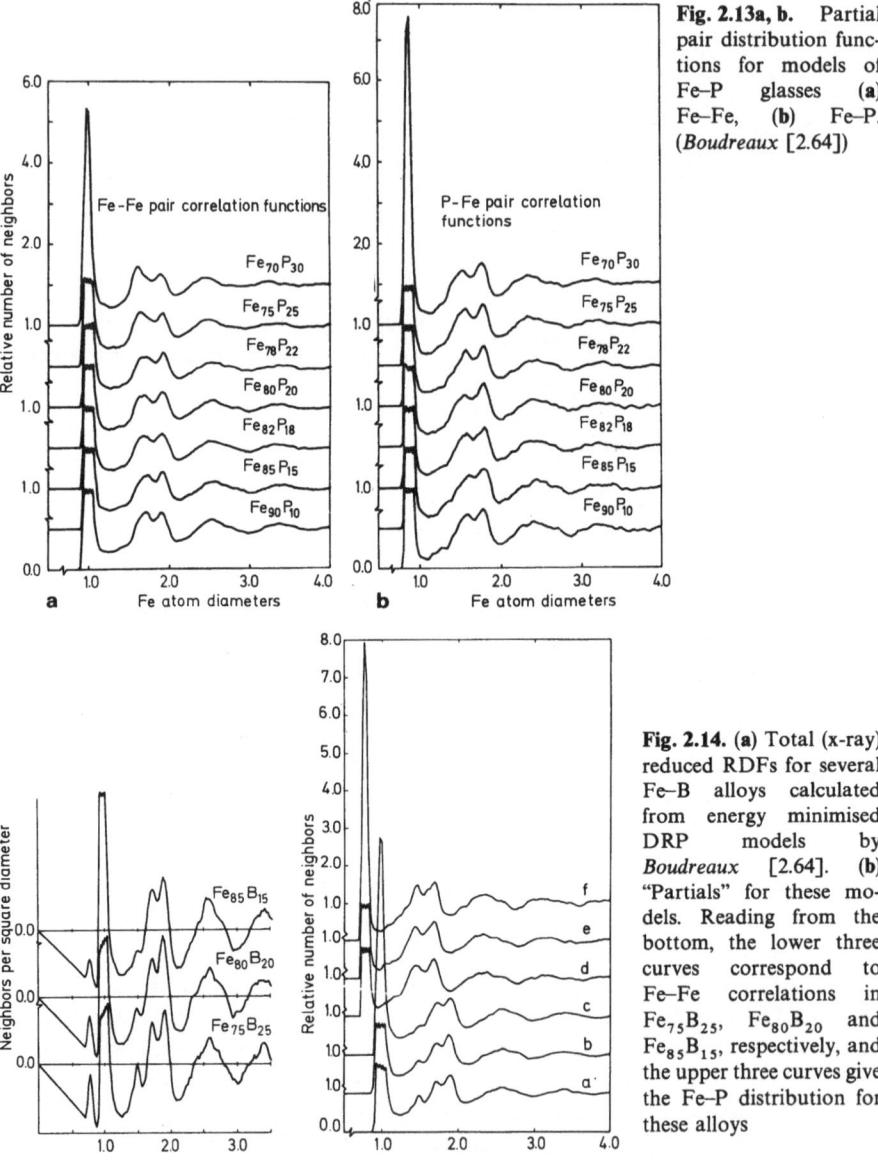

Fig. 2.13a, b. Partial pair distribution functions for models of Fe–P glasses (**a**) Fe–Fe, (**b**) Fe–P. (*Boudreaux* [2.64])

Fig. 2.14. (a) Total (x-ray) reduced RDFs for several Fe–B alloys calculated from energy minimised DRP models by *Boudreaux* [2.64]. (**b**) "Partials" for these models. Reading from the bottom, the lower three curves correspond to Fe–Fe correlations in $Fe_{75}B_{25}$, $Fe_{80}B_{20}$ and $Fe_{85}B_{15}$, respectively, and the upper three curves give the Fe–P distribution for these alloys

the second peak, whereas experimental second peak features are sharp in *Nold* et al.'s data (Fig. 2.4).

    Fujiwara and *Ishii* [2.65] have constructed 1600-atom models for Fe–P alloys, relaxed using a Morse potential. Total and partial distribution functions are given in Fig. 2.15 for a-$Fe_{76}P_{24}$. Agreement with experimental data for the total RDF is seen to be excellent (although the broadening function used is only

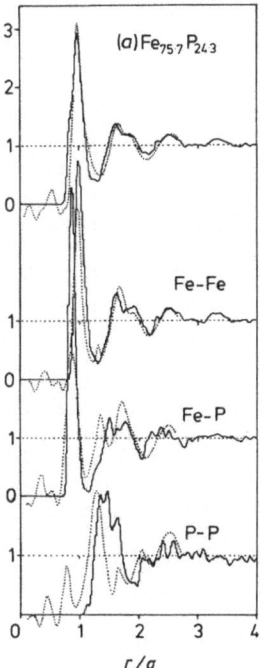

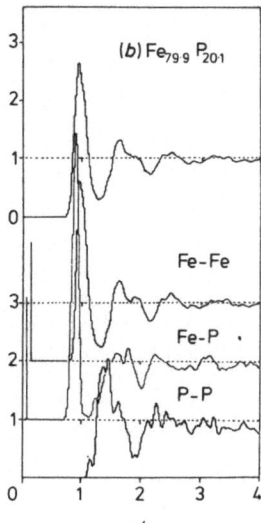

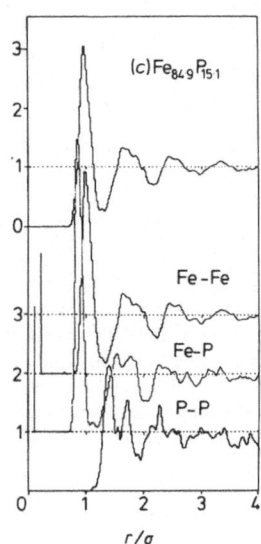

Fig. 2.15a–c. Total (top) and partial RDFs for Fe–P alloys (*Fujiwara* and *Ishii* [2.65]): (a) $Fe_{76}P_{24}$, (b) $Fe_{80}P_{20}$, (c) $Fe_{85}P_{15}$. Experimental data (dotted) are those of *Waseda* et al. [2.37]

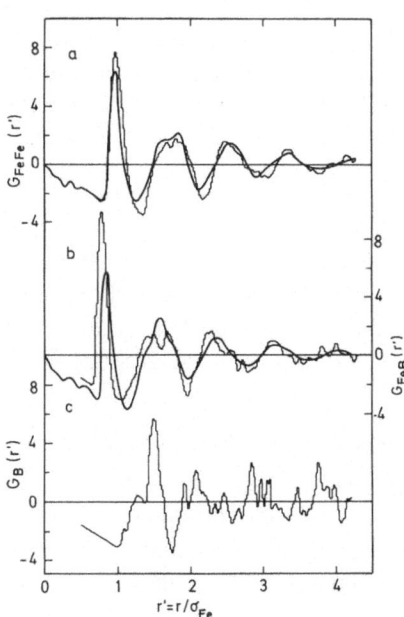

Fig. 2.16. Calculated and experimental partial distribution functions for a-$Fe_{85}B_{15}$. The histograms give functions for a relaxed DRP model (*a*) Fe–Fe, (*b*) Fe–B, (*c*) B–B. Experimental data should be compared with those given in Fig. 2.5

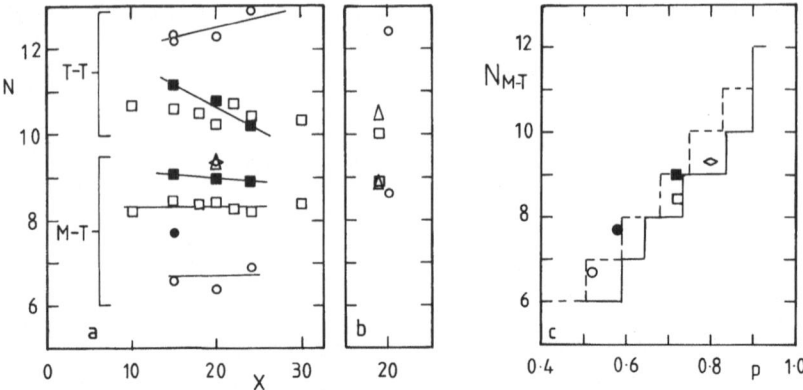

Fig. 2.17a, b. Calculated (**a**) and experimental (**b**) M–T and M–M coordination numbers for several amorphous alloy systems. Lines are drawn merely to connect points for compositionally-related alloys. (**a**) (○) Fe–B *Boudreaux* [2.64]; (●) Fe–B *Fujiwara* et al. [2.37]; (□) PdSi, Fe–P *Boudreaux* [2.64]; (■) Fe–P *Fujiwara* and *Ishii* [2.65]; (◇) $Pd_{80}Ge_{20}$ *Boudreaux* [2.64]; (△) $Pd_{80}Si_{20}$ *Gaskell* [2.59]. (**b**) (○) $Fe_{80}B_{20}$, (△) $Ni_{81}B_{19}$ *Lamparter* et al. [2.32]; (□) $Co_{81}P_{19}$ *Sadoc* and *Dixmier* [2.31]. (**c**) Comparison of calculated values of the M–T coordination numbers for DRP models and the "staircase" curve giving the number of spheres surrounding a central sphere of radius ratio p

about half that for termination broadening alone). The errors are more serious for the partial functions where, since $Q_{max} \simeq 7 \text{Å}^{-1}$, $\sigma_t \simeq 0.022–0.033$ nm compared with 0.005 nm used in the calculation.

Calculations for $Fe_{85}B_{15}$ have been reported by *Fujiwara* et al. [2.37]. The resulting partial PPDFs are shown in Fig. 2.16 and the general features show some agreement with Boudreaux's results (Fig. 2.14). The relative intensities of the components of the second peak are again reversed compared to Pd–Si and Fe–P alloys. Judgment of the validity of these calculations rests on the reliability of the conflicting x-ray data of Waseda and co-workers [2.37] on the one hand and *Fukunaga* et al. [2.35], *Nold* et al. [2.32], on the other. *Dini* et al. [2.38] have systematically examined the factors which could lead to such disagreement and conclude that differences in the thermal history of the Fe–B alloys during quenching could be a factor and that combinations of random experimental errors and systematic differences in data analysis techniques could be important. They do not state which they regard as the more reliable measurement, but suggest that the shape of the second peak may not be a particularly useful feature of the structure of Fe–B glasses on account of its sensitivity to experimental variables.

b) Coordination Numbers and Analyses of Interstice Distribution Functions

Boudreaux [2.64], *Jansen* et al. [2.25], *Boudreaux* and *Frost* [2.26] and *Fujiwara* and *Ishii* [2.65] each conclude that M–T coordination numbers, N_{M-T}, are relatively independent of concentration for a given alloy system. Their data are collected in Fig. 2.17a together with estimates for N_{T-T} which decreases slightly with increasing concentration. Experimental coordination numbers for

Table 2.3. Calculated and experimental density values for several amorphous alloys. Densities are calculated with the metal atom diameter corresponding to the experimental first peak in $G_{TT}(r)$, where available, or $G_{tot}(r)$ otherwise. (Experimental values for Fe–P are from *Waseda* et al. [2.78].) Values asterisked are calculated in this way; values below, in brackets, are the authors own estimates

Alloy	Density/kg m⁻³		Ref.
	Calculated	Experimental	
$Fe_{85}B_{15}$	7.82	7.4	[2.37]
$Fe_{100-x}B_x$	6.7 ± 0.3	$7.1 - 7.4$	[2.64]
$15 < \times < 25$			
$Fe_{85}P_{15}$	6.82	7.25	[2.65]
$Fe_{80}P_{20}$	6.84	7.13	[2.65]
$Fe_{76}P_{24}$	6.90	7.03	[2.65]
$Fe_{80}P_{20}$	7.35*	7.13	
	(7.1)		[2.26]
$Fe_{75}P_{25}$	7.65*	7.0	
	(6.84)		[2.26]
$Fe_{75}P_{25}$	6.0	7.0	[2.66]
$Pd_{80}Si_{20}$	10.4	10.25	[2.65]
$Pd_{80}Si_{20}$	10.4	10.25	[2.59]

the higher precision measurements quoted in Table 2.1 are also reproduced in Fig. 2.17b.

Calculated N_{M-T} values are 8–9 for alloy systems with large radius ratio (Pd–Ge, Pd–Si, Fe–P) and significantly lower for the borides – a feature which is not observed experimentally. Further work is needed on this important point, but evidence now available suggests that the larger metalloid atoms in DRP simulations force a large cavity size and *therefore* a high coordination number. The alternative, a variable cavity size with *constant* N_{M-T}, is not in evidence. In fact, N_{M-T} for all relaxed DRP models is approximately a linear function of p (Fig. 2.17b) and follows the "staircase" curve giving the maximum (integer) number of spheres which can surround a central sphere of radius ratio p (see *Blétry* [2.25]). Unrelaxed DRP structures exhibit even lower values of N_{M-T} (*Blétry* [2.25], *Jansen* et al. [2.26]).

Fujiwara and Ishii's analysis of the Voronoi polyhedron in Fe–P alloys shows that P is surrounded by trigonal prisms (44–58%), tetragonal dodecahedra (12–23%) and Archimedean antiprisms (11–16%). Boudreaux and Frost's analysis of Fe–B alloys suggests a mixture of uncapped trigonal prisms and octahedra, the precise numbers depending on the method used to analyse the cavity distribution. There are, in addition, a number of polyhedra not listed by Bernal such as the dihedron in which five Fe atoms surround B. This figure makes a significant contribution to the cavity type distribution in alloys at both ends of the composition range.

The environment around M atoms is found to be icosahedral by Boudreaux and Frost, whereas Fujiwara and Ishii observe polyhedra with a mean number of faces equal to 15 (an icosahedron has 20 faces).

c) Density

Density values calculated for several models are quoted in Table 2.3. Agreement within about 6% is achieved by most models. The difference in density of amorphous and crystalline phases is about half this value. More than any other simple property, perhaps, density predictions emphasise the inadequacies of current models for amorphous metals.

2.4.2 Stereochemically-Defined Models

a) Correlation Functions

Kobayashi et al.'s model [2.66] for a-$Fe_{75}P_{25}$ is the only example of an energy minimisation procedure using an orientation-dependent P–Fe interaction. A random collection of atoms is the starting structure and the algorithm attempts to force a cluster in which the phosphorus atom is surrounded by Fe atoms with trigonal prismatic symmetry. A degree of cooperative rearrangement of trigonal prisms is allowed, but it is not clear – from density values, for example – that this has been achieved. Reduced RDFs are shown in Fig. 2.18 without broadening apparently. The calculated total RDF is not quoted, which is unfortunate in view of the possible errors in the partials, but it seems unlikely that agreement with experiment is better than Fujiwara and Ishii's simulation in view of the poor agreement of $G_{Fe-Fe}(r)$, although the Fe–P PPDF (perhaps not surprisingly) does represent a distinct improvement.

Gaskell's model [2.57, 58] for $Pd_{80}Si_{20}$ was constructed initially as a random cluster of regular trigonal prisms with the Fe_3C, cementite, edge-sharing arrangement. The metalloid elements were "inserted" at the first stage of the energy-minimisation programme and an attempt was made to keep the metalloid in the centre of a distorted trigonal prism by defining two equilibrium Si–Pd distances as observed in c-Pd_3Si. Furthermore, distances between Si and the six Pd atoms which lie at the vertices of a given trigonal prism in the original model were included in the calculation of the elastic energy regardless of the distance they move subsequently (in retrospect, more attention could have been given to the inclusion of stronger Si–Pd force constants).

Total pair correlation functions and $S(Q)$ data are given in Fig. 2.19, 20. Gaussian broadening functions were used to attempt to simulate the breadth of the first peak and yet be reasonably consistent with estimated thermal and termination broadening parameters. The experimental data are terminated at $Q_{max} = 250$ nm^{-1}, thus $\sigma_t \simeq 1.6/250 \simeq 0.0072$ nm. Values of $\sigma = 0.0094$ and 0.014 nm were chosen for the Gaussian broadening parameters for Si–Pd and Pd–Pd correlations, respectively. The thermal broadening parameter is not known, but values of $\sigma_{th} = (\sigma^2 - \sigma_t^2)^{1/2}$ of 0.006 and 0.012 nm seems not unreasonable (σ_{th} for the Ge–Pd first neighbour peak is 0.008 nm [2.33] and less than 0.007 nm for Fe in a Fe–B glass [2.79]).

$S(Q)$ data are obtained as Fourier transforms of $G(r)$ functions broadened as described above. In this case, termination of the function and errors in $G(r)$

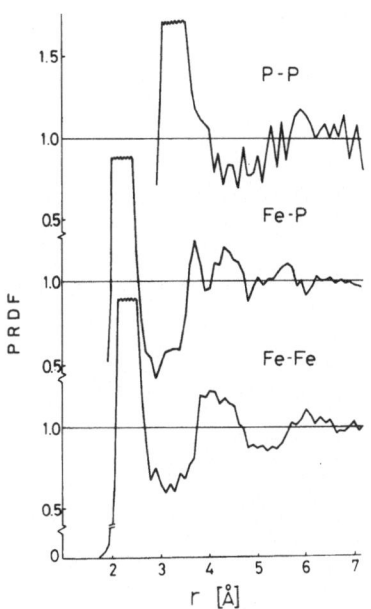

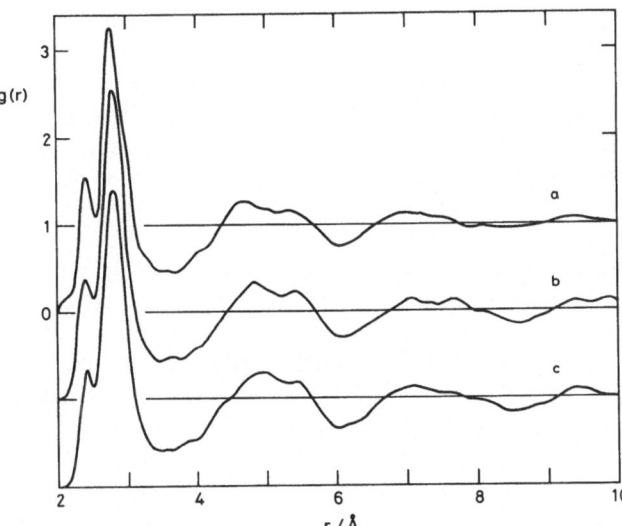

Fig. 2.18. Partial RDFs for an energy-minimised model of $Fe_{75}P_{25}$ using an angle-dependent force-field (*Kobayashi* et al. [2.66])

Fig. 2.19. Pair correlation functions $\varrho(r)/\varrho_0$ for a-$Pd_{80}Si_{20}$ [2.58]: *(Curve a)* Experimental data derived from neutron scattering measurements with $Q_{max} = 250\ nm^{-1}$ [2.35]. Curves (*b*) and (*c*) refer to spherical cores of the random trigonal prismatic model. (*b*) corresponds to a core radius of 0.86 nm and gives the better approximation to the near-neighbour distance distributions. (*c*) For a core radius of 1 nm the model is not fully dense but gives the best indication of $g(r)$ beyond about 0.5 nm where effects of finite model size become important

beyond about 0.8 nm introduced by the small size of the model were suppressed by applying a window function $\sin(\pi r/12)/(\pi r/12)$ to $G(r)$. This introduces some broadening into $S(Q)$, specifically, convolution by a Gaussian function for which $\sigma = 0.02\ nm^{-1}$.

Agreement between experimental and computed RDFs and $S(Q)$ is good. The splitting of second peaks in the RDF and $S(Q)$ are both adequately

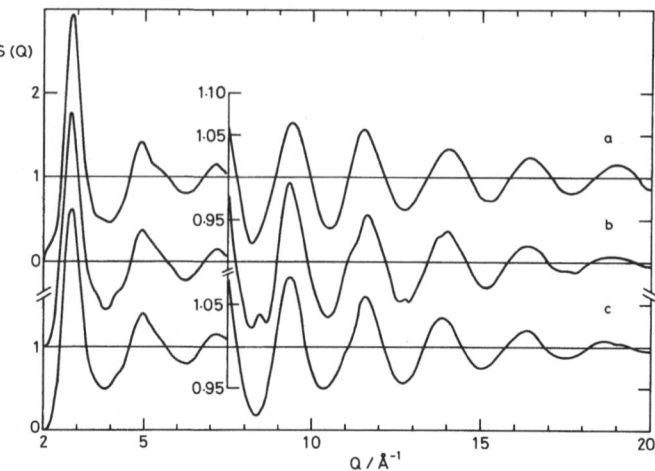

Fig. 2.20. (Curve a) Experimental structure factor for a-$Pd_{81}Si_{20}$ [2.58, 59]. **(Curves b, c)** Computed $S(Q)$ for the models with cores of radius 0.86 and 1.0 nm, respectively

reproduced. The width of the first peak (Si–Pd) in the RDF is less satisfactory. Had it been reasonable to do so, a reduced Si–Pd broadening parameter would have improved matters. This suggests that a real glass is probably *more* ordered than the model.

Calculated PPDFs [2.39] are shown in Fig. 2.21 and compared with the experimental data for $Pd_{84}Si_{16}$ of *Sadoc* and *Dixmier* [2.34]. Experimental partials were obtained from an analysis of only two measurements and it is therefore impossible to obtain all three PPDFs. The function labelled $G_{Si-Pd}(r)$ contains 17% of the Si–Si PPDF which is likely to have a strong peak near 0.36 nm and thus to contribute to the shoulder observed in the Si–Pd "partial" at this point. Contributions in other regions are likely to be very small. Functions are compared using the broadening parameters described earlier. Sadoc and Dixmier's neutron data are terminated at $Q_{max} = 200 \text{ nm}^{-1}$ but the figure for x-ray scattering is not quoted. The additional broadening required ($\sigma \sim 0.008$ nm) is unlikely to make any significant difference to the conclusions, especially in view of the uncertainty in experimental values. (See also the note to Fig. 2.21.)

The fit to experiment is clearly quite reasonable for the first two peaks and in view of the inherent difficulties in extracting PPDFs, it would be unrealistic to extend the comparison further.

b) Coordination Numbers, Density and Polyhedron Size Distributions

Kobayashi et al. [2.66] did not explicitly quote coordination numbers for a-$Fe_{75}P_{25}$ but it is clear from their data that values are generally lower than those observed in the crystal. This appears to be due to the anomalously low density of the model (6.0 compared with 7.0 kg m^{-3} observed experimentally).

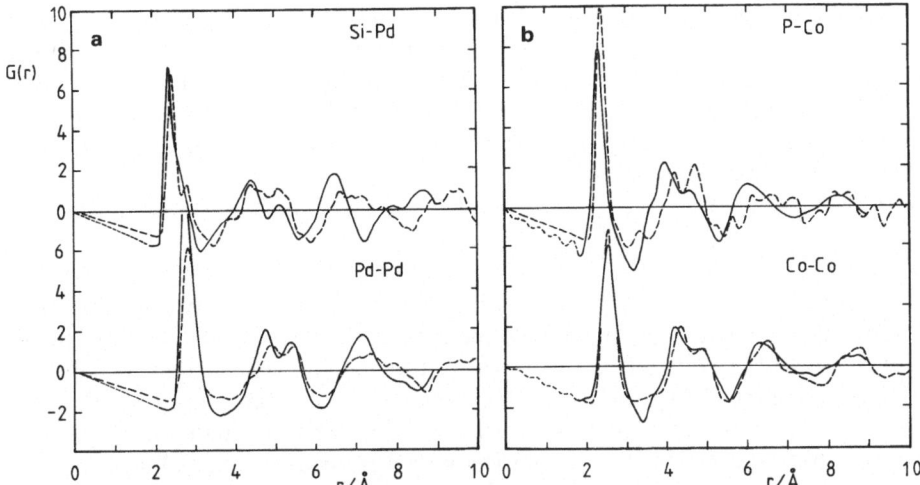

Fig. 2.21a, b. Experimental (solid lines) and calculated (dashed lines) partial-pair distribution functions for a-Pd–Si (**a**) and Co–P (**b**) alloys [2.39]. Experimental data are from *Sadoc* and *Dixmier* [2.31]. Both calculated distribution functions are based on the random trigonal prismatic model using cementite edge-sharing arrangements. Calculated and experimental data are in good agreement for Pd–Si but the P–Co PPDF shows a discrepancy near 4 Å indicating the possibility of "Fe_3P" edge-sharing arrangements in the glass. Due to differences in the definition of $G_{ij}(r)$, a small error is included in this figure. The computer PPDF's should be divided by the atom fractions of Pd and Co (0.84 and 0.81, respectively)

Coordination numbers estimated from their data are Fe–Fe: 10, P–Fe: 8, and P–P: 3.5.

Gaskell has determined the Si–Pd coordination number for the first sub-shell of Pd atoms around Si as 5.9 ± 0.3 with the distribution shown in Fig. 2.9. About 75% of the silicon atoms have at least two further (capping) Pd atoms within 0.34 nm giving a total of 9.2.

The density cannot be determined accurately in view of the small size of the model and the inevitable fluctuations in ϱ with radius. The calculated value, $10.4\,\mathrm{kg\,m^{-3}}$, lies within the range observed experimentally (10.25–$10.5\,\mathrm{kg\,m^{-3}}$).

Gellatley and *Finney* [2.68] have examined the distribution of polyhedra in this model using an analysis of Voronoi polyhedra and the "radical plane" technique. They argue that the latter gives more representative information than the more popular Voronoi technique for polyatomic alloys. Si and Pd polyhedra have an average of 12.1 and 14.5 faces according to the Voronoi analysis or 11.7 and 14.7, respectively, using the radical plane construction. A clear picture does not emerge from their analysis of polyhedron shapes. Regular trigonal prisms are rarely observed and this perhaps is not surprising as the prisms are extensively distorted even in crystalline Pd_3Si. However, some of the polyhedra observed in the model are distorted trigonal prisms. Nonetheless, the complexity of the distribution of shapes is surprising and until this has been examined in more detail, it will not be clear to what extent the trigonal prismatic character of the model has been preserved during energy minimisation. Stereo drawings of the model show that distortions of trigonal prismatic symmetry are extreme for atoms at the edge of the model (which have been excluded or given low weight in the calculation of RDFs, etc.) and this would be expected for a central force field. A high proportion of trigonal prisms occupy the central region, however.

c) Extension to Other T–M Alloys

It is important to recognise an important difference between Gaskell's stereochemically-defined model and the trigonal prismatic model of Kobayashi et al., *Gaskell* [2.57, 58], *Kobayashi* et al. [2.66] and, to a lesser extent, *Sadoc* et al. [2.19] attempted to produce a structure containing the local organisation around M atoms observed in the crystal. In all models – including DRP models where trigonal prismatic coordination may appear as a result of a particular choice of radius ratio or potential energy function – an element of randomness is included. Gaskell's model contains an additional constraint – edge-sharing designed to reproduce the structure observed in cementite. In Kobayashi et al.'s simulation, since the starting configuration is a *random* arrangement of M atoms, the manner in which units are packed is also random.

The question arises, does cementite-like packing adequately represent the structure of all amorphous T–M alloys? Whilst it was originally intended that the model for palladium-silicon should be generally applicable to other amorphous T–M alloys, the answer seems to be – no.

As discussed in Sect. 2.2.3, there are subtle but important differences in the distribution of T atoms around M in crystalline Fe_3C, Fe_3P and Ti_3P lattices. Such differences in the nearest-neighbour coordination shell appear to carry over into the amorphous state [2.39]. There are even more important and far-reaching differences at the level of the medium-range structure, specifically, in the way trigonal prisms are packed together. This is illustrated in Fig. 2.7. Generally, alloys with small radius ratios adopt the Fe_3C structure and, for large values of p, the Fe_3P or Ti_3P structure is preferred. However, this is only a crude guide and there are several important exceptions, see [2.44]. The differences can be characterised as packing of trigonal prisms around half octahedra (Fe_3C) (Fig. 2.7b) or around tetrahedra (Fe_3P and Ti_3P) (Fig. 2.7c). The two arrangements can be distinguished through differences in the distances from the metalloid to a second neighbour T atom, $r_{M-T(2)}$:

Tetrahedra/Tetrahedra, $r_{M-T(2)} = (p + 2p - 1/3)^{1/2} + (8/3)^{1/2}$

Tetrahedra/Octahedra, $r_{M-T(2)} = [p + 2p + 5 - 4(p^2 + 2p)^{1/2} \cos\beta]^{1/2}$,

where $\beta = \psi_1 + \psi_1$ and the angles ψ_1 and ψ_2 are dihedral angles for a half octahedron and a distorted tetrahedron:

$$\psi_1 = \cos^{-1}(3^{-1/2}) \qquad \psi_2 = \cos^{-1}[(3p + 6p)^{-1/2}].$$

Table 2.4. Calculated values of the second-neighbour metalloid-metal distance $r_{M-T(2)}$ on the basis of tetrahedral/tetrahedral (T/T) or tetrahedral/octahedral (T/O) packing of trigonal prisms observed in c-Fe_3P and Fe_3C, respectively. Distances are given in units of the metal atom radius [from the first peak in G_{TT})] and p is the metalloid radius ratio. Brackets indicate likely correlations of experimental and calculated data. Alloys, which in the crystalline phase adopt the Fe_3C structure, appear to be similarly packed in the glass

Alloy	P	$r_{M-T(2)}$		
		calc(T/T) (Fe$_3$P)	exp	calc(T/O) (Fe$_3$C)
$Co_{81}P_{19}$	0.83	3.05	3.11	3.26
$Fe_{80}B_{20}$	0.67	2.83	2.95	3.06
$Ni_{80}B_{20}$	0.67	2.85	3.03	3.07
$Pd_{84}Si_{16}$	0.74	2.93	3.19	3.15

Inserting values for the radius ratio calculated using experimental values obtained from PPDFs, the four alloys whose PPDFs are known most accurately, values of $r_{M-T(2)}$ in units of the metal radius are calculated and quoted in Table 2.4.

From this it is clear that Ni–B, Pd–Si alloys for which the cementite structure would be preferred are also more likely to have the Fe_3C tetrahedral/octahedral arrangement of prisms in the *glass* also.

A stable T_3M phase is not observed in the Co–P system but Fe_3P phases are found in P-rich ternary $Co_3(P, B)$ alloys [2.73]. The position of the second-neighbour M–T(2) distance in the amorphous alloy is more adequately represented by the Fe_3P-type structure, as Table 2.4 shows. For $Fe_{80}B_{20}$, the observed distance falls midway between the predicted levels.

It seems likely, therefore, that the medium-range structure in these amorphous alloys can be represented by two essentially different packings and that there may be some equivalence in the medium-range topologies of crystalline and amorphous lattices.

This conclusion is supported by attempts to make the stereochemically defined model for "Fe_3C-like" $Pd_{80}Si_{20}$ fit data for $Co_{81}P_{19}$. The model has been rescaled and relaxed under several sets of force constants. Computed Co–Co and P–Co PPDFs are given in Fig. 2.21 for the most successful calculation. Co–Co distances are reproduced adequately to about 0.8 nm so that the total RDF and $S(Q)$, which are dominated by this function, agree well with experiment. The profiles of the first P–Co peak and the position of the *second* component of the second peak at 0.49 nm agrees with experiment, but the first component at 0.39 nm is inadequately represented by a peak at 0.417 nm in the computed data. This difference, although small, is significant and represents the closest P–Co(2) distance in an essentially tetrahedral/octahedral arrangement. It seems likely that a new model using TP packing as in Fe_3P will be necessary to model successfully the experimental data for amorphous phosphides.

2.5 Concluding Remarks

The development and evaluation of atomic models for metallic glasses has now reached an interesting phase. Until recently it might be said that theory, simulation (and speculation) were advancing knowledge faster than experiment. Random sphere-packing models *appeared* to be good enough to explain the rather limited experimental information obtained from low resolution measurements. It was therefore natural to explore the properties of glasses *through* simulation and modelling has, for a time, become almost an end in itself. Experiment has now presented a new challenge which appears to have narrowed the range of possible structures in one sense and extended them in another. It now seems likely that fixed, geometry-independent metalloid coordination must be accommodated in models for T–M alloys along with M–M avoidance. At the same time, *none* of the existing models would predict anything resembling the form of $G_{BB}(r)$ observed in Ni–B (Fig. 2.5).

Further new experimental information of high quality is likely to emerge in the near future. EXAFS studies of the environment of low atomic number metalloids are now possible as shown by the recent work of *Flank* et al. [2.75] on a-Co–P and *Greaves* et al. [2.80] on Na and Si in silicate glasses. Analysis of x-ray near-edge structure (XANES) can, in principle, extend the information available in two important respects. Firstly, by providing the symmetry of the first-neighbour shell around the x-ray absorbing atom and secondly, by allowing higher-neighbour shells to be examined [2.36]. Almost certainly, great benefit will be derived from collecting structural information for the same set of samples using different techniques.

What kind of summary does *current* knowledge allow? What extrapolations to *new* knowledge can be envisaged?

a) Both DRP and SD models have been shown to be only *conditionally* acceptable as models for amorphous T–M alloys. *Some* DRP models are successful for alloys containing large metalloids (Fe–P, Pd–Si) where the radius ratio $p \simeq 0.75$ and where random packing leads to the observed ninefold coordination. Most of the properties of models for amorphous borides only qualitatively resemble experimental observations. SD models based on trigonal prisms have been shown to represent the properties of Pd–Si well but extension to borides has not been demonstrated (and has not been tried). Extension to other (high p) alloys will probably require a new physical model based on the medium-range structure of c-Fe$_3$P or Ti$_3$P lattices. All calculations to date have been conducted on models which are small and therefore subject to large errors.

b) There is reasonably strong evidence that the structure of T–M alloys requires models with greater complexity and sophistication than that contained in randomly-packed binary sphere models. Extrapolation of results for DRP models of monoatomic amorphous metals to binary alloys seems likely to give particularly misleading results for T–M alloys.

c) Experiments suggest a degree of structural complexity at the level of the *local structure*. It is not clear whether this represents an element of a more comprehensive organisational scheme, extending perhaps to other alloy systems, or to the medium-range structure or beyond. Evidence is accumulating for some degree of chemical short-range order in amorphous T–T alloys, for example, the neutron scattering studies of *Sakata* et al. [2.81] and complementary EXAFS data obtained by *Raoux* et al. [2.82] on Cu–Ti glasses and liquids. Moreover, studies by high resolution transmission electron microscopy provide persuasive evidence for positional order extending to perhaps 2 nm in thin amorphous metal films [2.83].

d) It is not obvious why the first coordination shell around the metalloid should be ordered nor why there may be an equivalence of local structures in amorphous and crystalline phases. Directional bonding could provide an answer but there is no obvious scheme of covalent bonding which would stabilize trigonal prismatic coordination of Fe (say) around boron, which has no d-electrons. Some calculations have been performed on the electronic structure of very small clusters containing boron [2.23], but the investigations are only at a pilot stage. An alternative proposition is that trigonal prismatic coordination in crystals represents a compromise between the requirements of close-packing for T and the need to create space to insert M atoms. This is achieved by inserting "chemical twinning planes". *Dubois* [2.84] has recently drawn attention to this point and stresses the possible importance of the concept for amorphous alloys. In this event it is difficult to see how the principle could apply unless order – positional and chemical – extends to perhaps 5 to 10 sphere diameters. The presence of the split first peak in $G_{BB}(r)$ for $Ni_{81}B_{19}$ and oscillation beyond 1 nm, together with high resolution electron microscopic results on Pd–Si alloys, support this view.

e) In order to meet the challenge of new experimental evidence, models must inevitably become more sophisticated in *concept* as well as technique. It seems likely that the emphasis may be forced away from sphere packing, or even random arrangement of clusters, towards models in which chemical twin planes, multiple twins dislocations or disclinations [2.59] observed in small metallic particles or distorted crystals are adapted to provide models for the amorphous state. It no longer seems inconceivable that the strange and beautiful amorphous cluster models for argon shown in Fig. 2.2 could have counterparts in structure for binary alloys.

2.6 Postscript   (July 1983)

This review was completed in April 1982. At that time, the results of *Nold* et al. for $Fe_{80}B_{20}$ and *Lamparter* et al. for $Ni_{81}B_{19}$ [2.32] were new and only partially digested (indeed the full version of the work on Ni–B alloys only appeared in November 1982). Since the results on Ni–B, at least, represent some of the most accurate PPDFs yet available, it was clear at the time of writing that a more discerning test of any model would be afforded by comparison of simulated and experimental data for $Ni_{81}B_{19}$ rather than Pd–Si or Co–P alloys. It was also clear that the split peak in $G_{BB}(r)$ and oscillations beyond 14 Å in $Ni_{81}B_{19}$

(Fig. 2.5) represented facts which would fit uncomfortably within the concepts of any of the models presented here as Sect. 2.5 indicates. Then, as now, there is little in the published literature to indicate the direction in which a more adequate solution could lie (apart from the speculations contained in Sect. 2.5) and while it would be difficult, and perhaps unethical, for a reviewer to anticipate future publications, the reader who has reached this point is entitled to some consideration. Is the evidence forcing us to continue to follow concepts of increasingly well-defined local- and medium-range order or should we be prepared to re-examine the more elemental sphere-packing models? If neither, is a radically new concept required?

The answer to these questions can only be given on the basis of personal conviction at the moment (although substantial evidence is now appearing in pre-prints). *Lamparter, Sperl, Steeb*, and *Blétry* [2.32] have constructed a soft sphere model, computed PPDFs for $Ni_{81}B_{19}$ and have compared these data with their experimental results. Agreement is good for the near-neighbour peaks of the Ni–Ni and Ni–B partials but the amplitude of oscillations in $G_{ij}(r)$ becomes much smaller than experiment beyond about 6 Å. Furthermore, while these calculations succesfully reproduce the observed splitting in the first peak of the B–B partial, the *relative* intensity of the two components is reversed and, again, $G_{ij}(r)$ is featureless beyond about 6 Å.

It seems likely that a more detailed fit to the distinctive features of the Ni–B data will require locally-defined coordination of the metalloid. Furthermore, the differences between the *experimental* partials for amorphous Fe– and Ni–B alloys, support the view that the medium-range structures of the two alloys are not identical and that these differences are related to different packing schemes in the corresponding crystal structures. This much, then, is likely to remain as true in 1984 as it was in 1982.

The new features, towards which experimental evidence may be forcing future thought, concern the nature of the preferred medium-range order and its spatial extent. It is easier to be more definite about the last point. Almost certainly the spatial extent of order in $Ni_{81}B_{19}$, at least, will be closer to ten nickel atom diameters than the two or three implied by a random arrangement of trigonal prisms. This, in turn, suggests correlated rather than random arrangements of local structural units. The nature of the medium-range structure, that is, the mode of packing of local structural units is more difficult to forecast. Probably some equivalence with the structure of the corresponding crystalline phase will continue to give a good guide, but the emerging evidence of *Dubois* and co-workers suggests that this may be only part of the answer and that new features which distinguish glasses and *stable* crystalline phases must also be considered.

It is not difficult to forecast that the pace of development of structural concepts will not slacken and that the next few years will prove to be at least as fascinating as the last.

Acknowledgements. Much of the thought contained in this article – if not the actual task of committing it to paper – was made possible by the award of a SRC Senior Fellowship. Generous past and present support from Pilkington is gratefully acknowledged. The manuscript has benefited from the critical comments of D. S. Boudreaux, J. M. Dubois, J. L. Finney, D. M. Glover, A.K. Livesey, and S. Steeb.

References

2.1 J.Hafner: In *Glassy Metals* I, ed. by H.-J. Güntherodt, H. Beck, Topics Appl. Phys., Vol. 46 (Springer, Berlin, Heidelberg, New York 1981) p. 93

2.2 J.Dixmier, K.Doi, A.Guinier: In *Physics of Non-Crystalline Solids*, ed. by J.A.Prins (North-Holland, Amsterdam 1965) p. 67
J.Dixmier, K.Doi: Comptes Rendu **257**, 2451 (1963)

2.3 G.S.Cargill: J. Appl. Phys. **41**, 12 (1970)

2.4 P.H.Gaskell: Phil. Mag. **32**, 211 (1975)

2.5 J.D.Bernal: Nature **185**, 68 (1960); Proc. Roy. Soc. A **280**, 299 (1964)

2.6 G.D.Scott: Nature **194**, 956 (1962)

2.7 J.L.Finney: Proc. Roy. Soc. A **319**, 479 (1970)

2.8 A.H.Boerdijk: Phillips Res. Rep. **7**, 303 (1952)
F.C.Frank: Proc. Roy. Soc. A **215**, 143 (1952)

2.9 G.S.Cargill: J. Appl. Phys. **41**, 2248 (1970)

2.10 J.L.Finney, B.J.Gellatly, J.Wallace: In *Metallic Glasses: Science and Technology*, ed. by C.Hargitai, I.Bakonyi, T.Kemeny (Central Res. Inst. Phys., Budapest 1980) **1**, p. 55

2.11 J.L.Finney: Nature **266**, 309 (1977)

2.12 D.E.Polk: Scripta Met. **4**, 117 (1970); Acta Met. **20**, 485 (1972)

2.13 M.R.Hoare, P.Pal: J. Cryst. Growth **17**, 77 (1972); Adv. Phys. **24**, 646 (1975)
M.R.Hoare: Ann. N. Y. Acad. Sci. **279**, 186 (1976)

2.14 J.A.Barker: J. de Physique **38**, Supplement C-2, 37 (1977)

2.15 A.L.Mackay: Acta Cryst. **15**, 916 (1952)

2.16 J.Farges, Thesis University of Paris-Sud (1977); J.Farges, M.F. de Feraudy, B.Raoult, G.Torchet: J. de Physique **38**, Supplement C-2, 47 (1977)

2.17 C.L.Briant, J.J.Burton: Phys. Stat. Sol. (b) **85**, 393 (1978)

2.18 J.A.Barker, M.R.Hoare, J.L.Finney: Nature **257**, 120 (1975)

2.19 J.L.Finney, J.Wallace: J. Non-Cryst. Sol. **43**, 165 (1981)

2.20 J.F.Sadoc, J.Dixmier, A.Guinier: J. Non-Cryst. Sol. **12**, 46 (1973)

2.21 A.C.Wright, R.N.Sinclair: In *Physics of Silica and its Interfaces*, ed. by S.T.Pantelides (Pergamon Press, Oxford 1978) p. 133

2.22 D.L.Griscom, E.T.Friebele, G.H.Sigel, R.J.Ginther: In *The Structure of Non-Crystalline Materials*, ed. by P.H.Gaskell (Taylor and Francis, London 1977) p. 113

2.23 M.J.Kelly, D.W.Bullett: J. Phys. C: Solid State Phys. **12**, 2531 (1979)

2.24 R.P.Messmer: Phys. Rev. B **23**, 1616 (1981)

2.25 J.Blétry: Rev. de Physique Appliquée **15**, 1019 (1980); Z. Naturforsch. **33a**, 327 (1978); **32a**, 445 (1977)

2.26 H.J.F.Jansen, D.S.Boudreaux, H.Snijders: Phys. Rev. B **21**, 2274 (1980)

2.27 D.S.Boudreaux, H.J.Frost: Phys. Rev. B **23**, 1506 (1981)

2.28 A.J.Leadbetter, M.W.Stringfellow: In Proc. of Grenoble Conf. (LAEA: Vienna) p. 501 (1972)

2.29 R.J.Temkin, W.Paul, G.A.N.Conner: Adv. Phys. **22**, 581 (1973)

2.30 J.F.Sadoc, C.N.J.Wagner: Chap. 2, this volume

2.31 J.F.Sadoc, J.Dixmier: Mater. Sci. Eng. **23**, 187 (1976)
J.Blétry, J.F.Sadoc: J. Phys. F: Metal Phys. **5**, L110 (1975)

2.32 E.Nold, P.Lamparter, H.Olbrich, A.Rainer-Harbach, S.Steeb: Z. Naturforsch. **36a**, 1032 (1981);
P.Lamparter, W.Sperl, E.Nold, G.Rainer-Harbach, S.Steeb: In *Rapidly Quenched Metals* IV, ed. by T.Masumoto, K.Suzuki, (The Japan Institute of Metals, Sendai 1982) p. 343;
P.Lamparter, W.Sperl, S.Steeb, J.Blétry: Z. Naturforsch. **37a**, 1223 (1982)

2.33 T.M.Hayes, J.W.Allen, J.Tauc, B.C.Giessen, J.J.Hauser: Phys. Rev. Lett. **40**, 1282 (1978)
2.34 J.F.Sadoc, J.Dixmier: In *The Structure of Non-Crystalline Materials*, ed. by P.H.Gaskell (Taylor and Francis, London 1977) p. 85
2.35 T.Fukunaga, M.Misawa, T.Masumoto, K.Suzuki: In *Rapidly Quenched Metals* III, ed. by B.Cantor (Metals Soc., London 1978) **2**, p. 325; Sci. Rep. Res. Inst. Tohuku Univ. A-**29**, 153 (1981);
      N.Hayashi, T.Fukunaga, M.Ueno, K.Suzuki: In *Rapidly Quenched Metals* IV (see [Ref. 2.32, p. 355])
2.36 P.H.Gaskell, D.M.Glover, A.K.Livesey, P.J.Durham, G.N.Greaves: J. Phys. C: Solid State Phys. (in press)
2.37 Y.Waseda, H.S.Chen: Phys. Status. Sol. **49a**, 387 (1978)
      T.Fujiwara, H.S.Chen, Y.Waseda: J. Phys. F: Metal Phys. **11**, 1327 (1981)
2.38 K.Dini, N.Cowlam, H.A.Davies: J. Phys. F: Metal Phys. **12**, 1553 (1982)
2.39 P.H.Gaskell: Nature **289**, 474 (1981)
2.40 S.Aur, T.Egami, I.Vincze: In *Rapidly Quenched Metals* IV (see [Ref. 2.32, p. 351])
2.41 P.Pannisod, D.Aliaga Guerna, A.Amamov, J.Durand, W.L.Johnson, W.L.Carter, S.J.Poon: Phys. Rev. Lett. **44**, 1465 (1980)
2.42 G.Czjek: Nucl. Inst. Methods **199**, 37 (1982)
2.43 E.Fruchart, A.M.Triquet, P.Fruchart: Ann. Chim. (Paris) **9**, 323
      J.Durand, M.Yung: *Amorphous Magnetism* II, ed. by R.A.Levy, R.Hasegawa (Plenum Press New York 1977) p. 275
2.44 P.H.Gaskell: Acta Metall. **29**, 1203 (1981)
2.45 A.L.Renninger, M.D.Rechtin, B.L.Averbach: J. Non-Cryst. Solids **16**, 1 (1974)
2.46 G.S.Cargill: In *Diffraction Studies on Non-Crystalline Substances*, ed. by I.Hargittai, W.J.Orville-Thomas (in press)
2.47 D.S.Boudreaux: In *The Magnetic Chemical and Structural Properties of Glassy Metallic Alloys*, ed. by R.Hasegawa (CRC Press Inc: Boca Raton, in press)
2.48 J.G.Wright: Inst. Phys. Conf. Ser. **30**, 251 (1977)
2.49 C.H.Bennett: J. Appl. Phys. **43**, 2727 (1972)
2.50 D.J.Adams, A.J.Matheson: J. Chem. Phys. **56**, 1989 (1972)
2.51 J.Blétry: Z. Naturforsch. **32a**, 445 (1977)
2.52 K.Maeda, S.Takeuchi: J. Phys. F: Metal Phys. **8**, L283 (1978)
2.53 T.Ichikawa: Phys. Stat. Sol. (a) **29**, 293 (1975)
2.54 T.Egami, K.Maeda, V.Vitek: Phil. Mag. A**41**, 883 (1980);
      D.Srolovitz, K.Maeda, V.Vitek, T.Egami: Phil. Mag. A**44**, 847 (1981)
      K.Maeda, S.Takeuchi: Phil. Mag. A**44**, 643 (1981);
      D.Srolovitz, K.Maeda, S.Takeuchi, V.Vitek: J. Phys. F: Metal Phys. **11**, 2209 (1981)
      D.Srolovitz, T.Egami, V.Vitek: Phys. Rev. B**24**, 6936 (1981)
2.55 W.Y.Ching, C.C.Lin: In *Amorphous Magnetism* II, ed. by R.A.Levy, R.Hasegawa (Plenum Press, New York 1977) p. 469
      G.D.Scott, G.J.Kovacs: J. Phys. D: Appl. Phys. **6**, 1007 (1973)
2.56 D.S.Boudreaux, J.M.Gregor: J. Appl. Phys. **48**, 156 (1977); **48**, 5057 (1977)
2.57 P.H.Gaskell: In *Rapidly Quenched Metals* III, ed. by B.Cantor (Metals Soc., London 1978) **2**, 277
2.58 P.H.Gaskell: Nature **276**, 484 (1978); J. Non-Cryst. Solids **32**, 207 (1979)
2.59 H.Koizumi, T.Ninomiya: J. Phys. Soc. Jpn. **49**, 1022 (1980)
      R.C.Morris: J. Appl. Phys. **50**, 3251 (1979)
2.60 R.Fletcher, C.M.Reeves: Comp. J. **7**, 149 (1964)
2.61 R.A.Johnson: Phys. Rev. A**134**, 1329 (1964)
      H.M.Pak, M.Doyama: J. Fac. Eng. U. Tokyo B**30**, 111 (1969)
2.62 L. von Heimendahl: J. Phys. F: Metal Phys. **5**, L141 (1975)
2.63 K.A.Gschneider: *Solid State Physics*, ed. by H.Ehrenreich, F.Seitz, D.Turnbull **16**, 275 (1964)
2.64 D.S.Boudreaux: Phys. Rev. B**18**, 4039 (1978)
2.65 T.Fujiwara, Y.Ishii: J. Phys. F: Metal Phys. **10**, 1901 (1980)
2.66 S.Kobayashi, K.Maeda, S.Takeuchi: Jpn. J. Appl. Phys. **19**, 1033 (1980); **48**, 1147 (1980)

2.67 R. Yamamoto, K. Haga, H. Shibutu, K. Doyama: J. Phys. F: Metal Phys. **8**, L179 (1978)
2.68 B. J. Gellatly, J. Finney: J. Non-Cryst. Sol. **50**, 313 (1982)
2.69 P. Chaudhari, J. F. Graczyck, S. R. Herd: Phys. Stat. Sol. **51**, 801 (1972); Phys. Rev. Lett. **28**, 425 (1972)
      N. J. Shevchik: Phys. Stat. Sol. **52K**, 121 (1972)
      A. Howie, O. L. Krivanek, M. L. Rudee: Phil. Mag. **27**, 235 (1973)
2.70 H. B. Sweet: B. A. Sc. Thesis (University of Toronto) [Quoted in T. W. S. Pang, U. M. Franklin, W. A. Miller: Mater. Sci. Eng. **12**, 167 (1973)]
2.71 B. E. Warren: *X-ray Diffraction* (Addison-Wesley: Reading, MA 1969) p. 116
      A. C. Wright: Phys. Chem. Glasses **17**, 122 (1976)
2.72 J. F. Sadoc, J. Dixmier, A. Guinier: J. Non-Cryst. Solids **12**, 46 (1973)
2.73 B. Aronsson, S. Rundqvist: Acta Crystallogr. **15**, 985 (1962)
2.74 P. Lamparter, E. Nold, G. Rainer-Harbach, E. Grallath, S. Steeb: Z. Naturforsch. **36a**, 165 (1981)
2.75 A. M. Flank, P. Lagarde, D. Raoux, J. Rivory, A. Sadoc: In *Rapidly Quenched Metals* IV, ed. by T. Masumoto, K. Suzuki (The Japan Institute of Metals, Sendai 1982) p. 393
2.76 J. M. Dubois, G. Le Caer: Nucl. Instrum. Methods **199**, 307 (1982)
2.77 G. S. Cargill, C. C. Tsuei: In *Rapidly Quenched Metals* III, ed. by B. Cantor (Metals Soc., London 1978) **2**, p. 337
2.78 Y. Waseda, H. Okazaki, T. Masumoto: In *The Structure of Non-Crystalline Materials*, ed. by P. H. Gaskell (Taylor and Francis, London 1977) p. 89
2.79 M. de Crescenzi, N. Motta, F. Comin, L. Innoccia, S. Mobilio, A. Balzarotti: In *Rapidly Quenched Metals* IV (in press), ed. by T. Masumoto, K. Suzuki (The Japan Institute of Metals, Sendai 1982)
2.80 G. N. Greaves, A. Fontaine, P. Lagarde, D. Raoux: Nature **293**, 611 (1981)
2.81 M. Sakata, N. Cowlam, H. A. Davies: J. de Physique, Colloque C-8 **41**, 190 (1980); J. Phys. F: Metal Phys. **9**, L235 (1979); **11**, L157 (1981)
2.82 D. Raoux, J. F. Sadoc, P. Lagarde, A. Sadoc, A. Fontaine: J. de Physique, Colloque C-8 **41**, 207 (1980)
2.83 P. H. Gaskell, D. J. Smith, C. J. D. Catto, J. R. A. Cleaver: Nature **281**, 465 (1979); P. H. Gaskell, D. J. Smith: J. Microscopy **119**, 63 (1980)
2.84 J. M. Dubois: Thesis, University of Nancy (1981); J. Phys. (Paris) C-9 **43**, 67 (1982)

3. X-Ray and Neutron Diffraction Experiments on Metallic Glasses

J. F. Sadoc and C. N. J. Wagner

With 16 Figures

Neutron and x-ray scattering experiments have been extensively employed to elucidate the structure of metallic glasses. Recent developments in both theory and experiment have enabled the evaluation of topological and chemical short-range order in binary alloys. The local atomic arrangement can be described either by the three partial interference functions which represent the Fourier transforms of the partial atomic distribution functions, by the number of j-type atoms per unit volume at the distance r from an i-type atom, or by the three number-concentration structure factors which are associated with the number-number (density), number-concentration, and concentration-concentration correlations. The three partial functions can be evaluated from three separate intensity measurements which are obtained by varying the atomic scattering amplitudes or lengths either through the combination of three different radiations (neutrons, x-ray, and electrons), the combination of nuclear and polarized neutron scattering, the utilization of the anomalous dispersion in x-ray scattering, isotopic substitution in neutron scattering, or isomorphous substitution in x-ray and neutron scattering. Using neutron diffraction it is also possible to obtain information on the local magnetic order. Local anisotropy can be determined and the sign of the first neighbors' magnetic interaction are derived from a magnetic interference function.

3.1 Introductory Comments

The atomic-scale structure of metallic glasses can be deduced either from their neutron, from x-ray, or from electron scattering patterns. However, the analysis of the scattering or diffraction pattern only yields a one-dimensional description of the three-dimensional distribution of the atoms in the metallic glass, i.e., the Fourier transform of the scattered intensity is related to the radial distribution function $4\pi r^2 \varrho(r)$ which represents the number of atoms in a spherical shell of radius r and thickness unity. This function is zero for values of r less than the hard-sphere diameter of the atoms and tends towards $4\pi r^2 \varrho_0$ for larger values of r, where ϱ_0 is the average atomic density of the glass. Otherwise, the radial distribution function (briefly called RDF) modulates about $4\pi r^2 \varrho_0$ and usually shows about four to five peaks before it merges with $4\pi r^2 \varrho_0$, within the error limits of the experiment. This is to say that short-range topological order exists in a metallic glass but no long-range order as observed in the corresponding crystal.

As it is very difficult to prepare pure amorphous metals, the structure investigations of amorphous metallic alloys are very important. The atomic distribution of the components in a binary alloy A_xB_{1-x} may be understood if the partial functions can be determined describing the AA, AB, and BB pair distributions. Thus, we need a set of three different experiments with different scattering factors. x-ray and neutron diffraction allow such determination for some alloys. In other cases some realistic assumption about the structure may be helpful in evaluating the partial functions.

In this chapter we shall discuss the recent advances in x-ray and neutron techniques, as well as the appropriate theoretical background, which will enable us to elucidate the topological and chemical short-range order in binary metallic glasses.

3.2 Diffraction Theory of Amorphous Materials

3.2.1 Interference Functions

a) Debye Formula

The scattered intensity by a disordered material has been calculated by *Debye* using two hypotheses [3.1–3]:
   (i) all interatomic vectors are isotropically distributed in space;
   (ii) every atom receives exactly the same electromagnetic wave (kinematic theory).

The intensity, i.e., the energy per unit cross section per unit time, scattered by N atoms is given by

$$I_N = I_0 \sum_i \sum_j f_i f_j \exp[i\mathbf{K} \cdot (\mathbf{r}_i - \mathbf{r}_j)], \tag{3.1}$$

where f_i is the scattering amplitude or length for atom i, $\mathbf{K}$ the scattering vector for the scattered plane wave, and $\mathbf{r}_i$ the position of the atom i. If we use the first Debye hypothesis, it is possible to average over all directions for a fixed length of the interatomic vector $\mathbf{r}_{ij} = \mathbf{r}_i - \mathbf{r}_j$, and we obtain the Debye formula

$$I_N(K) = I_0 \left[\overline{Nf^2} + \sum_{i \neq j} \sum_j f_i f_j \sin(Kr_{ij})/(Kr_{ij}) \right], \tag{3.2}$$

where r_{ij} is the distance between the atom i and j, $K = 4\pi(\sin\theta)/\lambda$ is the modulus of the scattering vector, and

$$\overline{f^2} = \sum_i f_i^2/N.$$

b) Monatomic Glasses

If we are looking at amorphous materials containing only one kind of atom, $f_i = f_j = f$ and

$$I_N(K) = I_0 f^2 \left\{ N + \sum_{i \neq j} \sum_j [\sin(Kr_{ij})]/(Kr_{ij}) \right\}. \tag{3.3}$$

If all atoms have the same local order, the double sum depends only on the distance between the i and j atoms, i.e., $r_{ij} = r_l$, where l extends over all interatomic distances in the sample

$$I_N(K) = N f^2 I_0 \left\{ 1 + \sum_l [\sin(Kr_l)]/(Kr_l) \right\}. \tag{3.4}$$

The interference function $I(K)$ is defined by the relation

$$I(K) = [I_N(K)/I_0]/N f^2. \tag{3.5}$$

The "minor interference function" $i(K)$ can then be written as

$$i(K) = I(K) - 1 = \sum_l \sin(Kr_l)/(Kr_l) \tag{3.6}$$

and is characteristic of the structure.

c) Polyatomic Glasses – Partial Interference Functions

We limit our study to binary alloys. Suppose we have $N_1 = c_1 N$ atoms of type 1 and $N_2 = c_2 N$ atoms of type 2 with respective scattering amplitudes f_1 and f_2 [3.4, 5].

The scattered intensity is

$$I_N(K) = I_0 N(c_1 f_1^2 + c_2 f_2^2) + f_1^2 \sum_{i \neq j} \sum \sin(Kr_{ij})/(Kr_{ij})$$

$$+ f_2^2 \sum_{i' \neq j'} \sum \sin(Kr_{i'j'})/(Kr_{i'j'}) + 2 f_1 f_2 \sum_i \sum_{j'} \sin(Kr_{ij'})/(Kr_{ij'}), \tag{3.7}$$

where i and j correspond to type 1 atoms and i' and j' correspond to type 2 atoms. It is possible to introduce three minor partial interference functions $i_{ij} = I_{ij}(K) - 1$, i.e.,

$$i_{11}(K) = [1/(Nc_1^2)] \sum_{i \neq j} \sum_j \sin(Kr_{ij})/(Kr_{ij}) = I_{11}(K) - 1,$$

$$i_{22}(K) = [1/(Nc_2^2)] \sum_{i' \neq j'} \sum_{j'} \sin(Kr_{i'j'})/(Kr_{i'j'}) = I_{22}(K) - 1, \tag{3.8}$$

$$i_{12}(K) = [1/(Nc_1 c_2)] \sum_i \sum_{j'} \sin(Kr_{ij'})/(Kr_{ij'}) = I_{12}(K) - 1.$$

The total scattered intensity will then be expressed in terms of the three partial interference functions

$$I_N(K) = I_0 N[(c_1 f_1^2 + c_2 f_2^2) + c_1^2 f_1^2 i_{11}(K) + c_2^2 f_2^2 i_{22}(K)$$
$$+ 2c_1 c_2 f_1 f_2 i_{12}(K)]. \tag{3.9}$$

If we want to define a total interference function depending only on the material structure and the scattering amplitudes, we have to normalize the intensity. There are two common ways to do this:
(i) the total interference function $S(K)$ is defined as

$$S(K) = I_N(K)/[I_0 N(c_1 f_1^2 + c_2 f_2^2)] = s(K) + 1 \tag{3.10}$$

so that

$$s(K) = [1/(c_1 f_1^2 + c_2 f_2^2)] \cdot [c_1^2 f_1^2 i_{11}(K) + c_2^2 f_2^2 i_{22}(K)$$
$$+ 2c_1 c_2 f_1 f_2 i_{12}(K)], \tag{3.11}$$

or (ii) define the total interference function $I(K)$ by

$$I(K) = [I_N(K) - I_0 N(c_1 f_1^2 + c_2 f_2^2) + I_0 N(c_1 f_1$$
$$+ c_2 f_2)^2]/[I_0 N(c_1 f_1 + c_2 f_2)^2]$$
$$= i(K) + 1 \tag{3.12}$$

so that

$$i(K) = [1/(c_1 f_1 + c_2 f_2)^2] \cdot [c_1^2 f_1^2 i_{11}(K) + c_2^2 f_2^2 i_{22}(K)$$
$$+ 2c_1 c_2 f_1 f_2 i_{12}(K)] \tag{3.13}$$

These interference functions $S(K)$ and $I(K)$ are also called the structure factors, mainly in neutron diffraction studies and in the theoretical evaluation of electronic transport properties.

There is no fundamental difference between the two definitions and in all cases, the partial functions are the same functions. The choice of one of these definitions must depend on the experimental condition. As we are going to see, definition (ii) is more useful for metallic glasses if it is difficult to know the position of the origin of the $I(K)$ function. This is namely the case in neutron diffraction because there is always a background which is difficult to estimate with a good accuracy. The first definition has to be used when $c_1 f_1 + c_2 f_2$ becomes zero which is possible for the "zero-alloys" consisting of isotopes with positive and negative scattering lengths.

If we use the second definition for $i(K)$, we can introduce the term

$$I_{LB} = I_0 N \cdot [(c_1 f_1^2 + c_2 f_2^2) - (c_1 f_1 + c_2 f_2)^2]$$ (3.14)

which is called the Laue monotonic background in powder diffraction.

d) Determination of the Partial Functions

Since the total $i(K)$ is a linear equation with three unknowns, (3.13), the partial functions can be obtained from scattering experiments which involve three independent sets of the atomic scattering factors f_1 and f_2:

(1) Three different isotopic compositions in three chemically identical samples. In this case the neutron scattering factors are different. This kind of experiment is very expensive, especially with amorphous material.
(2) A combination of x-ray and neutron scattering using different neutron magnetic scattering factors.
(3) Anomalous dispersion of x-rays which allows a change of scattering factors when the wavelength is modified.
(4) Chemical substitution of atoms by atoms of similar size and chemical properties.

Only through the partial interference functions or structure factors $I_{ij}(K)$ or $i_{ij}(K)$, (3.8) can we determine the local atomic arrangements in metallic glasses. In the case of binary alloys, three partial functions $i_{11}(K)$, $i_{22}(K)$ and $i_{12}(K)$ must be determined. This can be done quite readily, in principle, by measuring three independent scattering functions $I(K)$, (3.12), and solving for the three unknowns $i_{ij}(K)$, because $I(K)$ is the weighted sum of the three partial functions. The weighting factors, i.e., $c_i c_j f_i f_j$, depend upon the atomic concentration c_i and the scattering factor f_i of element i in the alloy. Since we cannot, a priori, assume that the partial functions $I_{ij}(K)$ are independent of concentration, the variation of c_i to obtain three different intensity expressions $I_N(K)$ should not be used as the first step to determine the three partial functions. Thus, we have to find ways to vary the scattering factors f_i. The methods employed so far can be grouped into two categories: (i) choice of different radiations (at least three radiations) to study the same sample material, and (ii) a change in the sample configuration by isotopic or isomorphous substitutions and employing a single radiation. However, the separation into the three partial functions is not straightforward because changes in the weighting factors are usually quite small.

Most calculations of the partial interference functions or structure factors in binary systems have involved the partial functions $i_{ij}(K)$ defined in (3.8). The three total interference functions $I(K)$ are the weighted sums of the three functions $i_{ij}(K)$ as shown in (3.13) which can be written in matrix notation as [3.6]

$$[T(K)] = [V(K)][P(K)],$$ (3.15)

where

$$T(K)] = \begin{bmatrix} I'(K) - 1 \\ I''(K) - 1 \\ I'''(K) - 1 \end{bmatrix} = \begin{bmatrix} T_1(K) \\ T_2(K) \\ T_3(K) \end{bmatrix}$$

$$[V(K)] = \begin{bmatrix} \alpha'_{11}(K) & 2\alpha'_{12}(K) & \alpha'_{22}(K) \\ \alpha''_{11}(K) & 2\alpha''_{12}(K) & \alpha''_{22}(K) \\ \alpha'''_{11}(K) & 2\alpha'''_{12}(K) & \alpha'''_{22}(K) \end{bmatrix} = \begin{bmatrix} V_{11}(K) & V_{12}(K) & V_{13}(K) \\ V_{21}(K) & V_{22}(K) & V_{23}(K) \\ V_{31}(K) & V_{32}(K) & V_{33}(K) \end{bmatrix}.$$

and

$$[P(K)] = \begin{bmatrix} i_{11}(K) \\ i_{12}(K) \\ i_{22}(K) \end{bmatrix} = \begin{bmatrix} P_1(K) \\ P_2(K) \\ P_3(K) \end{bmatrix}.$$

The weighing factors $\alpha_{ij}(K)$ are defined as

$$\alpha_{ij}(K) = c_i c_j f_i(K) f_j(K) / [c_1 f_1(K) + c_2 f_2(K)]^2. \tag{3.16}$$

The solution of (3.15), i.e., the functions defined in $[P(K)]$, is simply given by

$$[P(K)] = [V(K)]^{-1} [T(K)]. \tag{3.17}$$

A unique solution is found for $P(K)$ if the determinant of $[V(K)]$ is different from zero. However, in most of the experiments carried out so far to determine $[P(K)]$, the normalized determinant of the weighting factors V_{ij}, i.e.,

$$|V(K)|_n = \left| V_{ij}(K) \middle/ \left(\sum_j V_{ij}^2 \right)^{1/2} \right|,$$

is very small. Thus, (3.15) is rather ill-conditioned and the smallness of $|V(K)|_n$ is an indication of the difficulties to be expected when trying to solve (3.17). The problem is further aggravated by the fact that the experimental data $T(K)$ contain a certain amount of errors $\Delta T(K)$, usually taken to be about $t = \Delta T/T \simeq 3\%$. Therefore, we choose a function $T_e(K)$, i.e.,

$$T_e(K) = T(K) \pm \Delta T(K) = T(K)(1 \pm t)$$

which is derived from $T(K)$ by allowing it to vary between the limits $\pm t$.

As a first step in an iteration procedure, solutions are then sought for $P(K)$ subject to the conditions that each $i_{ij}(K)$ lies within a range of values chosen from a multitude of structure factors obtained by theory and experiment for metallic glasses, and that the three partial functions $i_{ij}(K)$ satisfy the inequalities [3.7]

$$c_1 + c_1^2 i_{11}(K) > 0,$$

$$c_2 + c_2^2 i_{22}(K) > 0,$$

$$c_2 + c_2^2 i_{22}(K) - \frac{(c_1 c_2)^2 [i_{12}(K)]^2}{c_1 + c_1^2 [i_{11}(K)]} > 0$$

which follow readily from (3.13), representing a positive quadratic form in f_1 and f_2. This process will yield $i_{11}(K)$, $i_{22}(K)$, and $i_{12}(K)$, each with a characteristic but mutually dependent error band.

In the second step of the iteration process, a smooth curve is drawn through $i_{11}(K)$ [corresponding to the largest weighting factor $W_{ij}(K)$ in the binary alloy] in such a way that the sum rule

$$\int i_{ij}(K) K^2 dK = -2\pi^2 \varrho_0 \tag{3.18}$$

is satisfied. The values of $i_{11}(K)$, obtained in this way, are taken as fixed and $i_{22}(K)$ and $i_{12}(K)$ are evaluated from two simultaneous equations in which there are only two unknowns. We choose the two equations yielding the highest normalized determinant whose value is usually an order of magnitude larger than the original determinant of the three simultaneous equations. Again, a smooth curve is drawn through $i_{12}(K)$ in such a way as to satisfy the sum rule (3.18). If it does, $i_{22}(K)$ can be evaluated immediately. Otherwise, $i_{11}(K)$ is adjusted and the entire process is repeated.

3.2.2 Radial Distribution Functions and Fourier Transform Techniques

a) Radial Distribution Functions

In the monatomic case, we have written the minor interference function as (3.5 and 3)

$$i(K) = (1/Nf^2) \sum_{i \neq j} \sum f^2 \sin(Kr_{ij})/(Kr_{ij}).$$

The double sum has $N(N-1)$ terms if r_{ij} take all the interatomic distances. We define a local density $\varrho(r)$, the number of atoms in an unitary volume at the distance r from the origin. Then, the number of atoms in a spherical shell of thickness dr is given by $4\pi r^2 \varrho(r) dr$. The function $4\pi r^2 \varrho(r)$ is usually called the radial distribution function. The minor interference function $i(K)$ can now be expressed as

$$i(K) = \int_0^\infty [4\pi r^2 \varrho(r) \sin(Kr)/(Kr)] dr. \tag{3.19}$$

If ϱ_0 is the mean atomic density of the material, it is possible to write

$$i(K)=\int_0^\infty 4\pi r^2[\varrho(r)-\varrho_0(r)]\,[\sin(Kr)/(Kr)]\,dr + \int_0^\infty 4\pi r^2\varrho_0[\sin(Kr)/(Kr)]\,dr\,.$$

$$(3.20)$$

The last sum is the scattering by an homogenous sample having the same shape as the amorphous sample. This term is different from zero only for very small values of K ($K\sim 1/R$ where R is the sample size). This term is not visible with an amorphous sample. Let $P(r)=\varrho(r)/\varrho_0\equiv g(r)$. Then, the interference function can be written as

$$I(K)=1+(4\pi\varrho_0/K)\int_0^\infty r[P(r)-1]\sin(Kr)dr=i(K)+1\,. \qquad (3.21)$$

The reduced interference function $a(K)=Ki(K)\equiv F(K)=K[I(K)-1]$ and the reduced radial distribution function $W(r)=r[P(r)-1]=G(r)/(4\pi\varrho_0)$ are related by

$$a(K)=4\pi\varrho_0\int_0^\infty W(r)\sin(Kr)dr=F(K)\,. \qquad (3.22)$$

The Fourier integral is defined by

$$\Phi(K)=(2/\sqrt{2\pi})\int_0^\infty f(r)\sin(Kr)dr\,.$$

By applying the Fourier theorem we have

$$f(r)=(2/\sqrt{2\pi})\int_0^\infty \phi(K)\sin(Kr)dK\,.$$

The $a(K)$ and $W(r)$ functions are related by Fourier transformation and

$$W(r)=[1/(2\pi^2\varrho_0)]\int_0^\infty a(K)\sin(Kr)dK=G(r)/(4\pi\varrho_0)\,. \qquad (3.23)$$

The radial distribution function (RDF) $R(r)=4\pi r^2\varrho(r)$ can be derived from $W(r)$ by

$$R(r)=4\pi r^2\varrho_0\{[W(r)/r]+1\}=rG(r)+4\pi r^2\varrho_0\,. \qquad (3.24)$$

b) Partial Radial Distribution Functions

In the case of binary alloys, the minor total interference function is

$$i(K)=\alpha_{11}i_{11}(K)+\alpha_{22}i_{22}(K)+2\alpha_{12}i_{12}(K)\,. \qquad (3.25)$$

The α_{ij} coefficients depend on the scattering factor and are defined in (3.16).

Let $\varrho_{ij}(r)$ be the number of j-type atoms in a unitary volume at a distance r from the i-type atom. Then

$$i_{ij}(K)=(1/c_j) \int\limits_0^\infty 4\pi r^2 \varrho_{ij}(r)\left[\sin(Kr)/(Kr)\right]dr. \qquad (3.26)$$

The probability of finding an atom of type j at a distance r from atom of type i is $P_{ij}=\varrho_{ij}/\varrho_0$. This probability tends towards c_j when r goes to infinity. Thus, it is convenient to use the partial atomic correlation function $P_{ij}(r)=\varrho_{ij}(r)/(c_j\varrho_0)$ which is normalized to 1. Then

$$i_{ij}(K)=(4\pi\varrho_0/K) \int\limits_0^\infty r[P_{ij}(r)-1]\sin(Kr)dr. \qquad (3.27)$$

The three partial reduced interference functions in (3.25) provide information about the structure of the binary amorphous alloy.

c) Total Radial Distribution Functions

As shown in (3.13, 25), the total interference functions $i(K)$ are the weighted sum of the partial interference functions $i_{ij}(K)=I_{ij}(K)-1$. In the case of nuclear neutron scattering, the weight factor α_{ij} (3.16) is independent of K, i.e.,

$$\alpha_{ij}=c_ic_jb_ib_j/(c_1b_1+c_2b_2)^2, \qquad (3.28)$$

where b_i is the coherent scattering length of element i. Then

$$a(K)=K[I(K)-1]=K\,i(K)=\sum_i\sum_j\alpha_{ij}a_{ij}(K). \qquad (3.29)$$

Its Fourier transform $W(r)$ is given by (3.23), i.e.,

$$W(r)=\alpha_{11}W_{11}(r)+\alpha_{22}W_{22}(r)+2\alpha_{12}W_{12}(r). \qquad (3.30)$$

In the general case, where $\alpha_{ij}(K)$ is a function of K, we must apply the convolution theorem, i.e.,

$$W(r)=\sum_i\sum_j A_{ij}(r)*W_{ij}(r), \qquad (3.31)$$

where

$$A_{ij}(r)=(1/\pi)\int \alpha_{ij}(K)\cos(Kr)dK. \qquad (3.32)$$

However, in most x-ray experiments involving metallic glasses, $\alpha_{ij}(K)$ is a slowly varying function of K so that $A_{ij}(r)$ is a sharp function, much sharper than $W_{ij}(r)$ which can be taken outside the convolution integral, i.e.,

$$W(r) = \alpha_{11}(0)\, W_{11}(r) + \alpha_{22}(0)\, W_{22}(r) + 2\alpha_{12}(0)\, W_{12}(r) \tag{3.33}$$

because

$$\int A_{ij}(r)\,dr = \alpha_{ij}(0).$$

It should also be noted that we could formally introduce an expression for the total atomic distribution function $\varrho(r)$ and the total correlation function $P(r)$. Then

$$W(r) = \sum_i \sum_j \alpha_{ij}(0)\, W_{ij}(r) = \sum_i \sum_j \alpha_{ij}(0)\, r[P_{ij}(r) - 1]$$

$$= r\left\{\left[\sum_i \sum_j \alpha_{ij}(0)\,[P_{ij}(r)] - 1\right\} = r[P(r) - 1] = r\{[\varrho(r)/\varrho_0] - 1\}. \tag{3.34}$$

On the other hand, the Fourier transform of $Ks(K)$, (3.11), yields the reduced distribution function

$$W_s(r) = \sum_i \sum_j [c_i c_j f_i f_j / \langle f^2 \rangle]\, W_{ij}(r) = \sum_i \sum_j [\alpha_{ij}(0) \langle f \rangle^2 / \langle f^2 \rangle]\, W_{ij}(r). \tag{3.35}$$

Thus, the differential correlation function $D(r)$ is given by

$$D(r) = 4\pi\varrho_0 \langle f \rangle^2 W(r) = 4\pi\varrho_0 \langle f^2 \rangle\, W_s(r) = 4\pi\varrho_0 \sum_i \sum_j c_i c_j f_i f_j W_{ij}(r). \tag{3.36}$$

For small values of r, i.e., $r < r_0$, $W_{ij}(r) = -r$. Thus,

$$[\langle f \rangle^2 W(r)]_{r \to 0} = [\langle f^2 \rangle W_s(r)]_{r \to 0} = -\langle f \rangle^2. \tag{3.37}$$

3.2.3 The Bhatia-Thornton Formalism

The partial interference functions (and the partial radial distribution functions) are very important in understanding the structure of a binary system, but they are often difficult to obtain as their evaluation needs three very accurate experiments.

Bhatia and *Thornton* [3.8] have defined the partial functions in a different way. In some alloys it is possible to approximate one or two of these functions and the remaining partials can then be obtained with only one or two experiments.

a) Number-Concentration Partial Functions

For homogeneous binary alloys made from atoms 1 and 2, the total scattered intensity may be written [3.8, 9]

$$I_N(K) = I_0 N[\bar{f}^2 S_{NN}(K) + 2\bar{f}\Delta f S_{NC}(K) + (\Delta f)^2 S_{CC}(K)], \tag{3.38}$$

where

$$\bar{f} = \langle f \rangle = c_1 f_1 + c_2 f_2 \quad \text{and} \quad \Delta f = f_1 - f_2.$$

From this equation it appears that $S_{NN}(K)$ is the total interference function defined in (3.10) if the two scattering factors f_1 and f_2 are equal. Thus, $S_{NN}(K)$ describes the overall structure and it has the same significance as the interference function in the monatomic case. The partial number-concentration functions are related to the partial interference functions by

$$S_{NN}(K) = c_1^2 I_{11}(k) + c_2^2 I_{22}(K) + 2c_1 c_2 I_{12}(K)$$
$$= c_1^2 i_{11}(K) + c_2^2 i_{22}(K) + 2c_1 c_2 i_{12}(K) + 1 = s_{NN}(K) + 1, \tag{3.39}$$

$$S_{NC}(K) = c_1 c_2 \{c_1 [I_{11}(K) - I_{12}(K)] - c_2 [I_{22}(K) - I_{12}(K)]\}$$
$$= c_1 c_2 \{c_1 [i_{11}(K) - i_{12}(K)] - c_2 [i_{22}(K) - i_{12}(K)]\}, \tag{3.40}$$

$$S_{CC}(K) = c_1 c_2 \{1 + c_1 c_2 [I_{11}(K) + I_{22}(K) - 2I_{12}(K)\}$$
$$= c_1 c_2 \{1 + c_1 c_2 [i_{12}(K) + i_{22}(K) - 2i_{12}(K)]\}. \tag{3.41}$$

The Fourier transform of $[S_{NN}(K) - 1] \cdot K$ is the reduced correlation function of local 1 and 2 atom number density fluctuations

$$r[P_{NN}(r) - 1] = (2\pi^2 \varrho_0)^{-1} \int_0^\infty [S_{NN}(K) - 1] K \sin(Kr) dK, \tag{3.42}$$

where ϱ_0 is the average number of atoms in a unit volume. $P_{NN}(r)$ is defined by $P_{NN}(r) = \varrho_{NN}(r)/\varrho_0$, where $\varrho_{NN}(r)$ is the global atomic distribution function, i.e.,

$$\varrho_{NN}(r) = c_1 \varrho_{11}(r) + c_2 \varrho_{22}(r) + 2c_1 \varrho_{12}(r) = c_1 \varrho_1(r) + c_2 \varrho_2(r),$$

where $\varrho_\alpha(r) = \varrho_{\alpha 1}(r) + \varrho_{\alpha 2}(r)$ is the number of atoms (type one or two) per unit volume at a distance r from an α atom.

$S_{NC}(K)$ is a function related to the correlation between fluctuations of the number of atoms in a volume and fluctuations of the atomic concentration in the same volume. $S_{NC}(K)$ oscillates around zero. Using the partial probability function defined in (3.27) we can write

$$S_{NC}(K) = c_1 c_2 (4\pi\varrho_0/K) \int_0^\infty [P_1(r) - P_2(r)] r \sin(Kr) dr, \qquad (3.43)$$

where $P_\alpha = c_1 P_{\alpha 1} + c_2 P_{\alpha 2}$ is the probability of finding any atom ($\alpha = 1$ or 2) in a unit volume at a distance r from an α atom. The quantity

$$N_1(r) - N_2(r) = 4\pi r^2 \varrho_0 [P_1(r) - P_2(r)] \qquad (3.44)$$

is the difference between the number of atoms in the spherical shell of radius r and unity thickness about a 1-atom and a 2-atom.

$S_{CC}(K)$ is a function related to the correlation between the kind of atom and the concentration around it. $S_{CC}(K)$ oscillates around $c_1 c_2$. A radial concentration function (RCF) $\varrho_{CC}(r)$ can be defined by

$$\varrho_{CC}(r) = c_2 [\varrho_{11}(r) - \varrho_{21}(r)] + c_1 [\varrho_{22}(r) - \varrho_{12}(r)]$$
$$= c_2 \varrho_1(r) + c_1 \varrho_2(r) - \varrho_{12}(r)/c_2.$$

Thus,

$$4\pi r^2 \varrho_{CC}(r) = c_2 N_1(r) + c_1 N_2(r) - N_{12}(r)/c_2 \qquad (3.45)$$

is the mean value of the difference between the number of α atoms and the number of β atoms in the spherical shell of radius r and unit thickness about an α atom (α and $\beta = 1, 2$). $\varrho_{CC}(r)$ is related to $S_{CC}(K)$ by

$$r\varrho_{CC}(r)/\varrho_0 = (2\pi^2\varrho_0)^{-1} \int_0^\infty K \cdot \{[S_{CC}(K) - c_1 c_2]/c_1 c_2\} \sin(Kr) dK. \qquad (3.46)$$

b) Properties of the Number-Concentration Partial Functions

To elucidate the meaning of these functions, we will discuss it for a simplified case in which all atoms have the same size. Then $\varrho_1(r) = \varrho_2(r)$, and in this case there is no correlation between the number and the species of atoms in a given volume. Thus the $S_{NC}(K)$ function vanishes.

The function $\varrho_{CC}(r)$ could be written [3.10, 11]

$$\varrho_{CC}(r) = \varrho_0 [P_{NN}(r) - P_{12}(r)]. \qquad (3.47)$$

If there is no chemical interaction between the 1 and 2 atoms they are randomly distributed, i.e., $P_{12}(r) = P(r) = P_{NN}(r)$ and $\varrho_{CC}(r) = 0$. For a distance with predominant heterocoordination, $P_{12}(r) > P_{NN}(r)$ and $\varrho_{CC}(r)$ is negative, and vice versa.

The number-concentration functions for small values of K (long-wavelength limit) are given by

$$S_{NN}(0) = \langle \Delta N^2 \rangle / N$$
$$S_{NC}(0) = \langle \Delta N \cdot \Delta C \rangle \qquad (3.48)$$
$$S_{CC}(0) = \langle \Delta C^2 \rangle,$$

where $\langle \Delta N^2 \rangle$ is the mean square fluctuation in the number of particles in the volume V, $\langle \Delta C^2 \rangle$ the mean fluctuation in the concentration, and $\langle \Delta N \cdot \Delta C \rangle$ is the correlation between the two fluctuations. In the case of a metallic alloy it is possible to neglect the compressibility. One obtains

$$S_{NN}(0) = \delta^2 S_{CC}(0)$$
$$S_{NC}(0) = -\delta S_{CC}(0), \qquad (3.49)$$

where $\delta = (N/V)(v_1 - v_2)$, v_1 and v_2 being the partial molar volume per atom of the two species.

For an ideal solution we find

$$S_{CC}(0) = c_1 c_2.$$

This corresponds to the Laue background, i.e., $I_{LB} \propto (\Delta f)^2 S_{CC}(0) = c_1 c_2 (f_1 - f_2)^2$.

In the general case, we can write

$$S_{CC}(0) = c_1 c_2 / (1 + 2 c_1 c_2 \omega k_B T),$$

where ω is an interchange energy such that if we start with two pure metals A and B and exchange two atoms A and B, the total decrease in energy is 2ω.

For large K values we have the asymptotic limits

$$S_{NN}(K) = 1, \quad S_{NC}(K) = 0, \quad S_{CC}(K) = c_1 c_2 \qquad (3.50)$$
$$K \to \infty \qquad K \to \infty \qquad K \to \infty$$

because $i_{ij}(K) = 0$, when $K \to \infty$.

c) Chemical Short-Range Order in Binary Alloys
[Generalized Warren Chemical Short-Range Order (CSRO) Parameter]

The atomic distribution function $\varrho_{NN}(r)$ describes the global or topological short-range order in the amorphous alloy. The term $\varrho_{NC}(r)$ is the correlation between density and concentration fluctuations and is zero when $\varrho_1(r) = \varrho_2(r)$ which is the case for A and B atoms of identical size. The last term $\varrho_{CC}(r)$ is of practical interest because it describes the concentration fluctuations.

It is convenient to express $\varrho_{CC}(r)$ in terms of the generalized Warren chemical short-range order parameter $\alpha(r)$ [3.12]:

$$\varrho_{CC}(r) = \varrho_w(r)\alpha(r), \tag{3.51}$$

where

$$\varrho_w(r) = c_2\varrho_1(r) + c_1\varrho_2(r) \neq \varrho_{NN}(r) \tag{3.52}$$

and

$$\alpha(r) = 1 - \frac{\varrho_{12}(r)}{c_2\varrho_w(r)}. \tag{3.53}$$

If we assume that the number of atoms about a 1-type atom is the same as the number of atoms about a 2-type atom, i.e., $\varrho_1(r) = \varrho_2(r)$, we find $\varrho_w(r) = \varrho_{NN}(r)$ and $\varrho_{NC}(r) = 0$. This assumption was originally introduced by *Keating* [3.4], and later applied by *Ruppersberg* and *Egger* [3.10], and is a good approximation for alloys consisting of atoms of similar size, as discussed above. In such a case, two functions $\varrho_{NN}(r)$ and $\alpha(r)$ are sufficient to describe the topological and chemical short-range order in a binary alloy, respectively.

The Warren CSRO parameter $\alpha(r)$ could, in principle, be determined from the three partial functions $\varrho_{11}(r)$, $\varrho_{12}(r)$, and $\varrho_{22}(r)$. However, it is more appropriate to evaluate an average Warren CSRO parameter α_p for the pth coordination shell.

The global or overall coordination number N^p can be calculated from the global radial distribution function, i.e., [3.13]

$$N_{NN}^p = \int_{r_{min}^p}^{r_{max}^p} 4\pi r^2 \varrho_{NN}(r)\,dr = c_1 N_1^p + c_2 N_2^p, \tag{3.54}$$

where

$$N_\alpha^p = \int_{r_{min}^p}^{r_{max}^p} 4\pi r^2 \varrho_\alpha(r)\,dr = \int_{r_{min}^p}^{r_{max}^p} 4\pi r^2 [\varrho_{\alpha 1}(r) + \varrho_{\alpha 2}(r)]\,dr. \tag{3.55}$$

Similarly, we can define the integral quantities:

$$N_{NC}^p = \int_{r_{min}^p}^{r_{max}^p} 4\pi r^2 \varrho_{NC}(r)\,dr = c_1 c_2 (N_1^p - N_2^p), \tag{3.56}$$

$$N_{CC}^p = \int_{r_{min}^p}^{r_{max}^p} 4\pi r^2 \varrho_{CC}(r)\,dr = c_2 N_1^p + c_1 N_2^p - N_{12}^p/c_2. \tag{3.57}$$

It is easily seen that the value α_p can be evaluated with the relation

$$\alpha_p = 1 - N_{12}^p/[c_2(c_2 N_1^p + c_1 N_2^p)] = N_{CC}^p/(c_2 N_1^p + c_1 N_2^p). \tag{3.58}$$

If the atoms have similar sizes, then $c_2 N_1^p + c_1 N_2^p \simeq N_{NN}^p$ and we obtain the relation

$$\alpha_p = N_{CC}^p / N_{NN}^p ,$$

first used by *Ruppersberg* and *Egger* [3.10].

Since $\varrho_w(r) = c_2 \varrho_1(r) + c_1 \varrho_2(r)$ is positive for all values of r, $\alpha(r)$ behaves like $\varrho_{CC}(r)$. Negatives values of α_p are found for like neighbor distances. α_p is zero for a random distribution of the elements in the alloy, i.e., when $N_1 = N_2 = N_{12}/c_2$.

The description of the binary alloy in terms of the N–C partial function is analogous to that of crystalline disordered solutions [3.14] which may be regarded as an ideal crystal consisting of "average" atoms with a superimposed "fluctuation" represented by the variations in the atomic scattering factors and static displacements due to the difference between the real atoms of the solutions and the "average" atoms.

It is often advantageous to consider the crystal as a periodic structure of these average atoms with the fluctuations due to different kinds and sizes of the atoms superimposed. The periodic structure of the average atoms produces regular or Bragg reflections, whereas the fluctuations are responsible for diffuse scattering [3.15, 16]. Fluctuations in chemical compositions, i.e., due to the kinds of atoms, produce a modulation of the "Laue diffuse scattering," whereas distortions due to the different sizes will result in diffuse "Huang scattering".

3.3 Experimental Diffraction Methods

X-ray and neutron diffraction methods have been used extensively in the past to determine the structure of amorphous solids or liquids. The diffraction pattern, i.e., the intensity scattered by N atoms in the amorphous sample, must be determined as a function of the length of the diffraction. This can be accomplished in two ways:

(i) Variable 2θ-method: measurement of the diffraction angle 2θ when using monochromatic radiation of wavelength λ. Then $K = (4\pi/\lambda) \sin \theta$.

(ii) Variable λ-method: measurement of the scattered photon energy E or neutron time of flight t when using polychromatic radiation (variable λ) at a fixed scattering angle $2\theta_i$. Then $K = (4\pi e/hc) (\sin \theta_i) E = 1.0135 (\sin \theta_i) E$ for photons when E is expressed in keV, and $K = (4\pi m/h) (\sin \theta_i) (L_0 + L)/t(\lambda)$, where L_0 and L are the flight paths of the incident and scattered neutrons, respectively.

The variable 2θ-method has been the conventional method for determination of structure. However, the K-range is limited by the wavelength of the radiation which is about $18-20 \, \text{Å}^{-1}$ when using $\lambda \sim 0.56$ to $0.71 \, \text{Å}$ available with conventional x-ray generators or neutron sources. However, much structure exists beyond the value of $20 \, \text{Å}^{-1}$ in the diffraction pattern of metallic glasses, and an extension of the K-range to values as high as 40 Å^{-1} is desirable. This can be accomplished with the newly established variable λ-method which yields values of $K = 40 \text{ Å}^{-1}$. These methods have been discussed in the literature [3.17–22].

3.3.1 Variable 2θ X-Ray Method

Any modern scattering experiment, based on the measurement of the scattered intensity as a function of the scattering angle 2θ, consists of a radiation source, a monochromator to select a narrow band of wavelengths and to suppress unwanted radiation, a diffractometer with sample holder to scan through the scattering angle 2θ, and a radiation detector with associated electronic equipment. The radiation source is usually an x-ray generator with 60 kV or less as an excitation potential, which limits the efficient use of a short wavelength radiation to Ag or Mo radiation. A 150 kV x-ray generator would permit us to use W K_α radiation with an energy of 60 keV. In general, the scattered intensities from an amorphous sample are relatively weak and intense radiation sources, such as rotating anode x-ray generators or even synchrotron radiation, would be beneficial.

In order to select a narrow band of wavelengths, usually a narrow band about the characteristic radiation of the x-ray target, a monochromator is most commonly used either in the primary or in the diffracted x-ray beam. Graphite is used as a monochromator crystal with conventional x-ray sources because of its high luminosity, and Si single crystals have been used with synchrotron radiation because of their good wavelength resolution. It is also possible to select a narrow band of diffracted x-rays with a modern solid state detector (Li-drifted Si or intrinsic Ge) in conjunction with a single channel pulse height analyzer. When using a Ge detector, it is necessary to consider carefully the perturbations of the x-ray background due to the escape peaks in Ge.

The x-ray detector should have high efficiency for the short wavelength radiation used in the scattering experiments of amorphous alloys. Scintillation counters, proportional counters with a 50 mm xenon path length at 2 atmosphere pressure and solid state detectors possess better than 80 % efficiency for Ag K_α radiation. However, the energy resolution $\Delta E/E$ for Ag radiation varies from ~ 30 % for the scintillation counter to ~ 15 % for the proportional counter and <2 % for the solid state detector. The high energy resolution of the solid state not only permits the elimination of unwanted fluorescence radiation from the sample, but also the partial removal of the incoherent or Compton scattering at higher scattering angles 2θ.

The conventional x-ray diffractometer scans sequentially through the scattering angle 2θ to register the scattered intensity with the help of a detector through a narrow receiving slit. If one adds either before or behind the receiving slit, an anti-scatter slit with a divergence equivalent to that of the divergence slit

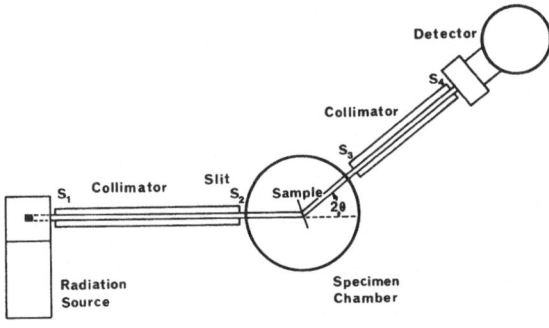

Fig. 3.1. X-ray transmission technique [3.12]

in the primary beam, much of the parasitic scattering due to the air, the window of the x-ray camera, etc., can be eliminated as shown in Fig. 3.1.

The old photographic film technique permitted the simultaneous registration of the scattered intensity over a large range in scattering angle, thus eliminating the scanning mechanism of the diffractometer. A modern development, however, promises to combine the advantages of both counter and film techniques. Position-sensitive detectors (PSD) have been developed for the registration of x-rays, as well as neutrons and electrons. The commercial x-ray position-sensitive proportional detectors consist of linear gas counters with a graphite-coated quartz fiber or a metal (stainless steel) wire and an active length of about 80 mm, a gas absorption thickness of about 5 mm and a position resolution <0.2 mm. When filled with Xe gas at 10 atmosphere pressure, these counters possess a counting efficiency of about 40% for Mo K_α radiation. The PSD must be used with a monochromatic radiation source obtained through the application of a monochromator in the primary beam. A balanced filter pair would also provide sufficient monochromatization but would require larger foils of uniform thicknesses. Provision must be made to eliminate the parasitic scattering. For example, the air scattering can be greatly reduced by evacuating the entire x-ray path or filling it with He gas.

Because of the fact that the x-ray position-sensitive proportional detectors presently available are linear detectors with finite thickness and length, a parallaxe will result when the detector wire is oblique to the incident x-ray beam and the angular range in 2θ is limited. For example, the angular range for an 80 mm detector at a sample to detector distance of 180 mm will be about 25° in $2\theta°$. The parallaxe at an oblique angle of incidence 12.5° or less, in a detector of 5 mm gas path length would correspond to a position uncertainty of $\leqq 0.8$ mm which corresponds to $\varDelta 2\theta \leqq 0.25°$, small enough to be of no consequence in the registration of scattered intensity from amorphous samples.

In spite of the fact that six or seven $2\theta_i$ positions must be chosen to cover an angular range of 125° including overlap, the application of the position-sensitive detector in the measurement of the scattered intensities from amorphous materials should greatly reduce the counting time and/or improve the counting statistics.

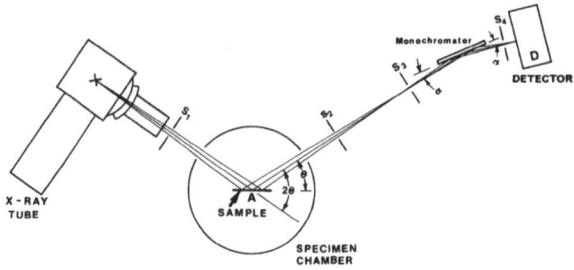

Fig. 3.2. X-ray reflection technique [3.12]

The variable 2θ-method can be applied in the reflection and transmission geometries. The most recent investigations of the structure of amorphous solids have been carried out using the reflection technique. Usually the focussing Bragg-Brentano geometry is employed because the absorption correction is relatively simple and is independent of the scattering angle 2θ for a flat, infinitely thick sample. The focussing geometry can be realized with the conventional $\theta - 2\theta$ diffractometer with a stationary x-ray source, or with the $\theta - \theta$ diffractometer with a stationary sample, as shown in Fig. 3.2 with the monochromator in the diffracted beam.

The transmission method is the standard technique when using neutrons as the radiation probe, but it has recently been used with x-rays for the study of liquids and amorphous solids. The diffraction geometry is shown in Fig. 3.1 employing a parallel or slightly diverging beam and a solid state detector to accomplish monochromatization of the scattered x-rays.

3.3.2 Correction of the Measured X-Ray Intensities

a) Measured Intensity

The coherent scattering per atom $I_a(K)$ is not directly accessible from experiment but must be deduced from experimental data by applying a number of different corrections which depend on the radiation and geometry used in the experiment.

The x-ray intensity $I_s(2\theta)$, scattered by an amorphous sample and measured by the variable 2θ technique, can be expressed as follows:

$$I_s(2\theta, E) = N\{I_0(E)A(2\theta, E, E)P(2\theta)[I_a(K) + I_{ms}^{coh}(K)]\}*M(E)$$
$$+ N\{I_0(E')A(2\theta, E, E')P(2\theta, E, E')I_{inc}(K')$$
$$\cdot \int h(2\theta, E', E'')M(E'')dE''\}*M(E'), \tag{3.59}$$

where $*$ represents the convolution product; N: number of atoms in the irradiated volume; $I_0(E)$: intensity of the primary beam; $A(2\theta, E, E')$: absorption of the scattered intensity of energy E and the primary intensity of energy E' in the sample; $P(2\theta)$: polarization factor; $I_a(K) = I_N(K)/N$: coherent scattering per atom as a function of K with $K = (4\pi/\lambda)\sin\theta = (4\pi e/hc)(\sin\theta)E$; $I_{\text{ms}}^{\text{coh}}(K)$: coherent multiple scattering as a function of K; $I_{\text{inc}}(K')$: incoherent (or Compton) scattering as a function of K', with $K' = (4\pi e/hc)(\sin\theta)E'$ being the initial energy of the incoming x-ray photon which is reduced to E after the incoherent scattering process, i.e., $E' = E + \Delta E = E/(1 - 0.00392 E \sin^2\theta)$; $M(E)$: normalized energy response function of the combination of a solid-state detector with the narrow window of a single channel analyzer, or the monochromator in the diffracted beam; $h(2\theta, E', E'')$: Compton profile, normalized so that $\int h(2\theta, E', E'')dE'' = 1$, E'' being the Compton shift of energy E', i.e., $E'' = E'/(1 + 0.00392 E'\sin^2\theta)$. The individual corrections have been discussed in detail in the literature [3.2, 3, 19–23]. We will only highlight some recent developments.

The response function $M(E)$ is centered about the energy of the K_α radiation used in the experiment. It is desirable to make it as narrow as possible to obtain monochromatic radiation. Since the K_α radiation is much sharper and larger than the narrow band $M(E)$ of continuous radiation, we can neglect the variation of the absorption correction $A(2\theta, E, E')$ and the polarization factor $P(2\theta)$ for the elastically scattered radiation with energy E. Thus, the convolution products in (3.59) can be replaced by the constant value $I_0(E_\alpha)$, where E_α is the energy of the K_α radiation.

Integration over the Compton profile can be accomplished with the approximation introduced by *Ruland* [3.24], i.e., by defining the Q-function:

$$Q(2\theta, E_\alpha) = \int M(E'')h(2\theta, E_\alpha, E'')dE'' \tag{3.60}$$

which can be evaluated from the measured $M(E)$ function and by assuming $h(2\theta, E_\alpha, E'')$ to be a Cauchy (or Lorentz) function. If one uses a monochromator in the primary beam and a wide energy window for the detector, $M(E) = 1$ and $Q(2\theta, E_\alpha) = 1$. The Q-function is a slowly decreasing function of 2θ and E_α, i.e., the diffraction vector K.

With this approximation, we can express the scattered intensity measured in the variable 2θ technique:

$$I_s(2\theta, E_\alpha) = NI_0(E_\alpha)\{A(2\theta, E_\alpha, E_\alpha)P(2\theta)[\langle f^2\rangle + \langle f\rangle^2 i(K) + I_{\text{ms}}^{\text{coh}}(K)]$$
$$+ A(2\theta, E_\alpha, E_c)P(2\theta)I_{\text{inc}}(K)Q(K)\}, \tag{3.61}$$

where $E_c = E''$ produced by E_α, the characteristic K_α energy.

b) Absorption Correction $A(2\theta, E, E')$ and Polarization Factor $P(2\theta)$

The absorption factor for a flat sample can be written for the nonfocussing reflection geometry as

$$A(2\theta, E, E') = \frac{A_0 \sin\beta}{\mu(E)\sin\beta + \mu(E')\sin\gamma}\left\{1 - \exp\left[-t\left(\frac{\mu(E)}{\sin\gamma} + \frac{\mu(E')}{\sin\beta}\right)\right]\right\}, \tag{3.62}$$

where A_0 is the cross section of the primary beam, γ and β are the angles between the incident and reflected beam, respectively, and the specimen surface, $\mu(E)$ is the absorption coefficient at energy E and t is the thickness of the sample. Equation (3.62) is valid as long as $A_0 < B \sin \gamma$, B being the width of the sample. If $A_0 > B \sin \gamma$, A_0 has to be replaced by $B \sin \gamma$ in the above expression. When $\mu(E) = \mu(E')$ and $\gamma = \beta = \theta$, one obtains the well-known expression $A(2\theta, E)$ for the focussing reflection geometry.

In the nonfocussing transmission geometry, the corresponding absorption factor is given by

$$A(2\theta, E, E') = \frac{C \sin \beta}{\mu(E') \sin \gamma - \mu(E) \sin \beta} \{ \exp[-\mu(E)t/\sin \gamma] \\ - \exp[-\mu(E')t/\sin \beta] \}, \tag{3.63}$$

where $C = A_0$ when $A_0 < B \sin \gamma$, and $C = B \sin \gamma$ when $A_0 > B \sin \gamma$. Remember that in the focussing transmission case, $\gamma = \beta = 90° - \theta$. Thus we obtain when $\mu(E) = \mu(E')$,

$$A(2\theta, E) = \frac{t}{\cos \theta} \exp[-\mu(E)t/\cos \theta]. \tag{3.64}$$

The polarization can be written as [3.25]

$$P(2\theta) = [1 + k \cos^2(2\theta)]/(1 + k'), \tag{3.65}$$

where $k = k' = 1$ when no crystal monochromator is used; $k = k' = \cos^2(2\alpha)$ when an ideally imperfect monochromator is used in the primary beam and $k = \cos^2(2\alpha)$ and $k' = 1$ when a monochromator is placed in the diffracted beam.

c) Atomic Scattering Factor and Multiple Scattering

The scattering factor f is calculated without considering resonance effects, but for wavelengths near the absorption edge of the sample elements there is a drastic change in the scattering factor, i.e.,

$$f(K, E) = f_0(K) + f'(K, E) + if''(K, E), \tag{3.66}$$

where f_0 is the high energy limit (tabulated in several books), and f' and f'' the real and imaginary parts of the anomalous dispersion term f' and f'' have been tabulated for characteristic x-ray targets but variations with K have not been determined; they are assumed to be independent of the scattering angle. For certain wavelengths there are some measurements and theoretical calculations [3.26, 27].

The anomalous scattering is a very attractive phenomenon for the determination of the partial functions because it is a way of charging the ratio of the scattering factors of two components systems. This possibility will be discussed in Sect. 3.4.1c but a lot of unknown properties of this effect remain to be uncovered in order to have reliable experimental results.

Multiple scattering contributes to background radiation and can be eliminated analytically [3.28]. To a first approximation it is sufficient to consider only doubly-scattered radiation.

d) Normalization of the Scattered Intensity

At large values of K, the reduced interference function $i(K)$ usually shows only small modulations about zero. Thus we can set $i(K)_{K \to \infty} = 0$ and obtain from (3.61) with $A(2\theta, E_\alpha, E_\alpha) \equiv A(2\theta, E_\alpha)$,

$$I_s(2\theta, E_\alpha) = I_s(K) = N I_0(E_\alpha) P(2\theta) \{A(2\theta, E_\alpha) [\langle f^2 \rangle + I_{ms}^{coh}(K)]$$
$$+ A(2\theta, E_\alpha, E_c) Q(K) I_{inc}(K)\}. \tag{3.67}$$

The quantity

$$\beta(K) = P(2\theta) \{A(2\theta, E_\alpha) [\langle f^2 \rangle + I_{ms}^{coh}(K)] + A(2\theta, E_\alpha, E_c) Q(K) I_{inc}(K)\}/I_s(K), \tag{3.68}$$

will modulate about a constant value at large values of K. We can replace it by its average value over a certain K range [3.21], i.e.,

$$\beta = \int_{K_{min}}^{K_{max}} \beta(K) dK/(K_{max} - K_{min}) \simeq [N I_0(E_\alpha)]^{-1}, \tag{3.69}$$

where K_{min} is the value of K above which the modulation of $\beta(K)$ are very small, and K_{max} is the upper limit of the scattering data.

It is also possible to apply the sum rule, i.e.,

$$\int_0^\infty K^2 i(K) dK = 2\pi^2 \varrho_0 \tag{3.70}$$

to evaluate $N I_0(E_\alpha)$ directly. Using (3.61) we can solve for $i(K)$, i.e.,

$$i(K) = (\{I_s(2\theta, E_\alpha)/[N I_0(E_\alpha)]\} - A(2\theta, E_\alpha) P(2\theta) [\langle f^2 \rangle + I_{ms}^{coh}(K)]$$
$$- A(2\theta, E_\alpha, E_c) P(2\theta) I_{inc}(K) Q(K))/[P(2\theta) A(2\theta, E_\alpha) \langle f \rangle^2]$$

which yields with (3.69)

$$[N I_0(E_\alpha)]^{-1} = \left(\int_0^{K_{max}} K^2 \{(\langle f^2 \rangle + I_{ms})/\langle f \rangle^2 + [A(2\theta, E_\alpha)/A(2\theta, E_\alpha, E_c)] \right.$$
$$\left. \cdot Q(K) I_{inc}/\langle f \rangle^2\} dK - 2\pi^2 \varrho_0 \right) /$$
$$\left(\int_0^{K_{max}} K^2 \{I_s(2\theta, E_\alpha)/[P(2\theta) A(2\theta, E_\alpha) \langle f \rangle^2]\} dK \right). \tag{3.71}$$

In principle, the values of NI_0 obtained from the sum rule should be equal to the value β^{-1} obtained by matching the measured intensity $I_s(2\theta, E_\alpha)$ to the independent scattering at large K values (high-angle method) and they usually are when the upper limit of K_{max} is larger than $12\,\text{Å}^{-1}$.

e) Spurious Effects Occurring in Fourier Transformation Termination Effects

The reduced interference $a(K) = Ki(K)$, (3.12), is not experimentally obtained in the interval $[0, \infty]$. It can only be determined from a minimum value K_m to a maximum value $K_M = 4\pi \sin\theta/\lambda$. It is possible by extrapolation to have values between 0 and K_m, but only the $W'(r)$ function can be obtained in the interval $(0, K_M)$, i.e.,

$$W'(r) = [1/(2\pi^2\varrho_0)] \int\limits_0^{K_m} a(K)\sin(Kr)dK. \tag{3.72}$$

Let

$$H(K) = 1 \quad \text{for} \quad |K| < K_M$$
$$H(K) = 0 \quad \text{for} \quad |K| > K_M.$$

We introduce the convolution product

$$W'(r) = [2\pi/(2\pi^2\varrho_0)]\left[(1/\sqrt{2\pi}) \int\limits_{-\infty}^{+\infty} a(K)\exp(iKr)dK\right]$$
$$* \left[(1/\sqrt{2\pi}) \int\limits_{-\infty}^{+\infty} H(K)\exp(iKr)dK\right].$$

In this relation the functions $W(r)$ and $a(K)$ are defined as antisymmetric functions. With

$$u(r) = \int\limits_{-K_M}^{K_M} \exp(iKr)dr$$

or

$$u(r) = 2K_M \sin(K_M r)/(K_M r),$$

we have

$$W'(r) = W(r)*u(r). \tag{3.73}$$

The $W'(r)$ function is the convolution product of the unknown function $W(r)$ and $u(r)$. This introduces a broadening of the peaks and oscillations with a pseudoperiod $2\pi/K_M$ called termination ripples.

Computer Programming Effect

In computer calculations, the integral in (3.72) must be replaced by a summation, i.e., we will calculate the function $W''(r)$ instead of $W'(r)$ [(3–14)]:

$$W''(r) = [1/(2\pi^2\varrho_0)]\Delta K \sum_{n=0}^{M} a(n\Delta K)\sin(n\Delta Kr), \qquad (3.74)$$

where ΔK is the interval in K space, $M = K_M/\Delta K$ and $K = n\Delta K$. Equation (3.74) can be written in the following form

$$W''(r) = [1/(2\pi^2\varrho_0)] \int_0^{K_M} a(K)\sin(Kr)dK \sum_{n=0}^{\infty} \Delta K\delta(K - n\Delta K).$$

It is easily seen that $W''(r)$ represents the convolution of the functions $W'(r)$, (3.72), and $f(r)$, the Fourier transform of $\sum \Delta K\delta(K - n\Delta K)$, i.e.,

$$f(r) = \sum_{n=0}^{\infty} \delta(r - n\Delta r),$$

where $\Delta r = 2\pi/\Delta K$.

Thus we can write

$$W''(r) = W'(r) * f(r) = W'(r) * \sum_{n=0}^{\infty} \delta(r - n\Delta r)$$

which reduces to

$$W''(r) = \sum_{n=0}^{\infty} W'(r - n\Delta r). \qquad (3.75)$$

If $\Delta r = 2\pi/\Delta K$ is large so that $|W'(r - \Delta r)| \ll W'(r)$, then we can write $W''(r) \simeq W'(r)$. This is usually the case when choosing $\Delta K = 0.05\,\text{Å}^{-1}$ which yields $\Delta r = 126\,\text{Å}$ because the atomic distribution function $W'(r)$ possesses characteristic oscillations only for $r < 20\,\text{Å}$.

Normalization Effect

The $W(r)$ function shows that it tends to $-r$ for r smaller than the first interatomic distance r_0 because $P(r) = 0$ for $r < r_0$. However, the first calculation of a Fourier transform of experimental data does not yield this result. Large oscillations are usually observed for small values of r if there is a misfit between gaseous scattering $N\langle f^2 \rangle$ and experimental, scattered intensity.

To illustrate this, we add a constant value C to the interference function $i(K)$. During Fourier transformation the function $\Omega(r) = \int_0^{K_M} CK \sin(Kr) dK$ is added to the function $W(r)$, i.e.,

$$\Omega(r) = C[\sin(K_M r) - K_M r \cos(K_M r)]/r^2$$

which has large oscillations at small values of r. By improving the normalization it is possible to reduce or eliminate these oscillations. If the misfit between $I(K)$ and the gaseous scattering is not a constant factor but a slowly varying function of K, it is possible to use the Fourier transform of the first oscillations of the function $W(r)$ as a means of estimating the misfit function.

3.3.3 Neutron Diffraction

a) Some Characteristic Properties of Neutrons

The neutron is a neutral particle with a mass $m = 1.675 \times 10^{-24}$ g and spin $I = 1/2$. As a consequence it has a magnetic moment γ ($\gamma = 1.913 \mu_{Bv}$, where μ_{Bv} is a nuclear Bohr magneton, i.e., $\mu_{Bv} = \mu_B/1826$, μ_B being the electronic Bohr magneton).

All neutrons with a velocity v have an associated de Broglie wave of wavelength $\lambda = h/mv$. Thermalized neutrons have a mean velocity given by $mv^{-2}/2 = 3kT/2$ which yields a value of $\bar{\lambda} = 1.44$ Å at a temperature $T = 300$ K. Neutrons are useful for diffraction because the wavelength is of the same order as the interatomic distances.

b) Neutron Interaction with Matter

Neutrons are scattered by the nuclei; the scattering amplitude or length is constant with the scattering angle because the nuclei are very small compared to the wavelength. The scattering amplitude does not follow a regular pattern when the atomic number increases. In addition, it changes with isotopic composition.

If the atoms possess magnetic moments, there is an interaction between neutron magnetic moments and electronic magnetic moments. This occurs with transition metals and rare-earth elements.

The magnetic scattering amplitude is

$$b_m = (r_0 \gamma/2) M f \sin \alpha, \tag{3.76}$$

where r_0 is the classical radius of the electron ($r_0 = e^2/mc$), γ is the neutron magnetic moment expressed in nuclear Bohr magnetons, M the atomic magnetic moment expressed in electronic Bohr magnetons, α is the angle between the scattering vector K and the atomic magnetic moment M, and f is a

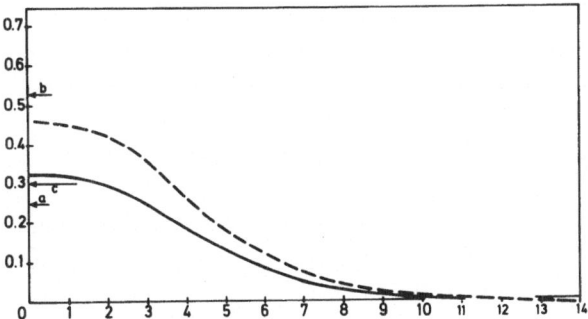

Fig. 3.3. Co and P neutron scattering factors. (*a*) Nuclear Co scattering factor b_{Co}. (*b*) Nuclear P scattering factor b_P. (*c*) Mean scattering factor $c_{Co}b_{Co} + c_P b_P$ for a-$Co_{81}P_{19}$. (- - -) Hexagonal Co magnetic scattering factor with ($n_B = 1.707\,\mu_B$). (——) Amorphous Co magnetic scattering factor obtained by affinity ($n_B = 1.2\,\mu_B$)

form factor decreasing with K because the electronic orbitals are comparable to neutron wavelengths.

3.3.4 Neutron Scattering Amplitude

a) Nuclear and Magnetic Scattering Amplitudes [3.29]

The amorphous CoP is a good example for studying all different neutron scattering terms.

The nuclear spin of cobalt is 7/2. Thus, there are eight different spin states of the Co nucleus and eight different scattering lengths. Experimentally, an averaged value is observed: it is called "the nuclear scattering length" $b_{Co} = 0.25 \times 10^{-12}$ cm. As there are no correlations between nuclear spins, the eight different scattering amplitudes are randomly distributed and a Laue background is observed: this term $(\overline{b^2} - \bar{b}^2)$ is improperly called "incoherent scattering". For Co this cross section is $\sigma_i = 4\pi(\overline{b^2} - \bar{b}^2) = 5.22$ barns. The scattering amplitude of phosphorous is almost "coherent": $b_P = 0.51 \times 10^{-12}$ cm.

In an a-CoP alloy, Co atoms have a magnetic moment. Thus, there is a magnetic scattering length b_m which decreases with a form factor f. This form factor has been investigated in pure Co metal [3.30] and found to have spherical symmetry in the reciprocal space. If we assume the d-orbitals to be the same in crystalline and amorphous cobalt, the form factor is the same for both materials and there is only a change in the b_m amplitude because the amorphous magnetic moment is smaller than the crystalline one ($\mu_B^a = 1.2\,\mu_B$ and $\mu_B^c = 1.707\,\mu_B$, respectively) (Fig. 3.3).

b) Polarized Neutron Beams

If all neutrons spins are parallel to a fixed direction, the beam is polarized. The scattering length, which we have to use in (3.9) in place of f_1, is $b_{Co}^\pm = b_{Co} + b_m$ for

Co atoms, and in place of f_2 we must use b_P for phosphorous atoms. $b_m = \pm (r_0\gamma/2) M f$ if the Co magnetic moment is perpendicular to the K vector, and b_m is positive $(+)$ if the magnetic moment is parallel to the neutrons' spin direction and negative $(-)$ if the moment is antiparallel.

c) Nonpolarized Neutron Beams

If the neutron spins are randomly distributed about two directions, the beam is unpolarized. Experimentally, we observe an averaged intensity when the Co magnetic moments are perpendicular to the K vector:

$$I_a(K) = I_0\{\langle[c_{Co}(b_{Co}^\pm)^2 + c_P b_P^2]\rangle + c_{Co}^2\langle b_{Co}^\pm\rangle^2 i_{CoCo}(K)$$
$$+ c_P^2\langle b_P\rangle^2 i_{PP}(K) + 2c_{Co}c_P\langle b_{Co}^\pm b_P\rangle i_{CoP}(K)\}$$

with

$$\langle b_m\rangle = 0.$$

We have

$$I_a(K) = I_0\{c_{Co}(b_{Co}^2 + b_m^2) + c_{Co}^2(b_{Co}^2 + b_m^2)i_{CoCo}(K)$$
$$+ c_P^2 b_P^2 i_{PP}(K) + 2c_{Co}c_P b_{Co}b_P i_{CoP}(K)\} \tag{3.77}$$

and

$$I_a(K) = I_n(K) + I_m(K),$$

where $I_n(K)$ is the nuclear scattered intensity (with the scattering factors b_{Co} and b_P) and $I_m(K)$ the magnetic scattered intensity (with b_m and 0 scattering factors). If Co magnetic moments are parallel to the K vector, the $I_n(K)$ function is measured because $b_m = 0$ which follows from (3.76) with $\alpha = 0$.

d) Polarization Analysis

The polarization analysis technique distinguishes scattering processes implying spin-flip and nonspin-flip of polarized neutrons [3.31].

For an atom, the corresponding coherent, scattered amplitudes are

$$b^{\pm\pm} = b \mp (r_0\gamma/2)M_z^\perp$$
$$b^{\pm\mp} = -(r_0\gamma/2)(M_x^\perp \pm iM_y^\perp), \tag{3.78}$$

where $M^\perp$ is the projection of a magnetic moment onto the plane perpendicular to the K vector. The z direction is chosen parallel to the polarization of the neutron. If the neutron polarization is chosen parallel to the K vector, $M_z^\perp = 0$ and the nonspin-flip process includes only nuclear scattering, and the spin-flip process includes only magnetic scattering.

Experiments using this technique are described in Sect. 3.4.3 with $ErCo_2$ amorphous alloy.

3.3.5 Correction of the Measured Neutron Intensity

a) Absorption Correction

This is similar to the x-ray absorption correction. The transmission geometry is mainly used with neutrons. Samples may be a plate, for which the absorption factors in (3.62–64) are valid, or a cylinder, for which computer programs were developed by *Mildner* et al. [3.32] and *Poncet* [3.33] for the absorption correction based on the formalism of *Paalman* and *Pings* [3.34].

b) Incoherent Scattering and Background

In neutron diffraction experiments one observes a neutron background measured without any sample in the beam. This background must be measured without the sample and corrected for sample absorption and then subtracted from the experimental data. The incoherent scattering must be determined. That needs an evaluation of $I(K) = I_{coh}(K) + I_{inc}(K)$ in order to have a normalization factor. For large values of K,

$$I_{\substack{a \\ K \to \infty}}(K) = I_0(\sigma_{coh} + \sigma_{inc}), \tag{3.79}$$

where

$$\sigma_{coh} = \sum_i b_i^2.$$

In the case of amorphous metals it is not too difficult to get a good evaluation of the background and the incoherent intensity if it is possible to use the hypothesis $I_{\substack{a \\ K \to 0}}(K) = I_{LB}$, I_{LB} being a Laue background defined in (3.14).

c) Multiple Scattering

Multiple scattering is not negligible in neutron scattering. It can be eliminated analytically [3.28, 35] but for amorphous materials it can be assumed that it is a constant and can then be included in the background.

d) Inelastic Scattering and Placzek Correction

A neutron collision with nuclei could be an elastic collision without a change in the neutron energy, or an inelastic collision with an energy transfer to phonons. This effect has been investigated by *Placzek* for liquid structures [3.36]. In amorphous materials it is smaller but not negligible. It produces a decrease in the coherent, scattered intensity for large K values [3.36, 37].

3.4 Structure of Metallic Glasses

X-ray diffraction techniques have been used extensively to investigate the structure of metallic glasses. However, as was pointed out in the previous sections, a single radiation experiment permits us to determine only the weighted sum of the partial interference functions $I_{ij}(K)$ or $S_{N-C}(K)$, (3.8, 39). As soon as larger amounts of glassy material became available (e.g., through the melt-spinning process), neutron diffraction experiments became feasible and have been carried out recently. This development opened up the possibility of determining more readily the partial interference functions, at least for binary alloys. As the structures are quite different for the two kinds of metallic alloys prepared in amorphous form, i.e., metal-nonmetal (metalloid) and metal-metal alloys, we present some examples of structural studies of the two types of amorphous alloys.

3.4.1 Partial Functions of Metal-Metalloid Glasses

a) Combination of X-Ray, Nuclear and Polarized Neutron Diffraction Data for the Evaluation of the Structure of an Amorphous Co_{80}–P_{20} Alloy

Four interference functions of an amorphous CoP alloy have been obtained from x-ray and neutron diffraction experiments [3.38] as follows:

$$I_x = i_x + 1 = (I_x - I_{LB})/[N(c_{Co}f_{Co} + c_P f_P)^2] \quad \text{for x-ray scattering}$$

$$I_n = i_N + 1 = (I_N - I_{NLB})/[N(c_{Co}b_{Co} + c_P b_P)^2] \quad \text{for neutron nuclear scattering}$$

$$I_n^\pm = i^\pm + 1 = (I^\pm - I_{LB}^\pm)/[N(c_{Co}b_{Co} + c_P b_P)^2] \quad \begin{array}{l}\text{for magnetic scattering} \\ \text{of polarized neutrons.}\end{array}$$

To obtain the partial functions, use is made of three functions

$$i_x, i_N \quad \text{and} \quad i^* = (i^+ - i^-) \cdot (c_{Co}b_{Co} + c_P b_P)/(c_{Co}b_m)$$

which give results to a good accuracy. The partial interference functions and reduced distribution functions are presented in Figs. 3.4, 5. From these functions it is possible to say that two phosphorous atoms are never first neighbors. Other features of these functions are well described by the tetrahedral close-packing model. The shoulder on the second peak on a Co–Co interference function is characteristic of that structure [3.39].

Using only neutron diffraction experiments, *Bletry* [3.40] has obtained the three partial functions $S_{NN}(K)$, $S_{NC}(K)$ et $S_{CC}(K)$ (Fig. 3.6). However, because of the fast decrease of b_m with K, only a limited K-range can be explored which is not sufficient for the evaluation of the partial atomic distribution functions.

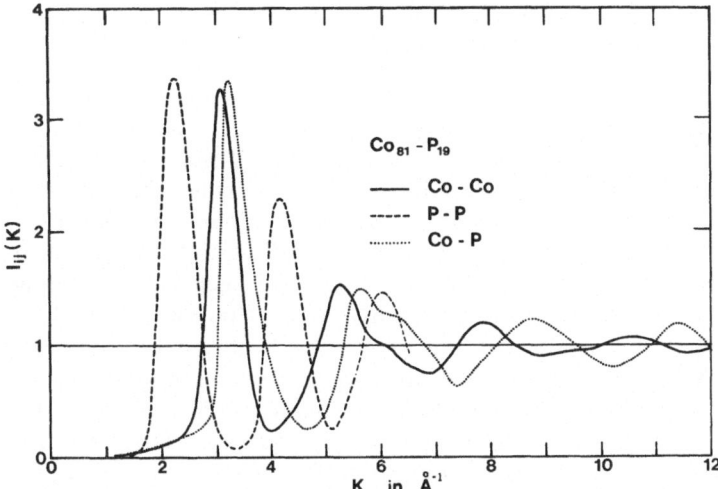

Fig. 3.4. Partial interference functions $I_{ij}(K)$ for a-CoP [3.38]. (——) Co–Co; ($\cdots$) Co–P; (---) P–P

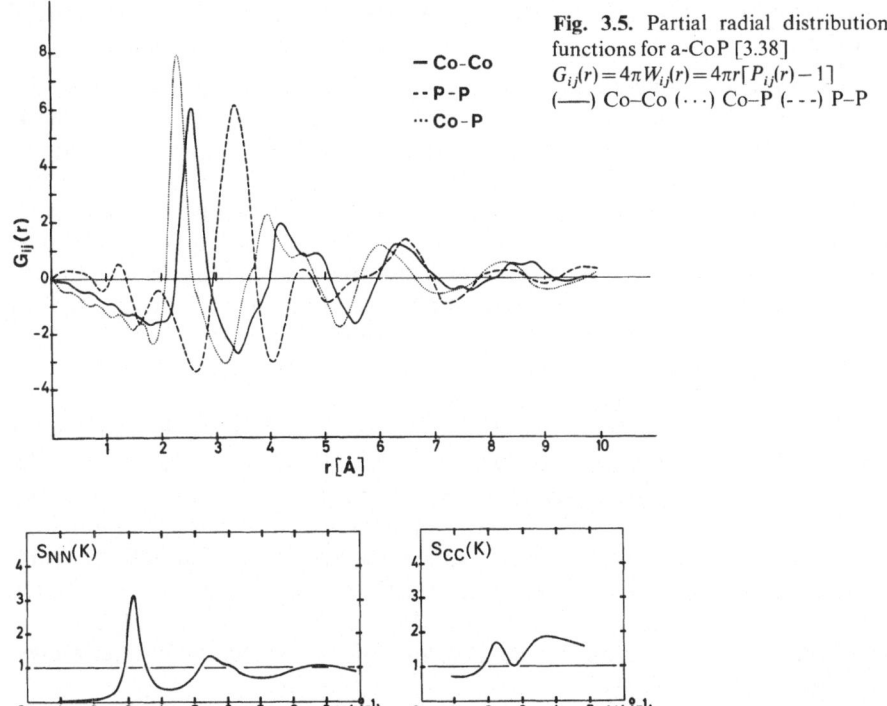

Fig. 3.5. Partial radial distribution functions for a-CoP [3.38]
$$G_{ij}(r) = 4\pi W_{ij}(r) = 4\pi r[P_{ij}(r) - 1]$$
(——) Co–Co ($\cdots$) Co–P (- - -) P–P

Fig. 3.6. Bhatia and Thornton partial structure factor of Co–P assuming $S_{NN}(K)$ to be a classical interference function for all amorphous pure metals. The first peak of $S_{CC}(K)$ is produced by chemical ordering in the CoP alloy [3.40]

b) Combination of X-Ray and Neutron Diffraction Data for the Evaluation of the Structure of an Amorphous Fe–B Alloy

Three interference functions of an amorphous $Fe_{80}B_{20}$ alloy have been obtained by *Lamparter* et al. [3.41] using x-rays and neutrons with naturally occurring Fe and with the isotope [57]Fe. As in the case of CoP, only the total neutron diffraction data of the [57]Fe alloy show a prepeak which must be related to the metalloid partial function, i.e., $I_{BB}(K)$. It cannot be produced by $S_{CC}(K)$ because it should be visible in both x-ray and naturally occurring Fe neutron diffraction data. Nevertheless, there must be chemical short-range order because the reduced atomic distribution function $G(r) = 4\pi\varrho_0 W(r)$ shows a splitting of the first peak, yielding the partial coordination numbers $N_{BFe} = 8.9$ and $N_{FeFe} = 11.5$. Using the three total interference functions, the individual partial functions have recently been evaluated by *Nold* et al. [3.41b], showing that B–B atoms are not nearest neighbors. Similar results have been obtained by *Lamparter* et al. [3.41c] for a $Ni_{81}B_{19}$ glass using the isotopic substitution of Ni in neutron experiments.

c) Combination of Three X-Ray Wavelengths Involving Anomalous Dispersion Corrections of the X-Ray Scattering Factors for the Evaluation of Partial Functions

Although this kind of experiment has not been done with amorphous metallic alloys, it is very important to describe a recent experiment on amorphous Ge–Se alloys utilizing synchrotron radiation and the anomalous scattering of x-rays for the determination of partial functions.

Ge and Se have very similar scattering factors for x-rays but unfortunately it is exactly the same thing with neutrons. So it is difficult to obtain partial functions combining x-ray and neutron scattering. Anomalous dispersion was the only way to vary the atomic scattering factors. In their work on amorphous GeSe, *Fuoss* et al. [3.42] employed synchrotron radiation with a high-resolution monochromator ($\sim 1\,eV$) in the primary beam to select wavelengths close to absorption edges of Ge and Se, and a medium-resolution LiF monochromator ($\sim 50\,eV$) in the diffracted beam to remove the fluorescent radiation from the sample. Because of this double monochromatization, the coherent intensities were relatively small and consequently, the accuracy in the partial functions is low. Nevertheless, they demonstrated that the anomalous dispersion method will yield valuable data when carried out with sufficient care.

Some experiments have been performed on metallic glasses using three different x-ray K_α radiations. *Waseda* and *Tamaki* [3.43] used molybdenum, copper and cobalt radiations for the determination of partial Ni–P interference functions.

In this case, the changes in f' and f'' are small as the x-ray energies are far away from the Ni absorption edge except for CuK_α radiation. Changes in the contribution of the partial functions to the three total interference functions are very small. This produces large uncertainties in the three partial functions.

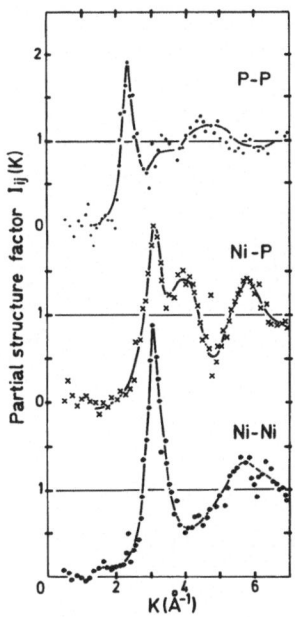

Fig. 3.7. The three partial interference functions of amorphous $Ni_{74}P_{26}$ derived from three different total interference functions obtained with Mo K_α, Cu K_α, Co K_α [3.43]

The three total functions are:

$$i(K) = 0.75\, i_{NiNi} + 0.02\, i_{PP} + 0.23\, i_{NiP},$$
$$Mo\,K_\alpha$$

$$i(K) = 0.70\, i_{NiNi} + 0.04\, i_{PP} + 0.26\, i_{NiP},$$
$$Cu\,K_\alpha$$

$$i(K) = 0.73\, i_{NiNi} + 0.03\, i_{PP} + 0.24\, i_{NiP}.$$
$$Co\,K_\alpha$$

The three partial interference functions presented in Fig. 3.7 are deduced from three functions which are very similar, so there is a large uncertainty in the results. This is so because the variations of $\alpha_{ij}(K)$ with different x-ray wavelengths are relatively small and the normalized determinant $|V(K)|_n$ becomes very small. The values of $\alpha_{ij}(K)$ used by *Waseda* and *Tamaki* [3.43] to determine the partial functions in an amorphous Ni–P alloy yielded a normalized determinant $|V_{ij}(K)|_n = 2 \times 10^{-4}$, a very small value indeed. This is compounded by the difficulty of normalizing the coherent scattering $I_a(K)$ when only a limited range in K (less than $7\,Å^{-1}$ with CoK_α radiation) is available, even when using the RDF normalization procedure [3.13].

Lastly, it should be remembered that fluorescent radiation excited by the white spectrum might be difficult to remove when using a monochromator in the diffracted beam or a solid state detector. This is particularly true when using

CuK_α with Ni alloys or MoK_α with Zr alloys. NiK_β and ZrK_β are too close to CuK_α and MoK_α, respectively, to be entirely removed by a monochromator in the diffracted beam. Unless a monochromator was used in the primary beam and care was taken not to excite $\lambda/2$ in the white spectrum when employing a graphite or LiF crystal, it is doubtful whether any of the anomalous dispersion data (obtained with conventional x-ray tubes) have yielded any reliable partial interference functions or structure factors $I_{ij}(K)$ so far.

The use of synchrotron radiation is a better method but a lot of practical and theoretical problems remain to be solved in order to obtain f' and f'' values. The x-ray wavelength must be defined with a high accuracy within two very narrow wavelength bands. In addition, the variations of the scattering factor are not yet well known.

3.4.2 Structure of Metal-Metal Glasses

a) X-Ray and Neutron Structure Investigation of Amorphous Transition Metal Rare-Earth Metal Alloys

It is possible to prepare metal-metal alloys in amorphous phases by vacuum deposition, sputtering or splat-cooling. But all these alloys are characterized by two metals with a large difference in atomic sizes. This is, for example, the case for Gd–Co [3.44, 45] and Gd–Fe alloys [3.44], whose reduced atomic distribution functions $G(r) = 4\pi\varrho_0 W(r)$, (3.23, 24), derived from x-ray experiments, are presented in Fig. 3.8.

It is possible to observe a splitting of the first maximum which can be attributed to the three first-neighbor distances between transition metal – transition metal, transition metal – rare-earth metal, and rare-earth – rare-earth metal. Partial functions using neutron and x-ray diffraction data have been obtained by *O'Leary* [3.46] who used the results of *Cargill* [3.44] and the neutron results of *Rhyne* et al. [3.47]. More recent investigations by *Givord* et al. [3.48] on Ni$_2$Y amorphous alloy show a small pre-peak in the neutron interference function (Fig. 3.9). It will be shown in Sect. 3.4.2b that the occurrence of a pre-peak might be an indication for chemical short-range order.

b) Isotopic Substitution in an Amorphous Cu$_{57}$Zr$_{43}$ Alloy for the Determination of the Partial Functions from Neutron Diffraction Data

Isotopes of Cu were originally used by *Enderby* et al. [3.7] to evaluate the partial interference functions in a liquid Cu–Sn alloy. Recently, the same isotopes of Cu, namely, ^{63}Cu and ^{65}Cu, have been used by *Mizoguchi* et al. [3.49], together with naturally occurring Cu, to prepare three amorphous samples of an amorphous Cu$_{57}$Zr$_{43}$ alloy. Three different total interference functions $I(K)$ were obtained from these three samples by neutron diffraction.

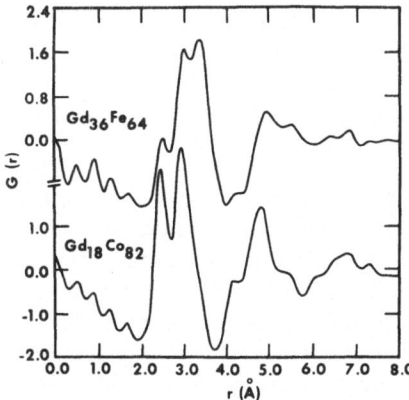

Fig. 3.8. Radial distribution functions for Gd–Fe and Gd–Co alloys. The first peak shows the three first distances [3.44]
FeFe $= 2.54\,\text{Å}$ GdFe $= 3.04\,\text{Å}$ GdGd $= 3.47\,\text{Å}$
CoCo $= 2.47\,\text{Å}$ GdCo $= 2.97\,\text{Å}$ GdGd $\sim 3.4\,$Å

▼ **Fig. 3.9.** Reduced interference function of amorphous $YNi_{1.85}$ obtained from x-ray and neutron diffraction data (a and c). The curves b and d are calculated functions from a model [3.48]

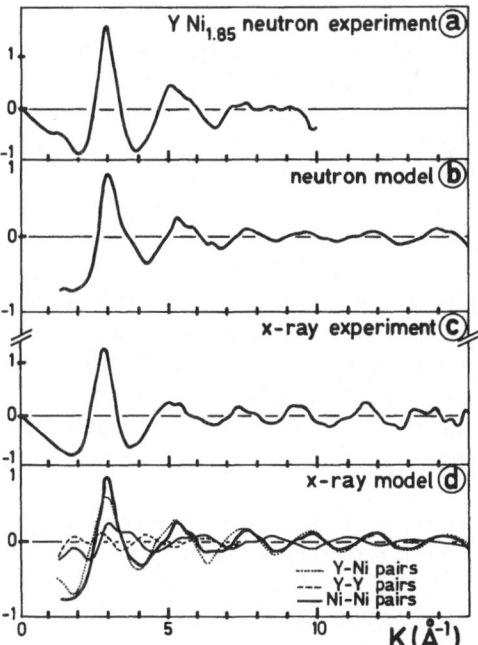

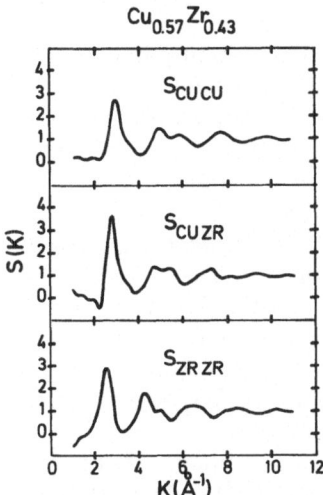

Fig. 3.10. Partial interference functions for CuZr alloy [3.49]

The partial interference functions shown in Fig. 3.10 were evaluated by the procedures discussed above. The reduced radial distribution functions $G_{ij}(r)$ $= 4\pi\varrho_0 W_{ij}(r)$ indicate that the distance $r_{CuCu} = 2.65\,\text{Å}$ is slightly larger than the size of the Cu atom ($2.56\,\text{Å}$ in diameter), whereas $r_{ZrZr} = 3.15\,\text{Å}$ is slightly smaller than the Goldschmidt diameter $3.20\,\text{Å}$ of Zr. The distance between Cu and Zr neighbors was found to be $2.80\,\text{Å}$. The Warren short-range order parameter α can be evaluated from the partial coordination numbers given by

Mizoguchi et al. [3.49] using (3.58) and the value $\alpha \simeq 0$ was obtained indicating that there is no chemical short-range order in amorphous Cu–Zr alloys.

c) Isomorphous Substitution of Zr by Hf in Amorphous $Cu_{60}Zr_{40}$ and $Ni_{35}Zr_{65}$ Alloys

In crystal structure determinations, it has been customary to make isomorphous substitution of light elements with heavy elements for the determination of the location of certain atoms in the unit cell. This approach should also be feasible for the elucidation of the amorphous structure as long as the substitution does not alter the structure. If the substituting element C has the same atomic size and chemical similarity to the element B which it replaces, it is to be expected that the structure of the metallic glass $A_x(BC)_{1-x}$ does not change. Examples of elements which can be substituted for each other are Zr and Hf and Nb and Ta in x-ray experiments, and Ni and Co in neutron experiments.

The x-ray diffraction patterns of amorphous $Cu_{60}Zr_{40}$ and $Cu_{60}Hf_{40}$ were determined by *Chipman* et al. [3.50] and used to calculate the partial coordination numbers. Their data indicate that there is preference for unlike nearest neighbors, in contrast to the result by *Mizoguchi* et al. [3.49] who found a random distribution of Cu and Zr in amorphous $Cu_{57}Zr_{43}$.

The interference functions $I(K)$ of seven $Ni_{35}Zr_{65-x}Hf_x$ with $x = 0, 5, 10, 15, 20, 25,$ and 30 at.% Hf were obtained from the x-ray scattering patterns [3.51, 52] and as an example, $I(K)$ is shown in Fig. 3.11 for the $Ni_{35}Zr_{65}$ alloy. The Fourier transforms of $K[I(K) - 1]$, i.e., the reduced atomic distribution functions $G(r) = 4\pi\varrho_0 W(r)$, exhibit a split first peak as shown in Fig. 3.12 for the same $Ni_{35}Zr_{65}$ alloy. The first maximum in $G(r)$, which decreases in height with increasing Hf content, lies at $r'_1 = 2.69$ Å, and the second of the split first peak is at $r''_1 = 3.19$ Å and becomes larger in height with increasing Hf concentration. This strongly suggests that the peak at r''_1 corresponds to the Zr–Zr interatomic distances.

The Bhatia-Thornton (N–C) partial structure factors were calculated from three $I(K)$ and are also shown in Fig. 3.11. It is readily seen that $S_{NN}(K)$ strongly resembles $I(K)$ because the factor $(f_2 - f_1)/\langle f \rangle \lesssim 0.5$ and the functions $S_{NC}(K)$ and $[S_{CC}(K) - c_1c_2]/(c_1c_2)$ are small, thus contributing little to $I(K)$ according to (3.38).

A similar conclusion can be reached when comparing $G(r)$ of the $Ni_{35}Zr_{65}$ alloy with $G_{NN}(r) = 4\pi W_{NN}(r)$ as shown in Fig. 3.12. Again, $G_{NN}(r)$ shows a splitting of the first peak with maxima at $r = 2.68$ and 3.20 Å. Also shown in Fig. 3.12 are the functions $G_{NC}(r)$ and $G_{CC}(r)$. It is readily seen that $G_{CC}(r)$ has a strong minimum at $r = 2.68$ Å indicating a preference for unlike nearest neighbors at that distance, which consequently represents the Zr–Ni nearest-neighbor separations. The maximum in $G_{CC}(r)$ at $r = 3.1$ Å represents like neighbor distances, i.e., Zr–Zr separation. The integrals were calculated as $rG_{NN}(r) + 4\pi r^2\varrho_0$, $rG_{NC}(r) = 4\pi r^2\varrho_{NC}(r)$, and $rG_{CC}(r) = 4\pi r^2\varrho_{CC}(r)$ in the range of

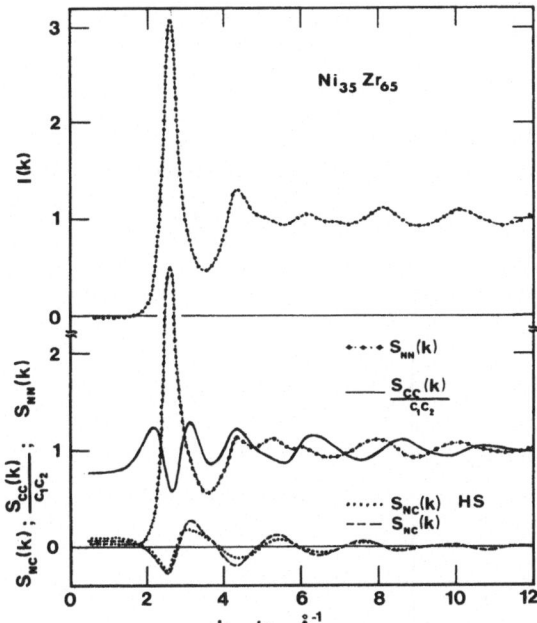

Fig. 3.11. Total interference function $I(K)$ of an amorphous $Ni_{35}Zr_{65}$ alloy. Also shown are the partial interference functions (partial structure factors) $S_{NN}(K)$, $S_{NC}(K)$, and $S_{CC}(K)/(c_1c_2)$ [3.51]

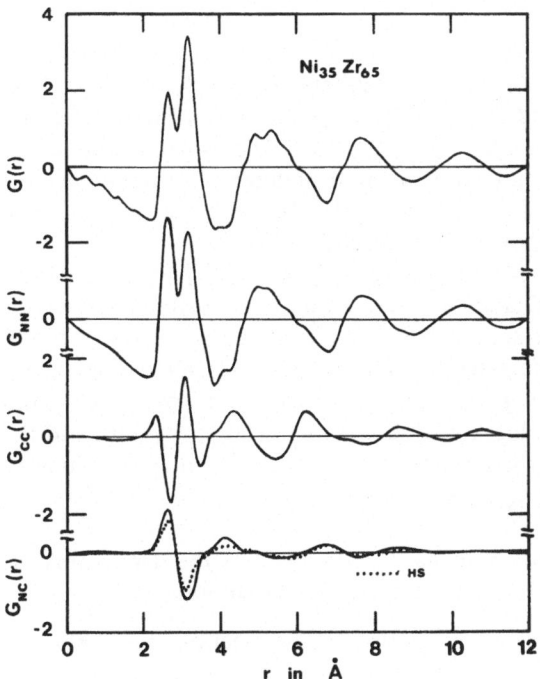

Fig. 3.12. Total reduced atomic distribution function $G(r)$ of an amorphous $Ni_{35}Zr_{65}$ alloy. Also shown are the partial reduced distribution functions $G_{NN}(r)$, $G_{NC}(r)$, and $G_{CC}(r)$ [3.51]

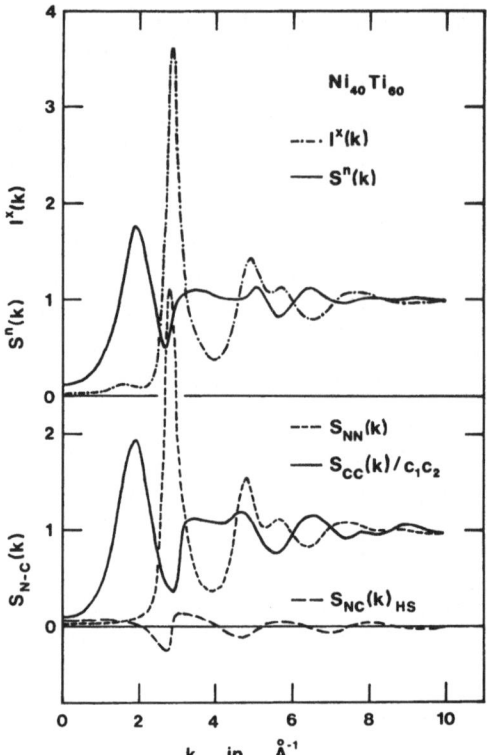

Fig. 3.13. Total interference function $I^x(K) = \{I_a(K) - [\langle f^2 \rangle - \langle f \rangle^2]\}/\langle f \rangle^2$ evaluated from x-ray data and $S^n(K) = I_a(K)/\langle f^2 \rangle$ obtained from neutron data of an amorphous $Ni_{40}Ti_{60}$ alloy. Also shown are the (N–C) partial functions (or structure factors) $S_{NN}(K)$, $S_{NC}(K)$, and $S_{CC}(K)/(c_1 c_2)$ [3.14, 51, 52]

$r = 0$ to $r = 3.9$ Å, yielding the coordination numbers $N^1_{NN} = c_1 N_1 + c_2 N_2 = 13.0$, $N^1_{NC} = c_1 c_2 (N_1 - N_2) = -0.5$, and $N^1_{CC} = (c_2 N_1 + c_1 N_2)\alpha_1 = 0.5$. From these values one obtains $N_1 = N_{Ni} = 11.6$ and $N_2 = N_{Zr} = 13.8$. The Warren short-range order parameter is $\alpha_1 = -0.04$, indicating a slight preference for unlike nearest neighbors in the first coordination shell, in agreement with the conclusion of *Chipman* et al. [3.50] for an amorphous $Cu_{60}Zr_{40}$ alloy.

In order to estimate the size effect term in (3.41), i.e., $S_{NC}(K)$, the hard sphere model [3.53, 54] was used to calculate it and its Fourier transform $4\pi r \varrho_{NC}(r)$ for a size ratio of 0.81. As can be seen in Fig. 3.11, good agreement was found between the H–S curves and the experimental functions $S_{NC}(K)$ and $4\pi r \varrho_{NC}(r)$, respectively. Thus, we may suggest that in cases where only two total interference functions are available, e.g., one obtained from x-ray data and the other from neutron scattering, the hard sphere $S_{NC}(K)$ curve may be added to evaluate the partial functions $S_{NN}(K)$ and $S_{CC}(K)$.

d) Combination of X-Ray and Neutron Scattering for the Evaluation of Topological and Chemical Short-Range Order in Ni–Ti Glasses

Natural Ti possesses a negative scattering cross section for neutrons ($b_{Ti} = -0.34$), whereas $b_{Ni} = 1.03$ so that the average scattering factor $\langle b \rangle$ for an

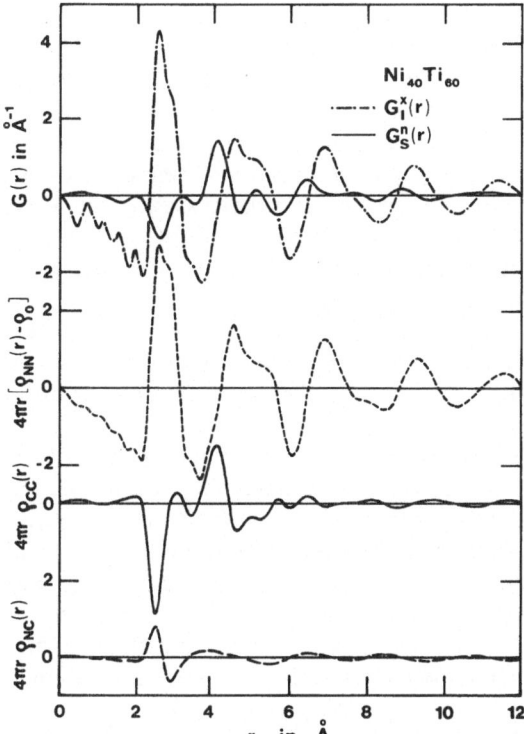

Fig. 3.14. Atomic distribution function $G(r) = 4\pi r \varrho_0 [g(r) - 1]$ evaluated from the x-ray interference function $I^x(K)$. Also shown are the (N–C) partial distribution functions $G_{NN}(r)$, $G_{NC}(r)$, and $G_{CC}(r)$ of an amorphous $Ni_{40}Ti_{60}$ alloy [3.14, 51, 52]

$Ni_{40}Ti_{60}$

---- $G_1^x(r)$

——— $G_S^n(r)$

amorphous Ni–Ti could be made very small. Amorphous $Ni_{40}Ti_{60}$ and $Ni_{35}Ti_{65}$ were prepared by the melt-spinning process. The neutron scattering patterns were measured by *Ruppersberg* [3.14] and the x-ray data were obtained by *Dokyol Lee* [3.52]. The total interference functions $I^x(K)$, (3.12), and $S^n(K)$, (3.10), are shown in Fig. 3.13 for the $Ni_{40}Ti_{60}$ alloy. The most striking feature is the small pre-peak in $I^x(K)$ which becomes the main peak in $S^n(K)$. This is readily understood when evaluating the weighting factors in (3.41) for the $Ni_{40}Ti_{60}$ alloy, i.e.,

	$\langle f \rangle^2 / \langle f^2 \rangle$	$2\Delta f \langle f \rangle / \langle f^2 \rangle$	$c_1 c_2 (\Delta f)^2 / \langle f^2 \rangle$
Neutrons	0.09	−1.15	0.91
x-ray	0.98	−0.49	0.02

The x-ray pattern $I^x(K)$ is dominated by the topological short-range order $S_{NN}(K)$ since the weighting factor $c_1 c_2 (\Delta f)^2 / \langle f^2 \rangle$ of $S_{CC}(K)/(c_1 c_2)$ is very small and the term $S_{NC}(K)$ is expected to be small because of the size factor $r_{Ni}/r_{Ti} = 0.86$ [3.53, 54]. The neutron scattering $S^n(K)$, however, is largely given by the term $S_{CC}(K)$ because $\langle f \rangle^2 / \langle f^2 \rangle$ is small.

The large modulation of the neutron structure factor $S^n(K) \sim S_{CC}(K)/(c_1 c_2)$ is a strong indication for the existence of chemical short-range order in the alloy. Nevertheless, the effect of chemical order is still visible in the x-ray pattern in terms of the pre-peak which falls at the same position in K as the main peak in the neutron structure factor $S^n(K)$ of the amorphous Ni–Ti alloy. Only when the average scattering length $\langle f \rangle$ is small can the chemical order be directly evaluated from the scattering pattern. In all other cases, i.e., in x-ray diffraction and also in neutron scattering, when the components of the alloy have scattering lengths of the same sign, only a pre-peak is observed when chemical order is present in the amorphous or liquid alloy.

As indicated above the size effect term $S_{NC}(K)$ was evaluated with the hard-sphere model (size ratio 0.86) and was then used together with the x-ray and neutron data of the $Ni_{40}Ti_{65}$ alloy to evaluate the three Bhatia-Thornton structure factors which are also shown in Fig. 3.13.

The Fourier transforms $G(r)$ of the total interference functions and of the partial structure factors $S_{N-C}(K)$ are shown in Fig. 3.14. The partial function $G_{NN}(r)$ shows a broad first peak resulting from the superposition of $G_{NiTi}(r)$ with a maximum at $r = 2.58$ Å [corresponding to the sharp minimum in $G_{CC}(r)$] and $G_{TiTi}(r)$ with a maximum at $r = 2.92$ Å. The first coordination number N_{NN} was evaluated from $4\pi r^2 \varrho_{NN}(r) = r G_{NN}(r) + 4\pi r^2 \varrho_0$ using (3.54) and found to be $N_{NN} = 12.8$. Similarly, the integral quantity N_{NC} was calculated with (3.55) and found to be $N_{NC} = 0.39$. With these two quantities, we can immediately calculate $N_{Ni} = N_{NN} + N_{NC}/c_1 = 13.85$ and $N_{Ti} = N_{NN} - N_{NC}/c_2 = 12.2$.

Lastly, the integral quantity $N_{CC} = -2.80$ was evaluated with (3.57) which yields, according to (3.58), the Warren chemical short-range order parametric $\alpha_1 = -0.21$, which corresponds to a Ni atom having 9.3 Ti neighbors instead of 7.7 for a random alloy.

e) Chemical Short-Range Order in Metal-Metal Glasses

All the amorphous systems which we have discussed are obtained by alloying two kinds of metallic atoms having a great difference in atomic size. Among all the studies, a large majority suggests a tendency towards hetero-coordination. This conclusion can be drawn from some of the results presented in the last paragraph, but also from the work by *Sakata* et al. on a-CuTi [3.55] and by *Sadoc* et al. on CuZr [3.56]. This effect may be explained by an increase in the packing density of systems with two atoms of different sizes when the number of first neighbor pairs consisting of big and small atoms are maximized.

3.4.3 Magnetic Structure Determination Using Neutron Diffraction

Polarization analysis experiments have been done by *Boucher* et al. [3.57] on an Er–Co$_2$ alloy, using the condition described in Sect. 3.3.4. The magnetic

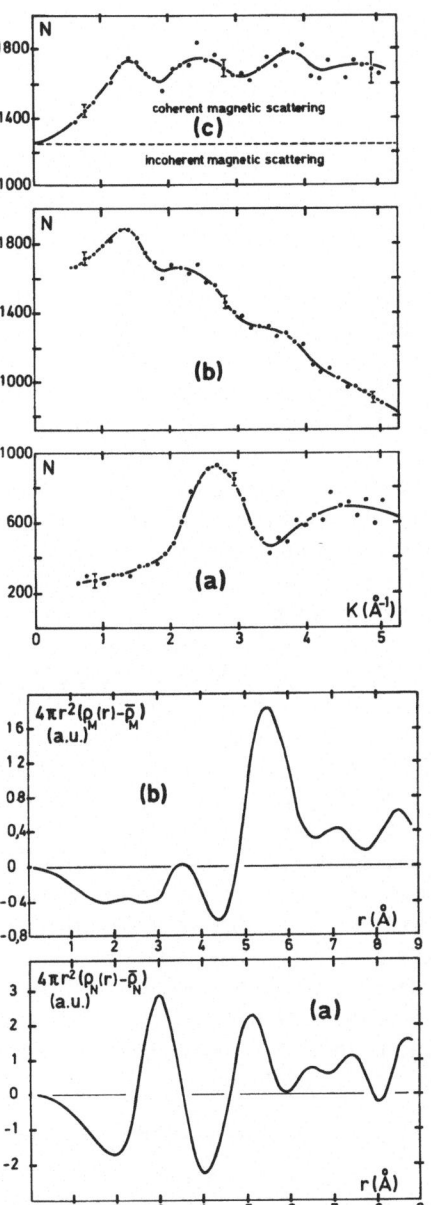

Fig. 3.15a–c. Scattered neutron intensities for ErCo$_2$ alloy. (**a**) Without spin flip. This diagram corresponds to the nuclear structure of the sample. (**b**) With spin flip. This diagram corresponds to 2/3 of the incoherent part of the nuclear scattering and to the magnetic scattering. (**c**) This curve is the same function as **b**, but corrected for the magnetic factors [3.57]

Fig. 3.16a, b. Radial distribution functions of amorphous ErCo$_2$ (**a**) Nuclear function $rG(r) = 4\pi r W(r)$ corresponding to the atomic structure. (**b**) Magnetic function. If all erbium moments were parallel and pointing in the same direction, the two curves would be proportional. If the erbium moments were parallel but pointing in the opposite direction, the first peak of $\varrho_M(r)$ would vanish. The experimental case is in between the two cases. *Boucher* et al. [3.57] proposed a model where the Er magnetic moments of neighboring atoms tend to be parallel with the coexistence of ferro and antiferro-magnetic configurations

moments were polarized by a 2 kOe field in the z direction. Only the x and y components perpendicular to the magnetization of the sample have an effect on the magnetic scattering. Scattered intensities are presented in Fig. 3.15. The two curves are very different and it is possible to obtain information on the magnetic local anisotropy as shown in Fig. 3.16.

With the experimental conditions used, there is no contribution to magnetic scattering from the magnetic long-range order. The existence of a coherent part clearly indicates that there are correlations between the directions of the easy magnetization axes of the erbium neighbors.

3.5 Conclusion

With the improvement in experimental techniques for the measurements of the interference function $I(K)$ or the structure factor $S(K)$, (3.12, 10), better data can be obtained and are very much needed in order to take advantage of the advancement in diffraction theory for amorphous (or liquid) binary alloys. Better experiments are possible because of the new developments of more intense radiation sources and efficient detectors.

Absorption measurements of x-rays or γ-rays yield valuable information about the local atomic arrangement in metallic glasses which can be directly compared with those occurring in the corresponding crystals. The measurements of the extended x-ray absorption fine structure (EXAFS) has provided support and confirmation of the topological and chemical short-range order in metal-metalloid and metal-metal glasses [3.58]. Mössbauer spectroscopy has been successfully employed to elucidate the nearest-neighbor interactions in Fe-base metallic glasses [3.59], whereas nuclear magnetic resonance (NMR) experiments using ^{11}B, and ^{71}Ga isotopes have shown that the local structures of amorphous and crystalline metal-metal alloys ($La_{75}Ga_{25}$) and metal-metalloid alloys ($Mo_{70}B_{30}$, $Mo_{48}Ru_{32}B_{20}$, $Ni_{78}P_{14}B_8$) are quite similar [3.60].

Another important point in structure determination is the modelling. It is very difficult to go from a diffraction experiment to a structure, but starting from a model and computing interference functions (which are compared with experimental functions) is a good way to test the validity of a hypothetical structure. In this case it is sometimes very surprising to see some very small details appearing in an experimental interference function which are confirmed by calculated interference functions. This is proof that these details are not experimental artifacts (such as termination ripples, or normalization oscillations), but conversely it also proves the validity of the model.

Acknowledgement. The research at UCLA has been supported by the grant DMR 80-07939 from the National Science Foundation

References

3.1   P. Debye: Ann. Physik **46**, 809 (1915)
3.2   A. Guinier: *Theorie de la Cristallographie* (Dunod, Paris 1976)
3.3   H. P. Klug, L. E. Alexander: X-*Ray Diffraction Procedure*, 2nd ed. (Wiley, New York 1976)

3.4   D.T.Keating: J. Appl. Phys. **34**, 923 (1963)
3.5   T.E.Faber, J.M.Ziman: Phil. Mag. **11**, 153 (1965)
3.6   F.G.Edwards, J.E.Enderby, R.A.Howe, D.I.Page: J. Phys. C**8**, 3483 (1975)
3.7   J.E.Enderby, D.M.North, P.A.Egelstaff: Phil. Mag. **14**, 961 (1966)
3.8   A.B.Bhatia, D.E.Thornton: Phys. Rev. B**2**, 3004 (1970)
3.9   J.Bletry: Z. Naturforsch. **31**a, 960 (1976)
3.10   H.Ruppersberg, H.Egger: J. Chem. Phys. **63**, 4095 (1975)
3.11   H.Ruppersberg, W.Knoll: Z. Naturforsch. **32**a, 1374 (1977)
3.12   C.N.J.Wagner: J. Non-Cryst. Sol. **31**, 1 (1978)
3.13   C.N.J.Wagner, H.Ruppersberg: Atomic Energy Review Supplement 1 (1981) p. 101
3.14   H.Ruppersberg, D.Lee, C.N.J.Wagner: J. Phys. F.**10**, 1645 (1980)
3.15   M.A.Krivoglaz: *Theory of X-Ray and Thermal Neutron Scattering by Real Crystals* (Plenum, New York 1969)
3.16   B.E.Warren: X-*Ray Diffraction* (Addison-Wesley, Reading, MA. 1969)
3.17   R.N.Sinclair, D.A.G.Johnson, J.C.Dore, H.H.Clarke, A.C.Wright: Nucl. Instrum. and Methods **117**, 445 (1974)
3.18   K.Suzuki, M.Misawa, K.Kai, N.Watanabe: Nucl. Instrum. and Methods **147**, 519 (1977)
3.19   T.Egami: J. Mat. Science **13**, 6587 (1978)
3.20   T.Egami: In *Metallic Glasses* I, ed. by H.J.Güntherodt, H.Beck, Topics Appl. Phys., Vol. 46 (Springer, Berlin, Heidelberg, New York 1981) p. 25
3.21   C.N.J.Wagner: *Liquid Metals, Chemistry and Physics*, ed. by S.Z.Beer (Marcel Dekker, New York 1972) p. 257
3.22   A.H.Compton, S.K.Allison: *X-Rays in Theory and Experiment* (MacMillan, London 1935)
3.23   *International Tables for X-Ray Crystallography*, Vol. III (1962), Vol. IV (1974) (Kynoch Press, Birmingham)
3.24   W.Ruland: Brit. J. Appl. Phys. **15**, 1301 (1964)
3.25   L.Jennings: Acta Cryst. A**24**, 472 (1968)
3.26   D.T.Cromer: Acta Cryst. **18**, 17 (1965)
3.27   S.Ramaseshan, S.C.Abrahams (eds.): *Anomalous Scattering* (Intern. Union of Crystallography and Munksgaard, Copenhagen 1965)
3.28   G.Malet, C.Cabos, A.Escande, P.Delorde: J. Appl. Cryst. **6**, 139 (1973)
3.29   G.E.Bacon: *Neutron Diffraction* (Clarendon Press, Oxford 1975)
3.30   R.M.Moon: Phys. Rev. A**195**, 136 (1964)
3.31   R.M.Moon, T.Riste, W.C.Koehler: Phys. Rev. **181**, 930 (1969)
3.32   D.F.R.Mildner, J.M.Carpenter, C.A.Pelizzari: Rev. Sci. Instr. **45**, 572 (1974)
3.33   P.F.J.Poncet: Intern. Scientific Rept. No. 78 P087s, ILL Grenoble (1978)
3.34   H.H.Paalman, C.J.Pings: J. Appl. Phys. **33**, 263r (1962)
3.35   G.H.Vineyard: Phys. Rev. **96**, 93 (1954)
3.36   G.Placzek: Phys. Rev. **86**, 377 (1952)
3.37   J.L.Yarnell, M.J.Katz, R.G.Wenzel, S.H.Koenig: Phys. Rev. A7, 2130 (1973)
3.38   J.F.Sadoc, J.Dixmier: Mat. Sci. and Eng. **23**, 187 (1976)
3.39   J.F.Sadoc, J.Dixmier, A.Guinier: J. Non-Cryst. Solids **11**, 381 (1973)
3.40   J.Bletry: Thesis, Grenoble (1979)
3.41a P.Lamparter, E.Nold, G.Rainer-Barbach, E.Grallath, S.Steeb: Z. Naturforschg. **36**a, 165 (1981)
3.41b E.Nold, P.Lamparter, G.Rainer-Harbach, S.Steeb: Z. Naturforschg. **36**a, 1032 (1981)
3.41c P.Lamparter, W.Sperl, S.Steeb, J.Bletry: Z. Naturforschg. **37**a, 1223 (1982)
3.42   P.H.Fuoss, W.K.Warburton, A.Bienenstock: J. Non-Cryst. Solids **35/36**, 1233 (1980)
3.43   Y.Waseda, S.Tamaki: Z. Physik B**23**, 315 (1976)
3.44   G.S.Cargill: In *Solid State Physics*, Vol. 30 (Academic, New York 1975) p. 281
3.45   C.N.J.Wagner, N.Heiman, T.C.Huang, A.Onton, W.Parrish: AIP Conf. Proc. **29**, 188 (1976)
3.46   W.P.O'Leary: J. Phys. F. **5**, L.175 (1975)
3.47   J.J.Rhyne, S.J.Pickart, J.A.Alperin: AIP Conf. Proc. **18** (1974)
3.48   D.Givord, A.Lienard, J.P.Rebouillat, J.F.Sadoc: J. de Physique **40**, C5–237 (1979)

3.49   T. Mizoguchi, T. Kudo, T. Irisawa, N. Watanabe, N. Niimura, M. Misawa, K. Suzuki: *Rapidly Quenched Metals III*, ed. by B. Cantor (The Metal Society, London 1978) p. 415

3.50   D. R. Chipman, L. D. Jennings, B. C. Giessen: Bull. Am. Phys. Soc. **23**, 467 (1978)

3.51   C. N. J. Wagner, D. Lee: J. de Physique **41**, C8–242 (1980)

3.52   Dokyol Lee: Ph. D. Thesis, Univ. California, Los Angeles (1980)

3.53   J. E. Enderby, D. M. North: Phys. Chem. Liquids **1**, 1 (1968)

3.54   N. W. Ashcroft, D. C. Langreth: Phys. Rev. **156**, 685 (1967)

3.55   M. Sakata, N. Cowlam, H. A. Davies: J. de Physique **41**, C8–190 (1980)

3.56   J. F. Sadoc, A. Lienard: *Rapidly Quenched Metals III*, ed. by B. Cantor (The Metal Society, London 1978) p. 405

3.57   B. Boucher, A. Lienard, J. P. Rebouillat, J. Sckweizer: J. Phys. F9, 1421 (1979)

3.58   J. Wong: In *Metallic Glasses I*, ed. by H. J. Güntherodt, H. Beck, Topics Appl. Phys., Vol. 46 (Springer, Berlin, Heidelberg, New York 1981) p. 45

3.59   I. Vincze, F. van der Woude: J. Non-Cryst. Solids **42**, 499 (1980)

3.60   P. Panissod, D. Aliaga-Guerra, A. Amamou, J. Durant, W. J. Johnson, W. L. Cartes, S. J. Poon: Phys. Rev. Lett. **44**, 1465 (1980)

4. Mössbauer Spectroscopy Applied to Amorphous Metals

U. Gonser and R. Preston

With 21 Figures

Amorphous metals are of great interest from a scientific as well as a technological point of view and, naturally, the whole arsenal of macroscopic and microscopic methods has been applied to the investigation of these materials. The microscopic methods are especially important for elucidating the atomic arrangements and magnetic structures of amorphous metals. In particular, Mössbauer spectroscopy has been able to increase our knowledge in these areas considerably because this technique singles out resonating atoms of a particular species and probes the nature of their immediate surroundings. From the metallurgist's point of view, it was especially kind of Nature to provide ^{57}Fe as the most suitable isotope for Mössbauer spectroscopy, given the fact that this is an isotope of iron, the most important constituent in a large variety of amorphous metals.

The first Mössbauer spectrum of an amorphous metal was published by *Tsuei* et al. in 1968 [4.1]. Up to the present time about 200 papers have appeared in which Mössbauer spectroscopy plays an essential role in the investigation of amorphous metals.

Of the various types of amorphous metals, the most promising technologically seem to be the alloys of a transition metal (T) and a metalloid (M) with a composition of about $T_{80}M_{20}$. In fact, the amorphous metals commercially available (Metglas®, Amomet®, Vitrovac®) are mostly of this approximate composition. In this chapter we shall restrict ourselves to these $T_{80}M_{20}$ alloys where typical examples have been chosen. The choice was dictated to a great extent by the authors' special interest in this field.

4.1 How Mössbauer Spectra are Measured

The principle of the Mössbauer effect has been treated at various levels of sophistication by a number of authors [4.2–10]. We shall, therefore, give only a brief discussion of the general features of Mössbauer spectroscopy, omitting much of the theoretical background. We believe this will enable the reader who is not already familiar with the Mössbauer technique to understand how it has been used to study amorphous metals.

A basic Mössbauer spectrometer consists of a source, an absorber, a detector with a counting system and a drive system for moving the source relative to an absorber, as shown in Fig. 4.1. In this apparatus the most common Mössbauer isotope, ^{57}Fe, is being used.

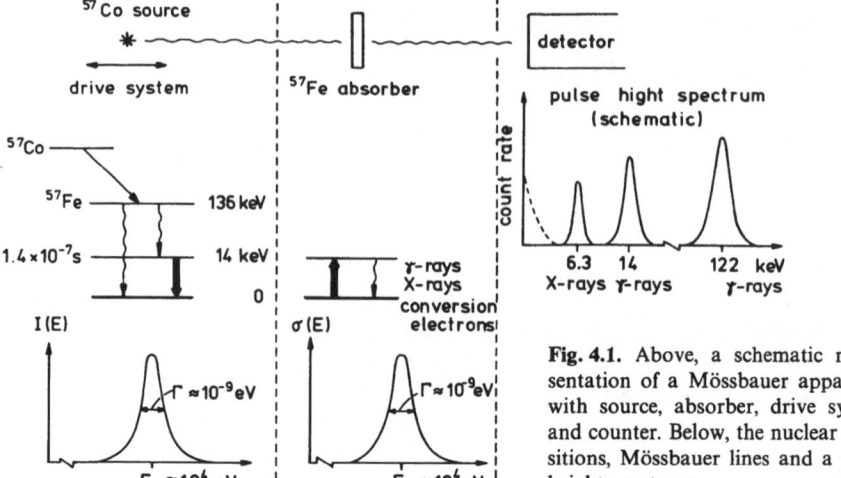

Fig. 4.1. Above, a schematic representation of a Mössbauer apparatus with source, absorber, drive system and counter. Below, the nuclear transitions, Mössbauer lines and a pulse height spectrum

4.1.1 Source

The source contains a few millicuries of ^{57}Co which decays with a 267-day half-life to the 14 keV excited nuclear state of ^{57}Fe, as shown in the nuclear energy-level diagram below the source in Fig. 4.1. The subsequent decay to the ground state of ^{57}Fe is the transition of interest. Mössbauer's discovery is basically the realization that the emission of the γ-ray from an atom fixed in a solid can occur in a recoil-free fashion, that is, without loss of energy; thus the γ-ray carries the total energy of the transition. The recoil-free event can be regarded as a typical quantum effect so that the lifetime of the 14 keV excited state determines the line width by the Heisenberg uncertainty principle. Since the lifetime is of the order of $\approx 10^{-7}$ s, the resulting line width Γ (full width at half maximum) turns out to be about 10^{-9} eV. Because the energy of the γ-ray is $E_\gamma \approx 10^4$ eV, the relative line width Γ/E_γ is about 10^{-13}. In other words, the angular frequencies present in 14 keV γ-radiation ($\approx 10^{20}$ Hz) tend to be concentrated within a range of about 10^7 Hz (the reciprocal of the 10^{-7} lifetime). By ordinary standards this is a very sharply peaked spectrum. The Lorentzian or Breit-Wigner shape of this spectrum is shown below the energy-level diagram for the source in Fig. 4.1.

4.1.2 Absorber

In the apparatus of Fig. 4.1, the 14 keV γ-rays from the source are allowed to pass through an absorber containing ^{57}Fe atoms whose nuclei are in the ground state. The ^{57}Fe atoms of the absorber have a chance of capturing incoming 14 keV quanta and thus to be excited to the 14 keV level, as shown in the energy-level diagram below the absorber. The resonance peak for absorption must be within a very narrow energy or frequency range – the same 10^{-9} eV (or 10^7 Hz) as for the source – to absorb the incoming radiation in a recoil-free transition. The recoil-free transitions of two resonating nuclei in the source and absorber are indicated in Fig. 4.1 by bold arrows. It has become customary to use the single word "Mössbauer" – the name of the discoverer of the method – in place of the expression "recoil-free nuclear resonance", as in "Mössbauer emission" or "Mössbauer absorption".

The Lorentzian cross section for Mössbauer absorption is shown below the energy level diagram for the absorber in Fig. 4.1. Ordinary electronic absorption also occurs but is usually of no interest because it is not a resonance phenomenon.

4.1.3 Detector

The detector may be of any standard type, such as a scintillation detector or proportional counter, if it has good efficiency for detecting Mössbauer radiation (14 keV in this case) with a moderate degree of energy selectivity. Part of the selectivity is obtained through proper choice of the materials and thickness of the detector, and part through the use of electronic pulse-height selection. A typical value of the overall resolution would be 10 %.

4.1.4 Drive System

Let us suppose that the source and absorber are nearly identical. For instance, the source might consist of aluminium containing a very low concentration of isolated substitutional ^{57}Co atoms, while the absorber might consist of aluminium containing a low concentration of isolated substitutional ^{57}Fe atoms. When a ^{57}Co atom decays to ^{57}Fe, its situation is then identical to that of any of the ^{57}Fe atoms in the absorber. The central frequency and shape of the emission spectrum of the source will then be exactly the same as the central frequency and shape of the absorption cross section of the absorber.[1] A certain amount of the 14 keV radiation incident on the absorber will be absorbed by ^{57}Fe nuclei which will thus be raised to the 14 keV level. Of course, each absorbing nucleus quickly returns to its ground state, but only a certain fraction of them do so by emitting a 14 keV γ-ray. The rest do so by emitting 7 keV internal conversion electrons (followed by low-energy x-rays) which are ignored by the detector and counting circuits when measured in the transmission mode illustrated in Fig. 4.1. Even in those cases where 14 keV γ-rays are re-emitted, only a fraction are emitted in the direction of the detector. Thus, the counting rate for 14 keV γ-rays is decidedly lower than it would be if the absorber consisted of pure aluminium or of aluminium containing the

---

1 This statement is correct if the source and the absorber are at the same temperature (Sect. 4.2).

same concentration of any other isotope of iron, for which there is no 14 keV nuclear level. In that case there would be no resonant nuclear absorption, and only ordinary, nonresonant electronic absorption would occur. But ^{57}Fe atoms are actually present so that the resonant nuclear absorption is relatively strong and the counting rate relatively low because the peak of the spectrum of the incident radiation coincides with the peak of the absorption cross section. However, as has been emphasized, the resonance is very sharply tuned and therefore easy to perturb. For instance, the source can be put into motion at a fixed velocity along the line connecting the source and the absorber. A relative velocity v will produce an ordinary (first-order) Doppler shift of the emitted radiation. That is, the energy of the radiation E_γ will be changed by an amount $\Delta E = (v/c) E\gamma$, where c is the velocity of light. In the case of 14 keV γ-radiation, a few mm/s will be sufficient to take the emitted radiation completely out of resonance with the absorber nuclei. At intermediate velocities (a few tenths of a mm/s), resonant absorption can still occur and the counting rate is still somewhat low, although not as low as when the relative velocity is zero.

This is the basic principle of the usual Mössbauer spectrometer which is an apparatus for measuring the counting rate at the detector as a function of the relative velocity between the source and the absorber. Usually the detected γ-ray pulses are stored in a multichannel analyzer where each channel corresponds to a specific relative velocity, between source and absorber.

4.1.5 Mössbauer Spectra

Figure 4.2 shows a typical single-line Mössbauer spectrum. In this case the source and absorber were not identical and therefore the resonance line is not centered at zero velocity. If the source and absorber are moving toward each other, the relative velocity is considered to be positive.

By using an appropriate constant of proportionality, the velocity scale may be converted to a frequency or energy scale, but this is usually not done. The full width at half maximum of the Mössbauer spectrum of Fig. 4.2 (in velocity, frequency or energy units) is approximately the sum of the full width at half maximum of the ^{57}Co–Cu source and the austenitic steel absorber. This is a simple consequence of the experimental folding of the emission and absorption spectra to obtain the Mössbauer spectrum.

The important Mössbauer parameters are calculated from the depths, widths and positions of the lines in the experimental spectrum. These parameters are obtained, in turn, from a least-squares fit of the experimental spectrum to a set of Lorentzian lines whose depths, widths and positions are adjusted for the best fit.

The points of an actual Mössbauer spectrum do not lie on a smooth curve. Instead they show a certain amount of scatter, as can be seen in Fig. 4.2, because of the usual statistical uncertainty in counting rates for nuclear events. These statistical fluctuations, expressed as fractions of the total number of

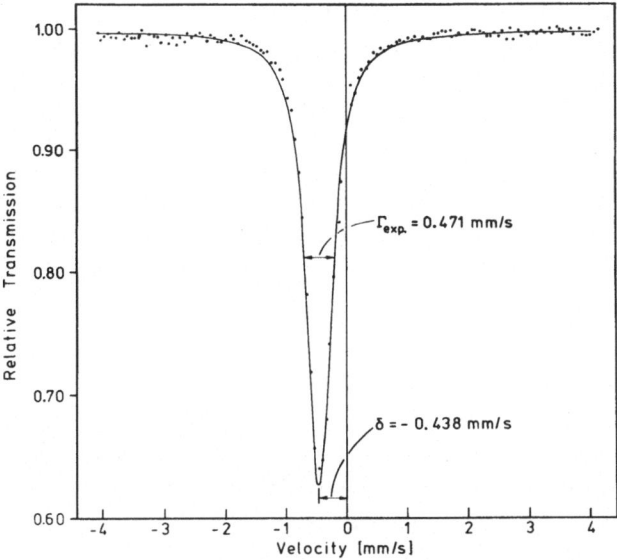

Fig. 4.2. Mössbauer spectrum obtained with a ^{57}Co–Cu source (80 K) and an absorber of austenitic steel (300 K). Γ_{exp} is the full width at half maximum of the dip. δ is the displacement of the centroid of the dip from the zero of velocity

counts, become smaller as the number of counts increases. As a result, the experimental spectrum becomes smoother in appearance and the parameters determined from the least-squares fitting become more reliable.

4.1.6 Scattering Geometry

The "transmission geometry" of Fig. 4.1 is in certain situations not very useful. For instance, in studies of surfaces and thin films or when the absorbing sample is very thin, it may be difficult to obtain a spectrum with resonance dips deep enough for reliable analysis. In such cases it may be preferable to locate the detector on the same side of the absorber as the source and measure the various radiations originating from the 14 keV excited state of the absorber (thin arrow in Fig. 4.1) after Mössbauer absorption has occurred (bold arrow in Fig. 4.1). This "scattering geometry" has the advantage that re-emitted γ-rays, x-rays associated with internal conversion and conversion electrons can all be measured. The counting rate in the detector is normally relatively low. When the resonance condition exists, however, the counting rate for the various radiations *increases* which results in a peak instead of a dip in the Mössbauer spectrum. Although the counting rate is small, the signal-to-noise ratio of the Mössbauer spectrum may be better than for the transmission geometry. That is, the resonance may stand out more clearly. The penetration depths of the various radiations (γ-rays, x-rays and electrons) are quite different. Thus, the spectra differ depending on which type of radiation is measured. This can be used to general advantage since the choice of one type of radiation selects a particular depth or range of depths in the sample.

The scattering geometry is particularly useful in probing with 7 keV conversion electrons. Since only the conversion electrons from nuclei near the surface can emerge from the sample and reach the detector, this method is applicable for selectively studying the atoms near the surface of a bulk sample. This technique is often abbreviated to CEMS (conversion electron Mössbauer spectroscopy).

4.2 Mössbauer Parameters

The Mössbauer resonance lines are determined by a number of parameters from which significant information can be obtained. Of importance are recoil-free fraction, thermal shift and hyperfine interactions.

4.2.1 Recoil-Free Fraction

Not all of the emitted and absorbed radiation is within the sharp resonance Lorentzian peak. This is a Doppler-shift effect due to the fact that the nuclei of the ^{57}Fe atoms are participating in the thermal motions of the solid at the same time that they are emitting the γ radiation. Because the thermal motion is oscillatory and at frequencies much higher than the 10^7 Hz width of the resonance, it turns out that the Doppler shifting does not broaden or shift the narrow Lorentzian peak at all, but instead removes a certain fraction of the emitted or absorbed radiation from the Lorentzian peak and throws it into a band of frequencies above and below that of the Lorentzian peak. This band is not broad by ordinary standards, but it is very broad in comparison with the Lorentzian peak and its intensity in any frequency interval of the order of 10^7 Hz is negligible. Thus, only the sharp, central, Lorentzian "Mössbauer peak" is normally observable in Mössbauer spectroscopy, and the fraction of the 14 keV radiation which is represented by this peak is called the *Mössbauer fraction*. It is also called the *recoil-free fraction* because a proper quantum treatment of the effect associates the loss of intensity from the Lorentzian peak with increases and decreases in the lattice vibrational energy due to the reaction force against the nucleus when it emits or absorbs a photon.

The fraction of the intensity which is removed from the Mössbauer peak increases with increasing amplitude (or energy) of the thermal vibrations. Thus, for a given source or absorber, the recoil-free fraction decreases with increasing temperature, while at a given temperature the recoil-free fraction is lower for sources and absorbers in which the ^{57}Co or ^{57}Fe atoms are more weakly bound in place and therefore oscillate with wider amplitudes.

4.2.2 Thermal Shift

A second effect of thermal vibrations is to shift the *frequency* of the Mössbauer peak. This is not a classical Doppler-shift effect, but a relativistic effect which is perhaps most easily understood as being a result of time dilatation, although there are other equivalent ways of viewing this phenomenon.

Time dilatation is the well-known slowing down of all processes within a moving object and it stems from the factor $\sqrt{1-v^2/c^2}$ in the Lorentz transformations, v being the velocity of the object and c the velocity of light. Because this factor contains only the square of v, the result of motion is always to decrease and never increase the rate of a process, regardless of the direction or sign of v.

The nuclei participate completely in the thermal motions of the atoms and therefore the frequencies of radiations emitted or absorbable by a nucleus are shifted downward by the time dilatation.

This shift is called the *thermal shift*, the *relativistic shift* or the *second-order Doppler shift*. The amount of the shift is proportional to $\langle v^2/c^2 \rangle$, where v is the instantaneous velocity of the atomic motion and the brackets signify the thermal average. At ordinary temperatures v is not large enough to be considered relativistic in the usual sense. That is, $\langle v^2/c^2 \rangle$ is much less than unity. Nevertheless, the recoil-free fraction of the γ-radiation is so sharply peaked that small but observable shifts can be produced by modest changes in the source (or absorber) temperature which, of course, alter the value of $\langle v^2/c^2 \rangle$ for emitting (or absorbing) atoms. Increasing $\langle v^2/c^2 \rangle$ shifts the recoil-free peak to lower frequencies.

4.2.3 Hyperfine Interactions

In addition to these intensity changes and shifts of Mössbauer spectra due to variations in the average motion of the Mössbauer atoms, there are other influences associated with changes in the electromagnetic environment of the emitting or absorbing nuclei. Usually the electromagnetic environment of a nucleus is at least partly determined by the electrons of the same atom, so that the effects of the electromagnetic environment are considered to be hyperfine effects.

a) Isomer Shift

Most atomic electrons have zero probability of being within the nucleus. The wave functions of *s*-electrons, however, do not have zero amplitude at the nucleus and therefore the probability density of an *s*-electron has a definite, small, but nonzero value within the nucleus. The negative charge associated with this small electron density within the nucleus interacts with the positive charge of the nuclear protons to shift the energy of the nucleus from its "bare nucleus" value. However, because the volume of the nucleus is different in the ground and excited states, the energy shifts are also different. Therefore, the energy of the *transition* between the two states is also shifted, and by an amount which depends on the density of electrons at the nucleus. That is, the energy of the Lorentzian line associated with the recoil-free emission or absorption undergoes a shift which is proportional to the *s*-electron density. This is called the *isomer shift* because it is the shift in energy difference between the ground state and the isomeric (14 keV) Mössbauer level. It is also called the *chemical shift* because a change in the chemical state of the atom has a direct or indirect effect on the *s*-electron wave functions and therefore on the density of electrons at the nucleus and thus, finally, on the resulting shift in the transition energy. If the source and absorber atoms are not identical chemically, there will be a shift of the Mössbauer spectrum away from zero velocity as seen in Fig. 4.2, even if the thermal shifts of the source and absorber are still identical. It is usually difficult to separate an observed shift unambiguously into a thermal-shift part and an isomer-shift part. Thermal shifts for iron, however, lie, in most cases, within a smaller range of values than isomer shifts. In practice, only *differences* in total shifts can be measured. These shifts are only relative. By custom, the zero of total shift is assigned to ^{57}Fe in metallic α-iron at room temperature. In an amorphous material there may be a distribution of isomer shifts because of the general inequivalence of sites for atoms of the same species.

b) Magnetic Splitting

When a ^{57}Fe atom is located in a magnetic field, the nuclear Zeeman effect splits the nuclear ground state into two sublevels and the 14 keV excited state into four sublevels. Of the eight possible transitions from the four upper sublevels to the two lower sublevels, only six are allowed in the absence of other perturbations. These are shown on the left side of Fig. 4.3. It is customary in

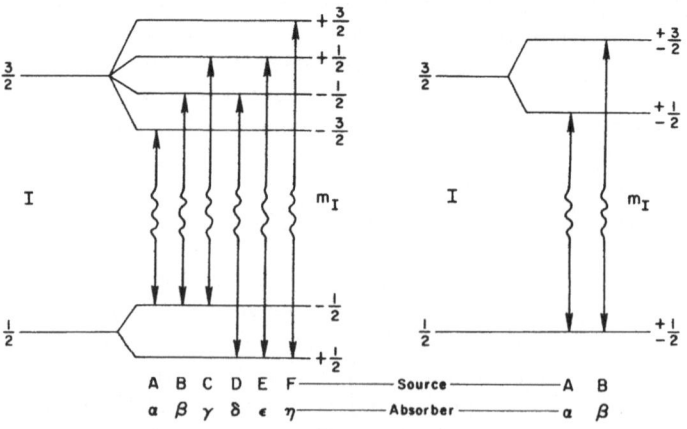

Fig. 4.3. Nuclear level diagram of ^{57}Fe for magnetic dipole and electric quadrupole interactions

Fig. 4.4a–c

Fig. 4.5a, b

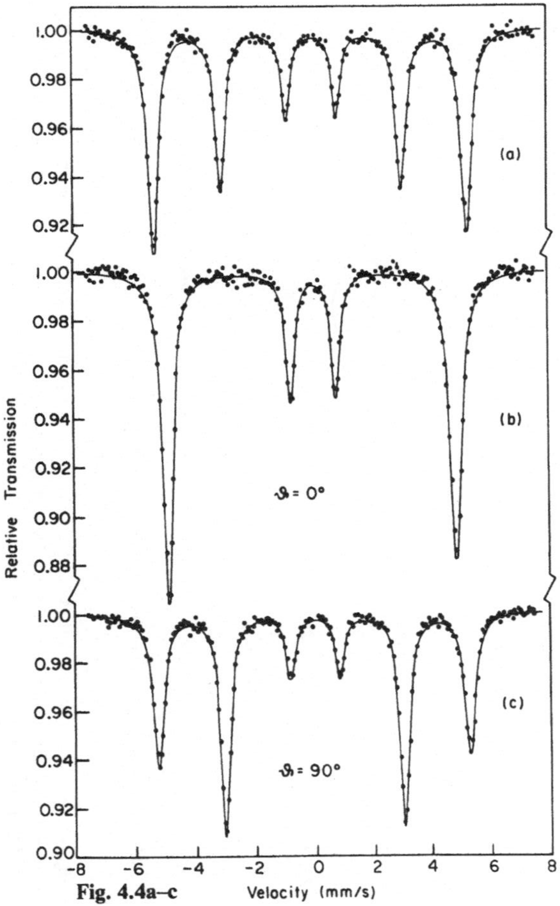

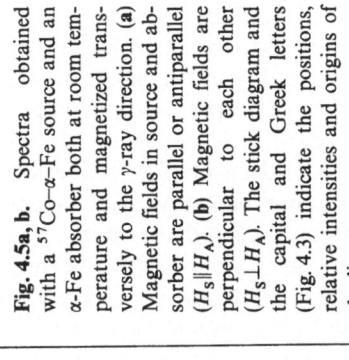

Fig. 4.4a–c. Mössbauer transmission spectra of α-Fe at room temperature: **(a)** $H_{ext}=0$; **(b)** $H_{ext}=50$ kOe, $\vartheta=0°$; **(c)** $H_{ext}=3.5$ kOe, $\vartheta=90°$. H_{ext} is the applied magnetic field and ϑ is the angle between the direction of H_{ext} and the direction of γ-ray propagation

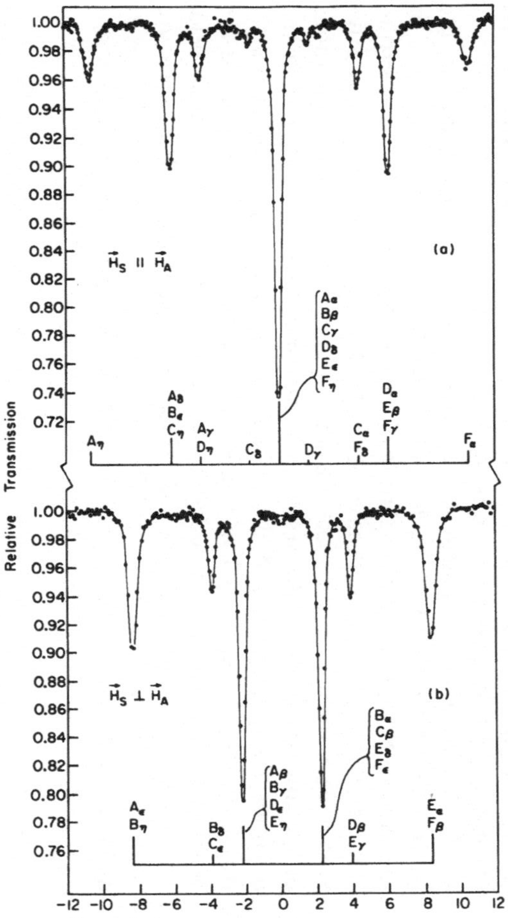

Fig. 4.5a, b. Spectra obtained with a ^{57}Co–α–Fe source and an α-Fe absorber both at room temperature and magnetized transversely to the γ-ray direction. **(a)** Magnetic fields in source and absorber are parallel or antiparallel ($H_S \| H_A$). **(b)** Magnetic fields are perpendicular to each other ($H_S \perp H_A$). The stick diagram and the capital and Greek letters (Fig. 4.3) indicate the positions, relative intensities and origins of the lines

Mössbauer spectroscopy to use a standard source (absorber) having a single, unsplit line to determine any possible splitting of the spectrum of the absorber (source) under investigation. Figure 4.4 shows typical Mössbauer transmission spectra for α-Fe at room temperature. If both the source and the absorber are in magnetic fields, a 36 line Mössbauer spectrum is produced by the absorption of each of the six emission lines of the source by each of the six absorption lines of the absorber at various velocities. Such 36 line spectra have been of some importance in the study of amorphous materials. Figure 4.5 shows spectra for a six-line ^{57}Co-α-Fe source and a six-line α-Fe absorber, both magnetized transverse to the γ-ray direction. Because of polarization and degeneracy, the number of lines is considerably less than 36. Due to polarization the occurrence of lines in the two spectra are mutually exclusive.

The splitting of the magnetic sublevels is directly proportional to the magnetic field present at the nucleus which is often called the *effective field* or the *internal field* H_i. Thus, the splitting of the lines of the Mössbauer spectrum is a direct measure of the effective or internal field. In addition, the relative intensities of the lines of the spectrum contain information on the orientations of the internal fields of the source and absober. This is because both the intensity and polarization of the radiation associated with each of the six transitions shown on the left in Fig. 4.3 depend on the angle ϑ between the propagation direction of the radiation and the direction of the internal field. The relative intensities of the first three transitions (the other three are equivalent) are shown in Fig. 4.6. When the directions of the internal fields at different atoms are isotropically distributed, the Mössbauer spectrum produced

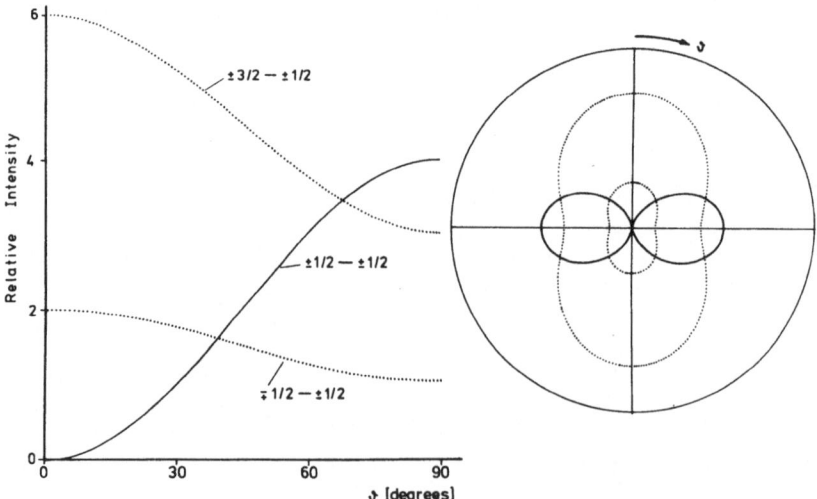

Fig. 4.6. Relative line intensities for the allowed transitions between states with spin 3/2 and 1/2 (Fig. 4.3) as a function of angle ϑ between the γ-radiation and the orientation of the magnetic field H_i. The sum of the contributions yields an isotropic total intensity as indicated by the circle

exhibits the $3:2:1:1:2:3$ intensity ratio shown in Fig. 4.4a. In all cases the sum of all the line intensities is isotropic as indicated by the circle. By making measurements of the angular dependence of the intensities and polarizations, it is possible to determine the direction and degree of alignment of the hyperfine fields in a sample.

If the magnetic field at the nucleus were entirely due to a uniform externally applied field, then it would have the same magnitude and direction at all nuclei. In paramagnetic materials the induced magnetism of the sample will make an additional contribution (positive or negative) to the field at the nucleus. If the material is spontaneously magnetic, however, there may be different orientations of the internal field associated with differences in domain orientation. In the absence of an applied external field, the internal field is a consequence of the magnetic ordering within the domain and its direction is determined by the domain direction. An applied external field then produces additional modifications of all the internal fields, first by contributing directly to the field at each nucleus and second by modifying the orientations and magnetizations of the domains so that the direction and magnitude of the domain contribution to the internal field is altered. This will be different for different initial domain orientations so that the hyperfine field will be inhomogeneous, even for crystalline materials. In an amorphous material there may be further variations in magnitude and direction of the internal field because of the general inequivalence of the atomic sites.

c) Quadrupole Splitting

If there is no magnetic field at the nucleus, but there is instead an electric field gradient (EFG) at that point, the quadrupole interaction of the 14 keV level of ^{57}Fe will cause a splitting into two sublevels while the ground state remains unperturbed, as shown in Fig. 4.3 on the right. This splits the emission or absorption spectrum into a quadrupole doublet. As an example, the spectrum of an $FeCO_3$ absorber made with a single-line source is shown in Fig. 4.7. If the orientations of the principal axes of the EFG for different nuclei are isotropically distributed, then the Mössbauer spectrum for an unsplit source and a quadrupole-split absorber (or vice versa) is split into a symmetric doublet. For anisotropic distributions of the principal axes, the intensities of the two lines may differ. In a single crystal with atoms in equivalent sites, the orientation and asymmetry parameter determine the relative intensities. In mixed crystals there may be complete randomness of orientation or there may be preferred directions (texture). In amorphous materials there may be additional variations in the magnitude and orientation of the quadrupole field due to the general inequivalence of atomic sites.

A uniform and sufficiently strong electric field gradient cannot be applied by external means, so that observed quadrupole patterns reflect just the influence on each nucleus of all the charged particles in the sample.

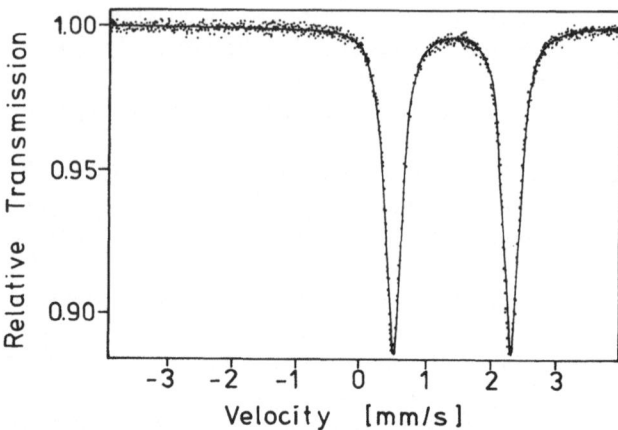

Fig. 4.7. Mössbauer spectrum of $FeCO_3$ (295 K) exhibiting quadrupole splitting. The zero of the velocity scale corresponds to the centroid of the α-Fe spectrum at room temperature

When magnetic dipole and electric quadrupole effects act simultaneously on the same nucleus, the splittings and intensities of the Mössbauer pattern may be difficult to predict or to interpret when observed. If one of these hyperfine effects is only a small perturbation on the other, both the prediction and the interpretation – at least qualitatively – are simpler.

4.3 Mössbauer Effect as a Microprobe

If two Mössbauer atoms in the same sample are in different local environments, their Mössbauer spectra will be different. If their motional properties are different, they will have different Mössbauer fractions and different thermal shifts. Likewise, their isomer shifts, magnetic dipole interactions and electric quadrupole interactions may all be different. For each type of environment there will be a different contribution to the total Mössbauer spectrum so that the total spectrum will be a superposition of all these contributions. It is this ability to "see" small differences in environments of a particular species of atom within one sample that makes Mössbauer spectroscopy an especially useful tool for studying amorphous metals, where a variety of environments must exist for each atomic species [4.1, 11–129].

4.4 Mössbauer Spectra of Amorphous Metals

In Fig. 4.8 a typical spectrum of an amorphous metal is shown. The dots represent experimental data. The solid curve represents a theoretical fit to the data. The dips in the solid curve are superpositions of the dips in the dashed curves. The dashed curves are a feature of the theoretical model used for fitting the data.

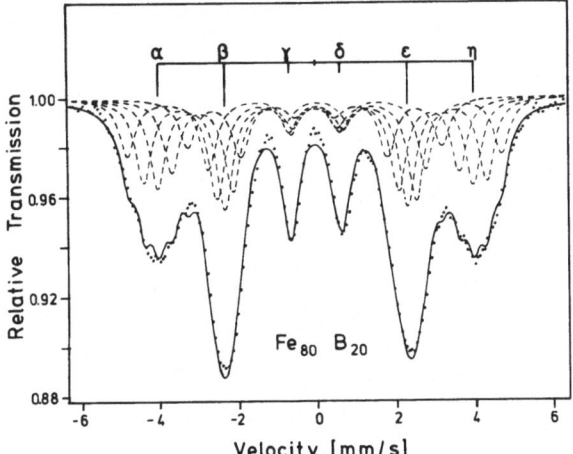

Fig. 4.8. Mössbauer spectrum of the amorphous metal $Fe_{80}B_{20}$. The stick diagram (α, β, γ, δ, ε, η) represents the average hyperfine pattern. The solid line represents the sum of the subspectra (Sect. 4.4.3)

It is remarkable that ferromagnetic amorphous alloys of the $T_{80}M_{20}$ type with different constituents exhibit very similar spectra. The lines are broad and usually the second and fifth lines are especially pronounced, indicating a preferred orientation of the spins within the plane of the ribbon. Most researchers in the field agree that the similarity in the spectra reflects similarities in the structure. Various attempts have been made to deduce magnetic and atomistic structures from these poorly resolved patterns.

4.4.1 Effects Due to the Variation of Metalloid Atoms

Substitution of one species of metalloid for another in amorphous $T_{80}M_{20}$ produces certain changes in the Mössbauer spectrum. Systematic studies of these effects have been made on the alloy systems [4.123]

$$Fe_{80}B_xM_{20-x} \quad (M=Ge, Si, C, P)$$
$$Fe_{80}P_xM_{20-x} \quad (M=Si, C, B).$$

Some results are shown in Fig. 4.9 where the average hyperfine field is plotted versus the concentration of the substituted metalloid. Other correlations have been found by considering the influence of the radius, electronegativity, concentration and other properties of the metalloid atoms on the isomer shift and internal field. However, due to the problem of resolution we are still far from a full understanding of these correlations. In particular, we are not able to predict or deduce the macroscopic behavior of the material from the measured microscopic parameters.

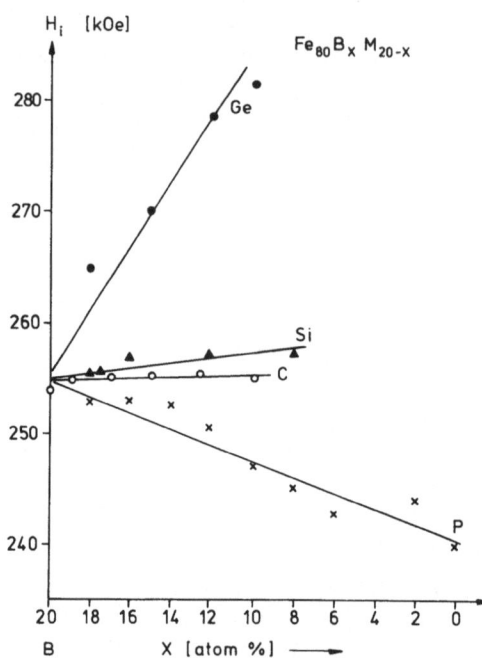

4.4.2 Effects Due to the Variation of Metal Atoms

Considerable work has been focussed on the system $(Fe, Ni)_{80}M_{20}$. Of special interest is a comparison of the amorphous alloys with the corresponding fcc crystalline states of the Ni_xFe_{1-x} system [4.106]. In Fig. 4.10 the hyperfine fields are shown at the top. The average hyperfine field of amorphous $(Ni_x, Fe_{1-x})_{80}P_{14}B_6$ is nearly invariant over the whole concentration range [4.39]. In recent work for $x < 0.91$ a cluster glass state was proposed [4.121]. In the crystalline Ni_xFe_{1-x} alloy the magnitude of the field changes significantly at the approach to the invar region [4.130, 131]. It has been suggested that ferro and antiferromagnetism coexist in the invar region. At lower concentrations of Ni, measurements are difficult because of the martensitic transformation to a *bcc* iron. In some cases the martensitic transformation can be circumvented by epitaxy and coherent precipitation. In the interpretation of the results, however, ferro as well as antiferromagnetism has been claimed to exist [4.132–135]. So far, the results on the magnetic ordering of fcc alloys are still inconclusive.

The Curie and Néel temperatures T_C and T_N are shown at the bottom of Fig. 4.10. Comparing again the "amorphous" curve with the "crystalline" curve, one finds that they are almost mirror images of each other. The high value of T_C for amorphous Fe-rich alloys indicates a large ferromagnetic Fe–Fe exchange parameter and a weak Ni–Ni exchange parameter. In the crystalline fcc alloys the opposite effect seems to dominate, that is, there is a negative (anti-

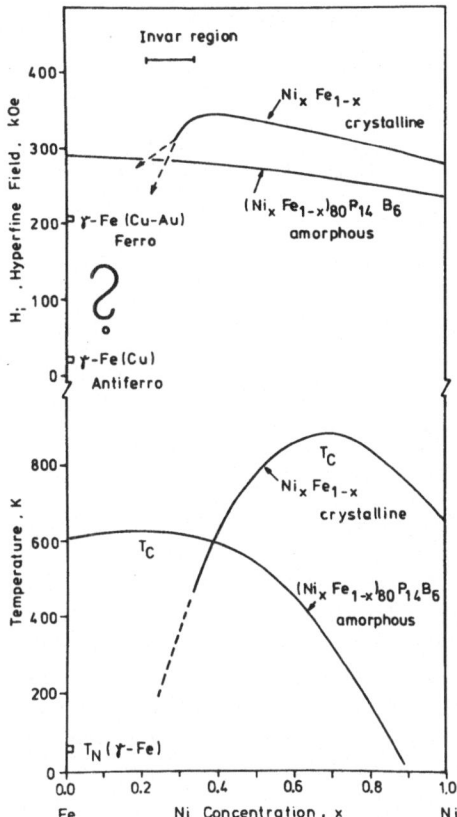

Fig. 4.10. Internal field H_i, Curie temperature T_C and Néel temperature T_N of amorphous $(Ni_xFe_{1-x})_{80}P_{14}B_6$ and crystalline fcc Ni_xFe_{1-x}. The question mark indicates the uncertainty of the value of H_i in this region

ferromagnetic) Fe–Fe exchange [4.136]. This behavior can be explained by the sensitivity of the magnetic exchange to interatomic distances as predicted by the Bethe-Slater curve. With increasing (decreasing) distance the Fe–Fe (Ni–Ni) exchange becomes positive and strongly ferromagnetic. In the amorphous Fe-rich alloys the increase in the separation of the Fe atoms is due to the presence of the metalloid atoms.

4.4.3 Hyperfine Field Distribution

There is general agreement that the broadening of the lines is mainly due to the presence of a distribution of hyperfine fields $P(H_i)$. However, a certain contribution of electric quadrupole interaction might also be present. If the principal axes of the electric field gradients are rather randomly oriented, the perturbation on the magnetic hyperfine fields will lead to a broadening of the lines while the spectrum as a whole will still appear to be more or less symmetric. The evaluation of the distribution of hyperfine fields has been done

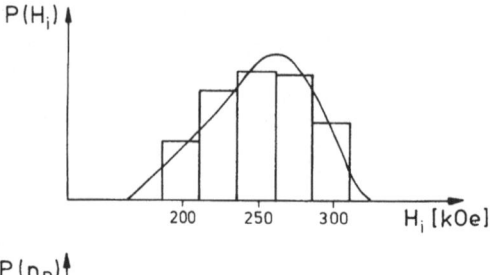

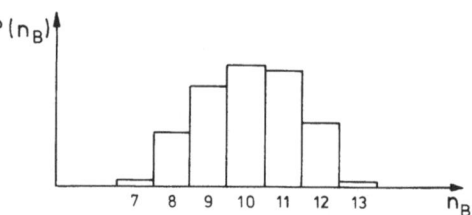

Fig. 4.11. Distribution of hyperfine fields $P(H_i)$ for the spectrum of Fig. 4.8. Evaluated by the method of *Window* [4.137] as a continuous function and also by fitting five discrete values of the field. The distribution of coordinations for the Bernal model $P(n_B)$ is shown below

by a Fourier analysis method [4.137], by an analytical method [4.138] and by least-squares fitting of a superposition of a finite number of subspectra [4.70].

Typical hyperfine field distribution curves $P(H_i)$ are shown in Fig. 4.11. These were evaluated from the Mössbauer spectrum in Fig. 4.8. The discrete distribution was obtained by a least-squares fitting in which it was assumed that five discrete values of H_i could completely account for the observed spectra. The five corresponding subspectra and their sum are shown in Fig. 4.8.

All methods of evaluating $P(H_i)$ are subject to difficulties associated with the thickness and texture of the sample because of dichroism. Dichroism is the change of polarization and therefore of the cross section for absorption of the γ-rays as they pass through the absorber, and it is more pronounced in a thick absorber than in a thin one. Furthermore, in a thick absorber the effect is more pronounced for the strong components of the absorption spectrum than for the weak ones. This causes a distortion of the derived $P(H_i)$ curve which cannot be corrected reliably since it is not known a priori how large the effect is for each value of H_i. The use of very thin absorbers minimizes this problem but makes it harder to obtain good experimental spectra.

A second difficulty is that there are small, random quadrupole perturbations of the magnetically split spectrum and these produce a further, unpredictable asymmetric broadening of all the lines. Thus, the derived $P(H_i)$ curve suffers a further distortion whose extent is not known a priori. In general, the perturbation of the magnetic hyperfine pattern by the quadrupole interaction and magnetic anisotropies tends to be underestimated [4.61]. The common assumption that the outer lines of the magnetically split spectrum are shifted by $\pm 1/2\, \Delta E_q$ relative to the inner lines is correct only in the special and extreme case where H_i is parallel to the principal axis of the EFG, the sign of the quadrupole coupling is unique and the EFG is axially symmetric. Since these conditions are not always met, the evaluation of $P(H_i)$ from Mössbauer spectra requires extreme care [4.103, 104, 139].

Indeed, a careful evaluation of $P(H_i)$ from the measured Mössbauer spectrum of an amorphous material would have to take into account not only the thickness of the sample, but also the distribution of isomer shifts, the distribution of EFG tensors, the texture of the EFG axes, the magnetic anisotropies, the spin texture and the correlations of all of these properties with each other[2]. Computer simulations (*Keller* [4.140] and H. Fischer, unpublished) show that spurious peaks on the low-H side of an experimental $P(H_i)$ curve can be produced by not taking proper account of the quadrupole interaction.

Lines 2 and 5 in a magnetic hyperfine pattern can be separated from lines 1, 3, 4 and 6 by a spectrum subtraction method [4.61]. However, the claim that the lines in the difference spectrum give more information about $P(H_i)$ than either of the two original spectra seems difficult to justify. In addition, the correct procedure for taking the difference has never been spelled out. In fact, the tendency for the derived $P(H_i)$ curve to be distorted, as discussed in the preceding paragraphs, might be further accentuated if determined from a difference spectrum.

The usual assumption is that the magnitudes of the hyperfine fields are the result of the individual microscopic surroundings of the resonant atoms, while the angular dependence of the magnetic hyperfine interactions is basically governed by the macroscopic domain structure which can be represented by an average spin orientation. This leads to the same relative line intensities for all contributions to the broadened hyperfine spectrum. The domain structure can be changed by the application of small magnetic fields or small mechanical stresses. This changes the relative line intensities through polarization effects but has little effect on the distribution of hyperfine fields $P(H_i)$.

It has also been found that substitution of other metal or metalloid atoms changes the average isomer shift and the average value of H_i. In most cases, however, the *shape* of $P(H_i)$ is nearly unchanged.

4.4.4 Can the Atomistic Structure be Deduced from the Spectral Structure?

The broad lines of the Mössbauer spectrum have inspired scientists to apply various models of the structure to interpret the spectra of ferromagnetic amorphous metals in terms of local variations in the environment of the Fe atoms. Three extreme viewpoints can be distinguished:

1) microcrystallites
2) elementary cells or molecular units
3) random dense packing (RDP)
   (Bernal structure)

---

2 Further problems can arise from the Goldanskii-Karyagin effect which is an asymmetry in the spectrum arising from anisotropic vibrations of Mössbauer atoms.

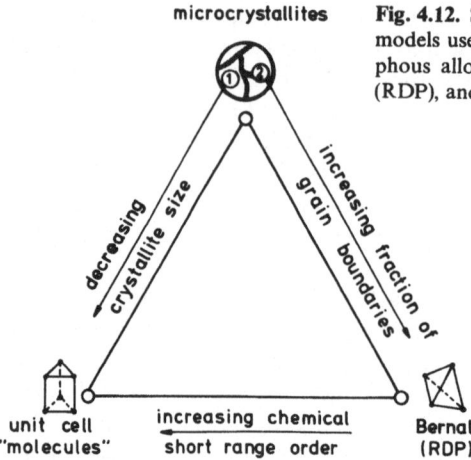

microcrystallites

Fig. 4.12. Schematic representation of three alternative models used to interpret experiments on $T_{80}M_{20}$ amorphous alloys: microcrystallites, random dense packing (RDP), and unit cells or molecules

unit cell
"molecules"   increasing chemical   Bernal
                short range order   (RDP)

The interrelationship of these three models is shown schematically in Fig. 4.12 where the corners represent the "pure" models. First we might consider the upper vertex of the triangle which represents the microcrystallite model, that is, small crystals separated by grain boundaries. Starting within a microcrystal at the point 1) in Fig. 4.12 and reducing the crystal size, we eventually arrive at the vertex on the lower left which symbolizes the unit cells as some kind of molecular units which are packed together in some fashion [4.141]. We might also choose a starting point within a grain boundary as at 2) in Fig. 4.12. In general, a grain boundary can be decomposed into a set of dislocations. As has been pointed out, these defects represent a kind of randomness similar to that found in the Bernal RDP structure [4.142] and it has been shown that the five typical classes of polyhedral holes of the Bernal structure are the core structure of dislocations [4.143, 144]. On going down the right side of the triangle, the fraction of the solid volume occupied by grain boundaries increases until we arrive at the vertex on the lower right which represents the Bernal structure, symbolized here by a tetrahedron. This symbol was chosen because the tetrahedron is the most dense arrangement of four atoms and it predominates in the Bernal structure [4.142].

If we adopt the view that microcrystals of carbides, borides, phosphides, etc., are present in these alloys, diffraction methods tell us that their linear dimensions cannot exceed 20 Å. The volume per Fe atom in cementite (Fe_3C) [4.145] is 13.0 Å^3 which is equivalent to a cube with an edge of 2.35 Å. If we assume, for simplicity, that the microcrystals have cubic shapes with edge lengths of ≈ 20 Å, each one would contain about 600 atoms. More than half of them would be surface atoms and the rest would represent the bulk. From a Mössbauer point of view, one would expect two different components in the spectra: from the bulk iron atoms for which the hyperfine field would be nearly unique, and from the surface atoms with their manifold of environments similar to the Bernal structure and with a distribution of hyperfine fields.

In considering the possibility that these alloys consist of elementary cells or molecular units, one faces the problem of visualizing and evaluating the multiplicity of arrangements and orientations of an assembly of these basic units. However, similar to the microcrystallite case, one would expect a fairly sharp spectral component for iron atoms as parts of the units and a variety of components for atoms in environments closer to associated "defects".

In the past the Bernal model has often served as a basis for interpreting experiments on amorphous metals. The Bernal structure can be realized physically by a random dense packing of macroscopic hard spheres and mathematically by computer simulation [4.146]. In the idealized topology of the Bernal structure one finds the following well-known holes with their percentage probabilities of occurrence given in parentheses: tetrahedron (86.7), octahedron (5.3), trigonal prism (3.8), archimedean antiprism (0.5), tetragonal dodecahedron (3.7). *Polk* [4.147] suggested that the relatively high stability of the amorphous metals in the vicinity of the composition $T_{80}M_{20}$ results from the filling of the larger holes by the smaller metalloid atoms. These holes can accommodate just about 20 at.- %.

All three of these models and variations have been used in attempts to interpret Mössbauer spectra. The preferred choice of the model depends greatly on the person you ask. Because in Mössbauer spectroscopy the point of observation is the Fe nucleus, one can hope to deduce the coordinations of the Fe atoms and the short-range arrangements of their neighbors. At present there is some disagreement about the reason for the distribution of ^{57}Fe hyperfine fields. Two seemingly contradictory assumptions are that the variety of fields is caused mainly by the neighboring transition metals T or mainly by the presence of the metalloid atoms B. It may be that this difference is, in part, merely semantic and that the two effects cannot be completely separated from each other. That is, a metalloid atom can perturb a Mössbauer spectrum not only by the influence of its electrons on the effective fields at neighboring Fe nuclei, but also because its presence perturbs the Fe–Fe separation of nearby Fe atoms. The Fe–Fe separation is also known to perturb the Fe hyperfine fields, an extreme example of this being ferromagnetic α-Fe and paramagnetic (or antiferromagnetic) γ-Fe.

In general, one has to realize that each of these starting models represents a rather idealized conception and in reality a variety of concepts ought to be taken into account: thermodynamics and kinetics of amorphization, metastable phases, stoichiometry, packing, size relationship, coordination, electronegativity, covalency, randomness, defects, distortion and possibly other factors. In fact, most of the terms in the preceding sentence should be read as if enclosed in quotation marks because they all have special meanings for the amorphous state. In fact, a "confusion principle" has been introduced. This principle states that the tendency toward amorphization is greater where there are more states and phases available to the system. The combined effect of all these is responsible for the real structures of amorphous phases which might then be represented by points somewhere within the triangle of Fig. 4.12. In the following we briefly describe the use of various models to evaluate the spectra.

As our first example we take the Bernal model which is easily realized physically and furthermore the correlation between model and (sub)spectra is rather transparent [4.70]. In this model, transition metal atoms with the following close contact coordinations n_B are present: 7–13 and their probabilities of occurrence $P(n_B)$ are shown in Fig. 4.11. When the spectrum of an amorphous metal of the type $T_{80}M_{20}$ is fitted by five subspectra, the resulting intensities correspond closely to the frequency of occurrence of the five nearest-neighbor coordinations (8–12) having the highest probability of occurrence according to Bernal's random dense packing of hard spheres. The five subspectra are indicated by dashed lines in Fig. 4.8. A plot of the hyperfine fields of the subspectra as a function of the five coordinations n_B turns out to be linear, which suggests that each close-contact Fe atom contributes the same amount to the internal field at the site of the resonating nucleus.

The model does not completely neglect the metalloid atoms which may stabilize the Bernal structure by occupying the bigger holes. The metalloid atoms will contribute to the total number of conduction electrons, thereby causing isomer shifts and, possibly, small quadrupole effects. Because the Fe hyperfine fields are sensitive to interatomic distances, the metalloid atoms might, because of their size, have some scaling effects on the magnitudes of the fields.

This model of amorphous alloys, which correlates the distribution of metal coordination (representing average numbers) with corresponding distributions of average hyperfine fields, continues to be useful when the constituents (metal and metalloid atoms) are varied or when temperature, magnetic field, stress, etc., are changed.

A similar approach was successfully used for a crystalline state in the case of the disordered γ-phase $Fe_{72}Pt_{28}$ [4.131]. In this case the expected distribution of Fe coordinations could be calculated by assuming completely random order. This fixed the relative intensity of the spectral component associated with each hyperfine field. In crystalline $Fe_{72}Pt_{28}$, every atom has 12 nearest neighbors. If the order is random, the coordinations n_d of the Fe nearest neighbors which occur with the largest probabilities $P(n_d)$ are 6 (6.2%), 7 (13.7%), 8 (22.0%), 9 (25.1%), 10 (19.4%), and 11 (9.1%) (Fig. 4.13). In the fitting it was assumed that these six nearest-neighbor coordinations n_d correspond to six subspectra, each having a characteristic hyperfine field H_i and a relative line intensity which was constrained to be proportional to $P(n_d)$. The Mössbauer spectrum of disordered crystalline $Fe_{72}Pt_{28}$ is shown in Fig. 4.14. The six subspectra are indicated by dashed lines. The solid line represents the summation. A plot of the best-fit values of the internal fields of the subspectra H_i versus the six frequently occurring coordinations n_d is a straight line, as seen in the lower part of Fig. 4.13. This suggests again that each Fe nearest neighbor adds a fixed contribution to the hyperfine field at the nucleus of the probe atom.

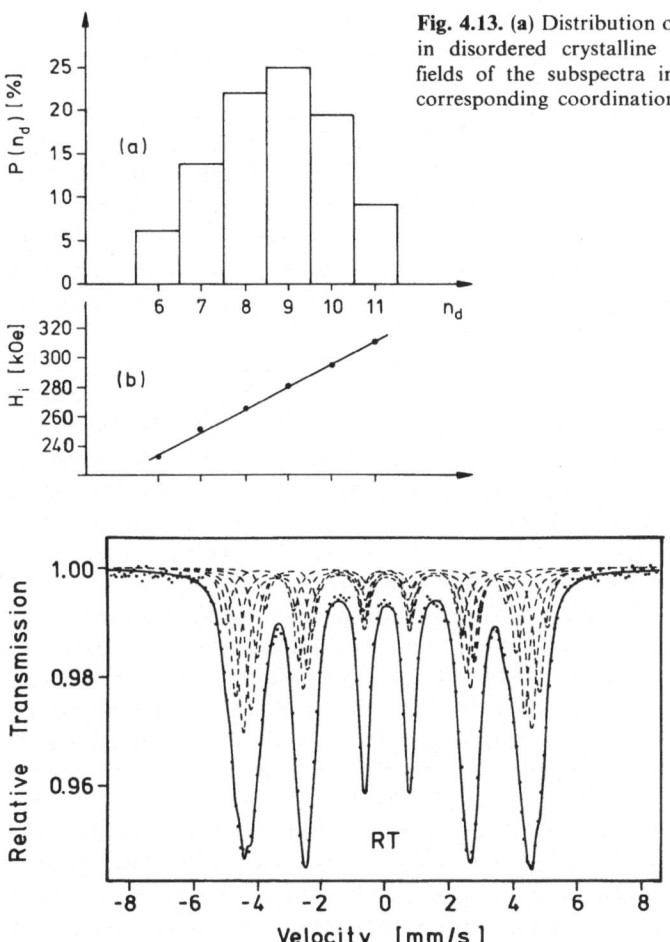

Fig. 4.13. (a) Distribution of Fe coordinations $P(n_d)$ in disordered crystalline $Fe_{72}Pt_{28}$; (b) hyperfine fields of the subspectra in Fig. 4.14 versus their corresponding coordination n_d

Fig. 4.14. Mössbauer spectrum of crystalline disordered $Fe_{72}Pt_{28}$ at room temperature. In the fitting, the relative intensities of the subspectra were fixed in accordance with the calculated distribution $P(n_d)$ shown in Fig. 4.13a

Another model lies somewhere between the microcrystallite model at the top of the triangle and the molecular unit or cell model at the lower left in Fig. 4.12 [4.116]. This model is based on x-ray diffraction evidence that crystalline order persists over about five atomic distances in these materials. It is therefore assumed that the immediate surroundings of each Mössbauer probe atom in amorphous $Fe_{80}M_{20}$ are very nearly the same as they would be in a random solid solution of $Fe_{80}M_{20}$ in a crystalline bcc lattice. Additional broadening of the spectral lines is introduced to simulate the effects of the inhomogeneities associated with the lack of real long-range order. This lack of long-range order is attributed to a high density of dislocations in the crystallite.

The calculated probabilities for various iron coordinations of iron sites for the disordered bcc lattice are surprisingly close to those for the seemingly quite different RDP model. They also give good fits to experimental data.

Two other models are based on the fact that in crystalline Fe–B compounds, the hyperfine field at a given Fe site appears to be strongly determined by the metalloid coordination n_M of boron atoms (rather than by the coordination of Fe atoms) for the site [4.91]. One model is an RDP model. The other is a quasi-crystallite model like the quasi-bcc model just discussed, except that it is based on a more complicated crystal structure. Both models have been used in attempts to interpret the Mössbauer data on various ordered, stoichiometric Fe–B compounds.

For their RDP model, the authors permitted each value of $H_i(n_M)$ to be broadened by an appropriate amount to account for inhomogeneities caused by the lack of real long-range order. The resulting $H_i(n_M)$ curves were then combined with calculated $P(n_M)$ values to obtain a predicted $P(H_i)$ curve by adjusting the only available parameter – the amount by which the individual $H_i(n_M)$ curves are broadened. The fits are not very good, but the authors believe that certain modifications of the assumptions made for the computer calculations of the $P(n_M)$ values might improve the fit.

For their quasi-crystallite model, these authors assumed that in the vicinity of each Fe atom there is short-range order based on the order existing in the metastable crystalline Fe_3B phase [4.91]. The $H_i(n_M)$ values just described were used again, and again some broadening was introduced to allow for inhomogeneities due to the lack of true long-range order. For iron concentrations greater than 75 % (Fe_3B), the excess iron atoms were assumed to be distributed randomly on B sites. The $P(H_i)$ curve predicted by this model for amorphous $Fe_{75}B_{25}$ agreed well with the $P(H_i)$ curve derived from the experimental Mössbauer spectrum. For higher Fe concentrations the agreement was poorer, but according to the authors, it is possible that the fit could be improved by adoption of an additional assumption that the excess iron atoms tend to form neighboring pairs on the B sites.

In a more recent paper [4.93], it was argued that the RDP model should be abandoned in favor of a quasi-crystallite model. $P(H_i)$ curves obtained by the difference method (Sect. 4.4.3) are found to agree with qualitative predictions based on a postulated hybridization of the $s-d$ band of the transiton metal T with the $s-p$ orbitals of the metalloid M in a quasi-crystalline material.

This section has to close on a sour note. While claims have been made that one model or another is the best, or closest to reality, the actual situation seems to be that a great variety of models – perhaps almost any model – can be "confirmed" by the fact that it correctly predicts the shape of the broadened hyperfine pattern. In determining the actual atomistic structures of these materials, two problems will have to be solved:
  a) how to determine the true distribution of hyperfine fields experimentally and
  b) how to interpret this distribution in terms of a unique structure.

Up to the present time, attempts to solve both of these problems have relied partly on guesswork.

4.5 Magnetic Properties

The magnetic properties of amorphous metals are of great potential significance from a technological point of view. Mössbauer spectroscopy has done its share in elucidating the nature and certain features of magnetic amorphous alloys. In the foregoing we were concerned only with the magnitude of the magnetic hyperfine fields; now we will focus our attention on the orientation of the hyperfine fields which can be deduced from the angular dependence of the intensities and polarizations of the hyperfine transitions as described in Sect. 4.2.3b.

4.5.1 External Static Fields

In crystalline ferromagnetic materials, the magnetocrystalline anisotropy normally governs the spin orientations along the easy direction of magnetization within the domains. In contrast, for amorphous metals, the shape, magnetoelastic and structural anisotropies become important. In general, two types of domain patterns seem to dominate: broad stripes about 25 μm wide whose spin orientation is close to the ribbon plane and patches of maze-type domains with closure structure and a width of about 3–5 μm (Fig. 4.15). The type of domain structure that develops depends greatly on internal stress and on the sign and magnitude of the magnetostriction coefficient. The spins can easily be aligned in the plane of the ribbon by magnetic fields applied parallel to the plane. But application of an external longitudinal magnetic field H_{ext} (along the γ-ray propagation direction and *perpendicular* to the ribbon plane) produces a number of interesting effects [4.124]. In Fig. 4.16 the Mössbauer spectra of amorphous $Fe_{83}P_5C_{12}$ in various applied longitudinal magnetic fields are shown. These fields were produced by a superconducting solenoid.

1) With increasing external magnetic field the internal field decreases. This indicates that the hyperfine interaction is negative, that is, that the nuclear and atomic moments are opposite in orientation.

2) In many weak ferromagnetic amorphous alloys a rather surprising effect is observed: the average orientation of the hyperfine field tends to align itself perpendicular to the externally applied magnetic field. This unexpected orientation effect is due to a corresponding orientation of the electron spins. As shown in Fig. 4.17, it can be followed by monitoring the relative line intensities which reflect the orientation of the hyperfine fields.

In the thin absorber approximation according to Fig. 4.6, the intensity ratios $I_2/I_3 = I_5/I_4$ of line 2 or 5 ($\Delta m = 0$; $\pm 1/2 \leftrightarrow \pm 1/2$) to line 3 or 4

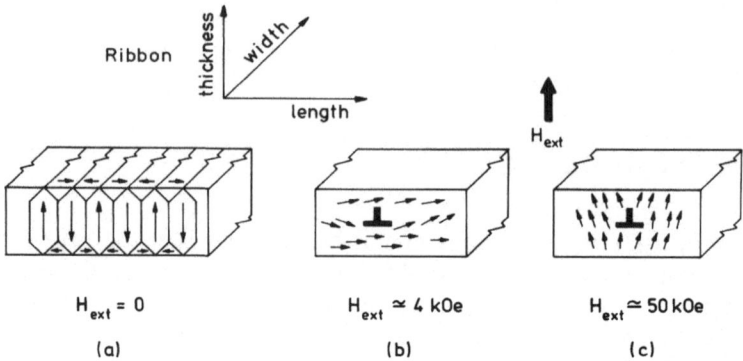

Fig. 4.15a–c. Schematic representation of maze-type domains with closure structure (**a**) and spin orientations (**b, c**) in amorphous $Fe_{83}P_5C_{12}$

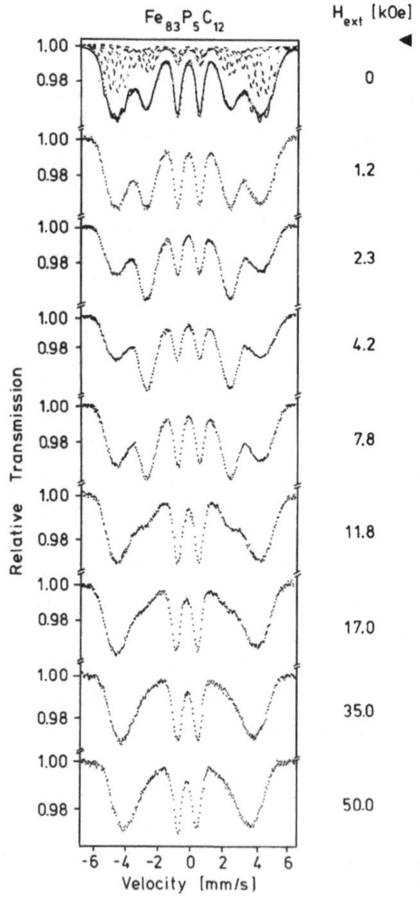

◄ **Fig. 4.16.** Mössbauer spectra of amorphous $Fe_{83}P_5C_{12}$ obtained in longitudinal magnetic fields H_{ext} oriented perpendicular to the ribbon plane

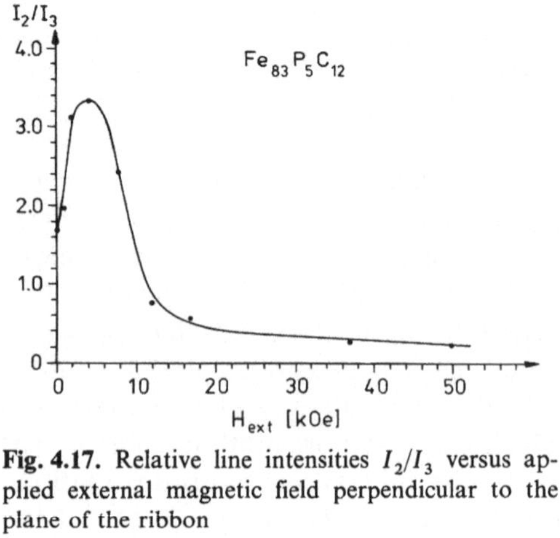

Fig. 4.17. Relative line intensities I_2/I_3 versus applied external magnetic field perpendicular to the plane of the ribbon

($\Delta m = \pm 1$; $\mp 1/2 \leftrightarrow \pm 1/2$) are 4 or 0 when all spins are aligned parallel ($\vartheta = 0°$) or perpendicular ($\vartheta = 90°$), respectively, to a longitudinal H_{ext}. The ratio is 2 in the case of $\vartheta \approx 55°$ and for random orientation.

In zero magnetic field the measured ratio I_2/I_3 is ≈ 1.5 for $Fe_{83}P_5C_{12}$. This value indicates that there is some kind of a preferred spin orientation (texture) perpendicular to the plane of the ribbon, in agreement with the assumption that closure domains predominate as suggested by Fig. 4.15a.

In small fields perpendicular to the ribbon plane ($H_{ext} \lesssim 4$ kOe), the ratio I_2/I_3 increases significantly which means that the spins orient themselves preferentially *along* the ribbon plane – *perpendicular* to the field direction H_{ext}. The following explanations can be given for this behavior. The external magnetic field is inducing a demagnetizing field which – in this geometry – completely compensates the applied external field. Under the usual experimental conditions, the lines of force are not precisely perpendicular to the ribbon plane. Thus, a small magnetic field component in the plane of the ribbon will be present. Such a field component will be effective because the demagnetizing field is negligible in orientations parallel to the ribbon plane. A similar anisotropy in well-annealed crystalline nickel measured by magnetoresistivity was observed as early as 1938 by *Bittel* [4.148]. While this effect was unexpected in crystalline materials, such behavior seems to be the rule in soft ferromagnetic amorphous metals. Mössbauer spectroscopy is suitable for analyzing this anisotropy quantitatively. With fields $H_{ext} \approx 4$ kOe, the ratio I_2/I_3 reaches a maximum of 3.5. Taking into account saturation effects and angular uncertainties, this value indicates that the average deviation from complete alignment of the spins in the plane of the ribbon is about 15° (Fig. 4.15b).

3) In larger fields ($H_{ext} \gtrsim 4$ kOe) the average spin rotates into the direction of H_{ext}. However, even in very large fields ($H_{ext} \approx 50$ kOe), the intensity of lines 2 and 5 has not completely disappeared which shows that complete alignment of the spins parallel to H_{ext} has not yet been accomplished. An explanation for this observation might be found in the presence of "defects", particularly "quasi-dislocations". *Kronmüller* et al. [4.149] estimate the number of such defects to be of the order of 10^{13} cm^{-2}. Such defects will cause inhomogeneous spin arrangements in their immediate vicinity (Fig. 4.15c).

4.5.2 External Dynamic Fields

In the past it has been shown that radio-frequency magnetic fields applied to weak crystalline ferromagnets can drastically influence Mössbauer spectra. Of interest were the collapse of the hyperfine pattern and the appearance of side bands in the Mössbauer spectrum.

Recently the influence of radio-frequency magnetic fields on the stability of amorphous metals was tested [4.122]. In these investigations, fields of 1–12 Oe with frequencies of 53 and 67 MHz were used. It was found that when an rf field is applied, the onset of crystallization occurs at temperatures lower than usual. This radio-frequency induced and enhanced crystallization is most likely connected with magnetostrictive vibrations of the metastable system. On the atomistic scale, this "shaking" assists the diffusive motion of atoms in overcoming potential barriers and promotes the nucleation and growth of microcrystals.

4.5.3 Surface Fields (Scattering Method)

It is an interesting question as to whether there are differences in the magnetic properties of surfaces as compared to the bulk [4.125]. Mössbauer spectroscopy offers the possibility of obtaining information on both regions simultaneously: on the bulk by measuring the γ-rays in transmission and on the surface by measuring the conversion electrons in scattering geometry as already discussed in Sect. 4.1.6. Figure 4.18 shows schematically the experimental arrangement using only one specimen, two independent counters and one multichannel analyzer. The conversion electrons are emitted from the absorber after a Mössbauer reasonance between source and absorber has occurred (Fig. 4.1). Because of the small penetration depth of electrons with energies of a few keV ($\approx 1000\,\text{Å}$), it is necessary to place the absorber inside the electron counter. The 2π geometry gives a reasonable count rate. The counts from these

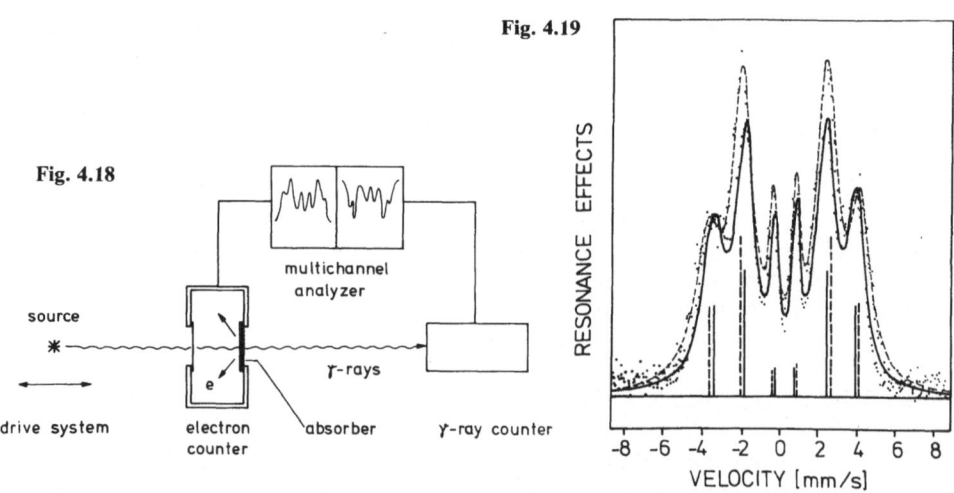

Fig. 4.18. Schematic representation of the setup for simultaneous measurement of absorption spectra in the transmission geometry and conversion electron spectra in the scattering geometry

Fig. 4.19. Superposition of the γ-ray absorption spectrum (——) and the conversion-electron scattering spectrum (– – –), as measured for amorphous $Fe_{40}Ni_{40}P_{14}B_6$. The stick diagram indicates the positions and relative line intensities

electrons are stored in the first half of the multichannel analyzer. The γ-rays passing through the electron counter and the sample are detected in a proportional counter and are stored in the second half of the analyzer. Figure 4.19 shows the conversion electron Mössbauer spectrum (CEMS) and the transmission spectrum (shown inverted to facilitate comparison) of amorphous $Fe_{40}Ni_{40}P_{14}B_6$. In comparing the two spectra, two effects are clearly seen:

1) Surfaces exhibit larger mean magnetic fields than the bulk. Similar effects have also been observed on surfaces of crystalline thin magnetic iron films; however, in the crystalline case the increased fields are restricted to a few atomic layers [4.150, 151].

2) The electronic moments in the surfaces are, on the average, aligned more nearly parallel to the plane of the ribbon sample than the magnetic moments in the bulk, as can be seen from the intensity ratio I_2/I_3. This suggests the presence of closure domains (in addition to the broad stripe domains) in the surface with stronger alignment along the plane of the ribbon than occurs in the bulk.

Recently the beginning of crystallization at the surface has been followed and an explanation for the concurrent anisotropy has been proposed [4.152, 153].

4.6 Stress

As already discussed in Sect. 4.5.1, the magnetic structure of ferromagnetic amorphous metals is rather sensitive to stress. In concluding this chapter we want to demonstrate stress effects by briefly discussing a more sophisticated variation of Mössbauer spectroscopy: polarized recoil-free γ-radiation.

4.6.1 Use of Polarized γ-Radiation

The relative line intensities of hyperfine split spectra are governed by the polarizations of source and absorber. This is well demonstrated by Fig. 4.5. In that arrangement we were concerned only with linearly polarized γ-rays because the γ-ray propagation direction was perpendicular to the magnetic fields of both source and absorber. Each line of the spectrum can be regarded as a Mössbauer polarimeter. By rotating the source (polarizer) magnetization relative to that of the absorber (analyzer), we change from opaque (absorption) to transparent. The intensity of each line follows a Malus curve as in optical polarimetry.

The spectra in Fig. 4.20 were obtained with a transversely magnetized ^{57}Co–α–Fe source and an amorphous $Fe_{40}Ni_{40}P_{14}B_6$ absorber whose ribbon direction was parallel to the source magnetic field. This is similar to the

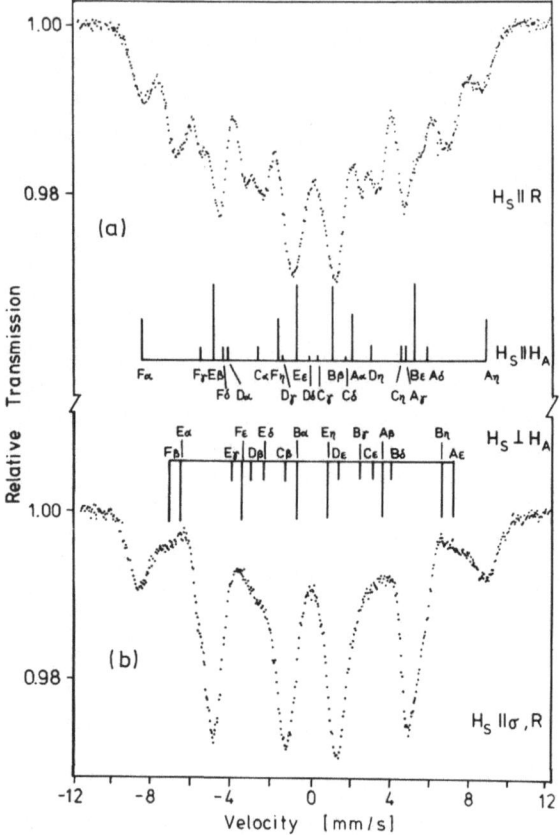

Fig. 4.20a, b. Spectra obtained with a transversely magnetized ^{57}Co–α–Fe source and an Fe$_{40}$Ni$_{40}$P$_{14}$B$_6$ absorber both at room temperature. (**a**) Magnetic field of the source and ribbon direction are parallel ($H_S \| R$); (**b**) tensile stress is applied parallel to R and H_S ($H_S \| \sigma, R$). The stick diagrams and the capital and Greek letters (Fig. 4.3) indicate the position, relative line intensities and origin of the lines

arrangement (^{57}Co–α–Fe vs α-Fe) which was used in obtaining the spectra of Fig. 4.5 except that here the source and absorber hyperfine fields are no longer equal and, in addition, the absorber lines are broad and rather complicated in appearance, as may be seen in the upper spectrum of Fig. 4.20. The two stick diagrams at the center of Fig. 4.20 indicate the line positions and relative intensities for the two special cases where the fields in source and absorber are parallel ($H_S \| H_A$) and perpendicular ($H_S \perp H_A$). Analysis of the upper spectrum of Fig. 4.20 shows that to a considerable extent it is a superposition of these two patterns, indicating that the absorber hyperfine fields have a tendency to line up parallel to the plane of the ribbon, as discussed in Sect. 4.5.3, with a preponderance of the spins aligned parallel to the long direction of the ribbon.

Upon application of tensile stress to the absorber ribbon, the spectrum changes considerably as can be seen in the lower spectrum of Fig. 4.20. Upon comparison of this spectrum with the stick diagrams, one recognizes that the spectral lines correspond completely to the upper stick diagram. This shows clearly that the spins have aligned themselves *parallel* to the applied stress [4.72, 85].

The degree of alignment of the electron spins can be determined quantitatively from spectra of the type shown in Fig. 4.20, after distribution of magnitudes of the hyperfine fields has been found by one of the methods of Sect. 4.4.3. The determination is simplified by the fact that the degree of alignment is essentially the same for all hyperfine fields which contribute to the broadened Mössbauer spectrum. This is because the direction of each electron spin is closely tied to the direction of domain magnetization and is little influenced by variations in the local atomic environment. Since the direction of the hyperfine field is in turn tied to the electron-spin orientation, the distribution of electron-spin or domain orientation is very nearly the same for all values of the hyperfine field. Using this fact and the already known distribution $P(H)$ of the magnitudes of the hyperfine fields, a basic six line absorption spectrum can be determined by a fit to the experimental spectrum. The relative intensities are the only adjustable parameters and they are the same for all contributions to the broadened spectrum because the distribution of field orientations is the same for all contributions.

An absorber polarization parameter α is defined which is one or zero in the limits of polarization, $H_S \| H_A$ and $H_S \perp H_A$, respectively. Random orientation is close to 0.5. The value of α is calculated directly from the relative intensities determined from the fit to the experimental spectrum. It is found that a tensile stress in the direction of the amorphous ribbon aligns the spins in the same direction. In Fig. 4.21 it is shown that relatively small tensile stresses ($16\,kg/mm^2$) are sufficient to produce this alignment.

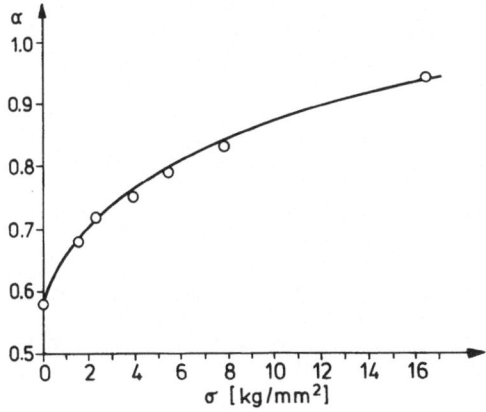

Fig. 4.21. Polarization parameter α versus external tensile stress σ

4.7 Conclusion

We hope that this brief treatment of some applications of Mössbauer spectroscopy to amorphous metals has given the reader a useful impression of the possibilities of the method and some of its achievements. Although our examples are a reflection of our own interests, the following list of references is a fairly complete bibliography of Mössbauer research on amorphous $T_{80}M_{20}$ type alloys as of the end of 1980. Concerning the last two years, the reader is referred to four international conferences where new detailed developments in Mössbauer spectroscopy as applied to amorphous metals are described in nearly 100 papers of the proceedings: 1) Rapidly Quenched Metals, RQM IV, Japan (1981) [4.126], 2) Amorphous Systems Investigated by Nuclear Methods, Hungary (1981) [4.127], 3) Application of the Mössbauer Effect, India (1981) [4.128], 4) International Conference on Magnetism, Japan (1982) [4.129].

Acknowledgement. This work was supported by the Deutsche Forschungsgemeinschaft. Discussions with Dr. M. Naka, S. Nasu, H. Fischer, M. Ghafari, and H.-G. Wagner are gratefully acknowledged.

References

4.1   C.C.Tsuei, G.Longworth, S.C.H.Lin: Phys. Rev. **170**, 603 (1968)
4.2   H.Frauenfelder: *The Mössbauer Effect* (Benjamin, New York 1962)
4.3   G.K.Wertheim: *Mössbauer Effect, Principles and Applications* (Academic Press, New York 1964)
4.4   H.Wegener: *Der Mössbauereffekt und seine Anwendungen in Physik und Chemie* (Bibliographisches Institut, Mannheim 1965)
4.5   I.J.Gruverman: *Mössbauer Effect Methodology*, Vols. 1–11 (Plenum Press, New York 1965–1976)
4.6   C.Janot: *L'Effect Mössbauer et ses Applications à la Physique du Solid et à la Métallurgie Physique* (Masson, Paris 1972)
4.7   N.N.Greenwood, T.C.Gibb: *Mössbauer Spectroscopy* (Chapman & Hall, London 1971)
4.8   U.Gonser (ed.): *Mössbauer Spectroscopy I and II*, Topics Appl. Phys., Vol. 5, and Topics Current Phys., Vol. 25 (Springer, Berlin, Heidelberg, New York 1975 and 1981)
4.9   R.L.Cohen: *Applications of Mössbauer Spectroscopy I and II* (Academic Press, New York 1975 and 1980)
4.10 A.H.Muir, Jr., K.J.Ando, H.M.Coogan: *Mössbauer Effect Data Index*, 1958–1965 (Interscience, New York 1966); J.G.Stevens, V.E.Stevens: *Mössbauer Effect Data Index*, 1966–1976 (Hilger and Plenum, London, New York)
4.11 V.V.Bondar, V.A.Povitskiy, Y.E.F.Makarov: Fiz. Metal. Metalloved **30**, 1061 (1970)
4.12 T.E.Sharon, C.C.Tsuei: Solid State Commun. **9**, 1923 (1971)
4.13 T.E.Sharon, C.C.Tsuei: Phys. Rev. B**5**, 1047 (1972)
4.14 C.C.Tsuei: In *Amorphous Magnetism* (Proc. Intern. Symp. Detroit, 1972) ed. by H.C.Hooper, A.M. de Graaf (Plenum Press, New York 1973) p. 299
4.15 C.L.Chien, J.C.Walker: AIP Conf. Proc., Magn. and Magn. Mater., 1974, ed. by C.D.Graham, Jr., G.H.Lander, J.J.Rhyne (American Institute of Physics, New York 1975) p. 127

Mössbauer Spectroscopy Applied to Amorphous Metals   123

4.16 F.E.Fujita: In *Mössbauer Spectroscopy*, ed. by U.Gonser, Topics Appl. Phys., Vol. 5 (Springer, Berlin, Heidelberg, New York 1975) p. 201
4.17 G.E.A.Bartsch, P.G.Glozbach, T.Just: Proc. "Rapidly Quenched Metals", 2nd Intern. Conf., ed. by N.J.Grant, B.C.Giessen (1975) Sect. 1, p. 343
4.18 C.C.Tsuei: Proc. "Rapidly Quenched Metals", 2nd Intern. Conf., ed. by N.J.Grant, B.C.Giessen (1975) Sect. 1, p. 441
4.19 C.L.Chien, R.Hasegawa: Proc. Intern. Conf. Mössbauer Spectroscopy Cracow, Poland (1975) p. 343
4.20 J.Sitek, S.Prejsa, P.Duhaj, M.Hucl, J.Cirak: Proc. Intern. Conf. Mössbauer Spectroscopy Cracow, Poland (1975) p. 345
4.21 R.W.Cochrane, R.Harris, M.Plischke, D.Zobin, M.J.Zuckermann: Phys. Rev. B12, 1969 (1975)
4.22 C.L.Chien, R.Hasegawa: AIP Conf. Proc., Magn. and Magn. Mater., 1975, ed. by J.J.Becker, G.H.Lander, J.J.Rhyne (American Institute of Physics, New York 1976) p. 214
4.23 C.L.Chien, R.Hasegawa: J. Appl. Phys. 47, 2234 (1976)
4.24 R.Hasegawa, R.C.O'Handley, L.I.Mendelsohn: AIP Conf. Proc., Magn. and Magn. Mater., 1976, ed. by J.J.Becker, G.H.Lander (American Institute of Physics, New York 1976) p. 298
4.25 C.L.Chien, R.Hasegawa: AIP Conf. Proc., Intern. Conf. of Structure and Excitations of Amorphous Solids, Williamsburg, Virginia, 1976, ed. by G.Lucovsky, F.I.Galenner (American Institute of Physics, New York 1976) p. 366
4.26 R.Hasegawa, C.L.Chien: Solid State Commun. 18, 913 (1976)
4.27 C.L.Chien, R.Hasegawa: J. Physique C6, 759 (1976)
4.28 C.L.Chien, R.Hasegawa: IEEE Trans. Mag.-12, 951 (1976)
4.29 F.E.Fujita: Bull. Japan Inst. Metals 5, 180 (1976)
4.30 P.Mangin, G.Marchal, M.Piecuch, C.Janot: J. Physique E9, 1101 (1976)
4.31 G.Marchal, P.Mangin, M.Piecuch, C.Janot: J. Physique C6, 763 (1976)
4.32 C.C.Tsuei, H.Lilienthal: Phys. Rev. B13, 4899 (1976)
4.33 P.Duhaj, J.Sitek, M.Prejsa, P.Butvin: Phys. Stat. sol. A35, 223 (1976)
4.34 L.A.Alekseev, V.T.Borisow, A.I.Dukhin, A.M.Durachenko, R.B.Levi, A.F.Prokoshin: Fiz. Metalloved 42, 887 (1976); Phys. Met. Metallurgy (USSR)
4.35 J.Logan, E.Sun: J. Non-Cryst. Solids 20, 285 (1976)
4.36 G.Marchal, P.Mangin, C.Janot: Solid State Commun. 18, 739 (1976)
4.37 C.L.Chien, R.Hasegawa: J. Appl. Phys. 47, 2234 (1976)
4.38 L.Takacs: Solid State Commun. 21, 611 (1977)
4.39 C.L.Chien, D.Musser, F.E.Luborsky, J.J.Becker, J.L.Walter: Solid State Commun. 24, 231 (1977)
4.40 D.K.Brown, I.Nowik, D.I.Paul: Solid State Commun. 24, 711 (1977)
4.41 K.Raj, A.Amamou, J.Durand, J.I.Budnik, R.Hasegawa: In *Amorphous Magnetism II*, ed. by R.A.Levy, R.Hasegawa (Plenum Press, New York 1977) p. 221
4.42 C.L.Chien, R.Hasegawa: In *Amorphous Magnetism II*, ed. by R.A.Levy, R.Hasegawa (Plenum Press, New York 1977) p. 289
4.43 A.S.Schaafsma, F. van der Woude: In *Amorphous Magnetism II*, ed. by R.A.Levy, R.Hasegawa (Plenum Press, New York 1977) p. 335
4.44 A.M. van Diepen, F.J. den Broeder: J. Appl. Phys. 48, 3165 (1977)
4.45 J.Sitek, J.Cirak, M.Presja, M.Hucl: Proc. Intern. Conf. Mössbauer Spectroscopy Bucharest, Rumania (1977) p. 233
4.46 U.Gonser, M.Ghafari, H.G.Wagner: Proc. Intern. Conf. Mössbauer Spectroscopy Bucharest, Rumania (1977) p. 237
4.47 A.S.Schaafsma, W.Venhaizen, F. van der Woude: Proc. Intern. Conf. Mössbauer Spectroscopy Bucharest, Rumania (1977) p. 239
4.48 I.Vincze, L.Takacs: Proc. Intern. Conf. Mössbauer Spectroscopy Bucharest, Rumania (1977) p. 421
4.49 I.Vincze, J.Balogh: Proc. Intern. Conf. Mössbauer Spectroscopy Bucharest, Rumania (1977) p. 421
4.50 F.E.Fujita, T.Masumoto, M.Kitaguchi, M.Ura: Jpn. J. Appl. Phys. 16, 1731 (1977)

4.51   C.L.Chien, R.Hasegawa: Phys. Rev. B**16**, 2115 (1977)
4.52   C.L.Chien, R.Hasegawa: Phys. Rev. B**16**, 3024 (1977)
4.53   C.L.Chien, R.Hasegawa: In *Proc. 4th Intern. Conf. Hyperfine Interactions*, ed. by R.S.Raghavan, D.E.Murnick (North-Holland, Amsterdam 1978) p. 866
4.54   C.L.Chien: *Proc. 4th Intern. Conf. Hyperfine Interactions*, ed. by R.S.Raghavan, D.E.Murnick (North-Holland, Amsterdam 1978) p. 869
4.55   P.Mangin, G.Marchal: J. Appl. Phys. **49**, 1709 (1978)
4.56   C.L.Chien, R.Hasegawa: J. Appl. Phys. **49**, 1721 (1978)
4.57   H.Franke, M.Rosenberg, F.E.Luborsky, J.L.Walter: IEEE Trans. Mag.-**14**, 952 (1978)
4.58   T.Tarnoczi, I.Nagy, C.Hargitai, M.Hosso: IEEE Trans. Mag.-**14**, 1025 (1978)
4.59   C.L.Chien: Phys. Rev. B**18**, 1003 (1978)
4.60   R.J.Birgeneau, J.A.Tarvin, G.Shirane, E.M.Gyorgy, R.C.Sherwood, H.S.Chen, C.L.Chien: Phys. Rev. B**18**, 2192 (1978)
4.61   I.Vincze: Solid State Commun. **25**, 689 (1978)
4.62   J.Balosh, I.Vincze: Solid State Commun. **25**, 695 (1978)
4.63   I.Vincze, E.Babic: Solid State Commun. **27**, 1425 (1978)
4.64   C.L.Chien, D.Musser, F.E.Luborsky, J.L.Walter: Solid State Commun. **28**, 645 (1978)
4.65   P.J.Schurer, A.H.Morrish: Solid State Commun. **28**, 819 (1978)
4.66   C.L.Chien, D.Musser, F.E.Luborsky, J.L.Walter: J. Physique F**8**, 2407 (1978)
4.67   C.L.Chien: Phys. Lett. **68**A, 394 (1978)
4.68   H.Franke, M.Rosenberg: J. Magn. Magn. Mater. **7**, 168 (1978)
4.69   D.K.Brown, I.Nowik, D.I.Paul: J. Magn. Magn. Mater. **7**, 182 (1978)
4.70   U.Gonser, M.Ghafari, H.G.Wagner: J. Magn. Magn. Mater. **8**, 175 (1978)
4.71   H.Franke, H.Herold, U.Köster, M.Rosenberg: J. Magn. Magn. Mater. **9**, 214 (1978)
4.72   H.Fischer, U.Gonser, R.S.Preston, H.G.Wagner: J. Magn. Magn. Mater. **9**, 336 (1978)
4.73   R.A.Manapov, A.V.Mittin, F.G.Vagizov: In *Magnetic Resonance and Related Phenomena*, Proc. 20th Congress Ampere, Tallinn, USSR, ed. by E.Kundly, E.Lippmaa, T.Saluvere (Springer, Berlin, Heidelberg, New York 1978) p. 259
4.74   R.Ingalls, C.M.Liu, K.V.Rao: J. Magn. Magn. Mater. **10**, 257 (1979)
4.75   B.Gantor (ed.): Proc. "Rapidly Quenched Metals III", 3rd. Intern. Conf., Brighton (1978), 2 vols.
        H.Franke, U.Herold, U.Köster, M.Rosenberg: In [Ref. 4.75, Vol. 1, p. 155]
        T.Kemény, I.Vincze, B.Fogarassy, S.Arajs: In [Ref. 4.75, Vol. 1, p. 291]
        Z.Wronski, J.Suwalski, H.Matyja: In [Ref. 4.75, Vol. 1, p. 397]
        A.S.Schaafsma, H.Snijders, F. van der Woude: In [Ref. 4.75, Vol. 1, p. 428]
        J.Durand, C.Thompson, A.Amamou: In [Ref. 4.75, Vol. 2, p. 109]
        H.G.Wagner, U.Gonser, A.Schertz: In [Ref. 4.75, Vol. 2, p. 333]
4.76   R.W.Cochrane, R.Harris, M.J.Zuckermann: Phys. Rep. **48**, 1 (1979)
4.77   D.Musser, C.L.Chien: J. Appl. Phys. **50**, 1571 (1979)
4.78   C.L.Chien, H.S.Chen: J. Appl. Phys. **50**, 1574 (1979)
4.79   C.M.Liu, R.Ingalls, J.E.Whitmore, K.V.Rao, S.M.Bhagat: J. Appl. Phys. **50**, 1577 (1979)
4.80   H.Ino, S.Nanao, T.Muto: H. Phys. Soc. Jpn. **46**, 63 (1979)
4.81   C.L.Chien: Phys. Rev. B**19**, 81 (1979)
4.82   C.L.Chien, H.S.Chen: J. Physique C**2**, 118 (1979)
4.83   A.Matsuzaki, S.Nanao, H.Ino: J. Physique C**2**, 104 (1979)
4.84   F.E.Fujita: J. Physique C**2**, 120 (1979)
4.85   U.Gonser, M.Ghafari, H.G.Wagner, H.Fischer: J. Physique C**2**, 126 (1979)
4.86   C.L.Chien, D.Musser, F.E.Luborsky, J.L.Walter: J. Physique C**2**, 129 (1979)
4.87   R.Oshima, F.E.Fujita, K.Fukamichi, T.Masumoto: J. Physique C**2**, 132 (1979)
4.88   H.Onodera, H.Yamamoto, H.Watanabe: J. Physique C**2**, 142 (1979)
4.89   M.Takahashi, M.Koshimura, T.Suzuki: J. Physique C**2**, 144 (1979)
4.90   R.Ingalls: J. Physique C**2**, 174 (1979)
4.91   I.Vincze, D.S.Bourdreaux, M.Tegze: Phys. Rev. B**19**, 4896 (1979)
4.92   C.L.Chien, D.Musser, E.M.Gyorgy, R.C.Sherwood, H.S.Chen, F.E.Luborsky, J.L.Walter: Phys. Rev. B**20**, 283 (1979)

4.93   T.Kemény, I.Vincze, B.Fogarassy, S.Arajs: Phys. Rev. B**20**, 476 (1979)
4.94   M.Takahashi, M.Koshimura: Japan. J. Appl. Phys. **18**, 685 (1979)
4.95   P.K.Tseng, S.Y.Chuang, H.S.Chen: J. Appl. Phys. **50**, 4292 (1979)
4.96   A.Amamou: Phys. Status Solidi A**54**, 565 (1979)
4.97   E.Babic, Z.Marohmic, F.Hajdu, M.Tegze, I.Vincze: Solid State Commun. **29**, 175 (1979)
4.98   L.D.Lafleur: Phys. Rev. B**20**, 2581 (1979)
4.99   T.A.Donnelly, T.Egami, D.G.Onn: Phys. Rev. B**20**, 1211 (1979)
4.100 A.S.Schaafsma, H.Snijders, F. van der Woude, J.W.Drijver, S.Radelar: Phys. Rev. B**20**, 4423 (1979)
4.101 P.J.Schurer, A.H.Morrish: Phys. Rev. B**20**, 4660 (1979)
4.102 D.Musser, C.L.Chien, H.S.Chen: J. Appl. Phys. **50**, 7659 (1979)
4.103 G. Le Caer, J.M.Dubois: J. Physique E**12**, 1083 (1979)
4.104 C.L.Chien, J.H.Hsu, J.P.Stokes, A.N.Bloch, H.S.Chen: J. Appl. Phys. **50**, 7647 (1979)
4.105 A.F.Prokoshin, B.V.Molotilov, Yu.A.Grotsianov, A.N.Zhelnov: Sov. Phys. JETP Lett. **29**, 621 (1979)
4.106 U.Gonser: J. Physique C**1**, 51 (1980)
4.107 J.M.Dubois, G. Le Caer, A.Amamou, U.Herold: J. Physique C**1**, 247 (1980)
4.108 P.J.Schurer, A.H.Morrish: J. Physique C**1**, 249 (1980)
4.109 J.Balogh, I.Dészi, B.Fogarassy, L.Granassy, D.L.Nagy: J. Physique C**1**, 253 (1980)
4.110 J.Balogh, G.Faigel, M.Tegze, T.Kemény, A.S.Schaafsma: J. Physique C**1**, 255 (1980)
4.111 I.Vincze, F. van der Woude, J.Balogh: J. Physique C**1**, 257 (1980)
4.112 E.Wieser: J. Physique C**1**, 259 (1980)
4.113 D.C.Price, S.J.Campbell, P.J.Back: J. Physique C**1**, 263 (1980)
4.114 L.Takacs: J. Physique C**1**, 265 (1980)
4.115 Z.Wronski, H.Matyja, J.Piekoszewski, J.Suwalski: Solid State Commun. **33**, 1155 (1980)
4.116 F.E.Fujita: In *"Structure and Properties of Amorphous Metals* II", Sci. Rept. 28 (Res. Inst., Tohoku Univ. 1980) p. 1
4.117 R.Oshima: In *"Structure and Properties of Amorphous Metals* II", Sci. Rept. 28 (Res. Inst., Tohoku Univ. 1980) p. 8
4.118 I.Vincze, F. van der Woude, T.Kemény, A.S.Schaafsma: J. Magn. Magn. Mater. **15–18**, 1336 (1980)
4.119 H.Franke, S.Dey, M.Rosenberg, F.E.Luborsky, J.L.Walter: J. Magn. Magn. Mater. **15–18**, 1364 (1980)
4.120 S.Bjarman, R.Kamal, R.Wäppling: J. Magn. Magn. Mater. **15–18**, 1389 (1980)
4.121 G.Hilscher, R.Haferl, H.Kirchmayr, M.Müller, H.-J. Güntherodt: J. Phys. F**11**, 2429 (1981)
4.122 M.Kopcewicz, U.Gonser, H.G.Wagner: Appl. Phys. **23**, 1 (1980)
4.123 H.G.Wagner, M.Ghafari, U.Gonser, M.Naka: J. Physique C**8**, 199 (1980)
4.124 U.Gonser, M.Ghafari, H.G.Wagner, R.Kern: J. Magn. Magn. Mater. **23**, 279 (1981)
4.125 S.Nasu, U.Gonser: J. Physique C**8**, 690 (1980)
4.126 Proc. "Rapidly Quenched Metals IV", 4th Int. Conf. Sendai, Japan (1981)
4.127 Intern. Conf. Amorphous Systems Investigated by Nuclear Methods, Balatonfüred, Hungary (1981), published in Nucl. Instr. Methods **199** (1982)
4.128 Proc. Intern. Conf. Application of the Mössbauer Effect, Jaipur, India (1981)
4.129 Intern. Conf. Magnetism, ICM 10, Kyoto, Japan (1982), be published in J. Magn. Magn. Mater. (1983)
4.130 C.E.Johnson, M.S.Ridout, T.E.Cranshaw: Proc. Phys. Soc. **81** (1963)
4.131 U.Gonser, S.Nasu, W.Kappes: J. Magn. Magn. Mater. **10**, 244 (1979)
4.132 U.Gonser, C.J.Meechan, A.H.Muir, H.Widersich: J. Appl. Phys. **34**, 2373 (1963)
4.133 U.Gradmann, W.Kümmerle, P.Tillmanns: Thin Solid Films **34**, 249 (1976)
4.134 J.G.Wright: Phil. Mag. **24**, 217 (1971)
4.135 W.Keune, R.Halbauer, U.Gonser, J.Lauer, D.L.Williamson: J. Appl. Phys. **48**, 2976 (1977)
4.136 J.S.Kouvel: *Magnetism and Metallurgy*, ed. by A.E.Berkowitz, T.E.Kneller (Academic Press, New York 1969)
4.137 B.Window: J. Phys. E**4**, 401 (1971)
4.138 J.B.Müller, J.Hesse, E.Hagen: Hyperfine Interact. **10**, 1189 (1981)

4.139 G. Le Caer, J.M.Dubois: Phys. State Sol. (a) **64**, 275 (1981)
4.140 H.Keller: J. Appl. Phys. **52**, 5268 (1981)
4.141 P.H.Gaskell: J. Phys. C**12**, 4337 (1979)
4.142 J.D.Bernal: Proc. Roy. Soc. A**280**, 299 (1964)
4.143 T.Ninomiya: Proc. Symp. on Structure of Noncrystalline Materials, Special Issue of Phys. Chem. Glasses (1977) p. 45
4.144 H.Koizumi, T.Ninomiya: J. Phys. Soc. Japan **44**, 898 (1978)
4.145 E.J.Fasiska, G.A.Jeffrey: Acta Cryst. **19**, 463 (1965)
4.146 R.Yamamoto, M.Doyama: J. Phys. F**9**, 617 (1979)
4.147 D.E.Polk: Acta Met. **20**, 485 (1972)
4.148 H.Bittel: Annalen Physik **31**, 219 (1938)
4.149 H.Kronmüller, M.Fähnle, M.Domann, H.Grimm, R.Grimm, B.Gröger: J. Magn. Magn. Mater. **13**, 53 (1979)
4.150 T.Shinjo: J. Physique C**2**, 63 (1979)
4.151 A.H.Owens, C.L.Chien, J.C.Walker: J. Physique C**2**, 74 (1979)
4.152 U.Gonser, M.Ghafari, M.Ackermann, H.-P.Klein, J.Bauer, H.-G.Wagner: In [Ref. 4.126, p. 847]
4.153 H.N.Ok, A.H.Morrish: Phys. Rev. B**23**, 2257 (1981)

5. Defects and Atomic Transport in Metallic Glasses

P. Chaudhari, F. Spaepen, and P. J. Steinhardt

With 31 Figures

In this chapter we review our current understanding of atomic transport in metallic glasses. We also summarize the results of computer simulation experiments about the role of point and line defects in amorphous solids simulated by simple pair potentials.

5.1 Background

The fundamental understanding of atomic transport in amorphous solids is not nearly as well developed as that of transport in crystalline materials. Atomic motion can be studied experimentally by applying either a mechanical driving force, as in viscous flow, or a chemical one, such as a concentration gradient in a diffusion experiment or a free energy difference in a phase transformation. The study of these processes is intrinsically more complicated in glassy materials since they are in a thermodynamically unstable state and hence undergo a continuous structural change through successively lower free energy states. This structural relaxation phenomenon not only introduces yet another type of atomic motion to be considered, but it also affects all properties of the material. Since it most strongly affects the atomic transport properties themselves, it requires special attention during experimental design and data analysis. In the first part of this chapter, the atomic transport data are critically reviewed, with special attention for possible effects of structural relaxation. The magnitude, temperature dependence and annealing kinetics of the various transport coefficients are compared. This allows some insight into the similarity, or lack thereof, of the underlying atomic scale processes.

In crystalline materials, atomic transport is governed by the presence and motion of well-characterized defects. The periodic lattice provides an obvious reference structure for the definition of point, line and planar defects. In amorphous materials it seems plausible that atomic motion can also occur more easily at certain special sites in the system, but the identification of these sites as "defects" is much less straightforward due to the difficulty in the definition of a defect-free reference state. The only unambiguous definition of such an "ideal" amorphous structure is the (metastable) *equilibrium* state of the amorphous solid at absolute zero temperature, which must have zero entropy and hence be fully configurationally ordered. This state is obviously not experimentally accessible. Although, as *Kauzmann* [5.1] has pointed out, the entropy difference between the amorphous (i.e., liquid at high temperatures) and crystalline state decreases considerably with decreasing temperature, the increasing sluggishness of the atomic rearrangements prevents the amorphous phase from reaching its metastable equilibrium state below some cooling rate-dependent temperature T_g (the glass transition temperature), thus forming a glass with a nonzero residual entropy. *Kauzmann* [5.1] also showed that an extrapolation of the equilibrium properties below T_g gave a vanishing amorphous-crystalline entropy difference at a nonzero temperature T_0. He postulated a "catastrophe" at this temperature, with the amorphous phase "spinodally" transforming to the crystalline one. This was based on the notion that full configurational order could only be obtained by crystal periodicity. However, since then it has become clear that there are a number of model structures that are fully amorphous (i.e., with no translational symmetry), but still have a high degree of configurational order. For example, the rigid coordination requirements of the tetrahedrally coordinated random network lead to a configuration entropy of less than $0.2\,k_B$ per atom (k_B: Boltzmann constant) [5.2]; an assembly of fully ordered amorphous clusters, containing several hundred atoms each [5.3, 4], would have a configurational entropy of the order of $10^{-2}\,k_B$; *Steinhardt* et al. [5.5] demonstrated the existence of bond orientational order in large amorphous clusters obtained by molecular dynamics simulation. It is conceivable, therefore, that further work along these lines could result in the construction of infinitely extendable, fully ordered amorphous models. Until this has been demonstrated, however, their use as "ideal" reference systems remains speculative.

The only remaining approach for model study of defects, therefore, is to make use of the amorphous models as they are presently available. Two alternative routes can be taken:

I) an existing model can be analyzed in the hope of identifying defects that have been "trapped" in it, or

II) one can attempt to introduce defects in an existing model and study their characteristic features, function, stability and motion.

Route I) has been taken by *Egami* et al. [5.6, 7] who analyzed the internal stresses in amorphous clusters and found regions where either hydrostatic pressure, hydrostatic tension or shear stress predominates. The identification of these regions as defects remains problematic, however, since it is not a priori clear which of these stressed regions would not be present in the (unknown) ideal reference structure. The same problem exists for the identification of the dilated and compressed regions with the free volume fluctuations in the theory of *Turnbull* and *Cohen* [5.8, 9], since this again requires knowledge of the zero-free volume reference state. The reference state problem is avoided when route II) is taken. Defects can be introduced in amorphous model systems either by deformation, e.g., of computer-built clusters [5.10], or two-dimensional bubble rafts [5.11], or by introducing point or line defects directly by appropriate removals and/or displacements of the atoms, e.g., in two-dimensional dynamic hard sphere systems [5.12] or in computer experiments. These last ones are the subject of the second part of this chapter where we shall report on attempts to introduce point defects in dense random packed (DRP) and tetrahedrally coordinated amorphous systems, and line defects in DRP systems using both static relaxation and molecular dynamics simulation. This is intended as a report of research in progress, not a full review of the field. Broad conclusions on the link with experimental evidence or on the significance of other modeling work would be premature.

5.2 Experimental Studies in Atomic Transport Properties

5.2.1 General Concepts

a) Types of Atomic Transport

The motion of atoms in a metallic glass can be observed in a number of different phenomena or experiments. In each of these cases, the atomic mobility is represented by a characteristic jump frequency, k. Four such cases will be considered here:

I) In a system with a shear viscosity η, each atom, when unbiased by a stress, performs a jump in a configuration that produces local shear at a frequency [5.13] $k_\eta = kT/\Omega\eta$, where Ω is the atomic volume.

II) In a system with a diffusivity D, each atom performs a diffusive jump of length λ_D with a frequency $k_D = D/\lambda_D^2 a_D$, where a_D is a constant that takes into account the dimensionality of the system and the correlation factor for the jump.

III) As will be discussed below, all metallic glasses undergo continuous structural relaxation. Most basically, this phenomenon can be described as an overall densification process. The fundamental jump frequency for this process can therefore be written as $k_v = (1/\Omega)d\Omega/dt$.

IV) Phase transformation processes in metallic glasses, such as crystallization, are characterized by the velocity u of the phase boundary. The corresponding characteristic jump frequency is then $k_c = u\lambda$, where λ is the interatomic distance.

An important part of the study of atomic transport is the development of an understanding of the relationships between these characteristic frequencies. For example, in liquids, where η and D are related by the Stockes-Einstein relation, k_η and k_D are the same – indicating that the same atomic configuration gives rise to both diffusional and shear-type rearrangements. On the other hand, if, for example, crystallization of an amorphous metal occurs by a fast interfacial reaction rather than being governed by the diffusion of impurities, the respective jump frequencies can be very different: $k_D \ll k_c$ [5.14].

b) Equilibrium and Isoconfigurational Properties

A liquid alloy is in stable equilibrium above the melting temperature T_M and in metastable equilibrium between T_M and the glass transition temperature T_g. In these states, the structural and thermodynamic properties of the system are well-defined and constant. Measurements of atomic transport, therefore, after possible initial transients, provide *equilibrium* values, i.e., are independent of time.

The glass transition temperature is defined as the highest temperature at which the system can no longer reach its equilibrium state within the time scale of the experiment. The value of T_g depends therefore on the nature of the experiment, but an often used convention is to choose the temperature where the viscosity is $10^{12}\,\mathrm{N s\,m^{-2}}$, corresponding to an atomic rearrangement frequency k_η of a few jumps an hour.

Below T_g, the system is not in equilibrium. Therefore, if there is any appreciable atomic mobility, it will *relax* towards its equilibrium state. As a result, its structure and thermodynamic properties, and hence also its transport properties, will change continuously until the equilibrium state is reached. If the temperature is sufficiently far below $T_g (T \lesssim T_g - 30\,\mathrm{K})$, this state cannot be reached within a practical timespan and the system will always remain in unstable equilibrium. The structural relaxation process is *irreversible*: the earlier configurational states cannot be reproduced as long as the system remains out of equilibrium.

Structural relaxation affects many of the physical properties of metallic glasses but for most of them the changes are small [5.15]. For example, the elastic moduli change on the order of 10% [5.16]; the features of the pair distribution function [5.7] and the Curie temperature of magnetic glasses [5.17] change on the order of a few percent; the electrical resistivity [5.18] and

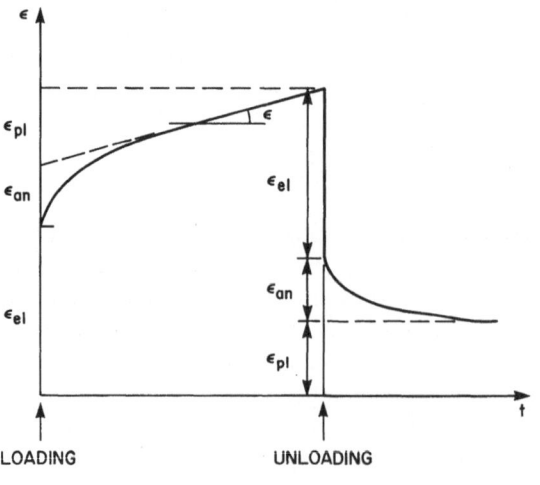

Fig. 5.1. Schematic creep curve, showing the elastic (ε_{el}), anelastic (ε_{an}) and plastic (ε_{pl}) contributions to the strain

the density [5.19] change only by a few tenths of a percent. On the other hand, atomic transport properties have been observed to change dramatically as a result of relaxation. For example, a five order of magnitude increase in viscosity has been reported in glassy $Pd_{82}Si_{18}$ [5.20].

Therefore, when measuring transport properties or comparing data, special care is required to take into account the structural state reached as a result of relaxation. For example, if transport properties are measured as a function of temperature, the only physically interpretable measurements are *isoconfigurational* ones. These can be performed if the relaxation is slow enough to ensure negligible structural changes during the time necessary for the measurements. Isoconfigurationality can be checked by cycling the temperature and reproducing the data. The changes in the transport properties under isoconfigurational conditions are therefore *reversible* ones, as opposed to those induced by structural relaxation.

5.2.2 Viscosity

a) Experimental Conditions and Techniques

The plastic (irreversible) deformation of metallic glasses occurs in two very different modes:

I) at high temperatures (around T_g) and at low stresses ($\tau < \mu/50$), they deform by *homogeneous* flow: every volume element of the specimen contributes to the strain;

II) at high stresses ($\tau \approx \mu/50$) and low temperatures ($T < 2/3\ T_g$), they deform by *inhomogeneous* flow: the deformation is localized in a small number of very thin shear bands.

Viscosity measurements can only be made in regime I); a meaningful viscosity can only be defined in the Newtonian viscous regime, i.e., at low enough stresses ($\tau < 10^{-2}\ \mu$). The usual technique is a tensile creep test: a load, corresponding to a tensile stress σ is hung from a wire or ribbon specimen and its elongation Δl, corresponding to a uniaxial strain $\varepsilon = \Delta l/l$, is measured as a function of time. Figure 5.1 is a schematic representation of such a creep curve. When the load is applied, the specimen elongates instantaneously by an elastic strain $\varepsilon_{el} = \sigma/E$ (E: Young's modulus). After an anelastic transient, the curve acquires a constant slope $\dot{\varepsilon}$, corresponding to the shear viscosity $\eta = \sigma/3\dot{\varepsilon}$. Extrapolation of this slope allows identification of the anelastic (ε_{an}) and plastic (ε_{pl}) components of the strain. This can be checked by unloading the specimen, whereupon the ε_{el} is recovered instantaneously and ε_{an} after a characteristic time; the remaining strain must be the irreversible plastic one, ε_{pl}. The shear viscosity is a measure of the rate of plastic deformation.

Although the creep test provides unique information about the plastic flow of materials, it is not the method of choice for studying the anelastic behavior, and certainly not for determining the elastic moduli. However, if sufficient care is taken, the elastic moduli and their temperature dependence can be obtained as a "by-product" of the creep tests; in $Pd_{82}Si_{18}$, for example, good agreement has been found with the results of ultrasonic techniques [5.21, 22]. Anelastic behavior is best studied by internal friction [5.23] but creep tests can provide useful additional information: in $Pd_{82}Si_{18}$ it was used to show stress-linearity of the anelastic response over a large stress range and in CuZr, a modified creep test involving strainrecovery after pre-straining and quenching was used to determine the activation energy spectrum for anelastic flow [5.24].

b) Equilibrium Measurements

Only a limited number of these measurements are available and none have been reported in the last nine years. The only data available for a glass-forming alloy melt are those from the viscometry measurements of *Polk* and *Turnbull* [5.25] on $Au_{77}Ge_9Si_{14}$. They exhibit a small activation energy, similar to most liquids at high temperatures. It may be useful to re-emphasize an important result of that investigation: the viscosity of an AuGeSi melt is *lower* than any reasonable extrapolation of the viscosity of pure Au to the same temperature; this argues against claims of extensive chemical clustering at these temperatures [5.26].

Around T_g, measurements have been reported for $Au_{77}Ge_9Si_{14}$ [5.27] and a number of Pd–Si-based glasses [5.28]. The apparent activation energy is very large (5–10 eV) and similar to that observed for crystallization processes in this temperature range [5.29]. Atomistically, this means that since the crystallization is diffusion-controlled, the local rearrangements controlling diffusion are similar to those governing flow. This has been confirmed by direct diffusion measurements [5.30] which showed the Stokes-Einstein relation to hold at this temperature (see also Sect. 5.1.3). *Chen* [5.31] has used this scaling relation to extract $\eta(T)$ from calorimetric studies of the crystallization process.

Experimentally, the equilibrium viscosity has a Fulcher-Vogel type temperature dependence: $\eta = A \exp[(B/(T - T_0))]$. It should be noted, however, that it is not possible to fit the entire temperature region $T_M \leftrightarrow T_g$ with a single set of $F - V$ parameters. Nevertheless, the behavior near T_g is theoretically rather well understood. The temperature dependence of the viscosity is dominated by that of the product $n_f k_f'$, where n_f is the concentration of sites that can produce local shear ("defects") and k_f' the stress-unbiased jump frequency of each defect [5.13]. Theories such as the free volume model [5.8, 9] or the configurational entropy model [5.32] can account for the $F - V$ dependence by the temperature dependence of the defect concentration n_f only. This implies that the jump frequency k_f' is probably only weakly temperature dependent.

c) Isoconfigurational Measurements

Early creep measurements [5.33, 34] did not take the structural relaxation phenomenon sufficiently into account. As a result, the structure of the specimens tested at higher temperatures was probably more relaxed than those

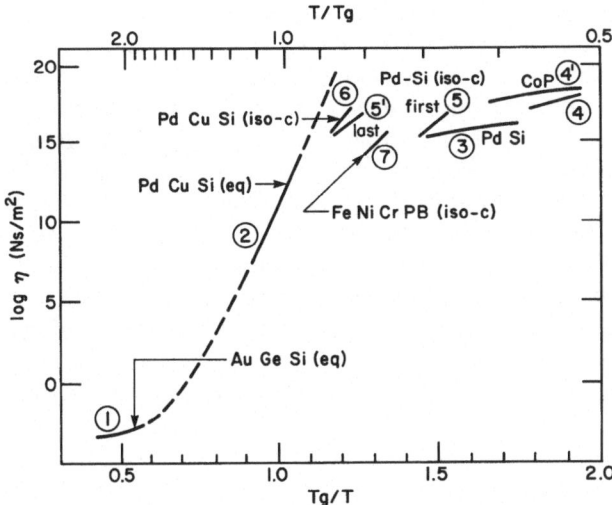

Fig. 5.2. Shear viscosity of a number of amorphous metals as a function of normalized inverse temperature

① $Au_{77}Si_{14}Ge_9$ melt (equilibrium) [5.25]
② $Pd_{77.5}Cu_6Si_{16.5}$ equilibrium [5.28]
③ $Pd_{80}Si_{20}$ not stabilized [5.34]
④ $Co_{75}P_{25}$ not stabilized [5.33]
④' $Co_{75}P_{25}$ not stabilized; rerun [5.33]

⑤ $Pd_{82}Si_{18}$ isoconfigurational; first measured state of annealing [5.20]
⑤' $Pd_{82}Si_{18}$ isoconfigurational; last measured state of annealing [5.20]
⑥ $Pd_{77.5}Cu_6Si_{16.5}$ isoconfigurational [5.36]
⑦ $Ni_{36}Fe_{32}Cr_{14}P_{12}B_6$ isoconfigurational [5.37]

of the specimens tested at lower temperatures. The viscosities at high temperatures are therefore too high, resulting in an artificially low apparent activation energy.

In order to perform isoconfigurational measurements it is necessary to let the specimens relax until the relative change in viscosity $[d(\ln\eta)/dt]$ is low enough to permit a number of measurements at different temperatures. The first isoconfigurational measurements were performed on $Pd_{82}Si_{18}$ [5.13, 35] and the results showed an Arrhenius-type temperature dependence with an activation energy of 2 eV. Figure 5.2 shows the isoconfigurational viscosities for the first and last of a series of structural states obtained by consecutive anneals on the same specimen, illustrating the five orders of magnitude change in viscosity due to structural relaxation. Isoconfigurational measurements have been reported also for $Pd_{77.5}Cu_6Si_{16.5}$ [5.36] and on $Ni_{36}Fe_{32}Cr_{14}P_{12}B_6$ [5.37], with activation energies of 2.5 eV and 2.6 eV, respectively. It should be noted that the data on the latter alloy are for a specific stress (9×10^8 Nm^{-2}), since the flow is not Newtonian viscous. Recent measurements on $Pd_{82}Si_{18}$ over a more extended temperature range have a non-Arrhenian temperature dependence, with the apparent activation energy at the higher temperatures being higher [5.38].

A theoretical understanding of this large activation energy has not been developed yet. The simplest interpretation of the isoconfigurational flow process simply assumed that, since the structure remained constant, n_f was constant and that hence the activation energy reflected the temperature dependence of k_f' [5.29]. However, a strong Arrhenius-type temperature dependence of k_f' would be inconsistent with the $F - V$ behavior of the equilibrium viscosity, as discussed above (Sect. 5.1.2). The pre-exponential is also much too small to be consistent with the simplest interpretation of a simply activated process for k_f' [5.38]. A new interpretation must therefore be found. In general, it can be said that the defect concentration n_f must change *reversibly* with temperature under isoconfigurational experimental conditions. One possible mechanism for this would be small, local reversible atomic position adjustments that redistribute the free volume change due to thermal expansion below T_g.

d) Stress-Dependence of the Flow Rate

The stress exponent m is defined from the stress (τ)-strain rate (γ) relation $\gamma \propto \tau^m$ or $m = \partial(\log\gamma)/\partial(\log\tau)$. In Newtonian viscous flow $m = 1$, and this has been observed for a number of systems [5.27, 28, 33, 35]. However, much larger values, up to $m = 10$, have also been reported [5.39]. This apparent contradiction has been resolved by *Taub* [5.40] who performed creep tests on $Pd_{82}Si_{18}$ over a large stress range on a well-stabilized specimen and observed an increasing deviation from linearity starting around 3×10^8 N m^{-2}.

This had been expected from theoretical models of the homogeneous flow in these materials [5.41–43]. Simple transition state theory gives for the relationship [5.13, 44]

$$\dot{\gamma} \propto \sinh(\tau\gamma_0 v_0/kT), \tag{5.1}$$

where v_0 and γ_0 are, respectively, the volume and the local shear strain of the defect governing flow. At low stresses this relationship is linear; at higher stresses it becomes exponential. A fit of (5.1) to *Taub's* data [5.40] gave $\gamma_0 v_0 = 48$ Å^3, or a little more than three times the atomic volume. Relation (5.1) also implies that

$$m = (\tau\gamma_0 v_0/kT)\coth(\tau\gamma_0 v_0/kT). \tag{5.2}$$

This means that m should depend on the stress range of the tests. Most of the reported values could be brought in agreement with (5.2) using a single value of $\gamma_0 v_0$. More recent determinations of $\gamma_0 v_0$ for Fe-based glasses gave even smaller values ($\sim$one atomic volume) [5.45].

The value of $\gamma_0 v_0$ can be interpreted in two ways: the flow defect consists either of a small number of atoms going through an extensive arrangement ($\gamma_0 = 1$), as in the free volume model, or it consists of a large number of atoms undergoing a small amount of shear [5.33]. An atomic-size dislocation ($b \approx \lambda$; λ: interatomic distance; b: Burgers vector) seems less appropriate since in this case, $\gamma_0 \approx v_0/\lambda \approx 1$ and $v_0 \approx \lambda^2 l$ (l: dislocation length), resulting in a dislocation length of only 3λ.

5.2.3 Diffusion

a) Experimental Techniques

Direct measurements of the diffusivity in glassy metals are difficult because they have to be performed below the temperature T_{kc} where rapid crystallization occurs, and for a time long enough to observe a change in the compositional profile. Usually $T_{kc} \approx T_g$, and at this latter temperature $D \approx 10^{-20} \, \mathrm{m^2 \, s^{-1}}$. Special techniques are therefore required.

One type of approach is micro-sectioning by sputtering. This was used by *Gupta* et al. [5.46] to determine the profile of implanted ^{110}Ag tracer atoms in $Pd_{81}Si_{19}$. *Cahn* et al. [5.47] used a similar technique (SIMS) to measure the concentration profile of ^{11}B in $Fe_{40}Ni_{40}B_{20}$ after sputter-deposition and annealing.

A different set of techniques is based on the use of nuclear particles to probe the concentration profile. *Chen* et al. [5.30] used Rutherford backscattering of α-particles to determine the profile of ion-implanted Au in $Pd_{77.5}Cu_6Si_{16.5}$. *Kijek* et al. [5.48] used the (p, α) nuclear reaction of ^{11}B to measure its concentration profile in Ni–Nb and Ni–Zr glasses by high energy proton irradiation. *Birac* and *Lesueur* [5.49] used the (n, α) reaction of ^{6}Li to measure its diffusivity in $Pd_{80}Si_{20}$ by neutron irradiation.

A third type of experiment, originated by *Cook* and *Hilliard* [5.50], for crystalline materials allows measurements of the interdiffusivity as low as $10^{-27} \, \mathrm{m^2 \, s^{-1}}$. *Rosenblum* et al. [5.51] and *Greer* et al. [5.52] have recently adapted it to amorphous metals. Using dual-target sputter deposition they prepared multi-layer compositionally modulated films (wavelength $\lambda \approx 25$ Å) consisting of alternating Pd-based and Fe-based amorphous metal layers. By monitoring the decrease in intensity of the low-angle x-ray scattering satellite as a function of annealing time, the interdiffusivity could be measured continuously as a function ot time. The small diffusion distances, corresponding to $\lambda \approx 25$ Å, allowed measurements of $\tilde{D}$ down to $10^{-26} \, \mathrm{m^2 \, s^{-1}}$.

b) Diffusivity of the "Smaller" Atoms

Some of the results of these diffusivity measurements are summarized in Fig. 5.3. *Kijek* et al. [5.53] have pointed out that it is useful to distinguish two general categories, depending on the diffusing atom species: "small" atoms (Li, small metalloids: B) seem to diffuse considerably faster (two orders of magnitude) than "large" atoms (metals: Au, Ag, Pd. Fe or large metalloids: P).

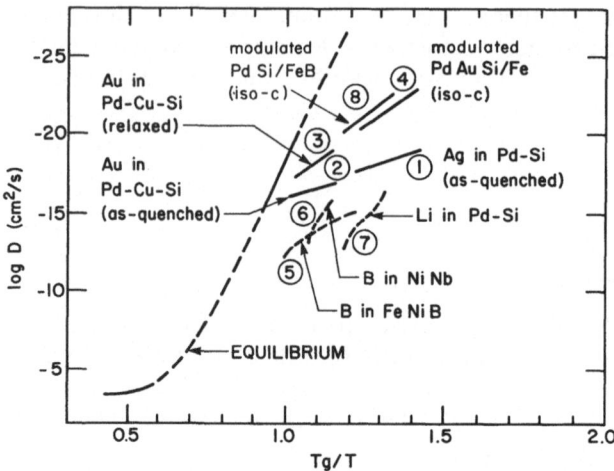

Fig. 5.3. Selected diffusivity data for amorphous metals. (——) "large" atom diffusivities; (---) "small" atom diffusivities. The equilibrium curve has been shown by scaling the equilibrium curve of Fig. 5.2 according to the Stokes-Einstein relation

① Ag in $Pd_{81}Si_{19}$; not stabilized [5.46]
② Au in $Pd_{77.5}Cu_6Si_{16.5}$; not stabilized [5.30]
③ Au in $Pd_{77.5}Cu_6Si_{16.5}$; annealed [5.30]
④ Interdiffusivity in a $(Pd_{80}Au_7Si_{13})_{70}/Fe_{30}$ modulated film; isoconfigurational [5.51]

⑤ B in $Fe_{40}Ni_{40}B_{20}$ [5.47]
⑥ B in $Ni_{59.5}Nb_{40.5}$ [5.48]
⑦ Li in $Pd_{80}Si_{20}$ [5.49]
⑧ Interdiffusity in a $(Pd_{85}Si_{15})_{50}/Fe_{85}B_{15})_{50}$ modulated film; isoconfigurational [5.52]

No clear Arrhenius-dependence has been observed for the first category, although in the case of B, the direct measurements seem to agree with the extrapolation of the Arrhenius plot of the diffusivities derived from crystallization experiments in Fe–B–C glasses [5.54].

It was also observed that the diffusivity of B in the Ni–Nb alloy did not change upon structural relaxation by annealing. It is not known whether this is a general feature of all the diffusivities in this category.

Kijek et al. [5.53] suggest that these small atoms diffuse interstitially through the skeleton formed by the large metal atoms. Such a mechanism is reminiscent of some of the original structural ideas about the structure of metallic glasses [5.55] and it seems plausible that it would give a rather high diffusion rate. Interstitial diffusion would probably also not be affected as much by structural relaxation of the skeleton as the diffusivity of the metal atoms themselves.

c) Diffusivity of the "Large" Atoms

The first measurement in this category was that of the diffusivity of Ag in $Pd_{82}Si_{18}$ [5.46]. Since structural relaxation had not been taken into account, the relatively low apparent activation energy (1.3 eV) is not physically meaningful. This was confirmed by later measurements on Au in $Pd_{77.5}Cu_6Si_{16.5}$ [5.30].

In as-quenched samples, the apparent activation energy was even lower (0.75 eV), but preannealing of the samples around T_g resulted in a decrease of the diffusivity by two orders of magnitude and an increase in the activation energy to 1.67 eV. This last set of measurements was probably close to isoconfigurational. This became clear from the work of *Rosenblum* et al. [5.51] on the interdiffusivity in Pd/Fe-based modulated films, since in this type of experiment the diffusivity can be monitored continuously. A continuous decrease of D, due to structural relaxation, was observed and the conditions for an isoconfigurational experiment could be checked specifically. The results were in agreement with the extrapolation of the Pd–Cu–Si(Au) data and exhibited an activation energy of 1.73 eV.

Since the rearrangements required for producing local shear must involve the metal skeleton of the metallic glass, one would expect the viscosity of these materials to be related to the diffusivities in this later category. The similarity between the two processes will be discussed in detail below (Sect. 5.1.5).

d) Scaling of Viscosity and Diffusivity

It is straightforward to show [5.13] that viscosity and diffusivity are related by

$$\eta D = kT/3\pi L, \tag{5.3}$$

where L is a characteristic length that depends on the concentration ratio, jump frequency ratio and the relative dimensions of the respective defects governing the two processes. When $L = \lambda$, (5.3) is the Stokes-Einstein (S-E) relation, which is known to apply to liquids at higher temperatures. $L = \lambda$ implies that a simple rearrangement involving only a few atoms governs both diffusion and flow.

Chen et al. [5.30], in their investigation of the diffusion of Au in Pd–Cu–Si, have observed that the S-E relation is obeyed also around T_g where the system is in equilibrium. Below T_g, this scaling has not yet been entirely established since it is difficult to compare experiments on different samples due to the different degrees of structural relaxation. However, recent diffusivity measurements in Pd–Si/Fe–B compositionally modulated films [5.52] seem to support the validity of the S-E relation below T_g. Since the results of this type of diffusion experiment are truly isoconfigurational, they can be compared to isoconfigurational viscosity data on Pd-based glasses in the same temperature range [5.20]. It is, therefore, significant that the activation energies measured in these two sets of experiments are the same (2.0 eV) and that the absolute values of η and D for similar annealing treatments scale according to the S-E relation. The isothermal relaxation rate of the inverse diffusivity $[d(1/D)/dt]$ with annealing is *constant* in time, similar to the viscosity relaxation rate $[d\eta/dt]$ (Sect. 5.1.4a), and at the same temperature the two ratios also scale according to the S-E relation. Since these relaxation ratios have been shown to be independent of thermal history [5.38], this last observation provides the clearest support for the validity of the S-E relation below T_g.

5.2.4 Structural Relaxation

a) Kinetics of the Viscosity Increase

The viscosity of a metallic glass increases *linearly* with time during structural relaxation below T_g. This was first demonstrated for $Pd_{82}Si_{18}$ (Fig. 5.4) [5.20] and later confirmed for $Pd_{77.5}Cu_6Si_{16.5}$ [5.20, 38] and some of the Fe-based glasses [5.56, 57]. As a result, the relative viscosity increase $[d(\ln\eta)/dt]$ decreases with time which allows isoconfigurational measurements to be performed.

The atomistic basis for structural relaxation in these materials is not fully understood. However, the most straightforward model is based on simple second-order reaction kinetics [5.58]. Since $\eta \propto n_f^{-1}$, a viscosity increase must correspond to a decrease in defect concentration. The rate of disappearance of the defects can be taken proportional to n_f itself and to the concentration of sites n_r, where such a defect can be annihilated (e.g., by collapsing out a free volume fluctuation at an appropriate stress concentration): $\dot{n}_f \propto n_f n_r$. If the concentrations of both these defects depend in the same way on the structural order of the glass such that $n_f \propto n_r$, then $\dot{n}_f \propto n_f^2$, which integrates to $n_f^{-1}(t) - n_f^{-1}(0) \propto t - t_0$, or $\eta(t) - \eta(0) \propto t - t_0$, which is the experimentally observed relation.

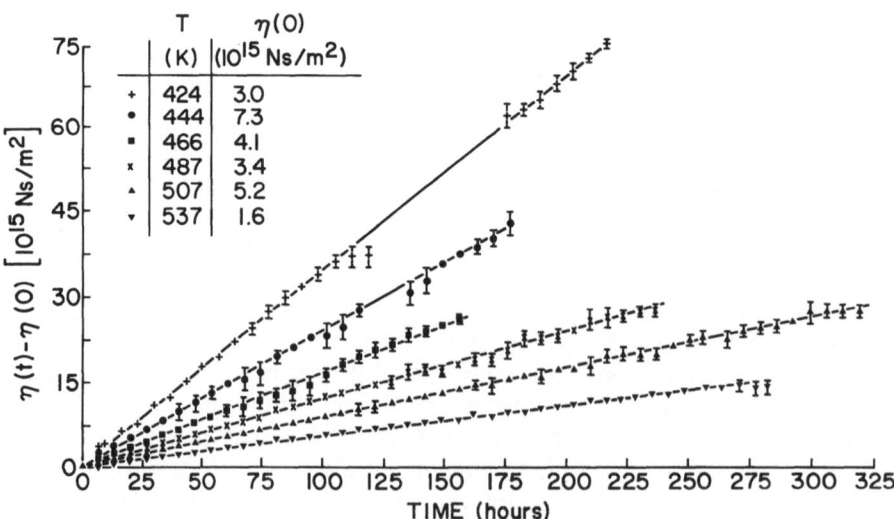

Fig. 5.4. Linear viscosity increase observed during successive isothermal annealing of amorphous $Pd_{82}Si_{18}$ at various temperatures. The initial viscosites (0) for each annealing temperature are tabulated [5.20]

b) Kinetics of the Diffusivity Decrease

As pointed out above (Sect. 5.1.2a), the diffusivity decreases inversely pro-portionally with time [5.52] with a rate that scales with the viscosity increase according to the S-E relation.

The atomic rearrangements governing the diffusivity decrease are probably quite different from those governing the diffusion process itself. In fact, it would not be possible to perform isoconfigurational experiments at all if the relaxation process was not considerably slower. This fact is also illustrated by the work of *Chen* et al. [5.20] on Au in Pd–Cu–Si (Fig. 5.3). In order to observe broadening of the composition profile, it is necessary that each atom performs at least a few hundred jumps. It is clear from Fig. 5.3 that all the measured diffusivities are still at least two orders of magnitude larger than the equilibrium diffusivity at the same temperature. This means that a few hundred atomic jumps are not sufficient to reach equilibrium. Atomistically, this could be understood if the concentration of sites for structural relaxation n_r were much less than the concentration of defects where diffusion can occur, n_D.

c) Relationship Between the Effects of Structural Relaxation on Different Properties

As pointed out in the introduction, structural relaxation affects many physical properties of metallic glasses, but so far very few links between these various property changes have been established. In one such recent attempt [5.58], the increase in the ferromagnetic Curie temperature T_c of binary metallic glasses such as $Fe_{80}B_{20}$ as a result of isothermal annealing has been linked to the viscosity increase expected during such an anneal. In a simple glass such as $Fe_{80}B_{20}$, there is good reason to believe that the increase in T_c is of topological origin. Using the free volume model for flow, the free volume decrease corresponding to the viscosity increase can be estimated. Since it is known that the nearest-neighbor distance changes very little during annealing (x-rays, sign of T_c change), this decrease in atomic volume corresponds to an increase in the average coordination member, and hence a proportional increase in T_c. Quantitative agreement between the changes in the two properties has been obtained. A similar attempt has been made to link the resistivity and viscosity changes upon annealing [5.59].

5.2.5 Similarity Between the Defects Governing Diffusion and Flow

In conclusion, it may be useful to summarize the observations concerning the similarity between the microscopic mechanisms governing atomic diffusion and viscous flow as follows:

i) the local volume strain $\gamma_0 v_0$ of the flow defects is small; since the defect governing flow requires only a nearest neighbor switch, it is presumably a small "point" defect; an extended flow defect (large $\gamma_0 v_0$) would obviously be very different, but a localized flow defect (small $\gamma_0 v_0$) can be similar.

ii) for $T \gtrsim T_g$, D, and η scale according to the Stokes-Einstein relation, which implies that the same defects govern both processes in this regime.

iii) for $T < T_g$, the isoconfigurational activation energy for η and D is the same in similar systems.

iv) the relaxation rates of η and $(1/D)$ are both constant in time and scale according to the Stokes-Einstein relation; if flow and diffusion were governed by qualitatively different defects, the annihilation rates of those defects would also be different.

5.3 Computer Simulation Studies of Defects and Atomic Transport

5.3.1 General Concepts

Simulations on a computer can be broadly classified as static or dynamic. This differentiation is based on the mode of relaxation of the atoms which interact via a prescribed potential and range. If the relaxation mode allows for thermal fluctuations, we shall call it dynamic relaxation. The extent of relaxation is limited by the amount of computer time available. This limitation generally favors static relaxation over dynamic if the number of particles is large. We have carried out both static and dynamic relaxations. Our approach to the problem of transport in amorphous solids is the following. We first examine the stability of those defects that are known to transport atoms in crystalline solids. These are point and line defects. Having established this basic issue, we then proceed to simulate atomic transport on a computer. (We will not be reviewing all aspects of defects in amorphous solids. In particular, we shall not present a discussion of the role of defects in describing or stabilizing the structure of amorphous solids. A brief discussion of this important issue has been given by *Chaudhari* [5.60].)

An ideal method for testing new notions concerning amorphous solids whether they be metallic or covalently bonded glasses is computer simulation. Computer-built models can be studied on both a microscopic (e.g., atomic position) and a macroscopic scale (e.g., temperature, stress field, etc.) simultaneously with absolute precision. In addition, experiments that cannot be performed on real glass systems in the laboratory can be performed on the computer models, e.g., removing a particular set of atoms and welding the material together again. All these properties of computer-built models have been especially useful in studying defects in amorphous systems.

On the other hand, any discussion of computer modelling must include the warning that computer-built models may not reflect the properties of real materials. The computer-built models have been built by a combination of mechanical and computer techniques which, at best, vaguely resemble the real material counterparts. In the case of metallic glasses, monatomic computer models of glasses have been used though no atomic transport data have been detailed on monatomic metallic glasses. One must always ask the question: in what way do both the computer model and/or the computer experiment reflect real laboratory conditions? For example, does the small size of the computer model affect the results? Does the method of relaxation sample all of phase space, given the time limitation and the method of relaxation? The ultimate test of the effectiveness of the computer experiment is whether from the study of computer models, real laboratory experiments can be suggested that will support the computer results. The computer results that will be presented are highly intriguing and suggestive, leading one to suggest various means of experimental verification. At the present time, only limited experimental support exists.

5.3.2 Computer-Built Models

For the computer investigations, models of both metallic glasses and of covalently bonded glasses were used. For almost all the studies, systems with only one atomic species were tested even though such systems cannot be produced in the laboratory. Various choices for the interatomic potential could have been employed but the studies were restricted to a Lennard-Jones potential for the metallic solids and a Keating potential for the covalently bonded random network models [5.61–63].

For the metallic solids, two kinds of models were studied which will be referred to as the Finney model and the Rahman model, even though the models are only derivatives of the models originally devised by *Finney* [5.64] and *Rahman* [5.65]. In both cases, the form of the Lennard-Jones potential utilized in the relaxation procedure was

$$U(r) = 4\varepsilon[(\sigma/r)^{12} - (\sigma/r)^6],$$

where the parameters for argon were used ($\sigma = 3.4$ Å, $\varepsilon/K = 120$ K, $m = 40$ a.m.u.). Actually, the full potential was not used; instead, the long-range potential was truncated at either 1.2 or 2.1 times the ideal atomic distance, depending on the particular model. Various smooth truncation procedures were employed to ensure that an artificial discontinuity was not introduced.

All of the static relaxations and some of the molecular dynamics relaxations were derived from the model originally constructed by Finney. The original model had approximately 8000 atoms and various subsets of it were relaxed under the Lennard-Jones potential. All the relaxations were carried out using the conjugate gradient technique. In order to verify our procedures and also to provide a reference, parallel computations were also carried out on a model of a face-centered-cubic crystal relaxed under the same potential.

For molecular dynamics studies of metallic glasses, the model originally constructed by Rahman and similarly constructed models were employed. The models were obtained by supercooling an equilibrium liquid which was equilibrated under periodic boundary conditions (see [5.66] for a review of dynamic relaxation techniques).

To study covalently bonded models, a 519-atom with tetrahedral coordination was utilized that was originally developed by *Polk* [5.67] and subsequently relaxed under a Keating potential by *Steinhardt* et al. [5.68]. The Keating potential was of the form [5.69]

$$V = \tfrac{3}{16}(\alpha/d^2)\sum_{l,i}(r_{li} \cdot r_{li} - d_2)^2 + \tfrac{3}{8}(\beta/d^2)\sum_{l(i,i')}(r_{li} \cdot r_{li'} + \tfrac{1}{3}d^2)^2 \, ,$$

where the sum is over all atoms with four neighbors i and with parameters chosen that are appropriate to amorphous silicon.

The models that have been relaxed using a Lennard-Jones potential have been tested for their elastic properties. It is important to understand these properties in order to be able to reliably relate the results of the computer investigations to real materials. The models have been shown to behave as elastic solids for small displacements. The upper plane of a computer-built solid was displaced with respect to the lower plane and held fixed as the model was relaxed. The energy of the model after relaxation was measured, corrected for changes in numbers of neighbors and plotted against the displacement. The results show (Fig. 5.5) that the energy difference rises with the square of the displacement for small displacements, as expected for an ideal solid. For

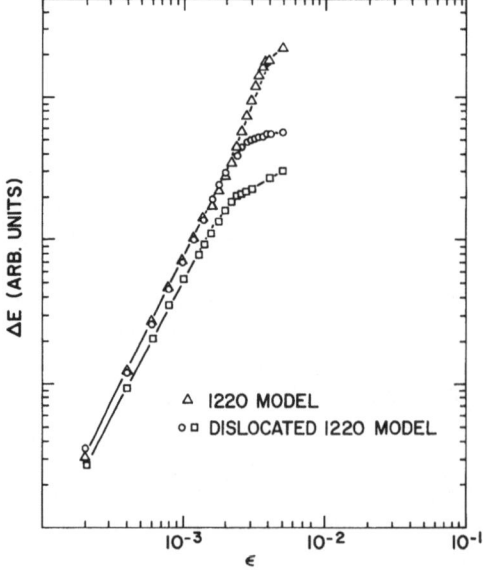

Fig. 5.5. Change in energy as a function of strain for a relaxed model ($\triangle$) and a dislocated model ($\square$, $\bigcirc$)

displacements of greater than 2% of an atomic diameter, there was a deviation from Hooke's law. For displacements of up to 5–10% of an atomic diameter the behavior was nonlinear, but elastic, i.e., once the bottom half of the solid was released, the elastic energy of the solid returned to its original value. Figure 5.5 also shows the results for a model with an edge dislocation introduced into the solid and then statically relaxed. Once again the solid obeys Hooke's law out to rather large displacements.

5.3.3 Point Defects [5.61]

a) Definition of a Point Defect in Glasses

In a crystalline solid a point defect (a vacancy or an interstitial) can be readily identified because it represents a deviation from the perfectly periodic lattice. A vacancy, in particular, maintains its local identity as a missing atom by being located at or near a site that is part of the periodic lattice (Fig. 5.6a). Once introduced into the lattice, the vacancies in a crystal anneal out very slowly. They can only disappear by having an atom jump into the vacant position; consequently, the vacancy moves one position (Fig. 5.6b). If this process is repeated often enough, the vacancy reaches a sink, for example, a free surface, an edge dislocation, a grain boundary, etc. However, at low temperatures, the probability of a jump is very small and, on the scale of times found in computer simulations [5.66], the vacancy remains forever.

In an amorphous solid the absence of a lattice makes it impossible to identify a vacancy site with an atomic site and the absence of symmetry makes it difficult to predict whether the vacancy will remain a localized unit upon relaxation. To test these notions, various clusters of atoms were removed from computer-built models and the models were allowed to relax. Two methods of tracking the vacant volume were used. In the first approach the model was sectioned into slices and the time evolution of the vacant volume was studied.

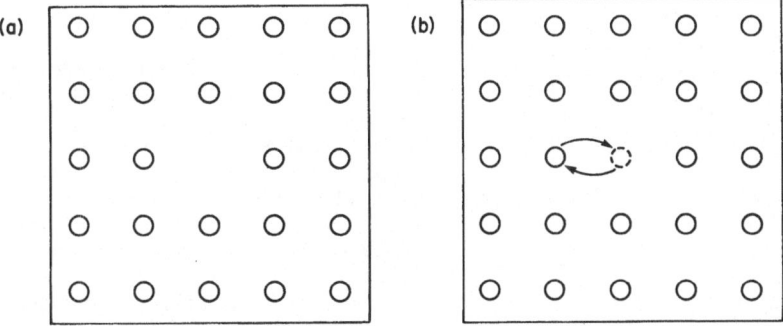

Fig. 5.6. (a) Vacancy in a crystal. **(b)** Transport of vacancy by motion of neighboring atom

In this approach the motion of the atoms surrounding the vacant volume could also be tracked. In the second approach, the volume of space between the atoms was measured. The volume containing the atoms was divided into equal test sections (with a volume small compared to the volume of single atom) and the "nearest-atom-distance", the distance from the center of the test section to the center of the nearest atom, was computed for each test section. A localized vacant volume manifests itself by a conglomeration of test sections with nearest-atom distances greater than a "Bernal length" (approximately 1.4 times the atomic radius). An integral over the volumes of such test sections yields a rough measure of the vacant volume. If the volume becomes delocalized, the vacant volume test sections with large nearest-atom distances disappear, and test sections with nearest-atom distances near the Bernal length become more numerous. The second approach is less subjective than the first and permits one to follow the annealing out of a vacancy and vacancy clusters with time in a molecular dynamics study.

b) Static Relaxation

The study of vacancies was first carried out for systems under static relaxation. The periodic boundary conditions of the 500-atom Rahman model were lifted and the model was relaxed at zero temperature under a Lennard-Jones potential by conjugate gradient methods. Three atoms were removed from the central part of the model to form a vacancy cluster and the remaining atoms were relaxed again. A plot comparing the positions of the remaining atoms with the positions of the atoms after the final relaxation is shown in Fig. 5.7. The vectors in the figure represent the displacement of atoms whose center, in either the initial or the final configuration, lies within a slice, of width equal to one atomic diameter, through the middle of the model. Solid circles and vectors represent atoms whose centers are in the middle slice both initially and finally.

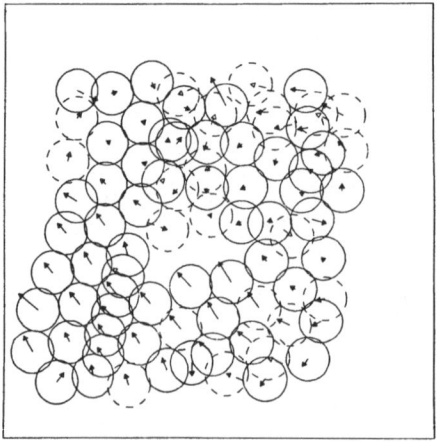

Fig. 5.7. A slice, with width equal to one atom diameter, through the Rahman model. In this figure the positions of the atoms before and after the introduction of three vacancies can be compared. Solid circles and vectors represent atoms whose centers are in the slice both initially and finally. The initial position is indicated by the circle. Dotted circles and/or vectors represent atoms whose centers either left or entered the middle slice, depending on whether the center of the circle lies at the foot or head of the vector

Dotted circles and/or vectors represent atoms whose centers either left or entered the middle slice depending upon whether the center of the circle lies at the foot or head of the vector. It is evident that the vacant volume has diminished and has been filled by a shuffling motion of a number of atoms. Not all atoms move uniformly towards the vacant volume. For the case shown, the bottom central right-hand group of atoms moved in to fill the vacant volume while the bottom left hand group moved towards the surface. The fraction of vacant volume left after considerable relaxation was found to be approximately 10%.

A second experiment with static relaxation was performed for covalently bonded systems using the Polk model. Unlike the previous case, a vacancy cluster formed by the removal of three atoms from the center of the continuous random network model did not disappear under static relaxation. The topological constraints imposed by the covalent bonding appear to be sufficient to stabilize the local vacant volume.

Several types of relaxations were made. First, three atoms were removed from the center of the 519-atom Polk model and the atoms were relaxed under the Keating potential. The plot in Fig. 5.8a shows the comparison in atomic positions before and after relaxation. Clearly, little movement has taken place and the vacancy is still present. Secondly, the bending force was set equal to nearly zero and the model was relaxed again. The resultant motion is shown in Fig. 5.8b. Both these results indicate that the combination of topological constraints – even without a bending force – plus the fact that no attempt is made to satisfy the dangling bonds in the center of the model, are sufficient to lead to stabilization of the vacancy cluster. It is not clear, however, to what degree this simulation reflects any experiment that could be done in the laboratory. A third test was to restore the bending force and relax again, but this time the dangling bonds were tied together in such a way as to insure that no rings of less than 5 atoms were formed by the tying process. Figure 5.8c shows that much more movement has taken place and the vacant volume is diminished; nevertheless, most of the vacant volume remains. A fourth experiment in which the dangling bonds are tied together and the bending force is set to nearly zero is shown in Fig. 5.8d. Little change from Fig. 5.8c was found.

As the change in the bond-bending forces makes only a small change in the positions of the atoms around the vacant volume, one can conclude that the topological constraints are more important in stabilizing vacancies in random network models. To test this hypothesis, another relaxation was performed in which the range of the interactions were increased to distances of 2.5 times the nearest-neighbor distance. The bond-bending force was kept at zero and the Lennard-Jones potential was used to relax the model. Under these conditions the network collapses and the resultant atomic packing is more characteristic of the metallic packing rather than the covalent packing. The number of nearest neighbors changes from a nominal four to eight. The interference functions of the original and collapsed network are shown in Fig. 5.9. In amorphous solids

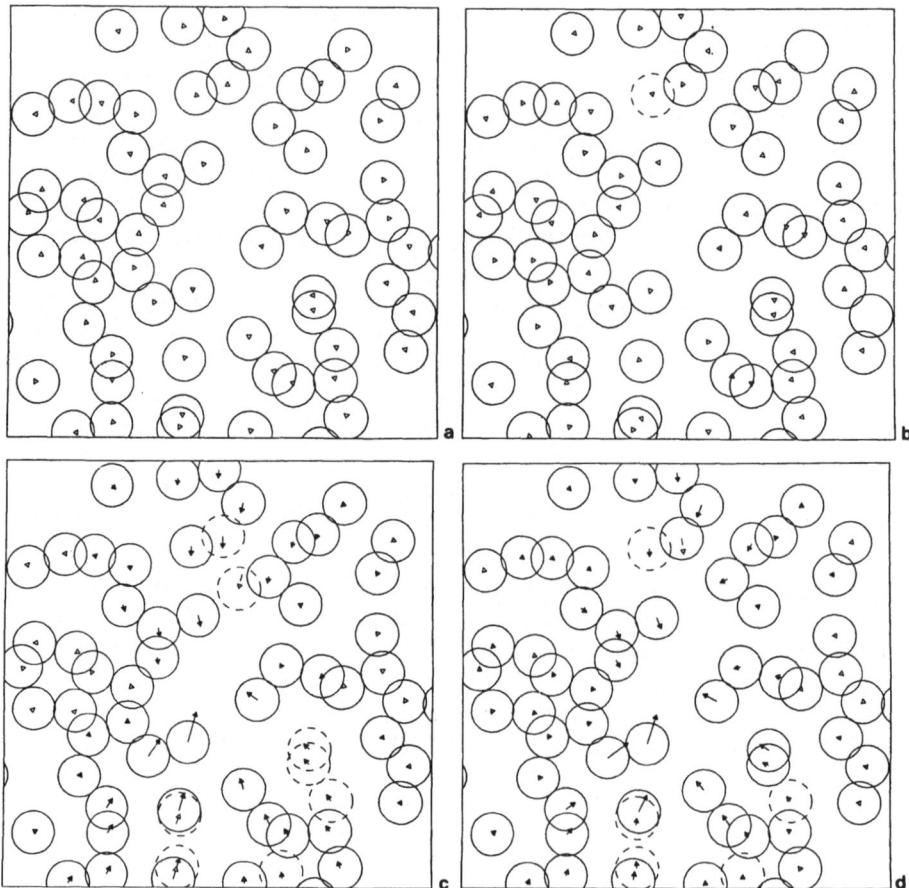

Fig. 5.8a–d. Plots showing initial and final positions of the atoms in a random network model with a three-vacancy cluster as a function of the relaxation conditions. (**a**) The positions after relaxation; (**b**) relaxation under the condition that bond-bending forces go to zero; (**c**) relaxation after the broken bonds have been tied together; (**d**) broken bonds tied together and bond-bending forces set to zero

such as Si or Ge, such a network collapse must indeed occur as the substances form not only covalently bonded glasses but also metallic liquids.

c) Dynamic Relaxation

A study of the time dependence of the annealing process for vacancies in a metallic glass was performed by molecular dynamics analysis of the Rahman model with periodic boundary conditions imposed. From one to five atoms were removed from the center of the 500-atom model. The atoms were confined in a periodic box with a size corresponding to a density of 1.589 and

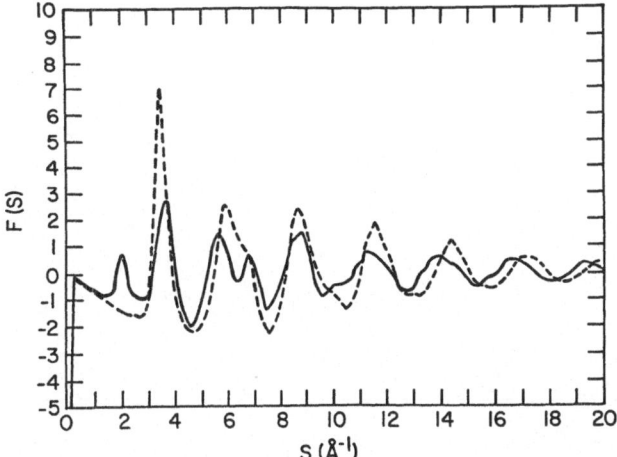

Fig. 5.9. Interference function of the random network before and after relaxation using a Lennard-Jones potential and no bending force. (———) initial random network model. (– – –) bending model relaxed with a Lennard-Jones potential

$1.602\ \text{g cm}^{-2}$, depending on the number of atoms that were removed. At low temperatures this density yields a negative pressure that would have stabilized a sufficiently large void. Nevertheless, in all cases that were examined, the vacant volume annealed out within 100 time units or less where a time unit corresponds to 2.2×10^{-12} s. For comparison, 100 time units corresponds to 10^{-15} jumps of a vacancy in a crystalline solid relaxed under similar conditions at a temperature of 12 K. Test section slices of the model containing five vacancies at time zero, an intermediate time and the final time are shown in Figs. 5.10a–c. The atoms or sections of atoms are indicated by solid circles. In Figs. 5.11a, b, plots comparing initial and final positions of atoms for the three vacancy case are shown where the same convention as in Fig. 5.7 has been employed. In the case of three vacancies, the vacant volume appears to be filled by a combination of general shuffling of surrounding atoms and certain large moves or jumps of individual atoms. In the case of the five-atom cluster, jumping of individual atoms into the vacant volume appears to dominate the annealing process and annealing is more complex. In both cases, certain pairs of atoms appear to interchange positions or rotate about their relative center of mass. Observations of such events are indicative of atomic diffusion during annealing. Histograms of the number of test sections found inside the model vs their nearest-atom distances are shown for early, intermediate and final times in Fig. 5.12a for the three-atom case and in Fig. 5.12b for the five-vacancy case. In both cases the number of test sections with large nearest-atom distances decreases with time. In Figs. 5.13a, b, the fractional vacant volume was found by integrating over the volumes of test sections with nearest-atom distances greater than 1.4 atomic radii. Once the vacant volume begins to decrease, one observes a combination of decay corresponding to large groups of atoms

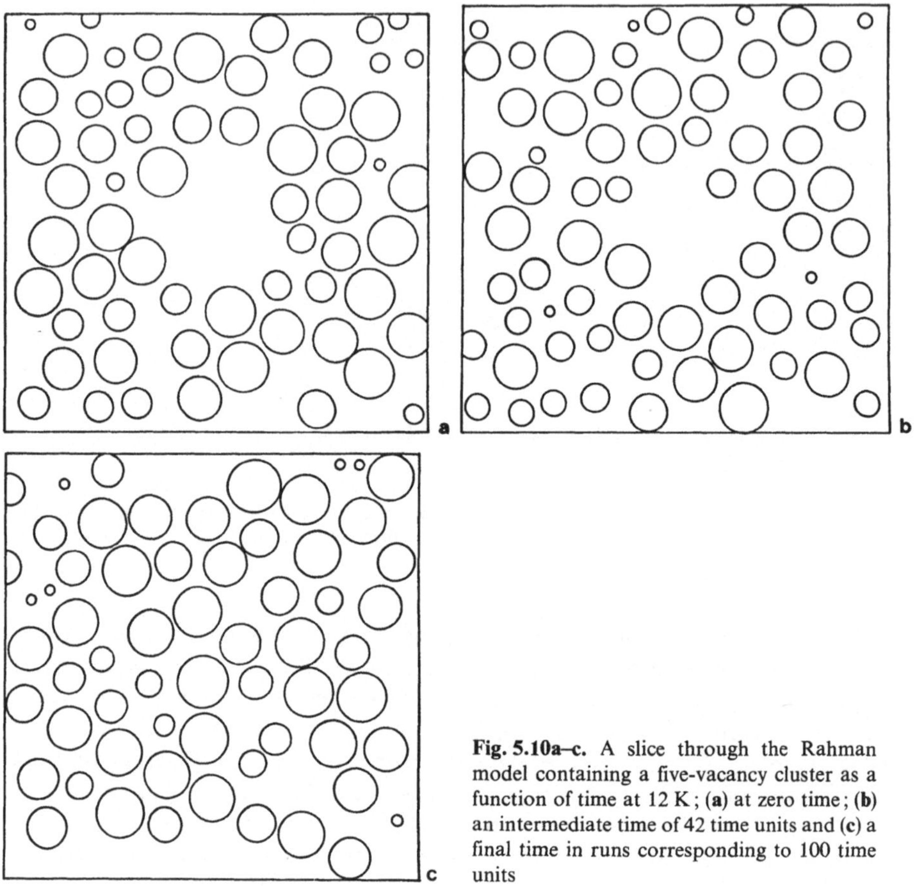

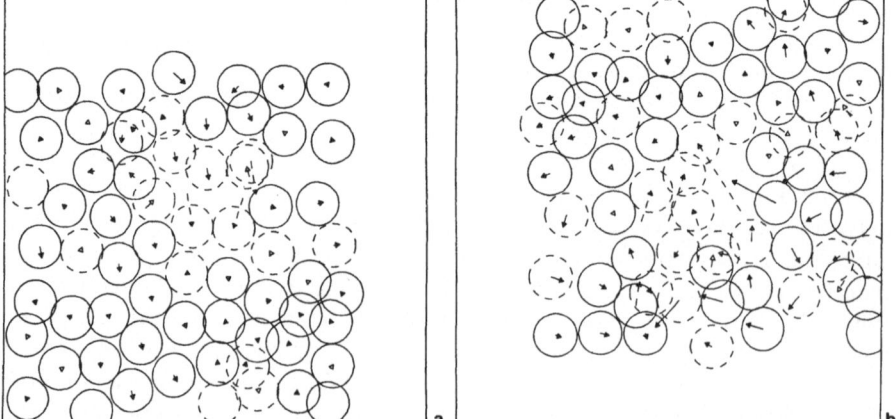

Fig. 5.10a–c. A slice through the Rahman model containing a five-vacancy cluster as a function of time at 12 K; (**a**) at zero time; (**b**) an intermediate time of 42 time units and (**c**) a final time in runs corresponding to 100 time units

Fig. 5.11a, b. Plots showing initial and final positions of atom in (**a**) a three-vacancy and (**b**) a five-vacancy cluster. Relaxation is carried out at 12 K. The notation is the same as that adopted for Fig. 5.6

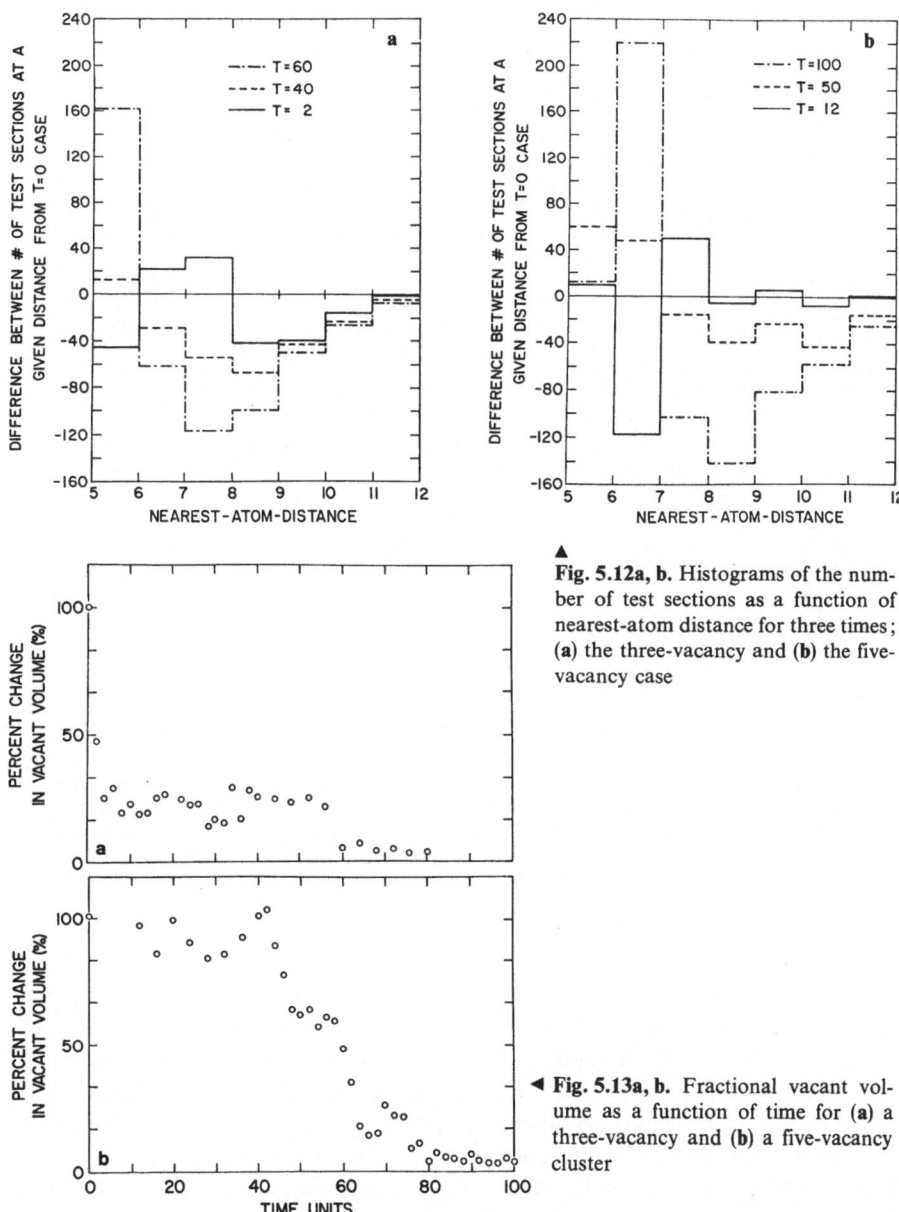

Fig. 5.12a, b. Histograms of the number of test sections as a function of nearest-atom distance for three times; (a) the three-vacancy and (b) the five-vacancy case

◄ Fig. 5.13a, b. Fractional vacant volume as a function of time for (a) a three-vacancy and (b) a five-vacancy cluster

shuffling in and sudden dips corresponding to individual atoms jumping into the void, as can be verified from computing lists of atoms closest to the center of the void as a function of time. It is also apparent from these figures that the localized vacant volume eventually redistributes itself uniformly over the sample.

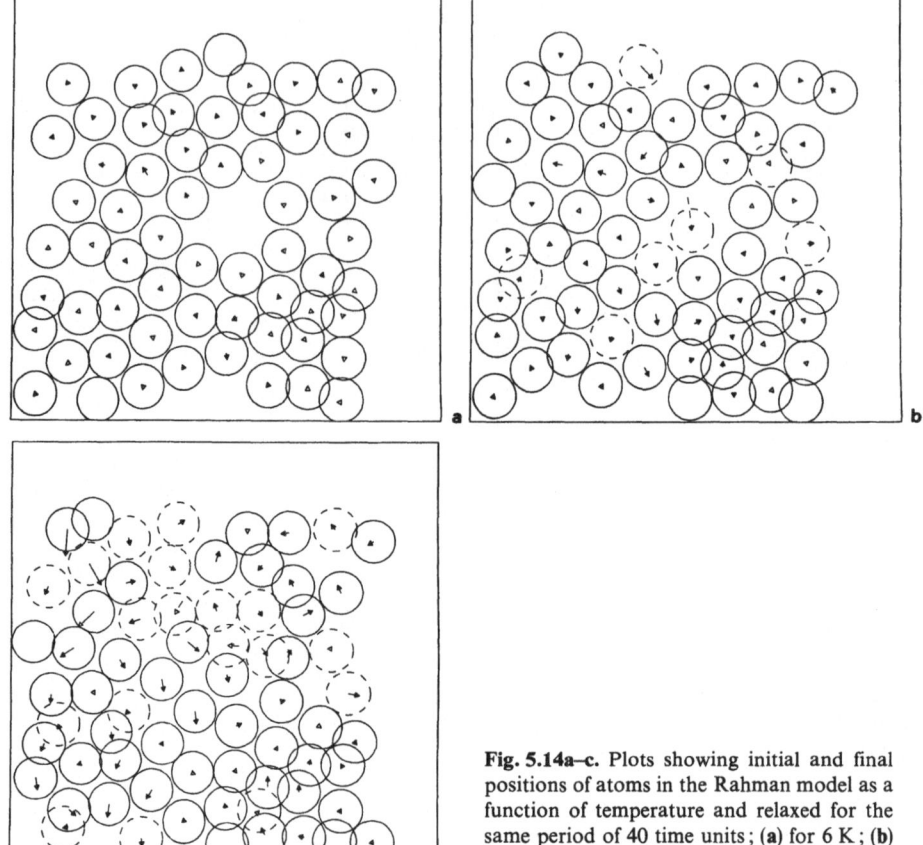

Fig. 5.14a–c. Plots showing initial and final positions of atoms in the Rahman model as a function of temperature and relaxed for the same period of 40 time units; (**a**) for 6 K; (**b**) for 12 K; (**c**) for 24 K

Molecular dynamics calculations carried out at higher and lower temperatures show that the annealing-out time is a function of temperature. As expected, lowering of the temperature increases the anneal time. In Figs. 5.14a–c, the results of the molecular dynamics studies at 6, 12, and 24 K on the annealing out of a vacant volume corresponding to the removal of one atom are shown. The plots were obtained after 40 time units of relaxation. At 6 K there is virtually no net displacement of the atoms. At 12 K there is considerably more motion and the slice at 24 K shows that most of the vacant volume has already annealed out. The effects of pressure were found to be similar to those from the temperature; increasing the pressure increased the rate of annealing.

Thus, under conditions of temperature and pressure at which vacancies in crystalline solids are unable to anneal out (due to the low probability of atoms jumping across the barrier to fill the void), vacancies in models characterized by

Lennard-Jones forces anneal out. Both the static and dynamic relaxations suggest that the potential barriers to vacancy motion are small in such solids. A similar observation has been observed in two-dimensional hard sphere simulations [5.12].

In contrast, vacancies have been found to persist after static relaxation in models characterized by a Keating potential. The topological constraints are sufficient to leave the vacancy in a metastable state, even when bond-bending forces are removed. Only by changing the neighbor topology by extending the range of the interaction does the persistence of the network structure and the vacancies cease.

5.3.4 Line Defects – Static Relaxation

a) Definition of Line Defects in Glasses

Although the definition of a defect-free glass may not exist, a point defect in a glass is recognizable as an extreme deviation in the local atomic density. However, line defects involve at most a very small deviation in the local density and so their study is much more difficult and controversial. Line defects in crystals are easy to recognize because they result in local deviations along a line of the perfect periodic structure. Although one need only be shown a single crystalline sample containing a defect and the defect can be readily recognized, in point of fact what is being done is that the sample is being compared to a defect-free perfect periodic array – a reference sample. If one is shown an atomic model of a glass, there is no reference sample with which to compare and so one cannot readily determine whether the model has a defect or not. More exactly, the strength of the line defect can be measured by taking a Burgers' circuit around the line defect measuring the displacement relative to a reference sample. Unless a definition of a defect-free glass can be obtained, this method of defining a line defect in a glass is not possible.

As a result, a common conclusion is that line defects in glasses are not sensible notions. One should keep in mind though that the classical studies of dislocations begin with continuous mechanics in which the material is treated as a continuum and the underlying atomic structure is ignored. Furthermore, there are various features of line defects which can be measured intrinsically, without reference to a defect-free system. Taking a given model of a glass and searching for line defects may prove too difficult, nevertheless, since the line may be highly twisted or tangled with other line defects. Therefore, the approach that has been attempted is to introduce a straight line defect in a computer-build model of a glass in much the same way as one would do for a crystalline solid, and then let the model evolve under either static or dynamic relaxation. Then one can check to see if, from intrinsic features of the resulting model, a more or less straight defect remains in the sample or whether the defect disappears as a localized entity. Thus, although the relaxation procedures used for these studies are the same as those used in the studies of point defects, methods of introducing the defects and methods of detecting them after relaxation are required.

b) Introducing the Defects in Glasses

The method of introducing dislocations which was found to be appropriate was to follow the concept introduced by *Volterra* [5.70] for continuous media. The edge dislocation resulted from cutting out a strip of material of width equal to one atom diameter (Burgers vector) (Fig. 5.15). The cut was "mended" by shifting atoms towards one another by about 0.3–0.45 atomic spacings so that the atoms on either side of the cut could sense the force of the other. Otherwise the cut would open and the internal stress would relax in an undesirable way. (In fact, the first results on edge dislocations yielded the result that they were

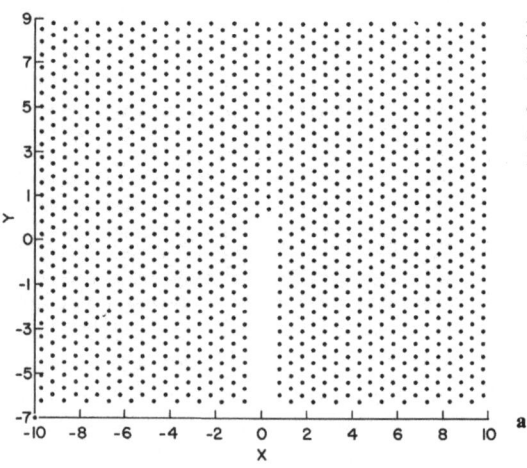

Fig. 5.15a–c. Introduction of edge dislocation into solid as shown through the projection of atoms. Introduction of cut in (**a**) crystal and (**b**) amorphous solid; (**c**) is the projection of atoms after mending and relaxation of the cut

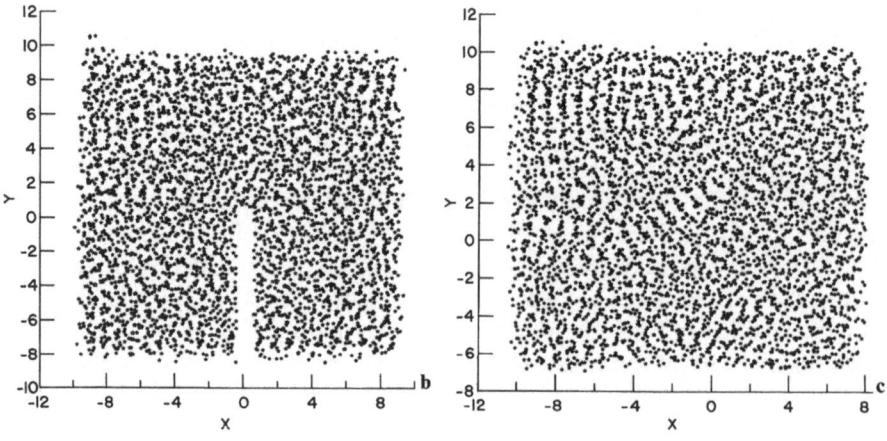

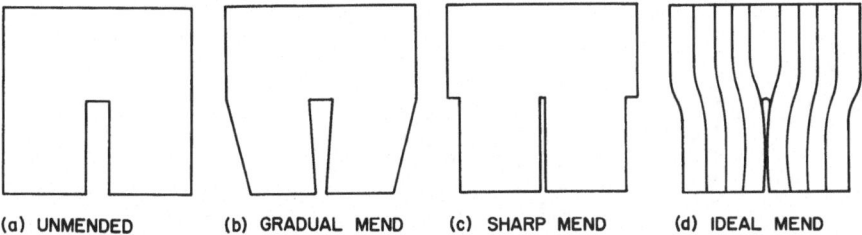

(a) UNMENDED   (b) GRADUAL MEND   (c) SHARP MEND   (d) IDEAL MEND

Fig. 5.16a–d. Mending procedures: (a) Remove a half plane of the atoms. Then displace atoms into the cut half of the solid by either (b) a graded displacement, (c) an equal displacement, or (d) a displacement computed for an ideal continuum solid

not stable entities in glasses [5.63]; this was because the cuts in the samples had not been properly mended and had opened up again.) Various mending methods were used (Fig. 5.16) but the results were found to be independent of the method so long as the cut did not reopen. For screw dislocations a cut was made halfway through a cylindrical model and suitable displacements in the atoms were introduced.

The particular model that was used for most of the studies was one that was derived from the Finney model. For the edge dislocations a section of the model containing approximately 4193 atoms, roughly $18 \times 18 \times 9$ atoms, was used. For the screw dislocation a cylindrical model containing 5390 atoms was used. Most of the studies that have been done to date involve static relaxation using conjugate-gradient techniques. In order to provide a check on the methods, similar calculations were carried out on a model of a face-centered-cubic crystal relaxed under the same Lennard-Jones potential as the glass samples.

c) Detection of Dislocation After Relaxation

A combination of four methods has been used to detect the dislocations after relaxation. The first approach was simply to project the atoms onto a plane perpendicular to the straight line defect. In Fig. 5.17 the results are shown for an edge dislocation in an fcc crystal. The presence of two partial defects is clearly marked by the deviation from strict periodicity and the location can be found by measuring **Burgers** circuits; both of these qualities can be measured because the system is nearly periodic and the perfectly periodic state can be used as a reference state. In Fig. 5.18 the same projection is shown for an edge dislocation placed in an amorphous solid. Figure 5.18a shows the solid after the cut has been introduced and the model has been mended but before the relaxation. The solid after some relaxation is seen in Fig. 5.18b; because the mending procedure brought two atoms on either side of the cut too close to one another, an atom has been cast out of the model by the relaxation procedure. Figure 5.18c shows a model after relaxation in which the mend was not sufficient and the cut never healed. In Fig. 5.18d, a well-healed and well-relaxed

154   *P. Chaudhari* et al.

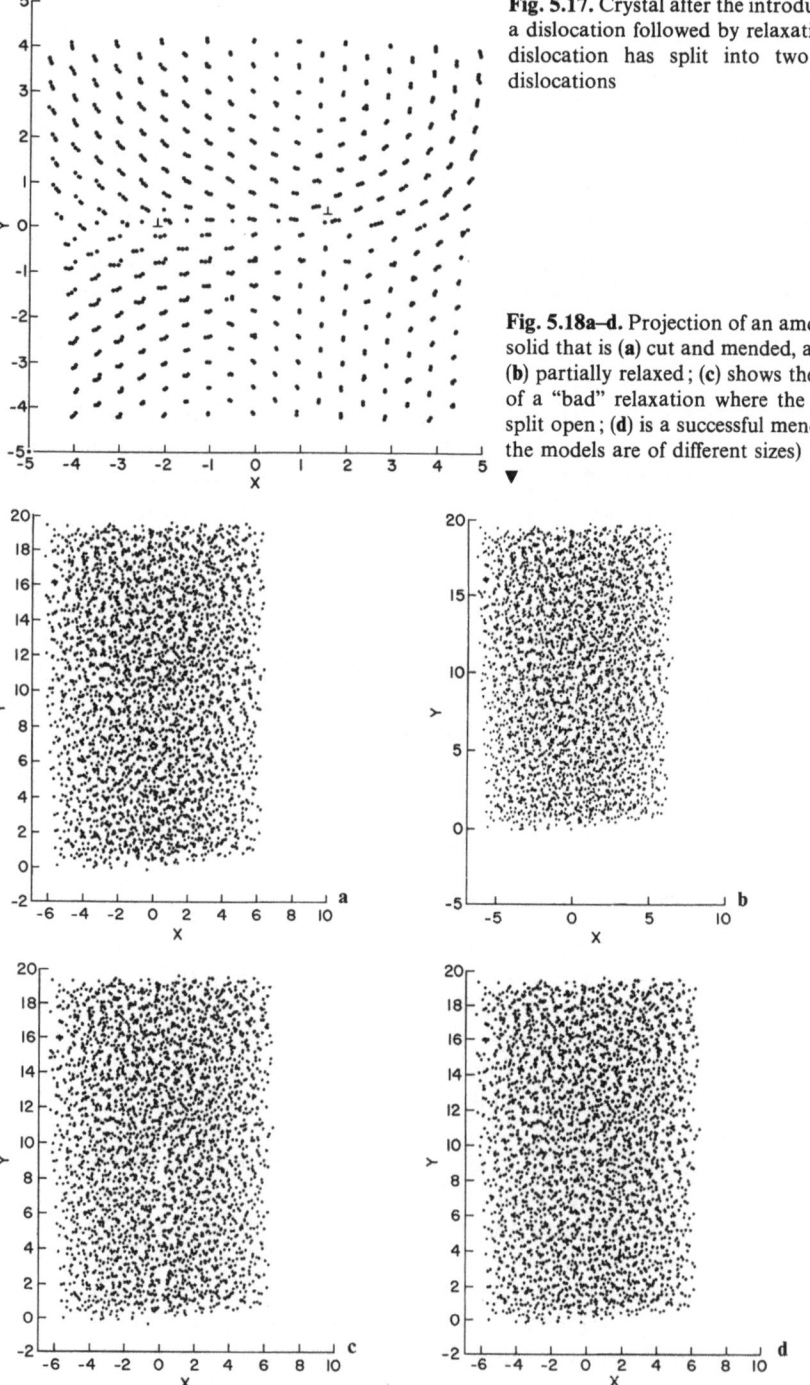

Fig. 5.17. Crystal after the introduction of a dislocation followed by relaxation. The dislocation has split into two partial dislocations

Fig. 5.18a–d. Projection of an amorphous solid that is (**a**) cut and mended, and then (**b**) partially relaxed; (**c**) shows the results of a "bad" relaxation where the cut has split open; (**d**) is a successful mend (note: the models are of different sizes)

model after the dislocation has been introduced is shown. Except for the fact that the model has a different shape due to the removal of some of the atoms, one cannot tell that the dislocation has been introduced or whether it has remained in the sample. This kind of projection is of very little use except as a check that the cut has properly mended after relaxation.

The second approach that has been utilized is a reference net procedure. Each atom in the cut, but unrelaxed and unmended solid is associated with the nearest point on an ideal three-dimensional lattice whose lattice spacing is on the order of the interatomic spacing. For the edge dislocation a cubic lattice is the most convenient and for the screw dislocation, a lattice with cylindrical symmetry is the most useful. After the model has been mended and relaxed, each point in the ideal lattice is displaced according to the average displacement of the atoms associated with it. In this way one can visualize how the displacements in the glass would look if they were imposed on a crystalline system. In Fig. 5.19 the results are shown for an fcc crystal. In Fig. 5.19a the dislocation has slipped out of the system upon relaxation and this is reflected by the presence of an extra half-plane of atoms on the edge of the sample. In Fig. 5.19b the dislocation has remained in the crystal and the lines in the reference net conform very closely with the shapes computed theoretically for

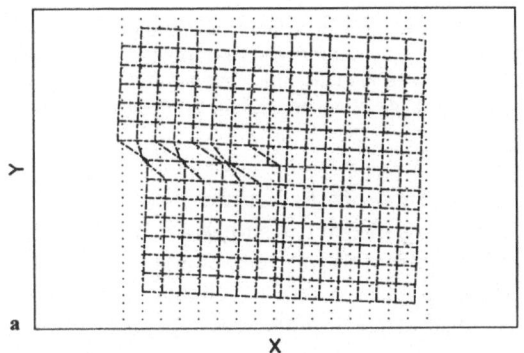

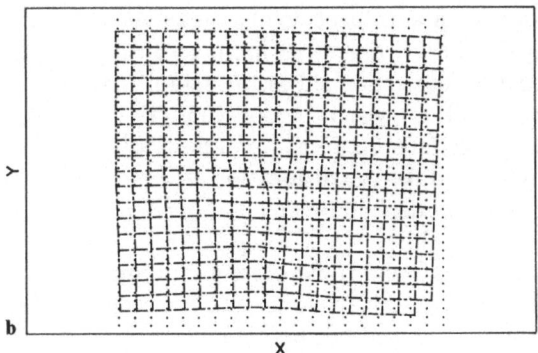

Fig. 5.19a, b. Reference net for the case of a dislocation that has been introduced into a crystal and which (a) has slipped out after relaxation and (b) has remained after relaxation

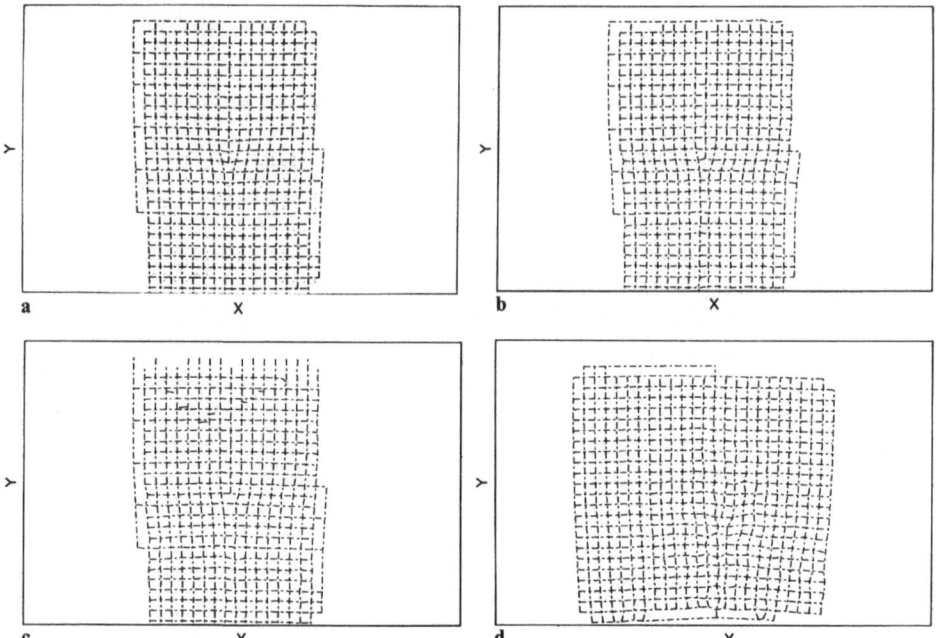

Fig. 5.20a–d. Reference net for a dislocation that has been introduced into an amorphous solid: (a) the sample has been cut and mended according to ideal continuum displacements; (b) the sample has been subsequently relaxed; (c) the result after relaxation of a sample that has been mended by graded displacements-cutoff = 1.2 atomic diameters; (d) same as (c) but cutoff is 2.1 atomic diameters

an infinite solid. The results of the reference net procedure for amorphous systems are shown in Fig. 5.20. In Fig. 5.20a the net is shown for a system in which the ideal displacements have been imposed (no relaxation) [5.71]. The results for two amorphous solids whose cuts have been mended by different procedures and then relaxed with a cutoff at 1.2 atomic spacings are given in Figs. 5.20b, c. Although there is some irregular motion in these compared to Fig. 5.20a, both cases conform rather closely with theory and with one another. In Fig. 5.20d the reference net is shown for a dislocation in a model relaxed with a cutoff of 2.1 atomic spacings. Again, there is regularity in the results. All in all, the results suggest that the dislocation has remained in the sample. However, experience has shown that this kind of visual proof can be deceiving. Furthermore, the reference net procedure requires foreknowledge of the orientation of the cut that was used to make the dislocation. A procedure for studying edge dislocations which does not depend on any sort of reference, be it real or imaginary, would be more powerful.

The reference net procedure has been found to be more useful for studying screw dislocations. Shown in Fig. 5.21a is the reference net for a cylindrical model after the dislocation has been introduced but before relaxation. A space

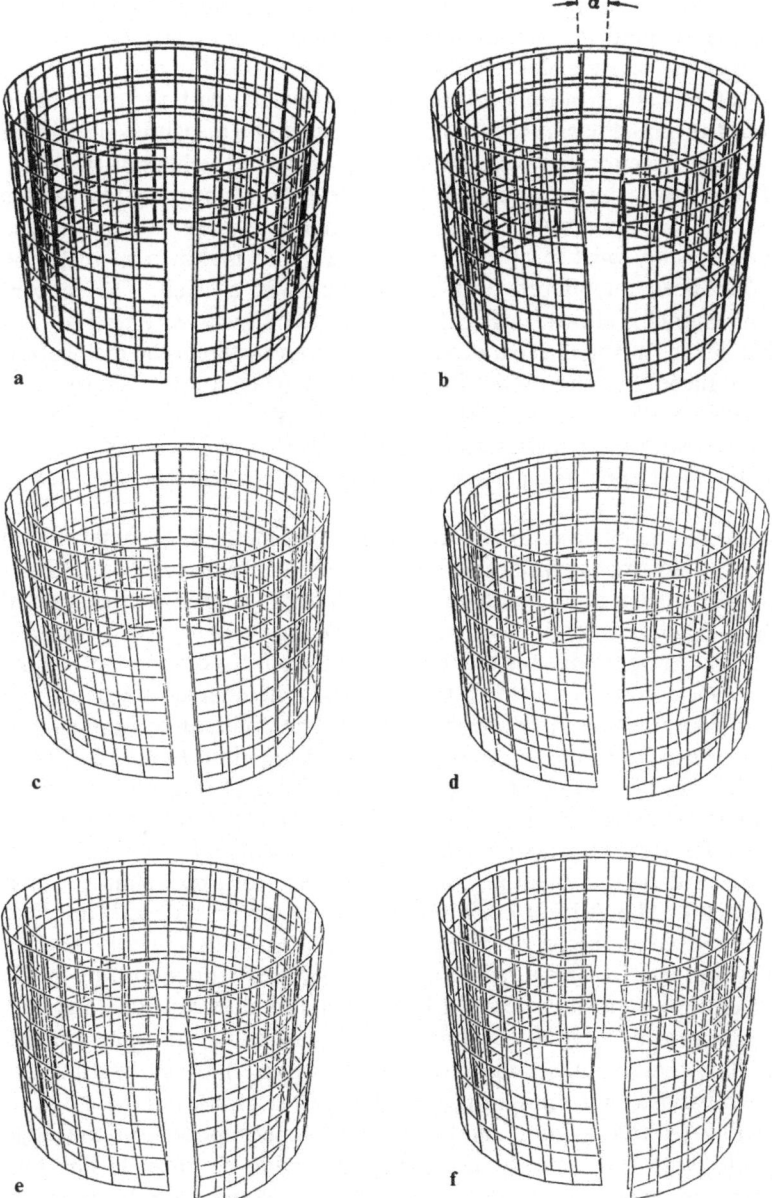

Fig. 5.21a–f. Reference net of the (**a**) unrelaxed and (**b**) relaxed amorphous model containing a screw dislocation. For clarity of presentation, only the two outer segments of the reference net are shown. The presence of the twist in (**b**) is indicated by the angle and measured between two lines lying in the front and back of the cylinder (**c**) Sample after dislocation and twist has been introduced but before relaxation. (**d**) Sample (c) after relaxation. (**e**) Sample relaxed and under constraint which prevents Eshelby twist. (**f**) Sample (e) after constraints are lifted and then the sample is relaxed

in the net has been left along which the slip plane lies. By following the lines of the reference net around the model, one finds that a helix is formed which is characteristic of the screw dislocation. Figure 5.21b shows the result after the model has been relaxed. Not only has the helix associated with the dislocation remained, but the model has a twist in it. The twist can be measured by finding the angle on the projection between the lines in front and the lines at the back. In Fig. 5.21a the angle is zero, but in Fig. 5.21b the angle is nonzero. The twist is due to the fact that the model is of finite extent and the relaxation has introduced an Eshelby twist [5.71]. In Fig. 5.21c a model is shown in which the dislocation has been introduced and the proper twist as well. The relaxed version of this model is shown in Fig. 5.21d and it conforms very well to that shown in Fig. 5.21b. In Fig. 5.21e a model can be seen in which the dislocation has been introduced, but the model itself has been relaxed under constraints along the z-axis and the $z-y$ plane that do not permit it to twist. This model was then relaxed with the constraints removed and that in Fig. 5.21f resulted. Once again, the twist appears as evidence that the dislocation has remained in' the model.

As has been emphasized repeatedly, an ideal method of determining whether dislocations exist in an amorphous system would be one that did not use any reference system at all. Such an approach that has been very successful in the study of edge dislocations has been to measure the Airy stress function for the solid. The straight edge dislocation has four nonzero components of the stress tensor associated with it. Rather than compute these separately, it is more convenient to compute the Airy stress function. The different components of the stress tensor can be derived from the Airy stress function. *Frank* [5.72] has shown how this can be computed for a model consisting of rods and pinjoints. The method has been recently extended [5.73] and applied to compute the Airy stress function for three-dimensional models held together by central forces, such as a Lennard-Jones potential, and with an axis of approximate translational symmetry. The Airy stress function for an fcc crystal with an edge dislocation is shown in Fig. 5.22. Typically, the stress function for an edge dislocation is characterized by a peak and a valley (of equal magnitude) in the potential located on opposite sides of a line through the dislocation core. The Airy stress function for edge dislocations relaxed inside an amorphous solid (please note how these results differ from those in [5.62]) are shown in Figs. 5.23a, b. The imbalance in the magnitudes of the peak and valley in these figures is probably due to a remaining dilatation core superimposed on the edge dislocation, although this issue is being studied further at present. The two figures represent the results of two different kinds of mending procedures. In the first case the cut has been mended by using the ideal theoretical form. For this particular example, the mending method was successful but, in general, atoms on opposite sides of the cut may be placed too close to one another and the relaxation process may fail. Therefore, the more generally used procedure has been to move the atoms by only 4/5 of the ideal displacement and then allow the relaxation procedure to complete the mend, as shown in Fig. 5.23b.

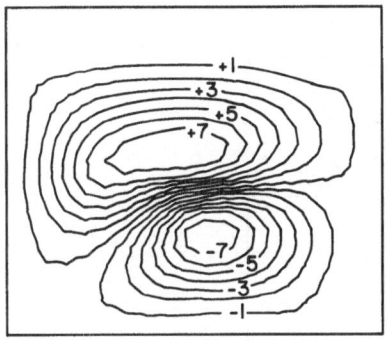

◀ **Fig. 5.22.** Airy stress function of a crystalline solid in which an edge dislocation has been introduced

Fig. 5.23a, b. Airy stress function of an amorphous solid in which an edge dislocation has been introduced (**a**) relaxed after an "ideal" and (**b**) relaxed after 4/5 times the ideal displacement ▼

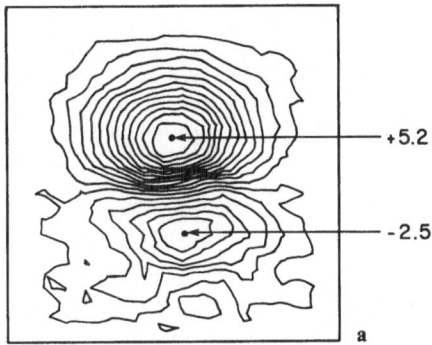

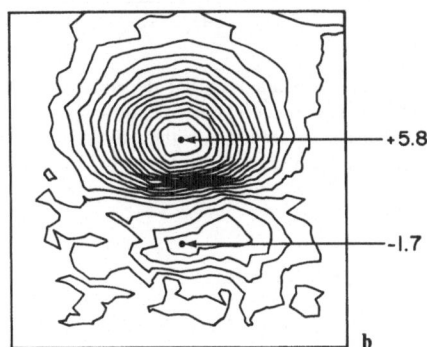

For the screw dislocation, the Airy stress function is not a successful way of testing for the presence of the dislocation because the nonzero components of the stress tensor are not associated with just the plane perpendicular to the line of the defect. The reason, of course, is that the Burgers vector is parallel instead of perpendicular to the line of the defect. For this case a method of detecting the dislocation that avoids the reference net is to measure the long-range elastic stress field associated with the dislocation. This can be done by computing the shear stress associated with the dislocation. In Fig. 5.24 the spatial dependence of the shear stress for amorphous and crystalline models is shown. The results are compared with elastic continuum theory and reasonable agreement is found.

From the observations of the reference nets and the stress field, one may conclude that the edge and screw dislocations are stable entities in Lennard-Jones solids at zero temperature. More recent efforts [5.63] have been directed towards measuring the features of these dislocations and the elastic properties of the models before and after their introduction. In addition, there has been a considerable effort in measuring the properties of the dislocations at finite temperatures.

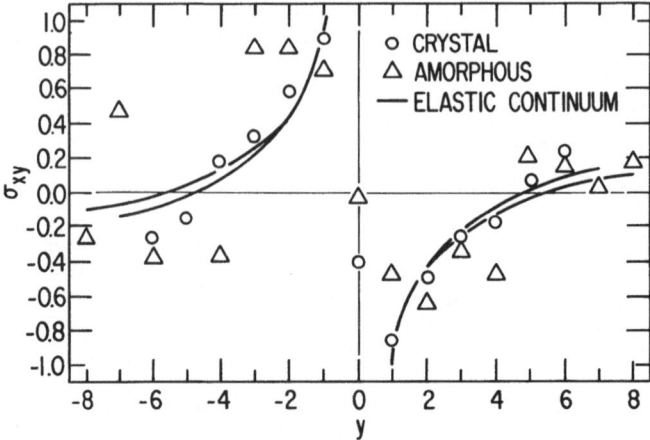

Fig. 5.24. Calculated spatial dependence of the shear stress of a screw dislocation in a finite cylinder. The two solid curves refer to the true diameters of cylinders. The computed shear stress for the models is averaged over the z-direction. The shear stress is in arbitrary units and the y-axis is in multiples of an atom diameter (equal to unity)

d) Burgers Vector

The first characteristic to be measured for the dislocations is the strength of the Burgers vector. In all cases the Burgers vector was computed by measuring displacements around a Burgers circuit of the reference net of the relaxed model. Fluctuations in the value were found along the length of the dislocation, but there were no systematic changes. Averages over the length of the dislocation are what are cited in this paper. For a crystalline solid the minimum strength is fixed by the lattice spacing of the periodic array. For the computer-built models the [111] strength can be measured by plotting the final displacement of the atoms compared to the initial displacement, as shown in Fig. 5.25 for an fcc crystal. Here the initial displacement was on a [111] plane and along the [$\bar{1}$10] direction. Below a certain critical displacement of approximately 0.3 atom diameters, the screw dislocation is unstable and the relaxation procedure eliminates it so that there is no final displacement. Above the critical displacement the screw dislocation stays in the solid. The final displacement along the [$\bar{1}$10] direction or, alternatively, the screw component of the Burgers vectors, was 0.5 atom diameters. If the initial displacement exceeds the critical value by a small amount, the final displacement is still found to be equal to the Burgers vector. This relationship of the initial and final displacement vector is what one expects for a crystal with a periodic arrangement of atoms.

A similar procedure was also carried out on an amorphous system, as shown in Fig. 5.26. Although not as sharp as in the crystalline solid, there is once again a threshold in the value of the initial displacement before the screw

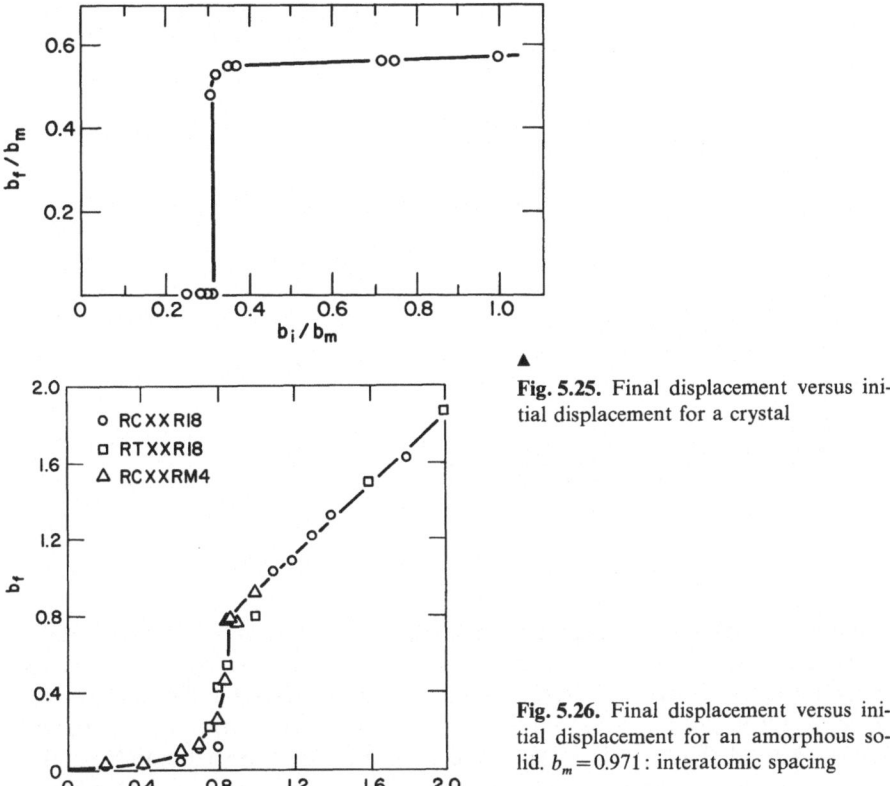

Fig. 5.25. Final displacement versus initial displacement for a crystal

Fig. 5.26. Final displacement versus initial displacement for an amorphous solid. $b_m = 0.971$: interatomic spacing

dislocation is stabilized. This value is approximately 0.83 times an atom diameter. Above this critical displacement the initial and final displacement relationship is linear and quite different from the behavior of the crystalline solid. The results have been shown to be independent of the size of the model. Below the threshold value the final displacement is small but nonzero except for initial displacements of less than 0.1 times an atom diameter. In Fig. 5.27 the angle of Eshelby twist vs the values of the final displacement is plotted. As expected for the screw dislocation, the relationship is linear. From these results one can conclude that a stable screw dislocation can be formed in a Lennard-Jones solid over a range of Burgers vectors. Although a threshold is observed, screw dislocations can exist both above and below this value.

To try to understand these results, an undislocated amorphous system was cut along a plane and one half of the model was displaced relative to the other. The change in nearest-neighbor assignments was examined both before and after relaxation and as the initial displacement increased, the number of assignment changes also increased. At a value of 0.8 atom diameters, the

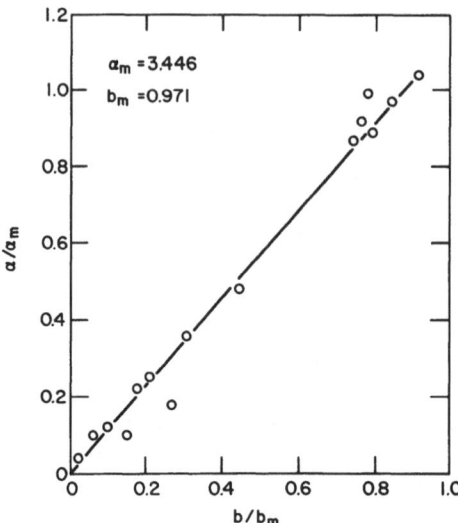

Fig. 5.27. Eshelby twist angle versus screw displacement for an amorphous solid

number of changes in assignments increased dramatically. The threshold for screw dislocations appears to correspond to the point where the nearest-neighbor pairs change. Parallel to the changes in nearest-neighbor assignments, the changes in nearest-neighbor spacing perpendicular to the shear plane was also measured. The spacing was found to increase by approximately 7% in a strip of one atom diameter on either side of the shear plane. This observation suggests that with deformation there is a decrease in density of material perpendicular to the slip plane and localized about it. This is clearly different from the case of a crystal where shear deformation in the manner described results in no change in density. The decrease in density on deformation suggests that the motion of dislocations in amorphous solids has a large frictional stress.

Thus, a shear displacement of one part of an amorphous solid with respect to another on a slip plane results in reversible deformation until the deformation reaches some critical value, which, for the models that have been investigated, is approximately 0.83 atom diameters. Deformation beyond this point is irreversible. Atoms on either side of the slip plane adjust locally to find positions of minimum energy. However, in the absence of thermal effects, these minima are not minima with respect to packing. There are a number of minima associated with the less densely packed plane and displacing the solid further results in moving from one shallow minimum to another. The shallowness of the minimum has been examined by calculating the reversible change in energy when two halves of the solid are displaced relative to one another by an amount of the order of 5% of an atom diameter. The change in energy with displacement is reduced by as much as fifty percent of the value obtained in the densely packed structures.

5.3.5 Dynamic Relaxation Studies

In order to investigate the effect of sample preparation conditions on transport properties and related phenomena, we have carried out molecular dynamic studies on liquids quenched at different rates to form glasses [5.74]. This study is still not complete. We shall report here some preliminary findings on the energy, pressure, density, diffusion constant and pair correlation function of a model containing 864 particles in a box with periodic boundary conditions. We have also investigated the stability as a function of temperature of an edge dislocation in a cylinder containing 3055 particles – the dislocation line is along the axis of the cylinder which is approximately 7 atomic diameters in radius and 20 in height [5.75].

In the smaller model the particles were initially arranged to form an fcc structure. The crystal was heated above its melting point of 84 K to 120 K to form a liquid. This liquid was then quenched by extracting kinetic energy from the system in a manner described by *Rahman* [5.65]. The quenching rate was of the order $\sim 10^{12}$ K s^{-1}. The energy of the system was computed as a function of temperature both under conditions of constant volume (density) and pressure. An example of the change in potential energy with temperature at a constant density of approximately 0.794 that of a crystal is shown in Figs. 5.28. There is a clear change in the slope of the curve at approximately 37 K which we shall call the class transition temperature $T_{g,\varrho}$ at constant density.

A similar change in slope of pressure with temperature is also observed in this sample at the same value $T_{g,\varrho}$. If the experiment is run at constant pressure rather than constant density, there is a change in the slope of the density with temperature which we shall call $T_{g,\varrho}$. This behavior is shown in Fig. 5.29. We have also measured the diffusion constant at constant pressure and volume.

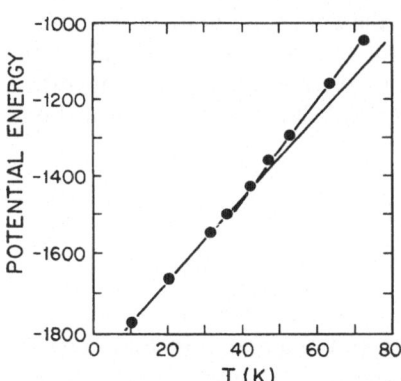

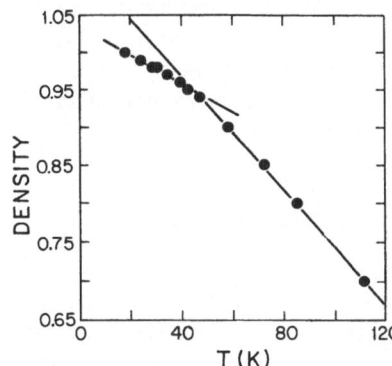

Fig. 5.28. Change in potential energy as a function of temperature. Density is 0.974 and melting temperature is 84 K. Note change in slope at ~ 40 K

Fig. 5.29. Density as a function of temperature at fixed pressure

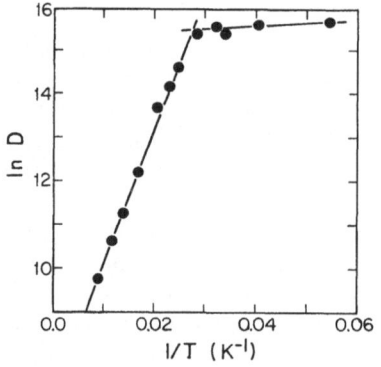
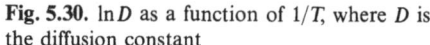

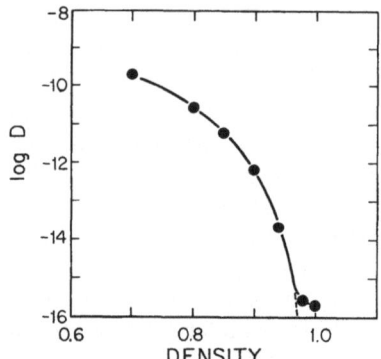

Fig. 5.30. ln D as a function of 1/T, where D is the diffusion constant

Fig. 5.31. ln D as a function of density, where D is the diffusion constant

Both of these diffusion constants show an abrupt change in slope at the glass transition temperature. An example is shown in Fig. 5.30. This data is obtained at constant pressure. The glass transition temperature at constant pressure and defined by the change in slope of the diffusion constant is higher than the above obtained from constant density simulations. The change in diffusion constant with density at constant pressure is shown in Fig. 5.31. The diffusion constant appears to be remarkably sensitive to the density of the sample at constant pressure.

The Airy stress function of the model containing the edge dislocation was measured as a function of temperature and time both below (24 K) and close to the glass transition temperature (36 K).

The Airy stress function was not measured on the sample at the temperature of annealing, but rather on samples which had been annealed at a given temperature and subsequently relaxed with the conjugate gradient technique. The low temperature anneal did not anneal out the dislocation up to the maximum running time of several hours. However, after the higher temperature anneal, the Airy stress function showed that the dislocation had annealed out.

5.4 Concluding Remarks

Although the results of these modeling studies are not definite enough yet to draw broad conclusions about defect stability and function, it is still of interest to discuss their significance in light of the experimental evidence and the new experiments and modeling studies they suggest. It is convenient to do this according to the type of transport.

I) The breakup of a vacancy in the monatomic dense random packed (DRP) system is significant for understanding the diffusion process. It should be checked whether it also occurs in a binary system with strong chemical interactions which is more representative of the glassy metal structure. If a full-size vacancy mechanism cannot operate, it should be investigated what type of local fluctuations are required to allow nearest-neighbor atom rearrangements. Molecular dynamics simulation seems to be the method of choice here. Given the continuity in the diffusion mechanism between the liquid and glassy state, the diffusive rearrangements can be studied above T_g where they are still frequent enough for convenient observation. Simulations as a function of temperature, above and below T_g, can be helpful in interpreting the changes in defect concentrations in equilibrium and isoconfigurational diffusion experiments. Diffusion of atoms that are much smaller than the host atoms (H, B, ...) probably occurs by an interstitial mechanism (Sect. 5.1.3b). Descriptions of possible paths through the DRP have been given by *Ahmadzadeh* and *Cantor* [5.8]. Molecular dynamics could provide more direct comparisons with the "substitutional" diffusion of the host atoms. The stability of a vacancy in a covalent random network suggests that a vacancy mechanism could be governing diffusion in amorphous Si and Ge. Measurements on these systems would therefore be highly desirable.

II) Given the experimentally indicated similarity between the defects governing viscous flow and diffusion, the kinematics of the diffusive rearrangements observed in the above simulations should be investigated in order to determine the local shear accompanying each. This should allow a check on the measured values for the volume strain $\gamma_0 v_0$.

III) The breakup process of the vacancy in the DRP system distributes the extra atomic volume more uniformly over the whole system, whose volume has been kept constant. At constant pressure, the extra volume would eventually be removed from the system by a process of *annihilation*, which is quite distinct from the *redistribution* process modeled so far. The most efficient way to remove the extra volume seems to be by elastic collapse [5.13] of the vacancy fragments. These annihilation processes are of great interest for the microscopic interpretation of the structural relaxation phenomenon since the extra volume increases the probability of forming sites where diffusive or shear rearrangements can occur. Molecular dynamics experiments can provide details of the annihilation processes and a quantitative check on the annealing kinetics.

IV) The simulations of the line defects in the DRP have established that under certain conditions of static relaxation or low temperature molecular dynamics, simulation stress fields characteristic of dislocations can be present. Some of their characteristics and their role in the plastic deformation of metallic glasses remain to be clarified. Dislocation-like defects have been postulated to explain the sharp shear bands that occur during plastic deformation at high stresses and low temperatures [5.76, 77]. However, no direct evidence for their occurrence has been given so far. In fact, the shear bands can be explained as macroscopic phenomena resulting from the lowering of the viscosity by disordering of the structure during shear [5.78]. A particularly striking illustration of the low viscosity in the shear band is the "vein pattern" in the fracture morphology which represents a typical failure mode for a fluid layer between two solid surfaces [5.79, 80]. This lowering of the viscosity and localization of the flow has also been explained by models that, instead of extended line defects, invoke more localized shear transformations accompanied by dilatation [5.41, 43]. This dilatation has also been confirmed by direct density measurements, after deformation [5.19]. The dilatation observed after shearing the entire DRP model is no doubt related to this, and merits a more quantitative investigation.

Since the diffraction conditions that make dislocation imaging possible in crystals do not exist in amorphous materials, different types of experiments must be devised to obtain direct evidence of the presence of atomic size dislocations. For example, it might be possible, under certain conditions of deposition or oxidation, to obtain an amorphous overlay on a crystalline substrate where the strain difference is taken up by "incoherency dislocations" in the interface; these could then be imaged on the crystalline side. Birefringence could also be used to image the stress fields in optically transparent amorphous materials, if a temperature and stress regime for localized flow could be found.

Acknowledgements. F. S. acknowledges research support in this area from the Office of Naval Research, Contract N 00014-77-C-0002. P..J.S. was supported in part by Dept. of Energy Grant No. E Y-76-C-02-3071.

References

5.1   W.Kauzmann: Chem. Revs. **42**, 219 (1948)
5.2   F.Spaepen: Philos. Mag. **30**, 417 (1974)
5.3   M.R.Hoare, P.Pal: Adv. Phys. **20**, 161 (1971)
5.4   M.R.Hoare: J. Non-Cryst. Sol. **31**, 157 (1978)
5.5   P.J.Steinhardt, D.R.Nelson, M.Ronchetti: Phys. Rev. Lett. **47**, 1297 (1981)
5.6   T.Egami, K.Maeda, V.Vitek: Philos. Mag. A**41**, 883 (1980)
5.7   T.Egami: In *Glassy Metals I*, ed. by H.-J.Güntherodt, H.Beck, Topics Appl. Phys., Vol. 46 (Springer, Berlin, Heidelberg, New York 1981) p. 25
5.8   D.Turnbull, M.H.Cohen: J. Chem. Chem. Phys. **31**, 1164 (1959)
5.9   M.H.Cohen, D.Turnbull: J. Chem. Phys. **34**, 120 (1961)
5.10 S.Kobayashi, K.Maeda, S.Takeuchi: Acta. Met. **28**, 1641 (1980)
5.11 A.S.Argon, H.Y.Kuo: Proc. 3rd Intern. Conf. on Rapidly Quenched Metals, ed. by B.Cantor (The Metals Soc., London 1978) Vol. 2, p. 269
5.12 F.Spaepen: J. Non-Cryst. Solids **31**, 207 (1979)
5.13 F.Spaen: in *Physics of Defects*, Les Houches Lectures XXXV, ed. by R.Balin et al. (North-Holland, Amsterdam 1981) p. 133
5.14 D.Turnbull: J. Phys. (Paris) **35**, C-4, 1 (1974)
5.15 T.Matsubara (ed.): *The Structure and Properties of Matter*, Springer Ser. Solid-State Sci., Vol. 28 (Springer, Berlin, Heidelberg, New York 1982)
5.16 A.Kursumovic, M.G.Scott, E.Girt, R.W.Cahn: Scripta Met., **14**, 1303 (1980)
5.17 A.L.Greer, J.A.Leake: J. Non-Cryst. Solids **38/39**, 379 (1980)
5.18 C.H.Lin, J.Bevk, D.Turnbull: Solid State Commun. **29**, 641 (1979)
5.19 N.Pratten, R.W.Cahn: Private communication

5.20 A.I.Taub, F.Spaepen: Acta Met. **28**, 1781 (1980)
5.21 A.I.Taub, F.Spaepen: J. Mat. Sci. **16**, 3087 (1981)
5.22 B.S.Berry, W.C.Pritchet: 1973, J. Appl. Phys. **44**, 3122 (1973)
5.23 B.S.Berry: In *Metallic Glasses*, ed. by J.J.Gilman, H.J.Leamy (American Society for Metals, Metals Park, Ohio 1978) p. 161
5.24 A.S.Argon, H.Y.Kuo: J. Non-Cryst. Solids **37**, 241 (1980)
5.25 D.E.Polk, D.Turnbull: Acta Met. **20**, 493 (1972)
5.26 J.J.Gilman: Philos. Mag. B**37**, 577 (1978)
5.27 H.S.Chen, D.Turnbull: Chem. Phys. **48**, 2560 (1968)
5.28 H.S.Chen, M.Goldstein: J. Appl. Phys. **43**, 1642 (1972)
5.29 F.Spaepen, D.Turnbull: In *Metallic Glasses*, ed. by J.J.Gilman, J.H.Leamy (American Society for Metals, Metals Park, Ohio 1978) p. 114
5.30 H.S.Chen, L.C.Kimerling, J.M.Poate, W.E.Brown: Appl. Phys. Lett. **32**, 461 (1978)
5.31 H.S.Chen: J. Non-Cryst. Solids **27**, 257 (1978)
5.32 G.Adam, J.H.Gibbs: J. Chem. Phys. **43**, 139 (1965)
5.33 J.C.Logan, M.F.Ashby: Acta Met. **32**, 1047 (1974)
5.34 R.Maddin, T.Masumoto: Mat. Sci. Eng. **9**, 153 (1972)
5.35 A.I.Taub, F.Spaepen: Scripta Met. **13**, 195 (1979)
5.36 A.I.Taub, F.Spaepen: Scripta Met. **14**, 1197 (1980)
5.37 A.L.Mulder, J.W.Drijver, S.Radelaar: J. Phys. (Paris) **41**, C-8, 843 (1980)
5.38 S.S.Tsao, F.Speapen: Proc. 4th Intern. Conf. on Rapidly Quenched Metals, ed. by T.Masumoto, K.Suzuki; Japan Inst. of Metals (1982) p. 463
5.39 J.C.Gibeling, W.D.Nix: Scripta Met. **12**, 919 (1978)
5.40 A.I.Taub: Acta. Met. **28**, 663 (1980)
5.41 F.Spaepen: Acta Met. **25**, 407 (1977)
5.42 F.Spaepen: Proc. 3rd Intern. Conf. on Rapidly Quenched Metals, ed. by B.Cantor (The Metals Society, London 1978) p. 253
5.43 A.S.Argon: Acta Met. **27**, 47 (1979)
5.44 S.Glasstone, K.H.Laidler, H.Eyring: *The Theory of Rate Processes* (McGraw-Hill, New York 1941) p. 480
5.45 A.I.Taub, F.E.Luborsky: Acta. Met. **29**, 1939 (1981)
5.46 D.Gupta, K.N.Tu, K.W.Asai: Phys. Rev. Lett. **35**, 796 (1975)
5.47 R.W.Cahn, J.E.Evetts, J.Patterson, R.E.Somekh, C.K.Jackson: J. Mat. Sci. **15**, 702 (1980)
5.48 M.Kijek, M.Ahmadzadeh, B.Cantor, R.W.Cahn: Proc. 4th Intern. Conf. on Rapidly Quenched Metals, ed. by T.Masumoto, K.Suzuki; Japan Inst. of Metals (1982) p. 593
5.49 C.Birac, D.Lesueur: Phys. Status Solids A**36**, 247 (1976)
5.50 H.Cook, J.E.Hilliard: J. Appl. Phys. **40**, 2191 (1969)
5.51 M.Rosenblum, F.Spaepen, D.Turnbull: Appl. Phys. Lett. **37**, 184 (1980)
5.52 A.L.Greer, C.J.Lin, F.Spaepen: Proc. 4th Intern. Conf. on Rapidly Quenched Metals, ed. by T.Masumoto, K.Suzuku; Japan Inst. of Metals (1982) p. 567
5.53 M.Kijek, M.Ahmadzadeh, B.Cantor, R.W.Cahn: Scripta Met. **14**, 1337 (1981)
5.54 U.Koster, U.Herold: In *Glassy Metals* ed. by H.-J.Güntherodt, H.Beck, Topics Appl. Phys., Vol. 46 (Springer, Berlin, Heidelberg, New York 1981) p. 225
5.55 D.E.Polk: Acta Met. **20**, 485 (1972)
5.56 P.M.Anderson, A.E.Lord: Mat. Sci. Eng. **44**, 279 (1980)
5.57 A.L.Greer, F.Spaepen: Ann. NY Academy Sci. **371**, 218 (1981)
5.58 K.F.Kelton, F.Spaepen: 1982, Proc. 4th Intern. Conf. on Rapidly Quenched Metals, ed. by T.Masumoto, K.Suzuki; Japan Inst. of Metals (1982) p. 527
5.59 J.J.Gilman: Philos. Mag. B**37**, 577 (1978)
5.60 P.Chaudhari: J. Phys. (Paris) C8, 267 (1980) p. 267
5.61 C.H.Bennett, P.Chaudhari, V.Moruzzi, P.Steinhardt: Philos. Mag. **40**, 485 (1979)
5.62 P.Chaudhari, A.Levi, P.Steinhardt: Phys. Rev. Lett. **43**, 1517 (1979)
5.63 P.Chaudhari, P.J.Steinhardt: "Displacement Vector and Energy of a Screw Dislocation in a Lennard-Jones Amorphous Solid" (1981), Philos. Mag. B (1982) submitted

5.64 J.L.Finney: Proc. R. Soc. London, Ser. A **319**, 479 (1970)

5.65 A.Rahman: J. Chem. Phys. **64**, 1564 (1976)

5.66 C.A.Angell, J.H.Clarke, L.V.Woodcock: In *Advances in Chemical Physics*, Vol. 48, ed. by I.Prigogine, S.Rice (Wiley, New York 1981) p. 397

5.67 D.Polk: J. Non-Cryst. Solids **5**, 365 (1971)

5.68 P.Steinhardt, R.Alben, D.Weaire: J. Non-Cryst. Solids **15**, 199 (1974)

5.69 P.N.Keating: Phys. Rev. **145**, 637 (1966)

5.70 V.Volterra: Cited by A.H.Love, *A Treatise on the Mathematical Theory of Elasticity* (Dover, New York 1944) p. 221

5.71 J.P.Hirth, J.Lothe: *Theory of Dislocations* (McGraw-Hill, New York 1968) p. 58
      J.D.Eshelby: J. Appl. Phys. **24**, 176 (1953)

5.72 F.C.Frank: Phys. Educ. **13**, 258 (1978)

5.73 P.J.Steinhardt, P.Chaudhari: "The Airy Stress Function in Atomic Models" (1981) J. Comp. Physics

5.74 M.Ronchetti, P.Steinhardt, P.Chaudhari: To be published

5.75 M.Ronchetti, P.Steinhardt, P.Chaudhari: To be published

5.76 J.J.Gilman: J.Appl. Phys. **44**, 675 (1973)

5.77 J.C.M.Li: In *Frontiers in Material Science*, ed. by L.E.Muir, C.Stein (Dekker, New York 1976) p. 527

5.78 P.S.Steif, F.Spaepen, J.W.Hutchinson: Acta Met. (1982)

5.79 F.Spaepen: Acta Met. **23**, 615 (1975)

5.80 P.G.Saffman, G.I.Taylor: Proc. R. Soc. London A**245**, 312 (1958)

5.81 M.Ahmadzadeh, B.Cantor: Proc. of 4th Intern. Conf. on Rapidly Quenched Metals, ed. by T.Masumoto, K.Suzuki; Japan Inst. Metals (1982) p. 591

6. Mechanical Properties of Metallic Glasses

H.-U. Künzi

With 16 Figures

The aim of this chapter is to review mainly the elastic and anelastic behaviour of metallic glasses and to discuss their relevance to structure, atomic dynamics and other properties. Since metallic glasses are usually only available in the form of thin foils (ribbons, splats or thin films adherent to a substrate), a brief description of the experimental techniques appears to be appropriate also. The chapter will be divided in different sections according to the different types of stress-strain behaviour exhibited by metallic glasses.

6.1 Overview

The mechanical properties of crystalline metals are well-known to be, in many respects, strikingly different from the behaviour of, say, a window glass which is commonly considered to be a typical representative for a glassy material. Glasses are usually very hard and brittle, whereas metals apart from a few exceptions are soft and ductile. Since the mechanical properties have their roots in electronic as well as structural properties, metallic glasses may be expected to present interesting behaviour.

But how do metallic and glassy properties combine in one material? Do we get materials with intermediate properties or do we get materials which show in some respects glass-like behaviour and in others metallic behaviour? From the point of view of basic science, this is certainly an interesting question which to a great extent is related to fundamental problems in the understanding of mechanical properties. The more application-minded engineer, on the other hand, may expect to find materials with a yet unknown combination of properties probably suitable for particular applications.

Furthermore, it is well-known from numerous investigations in crystalline materials that the mechanical properties are highly dependent on short-range order and the presence of lattice defects and impurities. This seems to be true also for amorphous solids with an almost equal sensitivity and therefore represents, in many cases, additional motivation to look into the mechanical properties of these new materials. This fact is in marked contrast with the behaviour of other physical properties such as the electrical resistivity which is known to be highly sensitive to deviations from a perfect order in crystalline metals; in amorphous materials the electrical resistivity merely reflects the disordered structure and is almost independent of small variations in such a structure. In metallic glasses, the investigation of structure-sensitive properties such as the mechanical properties, which at least can give indirect information on short-range order, are of particular value since classical diffraction experiments do not have sufficient sensitivity for determining the local arrangements of atoms. It is therefore hoped that studies on the mechanical properties, of interest in themselves, may help to give further insight into these problems.

In common with all other solids, metallic glasses behave in an essentially *elastic* manner at low temperatures and low stress levels. The elastic constants relevant for this mode of deformation will be discussed in the first part. The second part concerns the *anelastic* behaviour. This time-dependent elastic behaviour is a consequence of stress-induced relaxations operating within the material and in many cases has its origin in thermally activated atomic jumps. At low temperatures the stress-strain behaviour is still fully reversible. At higher temperatures close to the glass transition, the mechanical deformation becomes more and more irreversible. *Viscous flow*, which is a characteristic mode of deformation for all glassy materials at high temperatures, starts to dominate. Similarly at high stress levels *plastic deformation* sets in so that the shape cannot be recovered. Its occurence is, however, confined to the relatively high stress-levels prior to fracture. This mode of deformation is characteristic of crystalline metals and seems to distinguish metallic from nonmetallic glasses which usually fail by brittle fracture without prior plastic deformation.

6.2 Elasticity of Metallic Glasses

The action of external forces on an elastic medium sets up opposing stresses which, when the equilibrium is established, resists the displacement of atoms. For small displacements, the relationship between stress and deformation is for most materials, including metallic glasses, to a good approximation linear and reversible (Hooke's law). In the most general linear relationship between the stress (σ) and strain-tensor (ε), the elastic constants c are components of a fourth-order tensor. Of the 81 components of the 4th order elastic tensor for the lowest crystal symmetry, at most 21 are independent (this comes from the symmetry of the stress and strain-tensor and the existence of an elastic energy density which is a quadratic form of the strain components). The elastic constants can thus be represented by a symmetric 6×6 matrix $c_{\alpha\beta}$ (Voigt notation). The α and β indices from 1 to 3 refer to the normal stress and strain components, respectively, (11, 22, 33). From 4 to 6 they correspond to the shear components (23, 13, 12). The off-diagonal elements of the symmetric strain tensor (shear components) have to be multiplied by two to get the corresponding strains in Voigt's notation. Hooke's law is then given by

$$\sigma_\alpha = c_{\alpha\beta}\varepsilon_\beta .$$

(6.1)

The number of independent elastic constants $c_{\alpha\beta}$ depends on the crystallo-graphic point group symmetry. For monoclinic crystals, 21 constants are needed for a full description of their elastic behaviour. This number reduces to 3 in the case of cubic crystals and the matrix of the elastic constants takes the following form:

$$
c_{\alpha\beta} =
\begin{matrix}
c_{11} & c_{12} & c_{12} & 0 & 0 & 0 \\
c_{12} & c_{11} & c_{12} & 0 & 0 & 0 \\
c_{12} & c_{12} & c_{11} & 0 & 0 & 0 \\
0 & 0 & 0 & c_{44} & 0 & 0 \\
0 & 0 & 0 & 0 & c_{44} & 0 \\
0 & 0 & 0 & 0 & 0 & c_{44}
\end{matrix}
\tag{6.2}
$$

For amorphous solids which are generally asserted to be fully isotropic, the matrix keeps the same form but the additional relation

$$
c_{11} = c_{12} + 2c_{44} \tag{6.3}
$$

reduces the number of independent constants to two. For many purposes it is convenient to take as constants the bulk modulus B as the response to isotropic compression and the shear modulus as the response to elastic deformations at constant volume. The assumption of elastic isotropy is certainly justified for most of the metallic glasses. Notable exceptions occur, however, in strongly magnetic alloys. Here the magnetoelastic coupling effects, although formally second-order effects, may contribute significantly to the elastic constants and their anisotropy. In many cases these contributions are comparable to first-order effects.

6.2.1 Experimental Techniques

The elastic constants c_{11} and c_{44} have a direct meaning in terms of the sound velocities of longitudinal c_l and transversal c_t waves in an unbounded medium (i.e., wave length $\lambda \ll$ smallest geometrical dimension of the sample):

$$
c_{11} = \varrho c_l^2, c_{44} = \varrho c_t^2 ; \tag{6.4}
$$

ϱ is the mass density. In order to satisfy the boundary conditions $\lambda \ll d$ in metallic glass ribbons of thickness $d = 50\,\mu m$, frequencies well above $100\,MHz$ have to be applied. Such measurements using the ultrasonic pulse method have been made by *Weiss* et al. [6.1] on $Pd_{30}Zr_{70}$, but skillful preparation of the sample is then required which seems hardly worthwhile if no other measure-ments at high frequencies such as sound absorption are needed. At lower frequencies ($\lambda \gg$ thickness and width of specimen), the ultrasonic pulse technique can again be applied to metallic glass ribbons of several cm length to determine the velocity of extensional waves c_e, which is a direct measure of Young's modulus E

$$c_e^2 \cdot \varrho = E = \frac{(c_{11} - c_{12})(c_{11} + 2c_{12})}{c_{11} + c_{12}}. \tag{6.5}$$

Such measurements are relatively easy to perform and allow for a good accuracy. Since no geometrical dimension other than the length traversed by the pulse needs to be known, the experimental uncertainty is mainly limited by that of the mass density. Measurements using this technique have been performed by *Chen* et al. [6.2–4], *Davies* et al. [6.5], and *Kursumovic* [6.6].

In cases where only short (~ 1 cm long) specimens of amorphous metals are available, the impulse induced resonance technique [6.7] or the piezoelectric ultrasonic composite oscillator technique [6.8] may be applied. The latter method makes use of a driver and gauge transducer-unit onto which the sample has to be glued. The resonant frequency of the specimen is determined from the resonant frequencies of the driver gauge unit with and without the sample. In these experiments Young's modulus is determined from the resonant frequency for extensional waves. The experimental accuracies are intermediate ($\sim 3\%$) between the pulse-echo method and the vibrating-reed technique.

In the vibrating-reed technique, ribbon-like metallic glass samples of typically 1 cm length and 1–2 mm width are rigidly clamped at one end. Young's modulus and the shear modulus $G = c_{44}$ are determined from the frequencies of the resonant flexural and torsional modes, respectively. The absolute accuracy is mainly limited because the thickness enters the expressions relating the resonant frequencies to the elastic constants. For melt spun ribbons the thickness is usually not very uniform ($\pm 10\%$). The method offers, however, the advantage of high relative accuracy (a few ppm) combined with ease of measurement also at high temperatures. This method has been described in detail by *Berry* and *Pritchet* [6.9]. More recently, the vibrating-reed technique has also been adapted by the same authors to studies of thin films adhering to a substrate [6.10].

Apart from ultrasonic pulse and resonance techniques, static methods where the elastic deformation is directly observed as a function of the applied load have also been used to study elastic constants of metallic glasses [6.11–15]. A particularly simple method for measuring strains involves bonding strain gauges on ribbons of sufficiently large areas. Appropriate strain gauge units, having two orthogonally oriented planar sensing elements necessary for measuring lateral $\varepsilon_\perp$ and longitudinal $\varepsilon_\parallel$ (parallel to stress direction) strains, allow direct determination of Poisson's ratio v:

$$v = -\frac{\varepsilon_\perp}{\varepsilon_\parallel} = \frac{c_{12}}{c_{11} + c_{12}}. \tag{6.6}$$

According to studies made by *Chou* et al. [6.16, 17], corrections due to the stiffness of the strain gauge element itself may be kept within about 1 % of that of the underlying glassy specimen.

A direct determination of the bulk modulus B on metallic glasses does not appear to have been undertaken yet. But since the elastic behaviour of isotropic solids involves only two independent parameters, B can be calculated from two measured constants

$$B = \tfrac{1}{3}(c_{11} + 2c_{12}) = \frac{E}{3(1-2v)} = \frac{EG}{3(3G-E)}. \tag{6.7}$$

It should, however, be pointed out that the results of such calculations may turn out to be very precarious because of small denominators if the values of the two measured constants are not known with high accuracy. This also applies to other such relations between the elastic constants.

6.2.2 Elastic Constants

Table 6.1 gives a compilation of elastic constants for a large variety of glassy alloys in terms of the practical moduli which are the bulk modulus B, the shear stiffness G, Young's modulus E and Poisson's constant v. All values refer to room temperature and of course for each alloy, only two values are to be considered as independent. For comparison some values for polycrystalline metals are also given.

The elastic constants of solids are usually associated with the interatomic forces as given, for example, by the pair potential. Accordingly, the interatomic forces depend mainly on the *chemical nature* of the solid under consideration and its *density*, or more precisely, its interatomic distances. The chemical aspects are reasonably well documented in Table 6.1. The elastic constants of metallic glasses vary as in crystalline solids between elastically hard (Fe alloys) and elastically soft materials (Mg and MgZn).

The density aspect needs a more detailed discussion. The density of metallic glasses during crystallization usually increases by less than 1–2 % which corresponds to about half the changes that occur at the melting transition. We may, therefore, expect to have roughly identical cohesive forces in crystalline, amorphous, and liquid metals. The same conclusion can be drawn also from calorimetric measurements which always show that the internal energy decreases by about 5–10 % during crystallization near the temperature of the glass transition and slightly more during solidification at the melting temperature.

This similarity in the cohesive forces is, in fact, also reflected by the bulk moduli and its changes at the respective transitions. Table 6.2 gives some characteristic values together with the observed changes in the atomic volume. Notice that these values correlate with the changes in the moduli, as expected.

Table 6.1. Elastic constants of metallic glasses (B: bulk modulus, G: shear modulus, E: Young's modulus, and v Poisson's ratio)

Alloy	B [GPa]	G [GPa]	E [GPa]	v	Ref.
$Fe_{80}B_{20}$	141	64.9	168	0.30	[6.17]
$Fe_{40}Co_{40}B_{20}$	184	65.0	174.2	0.34	[6.17]
$Fe_{60}Ni_{20}B_{20}$	206	61.0	166	0.365	[6.17]
$Fe_{40}Ni_{40}B_{20}$	167	59.6	159.7	0.341	[6.17]
$Ni_{49}Fe_{29}P_{14}B_6S_2$	169	48.0	132	0.37	[6.16]
$Ni_{80}P_{20}$	161	36.7	103	0.394	[6.2]
$Co_{85}P_{15}$	–	39.0	120	–	[6.15]
$Co_{74}Fe_6B_{20}$	162	66.7	175	0.32	[6.16]
$Nb_{50}Ni_{50}$	–	–	132	–	[6.4]
$Cu_{50}Zr_{50}$	–	–	85.3	–	[6.5]
$Cu_{50}Ti_{50}$	–	–	96.7	–	[6.5]
$Be_{40}Ti_{50}Zr_{10}$	–	–	105	–	[6.5]
$Pd_{80}Si_{20}$	160	28.2	80.5	0.416	[6.2]
	182	35.5	66.7 [6.13]	–	[6.18]
	26	80 [6.23]		–	[6.22]
$Pd_{77.5}Cu_6Si_{16.5}$	165	31.3	88	0.41	[6.16]
	182	34.8	82.5 [6.14]	–	[6.19]
$Pd_{40}Ni_{40}P_{20}$			98.1		[6.5]
$Pt_{60}Ni_{15}P_{25}$	202	33.8	96.1	0.421	[6.2]
Sm_2Co_{17}	133	42.6	–	–	[6.20]
$Mg_{70}Zn_{30}$			35		
Ti (polycrystalline)	126	39.2	107	0.36	
Fe (polycrystalline)	169	82.4	211	0.28	
Pd (polycrystalline)	189	44.1	123	0.39	
Cu (polycrystalline)	136	45.5	123	0.35	
Mg (polycrystalline)	34	17.5	45	0.28	

Table 6.2. Bulk modulus B_a of glassy and liquid metals and its relation to the value B for the crystalline state. $\Delta V/V$ gives the observed decrease in the atomic volume during crystallization

	B_a [GPa]	B/B_a	$\dfrac{\Delta V}{V}$ [%]	Ref.
$Pd_{80}Si_{20}$	182	1.062	–	[6.18]
$Pd_{77.5}Cu_6Si_{16.5}$	182	1.065	1.6	[6.19]
Sm_2Co_{17}	133	1.045	1.6	[6.20]
Na (liquid at T_M)	54	1.08	2.5	[6.21]
Cs (liquid at T_M)	16	1.13	2.6	[6.21]

This clearly demonstrates the close similarity in the response of crystalline and disordered structures to compressive strains.

Previously, such results have also been obtained in studies of the compressibility in liquid metals (see, e.g., [6.21]). These authors convincingly showed that the compressibility of the crystalline and the liquid alcaline metals is

Table 6.3. Shear modulus G_a at amorphous and G_c of crystallized alloys

Alloy	G_a [GPa]	G_c/G_a	Ref.
$Pd_{80}Si_{20}$	35	1.34	[6.18]
$Co_{89.5}Sm_{10.5}$	42.6	1.36	[6.20]
$Co_{85}P_{15}$	39	1.34	[6.15]
$Fe_{80}B_{20}$	64.9	1.35	[6.17, 24]
$Ni_{76}P_{24}$	33.5	1.46	[6.12]
$Ni_{76}P_{24}$	35.0	1.29	[6.15]
$Pd_{77.5}Cu_6Si_{16.5}$	34.8	1.35	[6.19]
SiO_2	31	1.38	[6.18]

primarily a function of the atomic volume and depends to a much smaller extent on the structure. Differences in the compressibility between the liquid and solid state at the same atomic volume are about 1 % only. Accordingly, the changes observed during crystallization are primarily caused by the volume changes. *Fenkner* [6.22] also arrived at the same conclusion from a comparison of the sound velocities in liquid and crystalline metals.

Unlike the bulk modulus, the shear modulus of metallic glasses cannot be explained by using such simple arguments. This is already implied by the fact that for liquids, the low frequency shear modulus vanishes. Contrary to the case of compressive strain, shear strain demands no changes in volume and the atoms can respond by a convective flow. Metallic glasses, however, are solids and as such show a stiffness with respect to shear strains. Table 6.3 gives some values for the changes observed on crystallizytion near the glass transition. Typical changes for a variety of different alloys appear to be close to 35 %. The dispersion in the values recorded for the same material point to the fact that the glassy state of as-prepared materials is not always unique. In particular, this effect seems to be more strongly pronounced in electrolytically deposited materials (as the NiP alloy in Table 6.3) than in melt-quenched samples.

Very large annealing effects within the amorphous structure ($\Delta G/G \approx 1$) have been reported for an electrolytically deposited $Ni_{80}P_{20}$ sample by *Kiss* et al. [6.24].

Young's modulus E, which describes the behaviour with respect to uniaxial strain, is closely related to the shear stiffness G. For a sample submitted to a homogeneous uniaxial strain having a Poisson constant of 0.35 (typical for metallic glasses), 90 % of the deformation energy is stored in the shear mode and only 10 % is in the compressive mode. Consequently, E probes similar properties to G and in fact just like G, E also shows substantial changes during crystallization. Table 6.4 gives some values. In agreement with the above argument, the typical relative changes appear to be slightly smaller than those for the shear stiffness. Similarly to the shear modulus, thermal annealing treatments also have a considerable effect on Young's modulus. For the particular PdCuSi sample referred to in Table 6.4, the density increased by 0.5 %

and Young's modulus by 7% during the annealing treatment and by 1.6% and 24% during crystallization. More recently, *Kursumovic* et al. [6.6, 27] studied the kinetics of the changes in Young's modulus during isothermal annealing treatments and were thus able to monitor structural relaxations of metallic glasses.

Table 6.4. Young's modulus E_a of amorphous and E_c of crystallized alloys

	E_a [GPa]	E_c/E_a	Ref.
$Pd_{81}S_{19}$	80	1.26	[6.23]
$Fe_{80}B_{20}$	152	1.24	[6.8]
$Co_{80}P_{20}$	105	1.23	[6.25]
$Ni_{76}P_{24}$	95	1.32	[6.15]
as quenched	88	1.33	[6.26]
$Pd_{77.5}Cu_6Si_{16.5}$	94	1.24	[6.26]
annealed			

The observation of soft elastic shear constants finds its parallel in measurements of the specific heat at low temperatures. Here the lattice contribution to the specific heat of metallic glasses is notably enhanced with respect to the same contribution of the crystallized samples. Likewise the Debye temperatures are lower (see, for instance, [6.19, 28]). This indicates that the phonon density of states for low temperature phonons, presumably mainly for the transverse branches, is higher in the amorphous state than in the crystal.

The softening of the transverse mode extends, therefore, to much higher frequencies than those encountered for elastic studies. Indeed, also in elastic neutron scattering measurements *Suck* et al. [6.29, 30] show relatively soft, short, wavelength transverse collective modes in metallic glasses and an enhanced phonon density at low frequencies.

6.2.3 Theoretical Considerations

Theoretically, the softening of the shear elastic modulus in glassy metals has been anticipated by *Weaire* et al. [6.31]. These authors used pair potentials $V(r)$ to calculate the elastic constants. Since the pair potential approximation is quite a popular method for deriving energy expressions and the model presented helps to understand further features of the elasticity in metallic glass, a short description will be given. Accordingly, the adiabatic moduli M at $T=0$ were derived from the internal energy E_a given by the sum of energies of atoms taken in pairs:

$$M = \frac{1}{\Omega} \frac{\partial^2 E_a}{\partial \varepsilon^2} \quad \text{with} \quad E_a = \frac{1}{2} \int g(r, \varepsilon) V(r) d^3 r. \tag{6.8}$$

Here, Ω is the atomic volume and $g(r, \varepsilon)$ the pair correlation function of the strained atomic structure. Furthermore, for a given $V(r)$ and ε, the equilibrium condition $\delta E_a = 0$ (local minimum in E_a) imposes restrictions on the choice of $g(r, \varepsilon)$.

Obviously this condition demands that the atomic positions are considered as internal variables which arrange themselves (relax) in such a manner as to satisfy the local minimum condition for E_a. Such rearrangements depend on internally generated forces and therefore involve atomic displacements (internal displacements) which in general may differ from those imposed by the macroscopic strainfield. The overall effect of these displacements leads to a partial stress relaxation which finally finds its expression in relatively soft elastic moduli. In agreement with the observations made on the changes of the elastic constants, it appears also natural that such displacements are less likely to occur under isotropic compression than under pure shear deformation.

For the actual calculations, *Weaire* et al. [6.31] used computer simulation techniques to determine the relaxed pair correlation function of a cluster containing about 100 atoms. The results of these calculations yielded estimates of the changes in G and B in agreement with experiments.

Alternatively, elastic constants neglecting internal displacements were also calculated. In this case the strained pair correlation function was obtained by the principle of the local conservation of matter, that is, the atomic coordinates were changed according to the transformation prescribed by the macroscopic strain.

Since internal displacements are neglected, the changes in atomic coordinates depend on purely central forces (given by the pair potential). A structure described by a radially symmetric pair correlation function allows each atom to be a centre of symmetry. Consequently, this model satisfies the requirements necessary for the Cauchy relations. The Cauchy relations hold rigorously for a structure with a centre of symmetry at each atom, bonded by central forces. For an isotropic material $c_{11} = c_{12} + 2c_{44}$, the Cauchy relations demand in addition that $c_{12} = c_{44}$ which leaves us with one independent elastic constant only. In particular,

$$B = \tfrac{1}{3}(c_{11} + c_{12}) = \tfrac{5}{3}G. \tag{6.9}$$

This also means that fractional changes $\Delta B/B$ and $\Delta G/G$ occuring during crystallization have to be identical. Calculations have in fact yielded values typical of the changes encountered in the bulk modulus. It should, however, be pointed out that metallic glasses as well as all liquids and most crystalline solids do show substantial deviations from the Cauchy-relations.

An alternative explanation for the low shear stiffness in metallic glasses has been given more recently by *Cyrot-Lackmann* [6.32] in terms of soft atomic vibrations. The energy as a function of the atomic positions was calculated in a tight-binding model for the attractive part and a Born-Mayer type potential for the repulsive part. The bulk modulus was determined directly from the energy variation under compression and found to be similar in crystalline and amorphous metals. The shear modulus was derived by considering atomic vibrations in an amorphous structure. This is presumably allowed since the transverse softening extends over a large frequency range and also shows up in the specific heat and the Debye temperature.

Accordingly, the Debye frequency of an amorphous structure is reduced with respect to a crystalline structure with similar interatomic force constants. In a macroscopically polarized wave, individual atoms tend to vibrate in locally preferred directions. These preferred directions correspond to directions of the lowest Einstein frequencies (local vibration frequency with fixed neighbours) and have random orientation at different sites. Due to these preferred directions of individual atoms, the Debye frequency (collective mode) corresponds to the average of the minimum Einstein frequencies rather than the directional average of the Einstein frequencies at one site which with similar interatomic forces, would scarcely differ from the crystalline case. The softening of the shear stiffness is now explained by the fact that the Debye frequency depends mainly on the transverse sound velocity which in turn is directly related to the shear modulus.

A final remark in this section on the elasticity of metallic glasses concerns Poisson's ratio v for which several direct measurements have been carried out (Table 6.1). A direct comparison with crystallized alloys of the same composition does not appear to exist as yet. However, compared to pure crystalline metals, the values for the amorphous alloys are slightly larger. In fact from the changes observed in the other elastic constants and the relationship between v and these constants for isotropic solids, we expect v-values for the metallic glasses to be 3–7% larger than in the crystalline state.

As pointed out by *Köster* and *Franz* [6.33], the importance of Poisson's ratio lies in the fact that, in particular for metals, this number contains more direct information on binding forces than any other elastic constant. We have mentioned above that the elastic behaviour of an isotropic solid for which the Cauchy relation $c_{12}=c_{44}$ holds is described by one independent elastic constant only. For such a solid, Poisson's ratio is independent of the material

$$v = \frac{c_{12}}{c_{11}+c_{12}} = \frac{1}{2}\frac{c_{12}}{(c_{12}+c_{44})} = \frac{1}{4}. \tag{6.10}$$

Metallic glasses, however, violate the requirements for the Cauchy relations to hold. The atoms in a glass do not lie at centres of symmetry, or only rarely so, and noncentral forces contribute to a large extent, as in crystalline metals, to the cohesive energy. The main contribution is from the energy of the conduction electrons which depends on the volume rather than on the exact position of the atoms. Deviations from 0.25, the universal value of Poisson's ratio, provide therefore a measure for the importance of noncentral forces mainly due to the electron gas and internal displacements.

6.2.4 Magnetoelastic Behaviour

It is well known that the appearance of magnetic order not only shows up in the primary magnetic quantities but also affects most other solid state properties. This in particular is also true for the elasticity in ferromagnetic metallic glasses. The magnetoelastic phenomena are related to the occurrence of magnetostriction. In magnetism, the magnetostriction is generally considered to be a secondary effect. Since most magnetic metallic glasses are, however, extremely soft magnetically, such contributions are readily observable and may in certain cases dominate the first-order magnetic and elastic behaviour. With the onset of spontaneous magnetic order below the Curie temperature, the lattice assumed to be free of external constraints becomes slightly deformed both in volume and shape. For the following discussion it is assumed that whenever the temperature changes, the normal thermal expansion (at constant M) as observed in the paramagnetic state is always subtracted from the effective strain ε_{ij}. The remaining magnetostrictive deformations ε_{ij}^m in the magnetically ordered phase depend then on the direction and modulus of the spontaneous magnetization vector $M_s(T)$. The most general relationship between the components of the magnetostrictive strain tensor ε_{ij}^m and the components of magnetization M_i (assumed to be homogeneous throughout the sample) can be represented by a power series of homogeneous products $(M_k M_l)^n$ of order n ($n = 0, 1, 2, 3, \ldots$). Since on reversing the magnetization the strains must be left unchanged and furthermore, $\varepsilon_{ij}^m(M=0)=0$, only even orders $n \geq 2$ can appear. The lowest order terms, which in most cases already give a reasonable description of the experimental data, are therefore of the form

$$\varepsilon_{ij}^m = \Lambda_{ijkl} M_k M_l. \qquad (6.11)$$

This relation is of identical form to Hooke's law, which means that the symmetry of the fourth-order tensor Λ_{ijkl} of the magnetostrictive strain constants is identical to the symmetry of the fourth-order tensor of the elastic constants. For simplicity Voigt's notation can also be used. Consequently, for isotropic materials such as metallic glasses, we have at most two independent constants which similarly to the elastic constants are most conveniently taken as the constants describing the volume and shape changes. The direct significance of these constants becomes obvious when we separate the magnetostriction tensor into pure volume $\varepsilon_{ij}^{m,\,vol}$ and shape $\varepsilon_{ij}^{m,\,sh}$ strains

$$\varepsilon_{ij}^m = \varepsilon_{ij}^{m,\,vol} + \varepsilon_{ij}^{m,\,sh}$$

$$\varepsilon_{ij}^{m,\,vol} = \frac{\omega_s}{3}\delta_{ij}; \qquad \omega_s = \sum_i \varepsilon_{ii} = \frac{\Delta V}{V}$$

$$\varepsilon_{ij}^{m,\,sh} = -\frac{\lambda_s}{2}\delta_{ij} + \frac{3\lambda_s}{2}\frac{M_i M_j}{|M_s|^2}. \qquad (6.12)$$

It follows that ω_s is the relative change of volume $\Delta V/V$ due to the occurrence of spontaneous magnetization $M_s(T)$, and λ_s (saturation or linear magnetostriction) is the relative length change $\Delta l/l$ of the sample in the direction of the magnetization ($\vartheta = 0$) with respect to an imaginary reference sample of the same volume but with disordered magnetic moments:

$$\frac{\Delta l}{l} = \tfrac{3}{2}\lambda_s(\cos^2\vartheta - \tfrac{1}{3}). \tag{6.13}$$

Here ϑ is the angle between the direction of M_s and the direction in which $\Delta l/l$ is measured. Consequently, ω_s can only be observed when $|M_s(T)|$ changes, as is for instance the case when the temperature is varied, and the effect of λ_s can be observed when the direction of M_s at constant temperature is rotated.

For samples containing more than one domain, no such rigorous statements can be made. Assuming, however, homogeneous superposition of the strains and random orientation of the domain magnetizations in the demagnetized state: the saturation magnetostriction λ_s gives the relative change of length in the direction of the total magnetization between the demagnetized and saturated states. Similarly we can argue that within the same approximation, no volume changes occur when the sample is magnetized at constant temperature since the volume strain does not depend on the orientation of M_s.

a) Isotropic Effects

The volume effect can, in practice, be applied to synthesize materials with vanishing thermal expansion. In cases where the volume increases with $|M_s|$ ($\omega_s > 0$), the magnetic volume strain may compensate or even overcompensate the regular thermal expansion due to the anharmonicity of the atomic vibrations. Metals with a fully compensated thermal expansion coefficient are known as invar alloys. Several metallic glasses have been found to exhibit the invar effect. *Fukamichi* et al. [6.34–36] observed typical invar characteristics in the binary alloys Fe–B, Fe–P, and Gd–Co. In the Fe–B system, the room temperature coefficient of thermal expansion α is zero at 17% B, and below this concentration α exhibits a broad minimum around 14% B. For the Fe–P and Gd–Co alloys, the total coefficient of thermal expansion is always positive but strong anomalies occur below the Curie temperature. Invar characteristics with fully compensated thermal expansion coefficients have been reported by *Shirakawa* et al. [6.37] in $(Fe_xCo_yNi_z)_{90}Zr_{10}$ and by *Jagielinski* et al. [6.38] in $(Fe_xCo_yNi_z)_{78}Si_8B_{14}$. For the Fe-rich alloys the total magnetostrictive volume increase ω_s at 0 K is of the order of 1×10^{-2}. This is comparable to the volume decrease that occurs on crystallization of metallic glasses. A substantial effect on the coefficient of thermal expansion was found by *Fukamichi* et al. [6.39] after cold-rolling samples of Fe–B glasses. The authors claim that these modifications are due to anisotropic rearrangements in the short-range order (increased number of Fe–Fe pairs) in the direction of rolling which was also the direction of measurement.

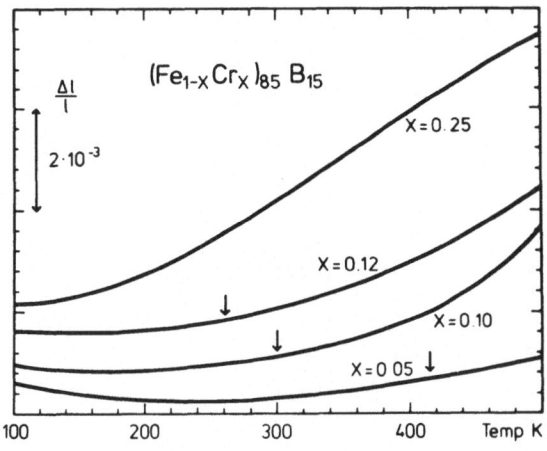

Fig. 6.1. Thermal expansion of $(Fe_{1-x}Cr_x)_{85}B_{15}$ glasses. To include all curves in the same figure they have been arbitrarily shifted. Alloys with low Cr contents have negative coefficients α of thermal expansion at low temperatures $[\alpha = \partial(\Delta l/l)/\partial T]$

Figure 6.1 shows the thermal expansion of glassy $(Fe_{1-x}Cr_x)_{85}B_{15}$. Well above the Curie temperatures (indicated by arrows) the magnetic contributions are negligibly small and the curves therefore show regular thermal expansion. At lower temperatures, however, strong magnetic contributions become evident and within limited temperature ranges give rise to temperature independent sample lengths. At even lower temperatures the regular thermal expansion is overcompensated and therefore the samples shrink with increasing temperature.

b) Anisotropic Effects

The shape effects (deformations at constant volume) stem from the fact that the magnetostrictive deformation is unaxially anisotropic when $\lambda_s \neq 0$. Accordingly, a spherical sample consisting of a single domain deforms when cooled below the Curie temperature (apart from isotropic volume changes) into a rotationally symmetric ellipsoid with the symmetry axis in the direction of the magnetization M_s. The strain along this direction measured with respect to a spherical sample of the same volume is λ_s and $-\lambda_s/2$ in all directions perpendicular to M_s. λ_s may be positive or negative and accordingly we speak of positive or negative linear magnetostriction. Numerous investigations of λ_s and its temperature dependence have been made in metallic glasses [6.38–64]. Values for some typical alloys are given in Table 6.5. For Fe-rich glasses, λ_s is of opposite sign and larger in magnitude ($\lambda_s = 20$–40×10^{-6}) than in polycrystalline iron ($\lambda_s = -9 \times 10^{-6}$) [6.64]. Co-rich alloys show negative magnetostrictions ($\lambda_s \approx -4 \times 10^{-6}$) but have lower values than polycrystalline Co ($\lambda_s \approx -62 \times 10^{-6}$) [6.64]. Exceptionally large magnetostrictions have been observed for sputtered samples of Tb–Fe$_2$ [6.50, 64]. In the crystallized state this alloy shows the largest positive magnetostriction known.

Apart from the Co alloy cited in Table 6.5, there are a number of similar magnetostrictive zero alloys known. In glassy Co alloys, the addition of Fe or

Table 6.5 Linear saturation magnetostriction λ_s os some amorphous alloys at room temperature

		λ_s	Ref.
$Fe_{80}B_{20}$		$+\quad 35 \times 10^{-6}$	[6.49]
$Fe_{80}B_{20}$	4.2 K	48	[6.49]
$Fe_{80}B_{20}$		39	[6.52]
$Fe_{40}Ni_{40}B_{20}$		14	[6.42]
$Fe_{40}Co_{40}B_{20}$		20	[6.42]
$Co_{80}B_{20}$		$-\quad 4$	[6.42]
$Co_{80}B_{20}$		$-\quad 6.2$	[6.52]
$Co_{75}Si_{15}B_{10}$		$-\quad 3.5$	–
$Co_{46.9}Ni_{35}Fe_{0.7}P_{17.4}$		$-\quad 12$	[6.40]
$Co_{70.4}Fe_{4.6}Si_{15}B_{10}$		$\sim\quad 0$	[6.58]
$Gd_{23}Co_{77}$		32	[6.53]
$TbFe_2$		308	[6.64]
$DyFe_2$		38	[6.64]
Crystalline metals			
$TbFe_2$		1753	[6.64]
Fe		$-\quad 9$	[6.64]
Ni		$-\quad 33$	[6.64]

Mn shifts the magnetostriction towards positive values. Depending on the metalloids in the Co glass, λ_s changes sign between 3–6 % Fe and at 7 % Mn [6.58–60]. Similar Fe and Mn concentrations are necessary in Co–Fe and Co–Mn based alloys when Ni [6.38, 58, 59, 61–63] or one of the metals V, Nb, Ta, Cr, Mo, W, Mn [6.63] is partly substituted for Co.

These particular alloys do not show any magnetoelastic shape effects since $\lambda_s = 0$, but they are of great technical importance because by virtue of the reciprocal action of the magnetoelastic coupling (on the magnetic properties), large magnetostriction constants usually have a negative influence on the soft magnetic properties. For metallic glasses this is of particular importance since the magnetically extrinsic properties (permeability coercitive field, core losses) are to a large extent controlled by the magnetoelastic stress anisotropy.

c) The ΔE Effect

The elastic manifestation of the magnetoelastic coupling effect on which we shall focus our interest now, results from the interaction of magnetostrictively deformed domains with an external stress field. For this purpose consider a deformed domain embedded in an elastic matrix subjected to a uniaxial stress. It is obvious that for $\lambda_s > 0$, the energetically most favourable orientation is the one in which the λ_s axis is parallel to the applied stress, assuming that there are no other constraints such as an external field H or internal stresses favouring other orientations. Since the magnetization vector M_s is "rigidly" fixed to the λ_s axis, M_s rotates together with this axis. The total strain ε_{ij} (at $H = 0$ or more generally $H = \text{const}$) resulting from the application of the stress σ is therefore

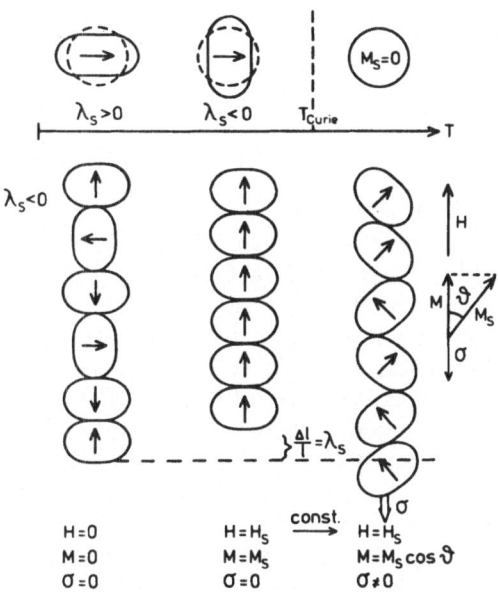

Fig. 6.2. Schematical representation of linear magnetostriction and magnetoelastic strain for a material with negative linear magnetostriction (for explanation see text)

the sum of the purely elastic strain ε_{ij}^{el} (at M_s = const) and the strain $\varepsilon_{ij}^{m,sh}$ due to the reorientation of the magnetization direction. This additional strain contribution leads to a softening of the elastic constants compared to the case at constant M_s.

Figure 6.2 summarizes schematically the effects discussed above for a material with negative linear magnetostriction λ_s. Initially the sample is in its demagnetized state. The magnetization directions in the deformed domains, represented by ellipsoids, have random orientations. Under conditions of no external and internal constraints on the orientations of the magnetization directions M_s, the sample shrinks ($\lambda_s < 0$) when an external field H is applied to magnetically saturate the sample. In the third step the field H is kept constant and the uniaxial traction σ is applied. The magnetization directions have now a tendency to arrange themselves along a plane perpendicular to the stress axis ($\vartheta \to 90°$). This equilibrium state is achieved by magnetization rotation and by domainwall motion. The interaction with the external $H(\mu_0 H \cdot M_s)$, however, opposes a full rotation and an intermediate angle ϑ depending on H and σ is an energetically more favourable orientation. The equilibrium angle ϑ, and therefore ε_{ij}^m, is obtained by minimizing the interaction energy of the magnetostrictively strained domain (corresponding to a magnetic and elastic dipole) with the external stress and magnetic fields. Neglecting interactions among domains and other possible magnetic anisotropies, this energy per unit volume of the domain is given by

$$g_m(\sigma, H) = -\frac{3\lambda_s}{2} \sigma \cos^2 \vartheta - |H| |M_s| \cos \vartheta, \tag{6.14}$$

and from the equilibrium condition

$$\frac{\partial g_m(\sigma, H)}{\partial \vartheta} = 0,$$ (6.15)

we obtain the angle ϑ_{eq}: $\cos\vartheta_{eq} = -|H||M_s|/3\lambda_s\sigma$.

Thermodynamically, $g_m(\sigma, H)$ is the magnetoelastic part of the magnetic free energy (σ and H are its natural variables). To get the total free energy g we have to add the elastic part $g_{el} = -(2\cdot E^M)^{-1}\sigma^2$ with the elastic compliance $1/E^M$ measured at constant M_s. The total strain is finally given by the first derivative of $g = g_{el} + g_m$ with respect to σ and the elastic compliance $1/E^H$ at constant H by the negative value of the second derivative. The softening of the elastic constant as a function of stress and field is described here by the contribution of g_m. Strictly speaking, only elastic constants related to shear deformations (E and G) are affected. An isotropic compressional stress has no effect on the orientation of the magnetization directions.

The expression derived above for the magnetic free energy cannot yet account for the experimental observations on the ΔE effect. As can be readily seen in the absence of a magnetic field H, our present approximation would predict a complete polarization of the λ_s axis for even the smallest uniaxial traction. The total magnetic moment of the sample may, nevertheless, vanish since M_s and $-M_s$ are equivalent orientations when $H = 0$. An applied field would give a certain stiffness for ε^m but not enough to explain the experimental data.

Since in metallic glasses the magnetocrystalline anisotropy averages out, internal stresses must play the dominant role. Such internal stresses impose locally preferred orientations for directions of the magnetization (see, e.g., *Kronmüller* [6.65, 66]) and therefore oppose an easy polarization of the λ_s axis.

Following the proposal made by *Becker* [6.67], the effect of a uniaxial internal stress σ^i having an angle ϑ_0 with the external uniaxial stress σ^{ex} can be described by assuming that in the absence of an external field H and stress σ^{ex}, the λ_s axis is polarized by σ^i. The effect of the external stress is to rotate M_s from the initial direction ϑ_0 (for $\lambda_s > 0$) into the new equilibrium direction ϑ. For $H = 0$, g_m is given by

$$g_m(\sigma^{ex}) = -\tfrac{3}{2}\lambda_s\sigma^{ex}\cos^2\vartheta - \tfrac{3}{2}\lambda_s\sigma^i\cos^2(\vartheta_0 - \vartheta).$$ (6.16)

For $\sigma^i \gg \sigma^{ex}$, the equilibrium direction ϑ, to be determined such that g_m is minimized, is close to ϑ_0 and the elastic compliance at $H = 0$ follows from the total free energy for $\sigma^{ex} \to 0$:

$$-\frac{\partial^2 g}{\partial(\sigma^{ex})^2} = \frac{1}{E^H} = \frac{1}{E^M} + \frac{3|\lambda_s|}{\sigma^i}\cos^2\vartheta_0\sin^2\vartheta_0$$ (6.17a)

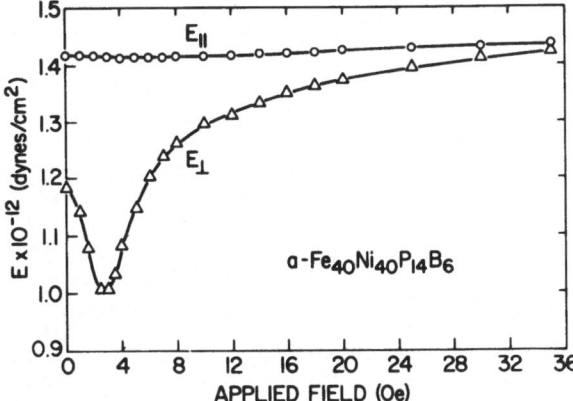

or

$$\frac{\Delta E}{E^H} = \frac{E^M - E^H}{E^H} = \frac{3E^M|\lambda_s|}{\sigma^i} \cos^2 \vartheta_0 \sin^2 \vartheta_0 . \tag{6.17b}$$

Accordingly, the ΔE effect parallels the behaviour of the linear magnetostriction λ_s and depends strongly on internal stresses which to some extent can be controlled by thermal annealing treatments. More generally, also other structural and magnetic anisotropies as well as all kinds of constraints opposing an easy domain motion may, if present, act in a similar way. In cases where anisotropies other than internal stress anisotropy ($\lambda_s \sigma^i$) are important, a ΔE effect proportional to λ_s^2 is to be expected rather than proportional to λ_s. The λ_s^2-dependence comes from the fact that the degree of polarization and the magnetic strain are both proportional to λ_s yielding $\Delta E \sim \lambda_s^2$. For the particular case of stress anisotropy, which seems to be the most important contribution in metallic glasses, the opposing forces are also $\sim \lambda_s$.

The orientation of the domain magnetization is also an important parameter. The maximum effect occurs when $\vartheta_0 = 45°$. In metallic glass, such angle dependent effects have been demonstrated by *Berry* and *Pritchet* [6.68] on field annealed samples. Such annealing treatments are known to induce an easy axis for the magnetization resulting mainly in 180° domain walls. Since for $H = 0$, $+M_s$ and $-M_s$ are equivalent orientations, the external stress can exert no force on the domain walls and when in addition the orientation is such that the domain magnetizations are mainly parallel or perpendicular ($\vartheta_0 = 0$ or 90°) to the applied traction. Then the ΔE effect can be expected to remain small. The situation may, however, change when some of the magnetization directions are rotated into a more favourable direction $\vartheta_0 \sim 45°$ by an external field H.

Figure 6.3 shows the results obtained by *Berry* [6.25] following magnetic annealing treatments parallel ($E_\parallel$) and perpendicular ($E_\perp$) to the ribbon axis. For the measurements of Young's modulus, the stress and the magnetic field

Table 6.6. Maximum ΔE effect for various metallic glasses at room temperature

	$\Delta E/E$	Annealing conditions			Ref.
		[min]	[°C]		
$Fe_{82}B_{18}$	0.7	120	370	–	[6.72]
$Fe_{80}B_{20}$	0.63	5	373	–	[6.77]
$Fe_{80}P_{13}C_7$	0.8	20	350	–	[6.71]
$Fe_{75}P_{15}C_{10}$	0.8	30	~300	$H_\perp$	[6.68]
	0	Crystallized		–	[6.68]
$Fe_{78}Si_{10}B_{12}$	1.75	5	373	–	[6.77]
$Fe_{78}Si_{10}B_{12}$	1.9	10	410	–	[6.44]
$Fe_{80}Si_5B_{15}$	4.5	2	350	$H_\perp$	[6.78]
$Fe_{71}Co_9B_{20}$	2.2	15	384	–	[6.77]
$Co_{70}Si_5B_{25}$	0.024	–	–	–	[6.54]
$TbFe_2$	$\Delta G/G = 0.12$ (at 10 MHz)				[6.84]

was applied along the ribbon axis. In contrast to the substantial minimum in Young's modulus exhibited by the transversly annealed sample, the longitudinally annealed sample shows only a very minor ΔE effect ($E^H \sim E^M$). For this material the maximum ΔE effect $\Delta E/E^H \approx (E_\parallel - E_\perp)/E_\perp$ is 0.4 which means that in the $E_\perp$ sample, the magnetic strain $\varepsilon^{m,\,sh}$ amounts to 40 % of the elastic strain. The magnitude of the elastic anisotropy obtainable is therefore very appreciable and provides an example of where amorphous materials do not always behave in an isotropic manner.

d) Experimental Results for the ΔE Effect

Numerous measurements of the ΔE effect and its temperature dependence have been performed in recent years (e.g., [6.25, 68–87]). Most of these papers deal with transition metal-metalloid alloys. Only a few studies were done on transition metal-rare earth metal alloys [6.84–87]. Table 6.6 gives some values for the maximum ΔE effect observed in various metallic glasses. It should be noted that contrary to λ_s, $\Delta E/E$ is an extrinsic elastic property and as such depends strongly on purity, method of production and sample history. The values quoted in Table 6.6 correspond to the fully relaxed difference between E^M and the minimum of E^H which occurs, in general, at a non zero external field H. For reasons to be discussed in Sect. 6.3, measurements intended to obtain the fully relaxed ΔE effect on ribbon shaped samples of thicknesses about 50 μm should be done below 10 kHz. Furthermore, to achieve these maximum values, rather well-defined (time and temperature) annealing treatments are necessary. In two cases the annealing was done in a magnetic field perpendicular to the direction of measurement. In particular, the results quoted for $Fe_{78}Si_{10}B_{12}$ and the similar alloy $Fe_{80}Si_5B_{15}$ show the strong enhancement of ΔE due to transverse field annealing.

The result given for $TbFe_2$ (Table 6.6) cannot be compared directly to the data of transition metal-metalloid alloys. This data was obtained from ultrasound velocity measurements at 10 MHz and may not correspond to the fully relaxed ΔE effect. Indeed, from the exceptionally large value for λ_s of $TbFe_2$ (Table 6.5) on the one hand, we might also expect a large ΔE effect. On the other hand, the sputtered transition metal-rich amorphous alloys also exhibit substantial anisotropies leading to large coercive fields at low temperatures.

e) Effect of Annealing on ΔE

The dependence of the ΔE effect on longitudinal and transverse field annealings has already been discussed in terms of the angular dependence between the directions of the applied stress and an intrinsic or induced uniaxial magnetic anisotropy. Annealings at zero field give intermediate curves for $\Delta E/E$. But by comparing total length changes as a function of the applied field between samples annealed in a longitudinal, transverse, and a zero field, *Berry* and *Pritchet* [6.68] found that the zero-field curve was surprisingly close to the longitudinal field curve and concluded that, probably due to shape anisotropy, the zero field annealed sample does not show a truly random orientation of the local domain magnetization M_s in its initially demagnetized state.

Taking advantage of the extrinsic nature of the ΔE effect, further information on domain motion opposing restraints can be obtained from thermal annealing treatments. As already mentioned, the magnetoelastic stress anisotropy appears to be the dominant pinning force for local magnetization directions in transition metal glasses with nonvanishing linear magnetostriction. The relation due to Becker then suggests that when λ_s is known, $\Delta E/E$ can be used to measure average values for the nonisotropic internal stresses.

Berry and *Pritchet* [6.69] applied this idea to study the annealing behaviour of such internal stresses in a $Fe_{75}P_{15}C_{10}$ glass. Their results indicate that with increasing annealing temperature, the internal stresses decrease linearly to almost the usually assumed minimum value $\lambda_s E^M$. Complementary to this decrease, $\Delta E/E$ increased. The minimum value $\lambda_s E^M$ corresponds to the self-imposed magnetostrictive stress and comes from the mismatch of the idealized magnetic strain ε^m at the domain walls.

This minimum stress was also used [6.25] to estimate the order of magnitude of the maximum possible ΔE effect. We have already noticed that the maximum response occurs when ϑ_0, the angle between stress and magnetization, is 45°. Inserting, furthermore, $\lambda_s E^M$ for σ^i gives $\Delta E/E \approx 0.75$. This upper limit is, however, quite far below the substantial value $\Delta E/E = 4.5$ observed for a field annealed sample of $Fe_{80}B_{15}Si_5$ by *Brouha* and *van der Borst* [6.78]. Obviously in field annealed samples, polarization of the λ_s axis in the demagnetized state is so strong that the idealized magnetic deformations ε^m can be superimposed in a more or less coherent manner leaving the internal stresses at a much lower level. A rather surprising feature of the magnetic annealing treatments is also the high degree of reversibility. Even after several consecutive treatments in a longitudinal and transverse field, the initial response can be closely reproduced [6.70].

The time and temperature conditions necessary to achieve the optimum ΔE effect are also rather interesting. In most cases [6.44, 70, 72, 77–79], the best results were achieved with annealing temperatures close to the crystallization temperature. At these temperatures the relaxations proceed quite fast so that after a few minutes (~ 30 min), equilibrium is achieved. Prolonged annealing times at high temperatures were found to have a negative effect whereas at low temperatures the optimum ΔE values cannot be achieved [6.44, 77]. A detailed analysis of the relaxation kinetics was done for $Fe_{40}Ni_{40}P_{14}B_6$ and $Fe_{75}P_{15}C_{10}$ [6.70]. An Arrhenius law with activation energies of 1.35 and 1.0 eV, respectively, was found to give a good description of the experimental data.

f) Temperature Dependence of the ΔE Effect

For metallic glasses where the magnetic anisotropy is mainly controlled by internal stresses, the ΔE effect should be, as discussed earlier, proportional to the linear magnetostriction. Indeed, measurements of the temperature dependence of $\Delta E/E$ for Fe-based glasses closely followed the temperature dependence of λ_s [6.73]. A detailed analysis of such curves gives information on the nature of the atomic interactions responsible for the magnetostriction [6.46, 47].

For practical purposes, the temperature dependence of the ΔE effect may be applied to control the temperature dependence of the shear or Young's modulus. Like all magnetostrictive effects, the ΔE effect also decreases with increasing temperature and finally vanishes at the Curie-temperature. Thereby the decrease of the magnetic softening compensates for the increasing thermal softening of Young's modulus E^H. The same applies to the shear modulus also. Materials with a temperature independent Young's modulus are called elinvar alloys. With the large ΔE effects reported for certain metallic glasses, it is relatively easy to find a proper alloy concentration and a suitable annealing treatment to arrive at a temperature independent shear or elastic modulus. Figure 6.4 shows the temperature dependence of several metallic glasses near room temperature at $H=0$. Prior to measuring the alloys were thermally annealed for two hours at 200 °C.

The practical usefulness of this effect to control the temperature coefficient of the elastic constants E and G is, however, limited because of its sensitivity to external magnetic fields. Of much greater value for this purpose are intrinsic effects which also depend on the magnetization but have already shown up in the elastic constants of magnetically saturated samples [i.e., $E^M(M)$]. Like the invar effect, the contribution of the intrinsic effect can only be defined with reference to Young's modulus of a hypothetical paramagnetic sample at the

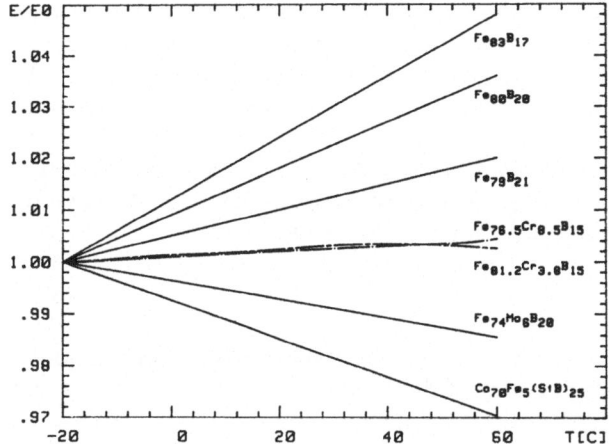

same temperature. Practically, these reference values correspond to the extrapolated curve of the paramagnetic state. The difference between the extrapolated curve and the effective value E^M is called the intrinsic ΔE effect. Sometimes both contributions taken together are referred to as the ΔE effect. In metallic glasses little is known about the intrinsic ΔE effect. *Kikuchi* et al. [6.81] claim that such an effect is present in $Fe_{83}B_{17}$ whereas no such anomalies were found by *Török* and *Hausch* [6.76] in Fe–Ni based glasses, even though the well-known crystalline Fe–Ni elinvar alloys depend mainly on the intrinsic effect.

The exact determination of the intrinsic ΔE effect in metallic glasses is in many cases difficult because the Curie temperature is often close to the crystallization temperature (e.g., $Fe_{83}B_{17}$) so that extrapolated curves become very uncertain. Furthermore, the vibrating reed technique, which is the preferred measuring technique when a high relative accuracy is needed, suffers from a major drawback when strongly magnetic materials have to be measured in an external field. Under these conditions the flexural and torsional vibrations are stiffened by external forces thus simulating also stiffer elastic constants. The effect was termed the pole effect by *Berry* and *Pritchet* [6.70, 88] who also calculated its magnitude. More refined calculations of this effect were given by *Wenger* and *Török* [6.89].

6.3 Anelasticity of Metallic Glasses

When discussing the elastic properties, we omitted how the deformation proceeded in time. For a given stress we were more interested in the equilibrium value of the deformation as given by Hooke's law. On closer examination, however, all solids show in addition to an immediate elastic response a time dependent creep and the equilibrium value of deformation is achieved only after a certain period of time. The same is true after releasing the load. A deformation is called anelastic if after a loading and deloading cycle the deformation is fully reversible, i.e., the shape of the sample is left unchanged [6.90].

In many cases this creep behaviour can be described by a single relaxation time τ (standard anelastic solid). Instead of Hooke's law the stress-strain behaviour of the standard anelastic solid is then given by

$$J_r\sigma + J_u\tau\dot{\sigma} = \varepsilon + \tau\dot{\varepsilon} , \tag{6.18}$$

where R_r and J_u are the relaxed and unrelaxed compliances (reciprocal moduli). Here σ and ε refer to simple modes of deformation only as, e.g., pure shear ($J = 1/G$) or uniaxial strain ($J = 1/E$). For more complicated deformations the tensor notation would of course be necessary. Hooke's law is obtained from this relation for the two limiting cases where τ is much smaller or much longer than the time (frequency) of measurement. The anelastic behaviour is caused by stress-induced relaxation processes which operate within the material and have a time-lag with respect to the direct elastic response. The mechanical stress may, therefore, be used as a sensitive experimental probe to investigate the kinetics of such relaxations.

6.3.1 Experimental Techniques

Basically the same experimental techniques as those already discussed for elastic measurements are also suitable for anelastic studies. The choice of the actual measuring method is, however, to some extent dictated by the length of the relaxation time τ involved in a particular relaxation process. For long relaxation times $\tau \cong 1$ s, the anelasticity can be studied directly by observing the time dependence of the creep after subjecting the sample to a constant stress σ_0 (elastic after effect). If the sample obeys the above relation for the standard anelastic solid, the strain increases as

$$\frac{\varepsilon(t)}{\sigma_0} = J_u + \delta J(1 - e^{-t/\tau}), \tag{6.19}$$

where $\delta J = J_r - J_u = \varepsilon_{an}/\sigma_0$ is called the relaxation strength. ε_{an} is the anelastic contribution to the strain. Figure 6.5 shows schematically the response of an anelastic solid for a loading and deloading cycle with a constant stress σ_0. Measurements on metallic glasses using this technique have been done by *Berry* [6.25], *Maddin* and *Masumoto* [6.91], *Kimura* et al. [6.92], and *Woldt* and *Neuhäuser* [6.93]. For faster relaxation times ($1 \mathrm{s} \gtrsim \tau \gtrsim 10 \,\mu\mathrm{s}$), the anelasticity can be measured indirectly by the internal friction Q^{-1} of an oscillating sample and for relaxation times at the lower end and below this range by

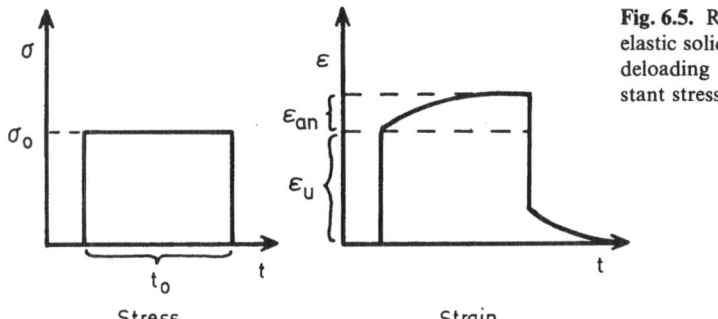

Fig. 6.5. Response of an anelastic solid for a loading and deloading cycle with a constant stress σ_0

ultrasonic attenuation. The internal friction is defined as the reciprocal value of the quality factor Q of an oscillating sample. Alternatively, Q^{-1} may be obtained from the loss tangent $\tan \delta$ [δ being the phase shift between $\sigma(t)$ and $\varepsilon(t)$] or the logarithmic decrement $\Delta = \ln(A_n/A_{n+1})$, where A_n and A_{n+1} are the amplitudes of two successive vibrations during the free decay of a natural vibration.

The following simple relationships between these quantities and the relative energy loss $\Delta W/W$ (ΔW is the energy dissipated per cycle and W is the total energy of the vibration) are valid for $Q^{-1} \lesssim 0.1$:

$$Q^{-1} = \tan \delta = \frac{\Delta}{\pi} = \frac{1}{2\pi}\frac{\Delta W}{W}. \tag{6.20}$$

Similarly, the ultrasonic attenuation α of a wave propagating through an anelastic solid is defined from the exponential decay of its envelope $U(x) = u_0 \exp(-\alpha x)$. α is related to the internal friction by $Q^{-1} = \alpha\lambda/\pi$, where λ is the ultrasonic wave length.

In dynamic experiments it is convenient to introduce the complex compliance $J^* = J_1 - iJ_2$. For harmonic excitation $\sigma(t) = \sigma_0 \exp(i\omega t)$, the strain is then given by $\varepsilon(t) = (J_1 - iJ_2)\sigma(t)$, where for a standard anelastic solid, (6.18)

$$J_1(\omega) = J_u + \frac{J_r - J_u}{1 + (\omega\tau)^2},$$

$$J_2(\omega) = (J_r - J_u)\frac{\omega\tau}{1 + (\omega\tau)^2},$$

and

$$Q^{-1}(\omega) = \frac{J_2}{J_1}\left(\simeq \frac{J_2(\omega)}{J_u} \quad \text{for} \quad Q^{-1} \ll 1\right). \tag{6.21}$$

The bell-shaped curve $Q^{-1}(\omega\tau)$ (Debye-Peak), which describes the energy dissipation, has its maximum for $\omega\tau = 1$.

In thermally activated relaxation processes, τ is a function of temperature so that the damping may achieve its peak value also at a fixed vibration frequency.

The relaxation processes which so far have been identified in metallic glasses are due to the thermoelastic effect, the magnetamechanical coupling (ΔE effect) and structural rearrangements involving atomic jumps. These will be discussed in turn below.

6.3.2 Thermoelastic Relaxation

Thermoelastic relaxation is a fundamental effect occurring in all materials and has, in fact, no direct relation to the amorphous nature of our samples. However, since in most internal friction measurements on thin samples the effect gives significant contributions, we have to be aware of its manifestations. Furthermore, experiments on thermoelastic relaxation allow for an almost isothermal determination of the thermal conductivity and the thermal expansion.

The thermal expansion of matter when heated is a well-known manifestation of thermoelastic interaction. The reciprocal phenomena, namely, heating of matter by applying a stress or a strain under adiabatic conditions, is an alternative expression of the same effect. Its contribution to internal friction in flexurally vibrating reeds arises from transverse thermal currents which flow as a consequence of the nonhomogeneous temperature distribution ΔT within the sample caused by thermoelastic heating and cooling of the sample:

$$\Delta T = -\frac{\alpha T}{c_p}\sigma. \tag{6.22}$$

As *Zener* [6.94, 95] has demonstrated, the internal friction Q^{-1} due to this effect can be described by a simple Debye peak:

$$Q^{-1} = \frac{\alpha^2 E T}{c_p}\frac{\omega\tau}{1+\omega^2\tau^2}. \tag{6.23}$$

α is the linear thermal expansion coefficient, E Young's modulus, c_p the specific heat and T the absolute temperature. The peak damping occurs at the frequency $\omega_0 = 1/\tau$, where τ is the relaxation time for the transverse heat exchange which in turn depends on the thickness d of the sample and the thermal diffusivity D of the material under consideration: $\tau = d^2/\pi^2 D$.

As melt quenched metallic glasses have a thickness of about 50 μm, thermoelastic damping becomes dominant in the low kilocycle range. The thermal diffusivity and conductivity can be directly obtained from the frequency ω_0 whereas the peak height is a sensitive measure for the thermal expansion coefficient α.

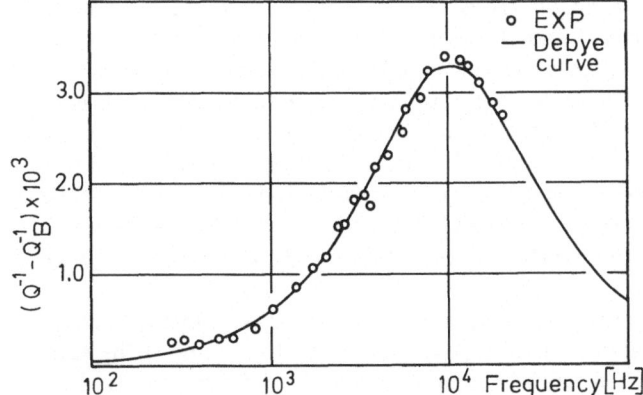

Fig. 6.6. Thermoelastic relaxation peak in a glassy $Mg_{70}Zn_{30}$ alloy at room temperature after correction for a constant background contribution $Q_B^{-1} = 0.5 \times 10^{-3}$

Figure 6.6 shows the thermoelastic peak in a glassy $Mg_{70}Zn_{30}$ alloy. The measurements were performed on a vibrating reed operated in the cantilever mode of flexural vibrations. Points represent experimental results obtained by exciting the fundamental as well as overtones of samples of different lengths (0.5–2 cm) and hence different resonance frequencies. The continuous curve shows the Debye peak which was adjusted for its position and height. The room temperature value obtained for the linear thermal expansion coefficient was $\alpha = 25.5 \times 10^{-6} \, K^{-1}$ and for the thermal diffusivity was $D = 6.0 \times 10^{-6} \, m^2/s$. The thermal conductivity λ calculated from D ($\lambda = 12.3 \, W/mK$) was found to be in reasonable agreement with the electronic contribution calculated from the Wiedemann-Franz law using the free electron value for the Lorenz ratio. Further examples of thermoelastic peaks are given in [6.25, 96].

6.3.3 Magnetoelastic Relaxation

The motion of domain walls and rotation of local magnetization directions under the influence of an external mechanical stress (as discussed in Sect. 6.2.4) leads to several types of losses. It is generally assumed that these magnetoelastic losses can be separated from other nonmagnetic losses by subtracting the losses observed in magnetically saturated samples. Under these conditions, all magnetic moments are polarized and the magnetoelastic interaction is blocked. It should, however, be checked that in cases where the thermoelastic damping is large, effects of nonisotropic thermal expansion can be neglected. Such an anisotropy is also present in metallic glasses when $\lambda_s \neq 0$ and the local magnetization direction does not have random orientations. The assumed nonmagnetic background depends then on the temperature coefficient of the linear magnetostriction and the degree of λ_s axis polarization. The latter is evidently different in conditions where magnetoelastic losses occur and in the magnetically saturated state.

The sources of magnetoelastic losses identified up to now are the magnetomechanical hysteresis effect and the eddy current relaxation. Eddy-current losses may be further divided into macroscopic and microscopic components. A second class of losses are due to the interactions of magnetostrictive stress fields with strainfields (elastic dipoles) of structural defects. The latter interactions are essentially identical to the structural relaxations to be discussed in the next section when we apply, instead of the magnetostrictive stress, an external stress. It should, however, be kept in mind that according to the above-mentioned separation procedure, the magnetostrictive stress induced component appears in the magnetoelastic losses.

Before discussing the properties of the individual loss contributions we would like to point out also that the above-mentioned losses are of the same type and are based on the same magnetic elementary processes as those occurring during magnetization reversals in an alternating magnetic field. This becomes immediately evident when we remember that both a uniaxial stress as well as an applied field tend to polarize the local magnetization vectors into certain directions.

The *magnetomechanical hysteresis* $\varepsilon^m(\sigma)$ results from the same source as the magnetic hysteresis $M(H)$ and describes the nonreversibility of the magnetostrictive strain when the applied stress is slowly cycled between positive and negative values. Both effects have a similar functional dependence on their respective variables and are caused by the fact that the domain wall displacement is not a single-valued function of the elastic stress or the applied field. The magnetomechanical hysteresis manifests itself primarily as a stress amplitude dependent loss. The energy ΔW dissipated per unit volume per cycle corresponds to the area enclosed by the $\varepsilon-\sigma$ loop. Just like the magnetic hysteresis loss, this loss component is frequency independent and appears to depend linearly on the strain amplitudes for crystalline and amorphous metals [6.69, 90]. These properties allow to separate the hysteresis loss to be separated from the relaxation losses. The latter correspond to the extrapolated loss at zero strain amplitude.

It should be noted here that the hysteresis loss is not the result of a relaxation but merely the expression of irreversible elementary processes and as such is not included in the stress-strain relation of the anelastic solid.

The eddy current losses arise from the work done on induced currents. A stress induced change in the magnetization of an electrically conductive sample produces a transient flow of induced eddy currents which in turn give rise to forces counteracting the applied stress. The full equilibrium strain (ΔE effect) can therefore develop only after the decay of the currents. Rate-limiting phenomena confining the full manifestation of the ΔE effect occur when the stress reverses sign before the currents have decayed. The physical principles governing the above process are the same as those involved in a generator. A torque (stress) applied on its axis induces currents in an external circuit (electrically conductive sample) which counteract the applied torque. The energy dissipated corresponds to the Joule heat produced by the currents. In a continuum generator, as represented by our magnetic sample, the currents flow in a path related to the geometry of the sample and the spatial extent of the current inducing regions.

These current inducing regions are to be considered as regions of coherent magnetization changes in time. In a ferromagnetic sample, two different types of regions can be distinguished: the entire sample and the ferromagnetic domains. This corresponds to a separation of the magnetization field into an average net magnetization and its local fluctuation given by the domains. This separation, even though not strictly necessary, is adequate since changes in the total magnetization and local changes give rise, in general, to effects occurring on widely different time scales. The losses related to changes in the net magnetization relaxation are called macroeddy losses and those caused by local magnetization changes (rotation and domain wall displacements) are termed microeddy losses. A demagnetized sample, for instance, can only show microeddy losses since elastically $+M_s$ and $-M_s$ are equivalent directions so that the application of a stress cannot result in a change of the net magnetization. Locally, however, two domains may rotate such that changes in the net magnetization cancel each other. The losses associated with each of these local movements do not cancel but add together.

In terms of the reciprocal quality factor as a function of frequency, the losses appear as peaks, similar to the relation given for the thermoelastic peak. At low frequencies $\omega \ll 1/\tau$, where the induced currents are homogeneously distributed over the sample, the losses increase linearly with frequency, and above the peak they decrease proportionally to $\omega^{-1/2}$. This is due to the skin effect limiting the current flow to a thin layer below the surface. Strictly speaking each magnetoelastic peak (microscopic and macroscopic) is the sum of several Debye peaks associated with different relaxation times (discontinuous spectrum of relaxation times). The relaxation times are the exponential decay times of the different (geometry dependent) modes into which the current distribution can be developed. For a detailed calculation of these magnetoelastic losses see [6.90].

For rod or ribbon-shaped samples excited longitudinally, the dominant relaxation time for the macroeddy current relaxation is given by the approximate relation

$$\tau \sim \frac{d^2}{D^2}, \quad D = \frac{\varrho}{\mu_0 \mu_r}. \tag{6.24}$$

Here d is the thickness or diameter of the sample, D the diffusion coefficient of magnetic flux, ϱ the electrical resistivity and μ_r the reversible permeability. Using the result that a relaxation peak has its maximum when $\omega\tau \sim 1$, we obtain the equivalent result that this occurs at a frequency where the skin depth $\delta_s \sim d$. At these frequencies the measuring conditions change from $H = \text{constant}$ to $B \sim M \sim \text{constant}$, causing the decline of the ΔE effect.

By analogy, the order of magnitude for the relaxation time of the microeddy current relaxation is obtained by inserting the typical linear dimension of the domains instead of the sample thickness.

For an average transition metal-metalloid ribbon d (sample) $= 50\,\mu m$, $\varrho = 100\,\mu\Omega\,cm$, $\mu_r \sim 10^3$–10^4, the peak damping may be expected to occur between 50–500 kHz for the macroeddy current damping. Assuming, furthermore, an average domain diameter of 10 μm yields a frequency of order 1 MHz for the microeddy current relaxation peak. For amorphous rare-earth iron alloys, the latter estimate is expected to be in the GHz range due to the much smaller domains (60–80 Å) operating in these materials [6.87].

At present there are only a few detailed studies on the mechanical losses in amorphous magnetic materials. The papers [6.24, 97–100] report on low frequency internal friction measurements in Fe and Co-based glasses but are mainly concerned with the high temperature structural relaxations. Measurements of the internal friction as a function of an external field H have been done in [6.25, 69, 70, 75, 79]. These measurements all show a maximum in the damping at an intermediate net magnetization level which corresponds to the field where the ΔE effect (Fig. 6.3) has its maximum. The field-dependent magnetic losses were always found to be proportional to the field-dependent values of $\Delta E/E$, indicating the common source of both effects.

A separation of the magnetoelastic relaxation effects into the eddy current and hysteresis components was discussed by *Berry* and *Pritchet* [6.69]. Even though for this particular sample in the frequency range studied the magnetic losses were not found to be large, substantial discrepancies were found between values calculated from eddy losses and the experimental data. The additional loss component was assumed to arise from interaction with structural defects. A clear manifestation of this type of interaction, even though with probable differences in the details, was observed in hydrogen charged samples of $Fe_{80}B_{20}$ and $Fe_{40}Ni_{40}P_{14}B_6$ [6.82]. Here the magnetic loss component (after correction for losses in the magnetically saturated sample) showed a large internal friction peak characteristic also of nonmagnetic metallic glasses charged with hydrogen.

Measurement of the frequency dependence of the magnetoelastic damping for the eddy current relaxation does not appear to have been done yet. The complementary decline of the ΔE effect has, however, been reported by *Arai* et al. [6.44, 71]. These authors studied the frequency dependence of the sound velocity at different external fields. Figure 6.7 shows their results. $\Delta t/t$ is the relative change in the signal transmission time between measurements at constant H and saturated samples and corresponds to $\sim 0.5\,\Delta E/E$. In agreement with our previous estimate, the relaxation frequency occurs in the vicinity of 100 kHz. From the complete disappearance of the ΔE effect above 1 MHz, we can conclude that the microeddy relaxation time must be close to the macroeddy relaxation time. This means that the average domain diameter is of the order of the ribbon thickness. The expected effect of an increase in the relaxation frequency $\omega_0 = \tau^{-1} \sim d^{-2}$ when thinner samples are used has also been demonstrated [6.44].

A practical application of the field dependence of the ΔE effect is the use of amorphous ribbons as ultrasonic delay lines. By applying relatively small

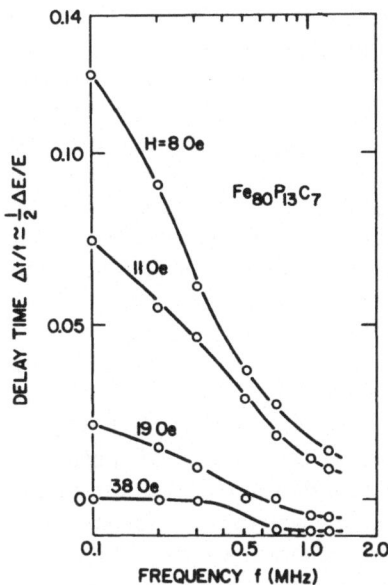

Fig. 6.7. High frequency decline of the ΔE effect in ribbons of $Fe_{80}P_{13}C_7$ as observed by ultrasonic pulse measurements of *Arai* et al. [6.71]

external fields, the delay time can be varied over a large range. In materials with a large ΔE effect, however, extremely large losses may be expected in the vicinity of the relaxation frequency. In terms of Q^{-1}, the peak damping [6.90] is of the order of

$$Q_{max}^{-1} \simeq 0.3 \frac{\Delta E}{E}, \tag{6.25}$$

which means that for an imposed harmonic vibration of frequency $\omega = 1/\tau$, the energy dissipated per cycles ΔW_s is of order $2(\Delta E/E)$ times the maximum stored energy W. With the large values for the ΔE effect in certain metallic glasses, it becomes evident that really giant damping effects are to be expected which would qualify these amorphous metals also as high damping materials.

Complementary observations on the decline of the impedance permeability due to magnetically induced losses give similar results for the relaxation frequency and its ribbon-thickness dependence (see, e.g., [6.101]).

6.3.4 Structural Relaxations

Crystalline metallic alloys provide many familiar examples for stress-induced structural relaxations. The well-known Snoek and Zener relaxations, for example, are associated with changes in the short-range order that occur through thermally activated atom displacements over one or two atomic distances. Studies on metallic and nonmetallic glasses have shown that similar relaxations may also occur in an amorphous structure [6.102].

The interaction of an external stress with the "structural variables" is most conveniently described by the concept of the elastic dipole [6.90]. The insertion of a point defect, for example, in an otherwise unstrained solid, produces local elastic distortions. Similar to the magnetostrictive deformation, the resulting strainfield may have an anisotropic component which can interact with an external stress. Anisotropic distortions are, in fact, to be expected when the local point group symmetry is lower than cubic, a situation certainly likely to occur in an amorphous structure at almost each possible defect site.

For simplicity we assume that a hypothetical spherical environment around a defect site deforms into an ellipsoid when the defect is introduced. It is again apparent that the interaction energy of the resulting strain field with an externally imposed uniaxial stress field is minimised when the longest ellipsoid axis points in the direction of the stress axis. In the absence of substantial internal displacements of the matrix atoms, this equilibrium orientation cannot, however, be achieved by a continuous rotation as was the case for magneto-strictively deformed domains. The only possible mechanism to reduce the energy of interaction comes through a diffusive jump of the defect to a nearby site associated with a strain field of a more favorable orientation with respect to the external stress.

Formally the strain ε_{ij}^{d} resulting from the insertion of a homogeneous distribution of defects, each with a defined orientation p, can be expressed by the following equation:

$$\varepsilon_{ij}^{d} = \sum_{p=1}^{n} \lambda_{ij}^{(p)} c^{(p)} . \tag{6.26}$$

Here $c^{(p)}$ is the concentration of defects (number of defects per matrix atom) with orientation p. In an amorphous structure the number of possible defect orientations is obviously large but for the moment we may assume that ε^{d} is reasonably well approximated by a fixed number of orientations. The second rank tensor $\lambda_{ij}^{(p)}$ characterizes the elastic dipole and corresponds to the strain per unit defect concentration with orientation p. Due to the homogeneous distributions of defects, the strain is assumed to be homogeneous throughout the sample. The $\lambda_{ij}^{(p)}$ tensors for different orientations of defects of the same type are related to one another by the standard equations for the transformation of second rank tensors:

$$\lambda_{ij}^{(p)} = \sum_{m=1}^{3} a_{im}^{(p)} a_{jm}^{(p)} \lambda_{m} . \tag{6.27}$$

λ_{m} are the three principal axes of the elastic dipole $m = 1, 2, 3$ describing the strain ellipsoid and $a_{im}^{(p)}$ is the direction cosine between the axis i and the mth principal axis for the dipole orientation p.

For a uniaxial stress σ_{11}, the total strain ε_{11} resulting from the purely elastic deformation $J_u\sigma_{11}$ and the effect of dipole reorientation achieved by defect migration is given by

$$\varepsilon_{11} = J_u\sigma_{11} + \sum_p \lambda_{11}^{(p)} [c^{(p)}(\sigma_{11}) - c^{(p)}(0)]. \tag{6.28}$$

Here $c^{(p)}(\sigma)$ represents the equilibrium concentration when the external stress is applied and $c^{(p)}(0)$ corresponds to the initial concentration.

The energy of interaction (free energy per defect) with the external stress is given by

$$\mu^{(p)} = - \int \sigma_{11} d\varepsilon_{11}^d dV = -\sigma_{11}\lambda_{11}^{(p)}\Omega, \tag{6.29}$$

where Ω is the atomic volume of the matrix. This energy depends on the orientation p of the defects and therefore leads to a splitting of their energy levels. To calculate the equilibrium values $c^{(p)}$ for the concentration at any given stress and temperature, we have to employ Boltzmann statistics. Accordingly, the occupation $c^{(p)}(\sigma, T)$ of the dipole orientation p with energy level $\mu^{(p)}(\sigma)$ is

$$\frac{c^{(p)}(\sigma T)}{c_0} = \frac{e^{-\mu^{(p)}/kT}}{\sum\limits_{p=1}^{n} e^{-\mu^{(p)}/kT}} \simeq \frac{1}{n}\left(1 - \frac{\mu^{(p)}}{kT} + \frac{1}{n}\sum_{p=1}^{n}\frac{\mu^{(p)}}{kT}\right). \tag{6.30}$$

To end up with a simple yet reasonable expression for the relaxation strength, we assume that two of the principal axes of the λ-tensor are identical $(\lambda_1 \neq \lambda_2 = \lambda_3)$ and that prior to the application of the external stress σ_{11}, all space directions are occupied with equal probability. Averaging finally over all dipole orientations, we obtain the fully relaxed compliance $J_r = \partial\varepsilon_{11}/\partial\sigma_{11}$ in the limit of small stresses $[\mu^{(p)}(\sigma) \ll kT]$ and the relaxation strength

$$\delta J = J_r - J_u = \frac{4}{45}\frac{c_0\Omega}{kT}(\lambda_1 - \lambda_2)^2. \tag{6.31}$$

A rough estimate for the case $c_0 \sim 0.01$ (1 % defects) $\Omega \sim 10^{-29} \text{ m}^3$, $T = 300 \text{ K}$, $(\lambda_1 - \lambda_2) \sim 0.1$ (i.e., 10 % difference in strain per unit concentration of defects in two principal directions) and $E = 1/J \sim 100 \text{ GPa}$ shows that $\Delta E/E \sim 2 \times 10^{-3}$ is much smaller than the corresponding magnetic ΔE effect unless both the total concentration of defects c_0 and the anisotropy $(\lambda_1 - \lambda_2)$ of the defect related strains are substantially larger. Observations at very low temperatures would also increase the effect but this is usually not practicable for point defect relaxations since the kinetics which we shall discuss next may be prohibitively slow. As already mentioned, the stress induced structural rearrangement involves atomic jumps over one or several interatomic distances. With the exception of very low temperatures where quantum mechanical tunneling may become important, the kinetics for structural rearrangements is usually well described by classical thermally-activated atomic jumps across energy barriers E_A. Accordingly, the jump frequency ω or its reciprocal value, the relaxation time τ, is given by an Arrhenius law

$$\frac{1}{\tau} = \omega = \omega_0 \, e^{-E_A/kT}, \tag{6.32}$$

where E_A is the height of the potential well separating the energy levels $\mu^{(p)}$ of the defects. The pre-exponential factor ω_0 (or its corresponding relaxation time $\tau_0 = 1/\omega_0$) depends on the characteristic frequency of the atomic vibrations and a factor taking into account the entropy of activation. For point-like relaxation centres, ω_0 is typically of the order 10^{13}–10^{15} Hz whereas for stress induced rearrangements of line defects (observed with certainty up to now only in crystalline materials), values of the order 10^{11}–10^{13} Hz are characteristic.

In contrast to the cases of thermoelastic and magnetic damping previously discussed, the relaxation kinetics for structural relaxation is strongly temperature dependent. Near room temperature an increment of $\Delta T \sim 40$ K, for example, at an activation energy of 0.5 eV decreases τ by an order of magnitude. Since all relaxation effects depend only on the compound variable $\omega\tau$ where ω is the frequency of measurement, observations on the damping peak and the modulus relaxation can be made by varying the temperature instead of the frequency. As a function of temperature the internal friction peak $Q^{-1}(T)$ is given by

$$Q^{-1} = \frac{1}{2} \frac{\delta J}{J} \frac{1}{\cosh\left[\dfrac{E_A}{kT}\left(\dfrac{1}{T_p} - \dfrac{1}{T}\right)\right]}, \tag{6.33}$$

where T_p is defined by the relation

$$\omega\tau = \omega\tau_0 \, e^{E_A/kT_p} = 1 \tag{6.34}$$

and corresponds to the temperature where the damping achieves its maximum.

It is clear that in particular for metallic glasses, the theoretical considerations given above are based on rather idealized assumptions more likely to occur in crystalline materials. The term defect, although often used, is not well defined in a metallic glass. Lattice vacancies, for instance, which are known to be present in thermal equilibrium in crystals, do not appear to be stable in metallic glasses [6.103]. They disintegrate into smaller portions of volume with different sizes and instead of the vacancy, which is totally identical at each lattice site, the more generalized concept of free volume is used. It is obvious that the kinetics of structural relaxations involving displacement of free volume depend on the local arrangement of atoms and involve a distribution of relaxation times rather than just one single parameter. Similar arguments evidently also hold for other types of defects. The effect of a distribution in the relaxation times is to broaden the width of the temperature and frequency range in which the relaxation step in the real part of the compliance and the damping peak in the loss component occur. Formally such a broad peak can be separated into a sum of single peaks, each one associated with a narrow interval of relaxation time τ and each contributing an incremental relaxation strength $\delta J(\tau)$. Since the relaxation times are indirectly related to the atomic short-range order, such studies contain important information on the structural variability of an amorphous structure. It should, however, be remembered that a distribution in τ may result from a distribution in τ_0 and/or a distribution in the activation energies E_A. This, unfortunately, is the source of considerable uncertainty in the interpretation of experimental data.

a) Experimental Results

A common feature of all glasses, whether metallic or nonmetallic, is the occurance of a glass transition. As in all glassy materials, the internal friction most probably associated with precursory effects of this transition shows an apparently unbounded exponential increase in the temperature range starting at about $200\,°C$ below the glass transition. In Fig. 6.8, this behaviour is represented together with Young's modulus for a Co-based glass. In the lower temperature range the internal friction is dominated by the thermoelastic effect. Accordingly, the temperature dependence of Q^{-1} is given by

$$Q^{-1} \sim \frac{\alpha^2(T)T}{c_p(T)} \sim c_p(T)T, \tag{6.35}$$

where for the latter step, the Grüneisen relation $\alpha = 1/3B \cdot \gamma c_v$ (with γ being the Grüneisen parameter and $c_v \simeq c_p$) was used. Above about $200\,°C$ the curve departs upwards from the thermoelastic behaviour and at the same time starts to exhibit frequency dependence (not shown in Fig. 6.8) characteristic of a thermally-acticated process. Interestingly, the onset of a substantial dissipation has no perceptible effect on the temperature dependence of the modulus.

At low stress levels and at temperatures not too close to the glass transition, quasistatic measurements [6.25] of the elastic after-effect in PdSi glasses indicate that the stress strain relation is still linear and reversible, i.e., truly anelastic, while at higher stress levels [6.91, 92], nonlinear and viscoelastic irreversible contributions become important. An analysis of the internal friction behaviour (low stress regime) in terms of a continuous spectrum of anelastic relaxation times [6.25] shows that an extremely wide distribution of relaxation times is needed to explain the observed data. This in turn signifies that the relaxation is governed by a relatively broad spectrum of activation energies $(1 \sim 4\,eV)$. Similar results have also been obtained on other metallic glasses [6.97, 99, 100].

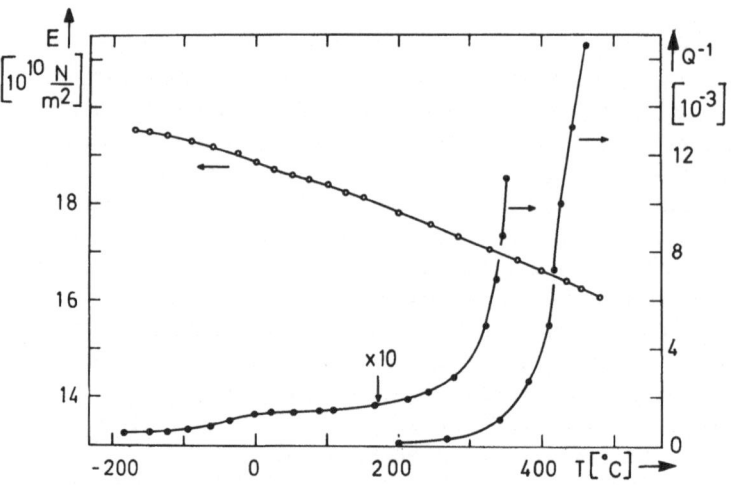

Fig. 6.8. Temperature dependence of Young's modulus and internal friction for glassy $Co_{70.4}Fe_{4.6}Si_{15}B_{10}$. The frequency of measurement was 140 Hz. The exponential increase of the internal friction below the glass transition temperature, here at 500 °C, is characteristic of all glassy materials

Obviously this enormous width of the activation energy distribution makes it difficult to draw convincing conclusions as to the exact nature of the relaxation centres operating in this temperature and frequency range. The occurence of this behaviour in all amorphous materials indicates that this is an intrinsic effect. From the proximity of this effect to the glass transition temperature, generally associated with a gradual softening of the amorphous structure (but apparently not reflected in the elastic constants), we may speculate that the relaxation arises from stress induced migration of free volume or loose sites. This conclusion is further supported from the annealing behaviour of freshly prepared samples. Large irreversible changes (up to an order of magnitude) may occur in the internal friction during the initial stabilisation of metallic glasses leading at the same time to a perceptible increase in density, or vice versa to a decrease in free volume.

The kind of motion involved in this relaxation process should not, however, be confused with motions responsible for irreversible viscoelastic behaviour. The reversibility of the effect established, at least in the lower temperature range, indicates that the moving object remains "bonded" to its initial site to which it finally returns after stress relief. In conclusion, it should also be pointed out that the strong internal friction increases in amorphous materials near the glass transition which is rather analogous to the behaviour of crystalline solids. Here it is also generally observed that the losses drastically increase at temperatures higher than about half the melting temperature. The idea to finally consider the temperature of the glass transition as the melting point of an amorphous structure is in fact not new and has been expressed in other contexts also.

More conclusive information on the behaviour of metallic glasses than from the "high temperature" effect considered above has been gained from observations at intermediate temperatures. Many of the earlier works on internal friction in metallic glasses were initiated to search for Snoek-like relaxation peaks in metallic glasses. The Snoek relaxation, originally detected in crystalline Fe containing C atoms as an impurity, arises from stress induced redistribution of small (mobile) atoms on the equilibrium interstitial sites of the crystal lattice. The idea to search for a similar effect in glassy transition metal-metalloid alloys which already intrinsically contain metalloids as small atoms is tempting. Up to now, however, no conclusive evidence has been reported which might convincingly demonstrate an enhanced mobility of such atoms comparable to crystalline metals. This is probably due to higher activation energies ($E_A \gtrsim 1.2\,\mathrm{eV}$). The internal friction peaks should then occur in the temperature range close to the glass transition where other kinds of losses become dominant. In fact very recently, *Chambron* et al. [6.104] observed that the induced magnetic anisotropy which develops in ferromagnetic alloys, as a result of thermal treatments in a magnetic field, shows a notable difference between a $Fe_{81}N_{13.5}Si_{3.5}C_2$ and the same alloy containing no carbon. This difference was attributed to mobile species of carbon atoms moving with an activation energy of $1.6\,\mathrm{eV}$.

The first atom found to have a sufficiently high jump frequency was hydrogen (present as an impurity in a sputtered sample of $Nb_{75}Ge_{25}$ and $Nb_{75}Si_{25}$ [6.105, 106]). Hydrogen is known to be soluble in abundant quantities in many crystalline metals such as Pd, Ti, Zr, V. Comparable saturation solubilities have also been found in chemically similar metallic glasses [6.107–109].

Numerous studies of the volumetric expansion during hydrogen absorption in crystalline metals have shown that even though the hydrogen atoms occupy interstitial sites, the lattice expands by about $3\,\text{Å}^3$ per hydrogen atom [6.110]. This value is almost universal for a large class of widely different materials and therefore similar values may be expected for metallic glasses also. The anisotropic part of this strain (which we may expect not to vanish when the local defect symmetry is lower than cubic), a condition certainly satisfied at almost all sites in an amorphous structure, forms the elastic dipole capable of a directional interaction with an external stress.

Even though the saturation solubility of H in a $Pd_{80}Si_{20}$ under normal conditions (one atmosphere H-pressure at room temperature) is only about 1–2 at. % H, this alloy was used for many studies [6.111, 115] dealing with hydrogen related problems.

Figures 6.9, 10 show the temperature dependence of the effect of H and D on the internal friction and Young's modulus in an amorphous $Pd_{80}Si_{20}$ sample operated in the cantilever mode of flexural vibration. The flat curve (base line) in Fig. 6.9 corresponds to the internal friction of the well annealed and hydrogen free sample. This curve follows the behaviour expected for thermoelastic damping (Sect. 6.4.1) nicely. Above about 200 K where the specific heat

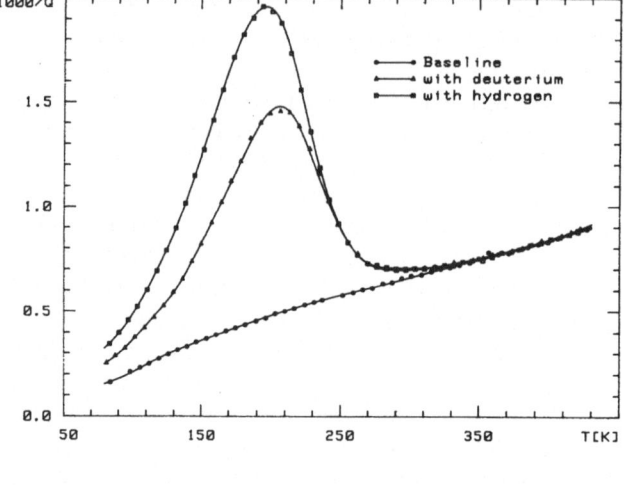

Fig. 6.9. Temperature dependence of the internal friction of an amorphous $Pd_{80}Si_{20}$ alloy charged with H or D. The lowest curve (base line) shows the behaviour of the well annealed and fully desorbed sample. Differences with respect to this curve show the effect of subsequent H or D dopings. The frequency of measurements was 2650 Hz

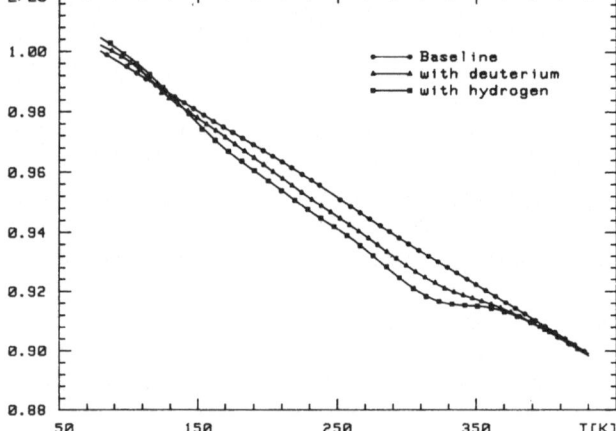

Fig. 6.10. Temperature dependence of Young's modulus for the same sample and treatments as described in Fig. 6.9. The changes occuring near 350 K correspond to the desorption of H or D

becomes temperature independent, the internal friction increases linearly with temperature. The sample was then allowed to absorb H until it was saturated. After charging the sample with H, a well-resolved and relatively narrow internal friction peak showed up. Charging was done in situ under approximately 1.6 atmospheres of H at ambient temperature. The saturation concentration (about 1 at. %) was achieved after about one hour. Since above 350 K H rapidly desorbs (the internal friction follows the base line again while cooling to low temperatures), the same sample was used for the experiments with D. Charging was done under the same conditions. As with H, D also gave rise to a peak. The peak positions of the D-peak at the same measuring frequency of $f = 2650$ Hz occured at a temperature of about 9 K higher than for the H-peak.

The peak heights as measured from the base line differed by a factor of 1.4. As discussed earlier, the height of an internal friction peak depends on the concentration of relaxation centres and the size of the elastic dipole $(\lambda_1 - \lambda_2)$. Unfortunately, the precise saturation concentrations for H and D are not known. We cannot yet conclude, therefore, which of the two quantities affects the peak height. It is, however, known [6.116] that crystalline Pd dissolves larger quantities of H and D at the same pressure. This is mainly due to the difference of the zero-point energies of the H_2 molecules and the H-atoms in the metal. For hydrogen, this difference is larger than for deuterium. It is therefore likely that the differences in the peak heights result from different concentrations.

Parallel to the occurrence of the internal friction peak, Young's modulus exhibits relaxation decline (Fig. 6.10). Annealed and desorbed samples (base line) show the usual temperature dependence of Young's modulus. Deviations from this curve give the effect of H and D. At low temperatures the H and D atoms have low mobilities. The relaxation time τ for atomic jumps is smaller than the period of sample vibration. With increasing temperaure, τ rapidly decreases, and consequently, the anelastic strain (due to partially oriented dipoles) becomes more and more important. The peak occurs at a temperature where $\omega\tau = 2\pi f\tau = 1$. f is the frequency of measurement. At temperatures higher than the relaxation peak, the relaxation time is smaller than the period of vibration; H and D atoms follow the applied stress immediately and the degree of orientation attains its equibrium value. At about room temperature the mobility is so large that H and D diffuse out of the sample during the measurements. After full desorption, Young's modulus and the internal friction follow their respective base line again.

The relaxation can be further characterised by measurements at different frequencies (overtones of the fundamental vibration frequency). Figure 6.11 shows the results for the H-peak after substruction of the base line. Since the peaks are relatively sharp, we may expect that the model of the standard anelastic solid together with the Arrhenius law gives an approximate de-

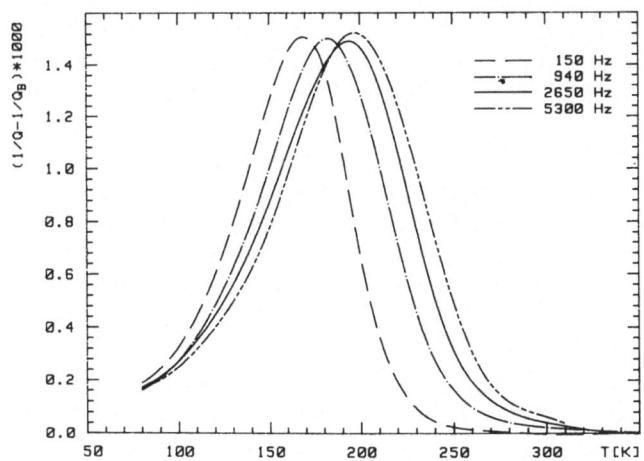

Fig. 6.11. Internal friction peaks due to H in glassy $Pd_{80}Si_{20}$ measured at different frequencies

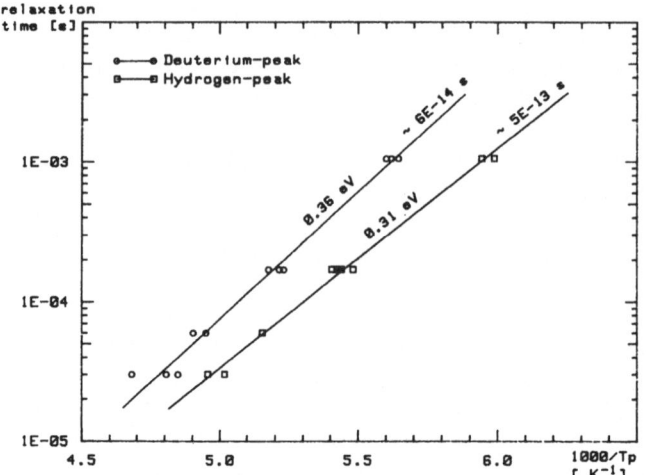

Fig. 6.12. Arrhenius representation of the relaxation time τ as obtained from peak shifts with frequency. Entries give activation energies E_A and pre-exponential factor τ_0

scription of the relaxation. Hence, the relaxation times τ at temperatures where the peaks achieve their maxima can be obtained from the relation $\omega\tau = 1$.

Figure 6.12 shows the Arrhenius plot for the relaxation time $\tau = \tau_0\exp(E/kT)$ together with values obtained for the pre-exponential factors τ and the activation energies E. The activation energies as well as τ_0 differ markedly for H and D.

From simple classical arguments we would expect that $\tau_0 \sim \sqrt{m}$, where m is the H or D atomic mass. Therefore, the two characteristic relaxation times τ_0 should differ by a factor of $\sqrt{2}$. Unfortunately, the accuracy obtained in the τ_0 values from the Arrhenius plot is rather poor and in fact the results obtained would indicate that D vibrates faster than H. The relative position of the H and D-peak, however, indicate the expected trend for the pre-exponential factors; but quantitatively, the shift is mainly due to the different activation energies. Mass dependent differences in the activation energy may occur when quantum mechanical tunneling starts to compete with thermally activated atomic jumps. A notable difference in the activation energy may, however, also occur when the concentrations of dissolved H and D are different. This effect will be discussed later on (Fig. 6.14).

From the overall behaviour of the H and D relaxations in $Pd_{80}Si_{20}$ and from parameters of the same H relaxation in various other metallic glasses compared in Table 6.7, we can draw the following important conclusions. The phenomenon seems to be quite general in metallic glasses and appears to occur whenever a small concentration of hydrogen can be dissolved. Moreover, the parameters governing the relaxation in even very different alloys nicely correlate on the Wert and Marx plot (Fig. 6.13). This plot shows the activation energy versus the peak temperatures T_p at 100 Hz. The line drawn corresponds to a pre-exponential factor of $\tau_0 = 1.5 \times 10^{-14}$ s which is of the order expected for a single atomic jump, i.e., $\tau_0 \cong 1/2\pi f_0$, with $f_0 = 10^{13}$ Hz as a typical atomic

Table 6.7. Parameters of hydrogen relaxation in metallic glasses

Alloy	E [eV]	τ_0 [s]	T_p [K at 100 Hz]	Relative width	Ref.
$Pd_{80}Si_{20}$	0.31	5×10^{-13}	156	4.5	[6.115]
$Pd_{80}Si_{20}$	0.33	1×10^{-14}	148	3.8	[6.113]
$Pd_{80}Si_{20}|D|$	0.36	6×10^{-14}	166	4	[6.115]
$Pd_{77.5}Cu_6Si_{16.5}$	0.31	3×10^{-13}	160	3.4	–
$Ni_{64}Zr_{36}$	0.42	8×10^{-15}	187	3.6	–
$Ni_{24}Zr_{76}$	0.44	5×10^{-14}	211	3.0	[6.118]
$Nb_{40}Ni_{60}$	0.47	3×10^{-14}	220	2.8	[6.113]
$Nb_{75}G_{25}$	0.55	3×10^{-14}	258	2.8	[6.113]
Metglas 2826 2826 A 2605	0.5 to 0.6	10^{-14} to 10^{-16}	230 to 260	4	[6.113]

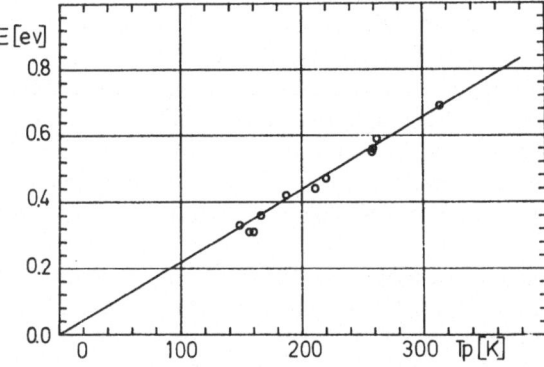

Fig. 6.13. Wert and Marx plot for hydrogen relaxation peaks in metallic glasses. The full line corresponds to a pre-exponential factor of $\tau_0 = 1.5 \times 10^{-14}$ s. The peak temperatures T_p are given for a vibration frequency of 100 Hz

vibration frequency. This is clearly indicative of a short-range reorientation mechanism. Also the activation energies are of reasonable magnitude. They correspond to values characteristic of the diffusion of small interstitial atoms in crystalline metals. The activation energies for the diffusion of H at low concentrations are, for instance, 0.23 eV in Pd, 0.4 eV in Ni, and 0.5 eV in Zr [6.117].

The most surprising feature of these hydrogen peaks in metallic glasses is certainly their relative sharpness. In an amorphous structure we might expect to encounter a large variation in the local arrangement of atoms which in turn would result in large local differences in the activation energy and the pre-exponential factor τ_0. Under such conditions the internal friction may be expected to parallel the behaviour at high temperatures discussed earlier. In fact, the form of the relaxation peaks cannot be quantitatively explained by a single Debye peak, i.e., by a single activation energy and a single τ_0-value. The observed peaks have a width corresponding to 3–4 times the width of a single peak (Table 6.7). In terms of a continous spectrum of activation energies, this means that the distribution width extends from about 0.75 to 1.15 of the average activation energies given in Table 6.7. All the H peaks observed up to now show a notable asymmetry. They start to increase relatively slowly at the low temperature side and fall off quite fast above the peak temperature. Calculated activation energy spectra extend further, therefore, to the low than to the high energy side.

From this result we may conclude that most of the H atoms sit in more or less well-defined "cages" which do not contain large holes allowing an easy escape. Only occasionally do we find moderate holes giving rise to smaller activation energies and hence an asymmetry in their distribution function. First results on direct studies of the atomic structure in hydrogenated amorphous samples using x-ray scattering and EXAFS measurements [6.119] show that the first peak in the radial distribution function in Nb_3Ge splits into two maxima separated by 0.35 Å when hydrogen is dissolved. This splitting is explained by pairs of host atoms having increased separations caused by the incorporation of hydrogen atoms into interstitial sites of the amorphous structure. The fact that hydrogenated samples show two well resolved peaks rather than a continuously smeared out first maximum is further indicative of more or less identical distortion of the atomic arrangement around occupied hydrogen sites. It should, however, be noted that this applies only for H concentrations not exceeding a few at.%. At higher concentrations the initially narrow peaks in internal friction not only increase in height, but also become much broader and shift to lower temperatures. This is shown in Fig. 6.14 for a $Ni_{24}Zr_{76}$ glass.

The times indicated in hours correspond to the total charging time at approximately one atmosphere of H_2 at 200 °C. The indications can only be considered as a relative measure of the H-concentration since the absorption kinetics may depend heavily on the surface conditions. The diffusion of H within the sample is much faster than the absorption so that we can assume a

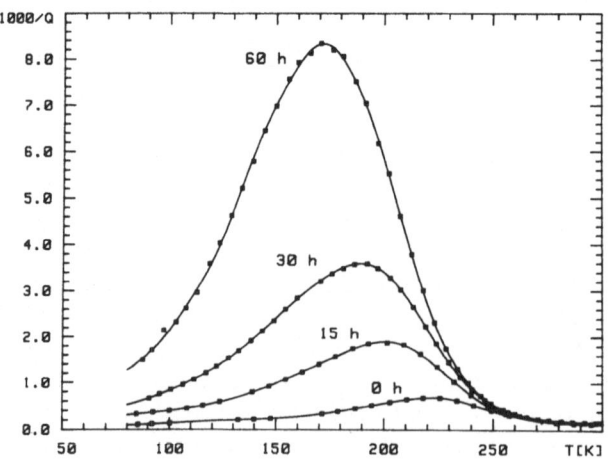

Fig. 6.14. Variation of the hydrogen related internal friction peak with increasing hydrogen concentration in a glassy $Ni_{24}Zr_{76}$ alloy. The frequency of measurement was 270 Hz

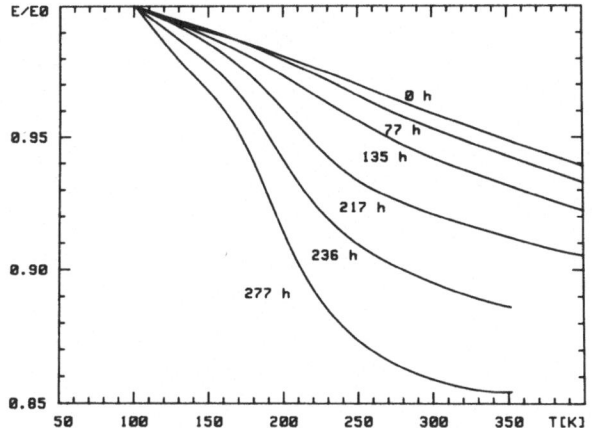

Fig. 6.15. Young's modulus relaxation in glassy $Pd_{35}Zr_{65}$ alloys containing different amounts of H at a vibrating frequency of 1070 Hz

homogeneous distribution of the hydrogen. The curve indicated with 0 h corresponds to a vacuum annealed sample (24 h at 250 °C) which was not brought into contact with hydrogen after its preparation. Obviously, freshly prepared materials may contain hydrogen as impurities.

The corresponding measurements of the relaxation strength $\Delta E/E$ in the same sample (after 60 h) was found to be 8 %. An even larger effect was observed in PdZr alloys which otherwise show exactly the same behaviour. Young's modulus relaxation for $Pd_{35}Zr_{65}$ is shown in Fig. 6.15. Here a relaxation strength of up to 10 % was observed indicating that the anelastic strain due to the reorientation of hydrogen atoms amounts to 10 % of the purely elastic strain. This effect does not yet correspond to values of hydrogen saturated samples. At large concentrations, measurements become extremely difficult since the quality factors decrease to small values and the samples usually start to curl.

The Arrhenius representation of the relaxation time τ, determined from the relation $\omega\tau = 1$ at the peak temperaure, is shown in Fig. 6.16 for glassy $Ni_{24}Zr_{76}$. The results show that with increasing hydrogen concentration, the activation energies shift to lower values, whereas the pre-exponential factors τ_0 increase. Strictly speaking, the relaxation peaks now under discussion are characterized by a relatively broad spectrum of activation energies. The activation energy obtained by fitting an Arrhenius law therefore corresponds to an average value. The same is true for the pre-exponential factor when a broad distribution in τ_0 values has to be considered. Furthermore, the relaxation times determined from the frequency shift of the peak temperature do not need to follow the Arrhenius law exactly, even though the individual relaxation centres may obey this law. Activation energies determined by the above simple procedure are slightly too large and pre-exponential factors are too small compared to the true average values of the respective distribution functions.

relaxation
time [s]

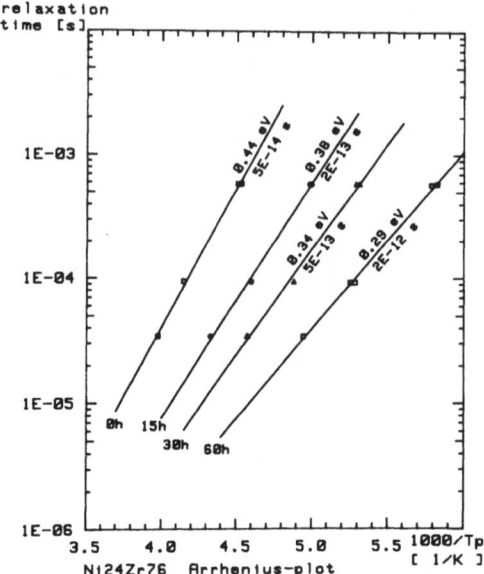

Fig. 6.16. Arrhenius representation of the relaxation time τ in glassy $Ni_{24}Zr_{76}$ for different H-concentrations. Entries give activation energies E_A and pre-exponential factors τ_0

The observed variations in E and τ_0 with hydrogen concentration are, however, much larger and of opposite sign.

In view of the amorphous nature of the atomic structure, a shift in the average activation energy becomes understandable when we argue that within this structure, the hydrogen atoms may occupy a great variety of structurally and hence also energetically different sites. It appears reasonable, furthermore, that sites with low energy levels (deep potential wells) are associated with higher activation energies than those with shallow potentials. Since with increasing concentration these sites will be occupied starting at the low energy end, the width of the spectrum of activation energies contributing to the relaxation will extend towards lower energies. This is phenomenologically just the behaviour necessary to explain the observation made in Fig. 6.14. The extension of the peak width towards lower temperatures corresponds to an extension towards lower relaxation times and hence lower activation energies. Obviously, this picture corresponds to a "rigid band" approximation of the hydrogen levels and neglects effects due to the interaction of the relaxation centres among themselves. But from the concentration independence of the upper end of the peak (~ 250 K in Fig. 6.14), we may conclude that this interaction does not give rise to a substantial shift of the hydrogen levels with respect to the saddle points of the potential experienced by the hydrogen atoms. This in turn signifies that the vibration frequency and hence the intrinsic pre-exponential factor τ_0^i for the hydrogen jump frequency cannot depend strongly on the concentration. From the interpretation of the relaxation peak (cf. Fig. 6.15) we concluded, however, that the pre-exponential factor describing the frequency shift of the peak strongly increases with hydrogen concentration. This behaviour finds a simple explanation when we assume that at high concentration, many of the appropriate sites in the neighbourhood of a given hydrogen atom which could lead to a relaxation are already occupied. The hydrogen atom under consideration has then to probe sites within a more extended region. Consequently, assuming diffusive motion with a diffusion coefficient D, the effective relaxation time τ will be $\tau \sim d^2/D$, where d is a concentration dependent average distance within which the hydrogen atom finds a new nonoccupied equilibrium site. Therefore, the elastic relaxation measures, instead of the intrinsic τ_0^i valid at low concentrations only, the corresponding pre-exponential factor $\tau_0 \cong d^2/a^2 \cdot \tau_0^i$ where a is the elementary distance for a single diffusive jump process. The results obtained for the highest concentration suggest that here d is of the order of 10 interatomic distances.

The above argument already makes use of the equivalence of the elementary jump processes giving rise to elastic relaxation and the long-range diffusion of hydrogen. This is, in fact, not a condition which needs to be satisfied for the occurrence of an internal friction peak. Here one might conceive relaxation centres which always remain confined around a spatially fixed structural irregularity such as, for instance, a hydrogen atom that can jump easily between a few sites located within a possibly particular atomic grouping but only rarely jumps out of this region. The first evidence that the short-range rearrangements responsible for the hydrogen relaxations do not depend on the presence of particular irregularities in the amorphous structure but are rather related to the same elementary jump processes that give rise to long-range diffusion has been obtained from studies of hydrogen absorption and desorption kinetics in $Pd_{80}Si_{20}$ and $Pd_{77.5}Cu_6Si_{16.5}$ glasses [6.112]. In these alloys, as is also well known in crystalline Pd, the absorption and desorption kinetics are mainly controlled by the bulk diffusion. A number of independent measurements of kinetics by volumetric methods and by the observation of the changes in Young's modulus during desorption (see Fig. 6.10 near room temperature) gave an activation energy for the hydrogen diffusion of 0.25 eV for the diffusion coefficient. This value is slightly lower than the corresponding average value from the internal friction measurement but certainly lies within the distribution width of the activation energies contributing to the elastic relaxation. In fact, it appears reasonable that the path taken by the hydrogen atoms on their long-range movement is dictated by the small activation energies (fast jump rates), whereas the jump direction in the anelastic measurements are governed by the stress induced splitting of the hydrogen energy levels on the neighbouring sites.

These results and conclusions have also been confirmed by measurements of the Gorsky relaxation in $Pd_{80}Si_{20}$ [6.114]. The Gorsky effect is a phenomenon associated with mobile defects which act as centres of dilatation in a host material. The relaxation is excited by a macroscopically inhomogeneous stress which in the present case, changes locally the equilibrium concentrations of the dissolved hydrogen. The transient flow of hydrogen atoms which occurs following a sudden application of the stress is obviously rate-controlled by long-range diffusion and can be observed externally by the anelastic creep response. Formally, the manifestations and driving forces in the Gorsky relaxation are closely analogous to the thermoelastic relaxation (discussed earlier) when instead of the heat currents we consider the flow of hydrogen. In particular, the relaxation time τ is obtained by simply replacing the thermal diffusivity by the diffusion coefficient D of hydrogen. Accordingly, for strips of thickness d subject to sudden bending we have $\tau = d^2/\pi^2 D$. Experimentally, creep relaxation times of the order of minutes were observed near 250 K in 40 μm thick strips of glassy $Pd_{80}Si_{20}$. As opposed to the case of reorientation kinetics, the Gorsky effect involves only a single relaxation time. Although these observations were made in an amorphous material, this behaviour is to be expected since the long-range diffusion serves to average the distribution of local jumps into a single-valued diffusion coefficient D finally responsible for the magnitude of τ.

6.4 Plastic Deformation of Metallic Glasses

Many experimental and theoretical studies have been carried out in the field of plastic deformation and fracture mechanics of metallic glasses. A detailed discussion of this work is not possible within this paper. A few remarks should, however, give an insight into the fundamental problems encountered in this field.

In crystalline metals it is well established that plastic deformation proceeds by the movement of dislocations along certain crystal planes. In contrast, little is known about the plasticity and the properties of dislocations in amorphous solids. Experimentally, the surfaces of plastically deformed metallic glasses show shear bands as is typical for crystalline metals. In crystals this step-like surface offsets are explained by positions where several dislocations of a slip plane appeared at the surface. Indeed, the concept of dislocations does not need the presence of a periodic crystal lattice, as might be wrongly inferred from illustrations in text books. This is not saying that these illustrations are incorrect. We would rather like to point out that in an amorphous structure, a dislocation can not be made visual by simply drawing the position of atoms. The internal stress or strain field associated with a dislocation is, however, an equally good and unique description of a dislocation and there is, in fact, no objection to producing such stress fields in amorphous solids also. The discrepancy in opinions arises only when the stability of such fields is discussed. *Spaepen* [6.120] argues that due to structural relaxations, the core of the dislocation (region of high stresses) would become very diffuse and the line itself might break up into small segments which under the influence of an external stress, would move in an uncorrelated way as individual defects, rather than coherently as would be necessary for the glide of a dislocation. To clarify this point, *Chaudhari* et al. [6.121] studied such structural relaxations using

Table 6.8. Tensile strength of metallic glasses and steels

	G [GPa]	σ_y [MPa]	G/σ_y
$Fe_{80}B_{20}$	64.9	3,630	17.8
$Ni_{49}Fe_{29}P_{14}B_6Si_2$	48.0	2,380	20.1
$Pd_{77.5}Cu_6Si_{16.5}$	31.4	1,470	21.3
Fe whiskers		12.000	6.6
Fe and steel		300–2000	40–260
18/8 CrNi steel		570	140

computer simulation techniques. Their results indicate that edge dislocations are not stable whereas screw dislocations are so and have a stress field corresponding to the one in a crystalline solid.

Experimentally, this problem has not yet been solved rigorously. *Kronmüller* [6.66], however, concluded from measurements of the magnetization curve in magnetoelastic amorphous alloys that internal stress fields, decaying with distance r from its centre as $1/r^2$, are present. Such stress fields would correspond to dislocation dipoles. The stress field of an isolated dislocation decreases as $1/r$.

Since metallic glasses have no underlying periodic lattice, the mobility of dislocation may be anticipated to be very low. In fact, as Table 6.8 shows, the yield stresses σ_y are close to the theoretical values for dislocation free crystals (Fe whiskers) and consequently, metallic glasses have G/σ_y ratios which are matched by only few crystalline metals.

More recently, *Rivier* [6.122], and *Rivier* and *Duffy* [6.123, 124] have claimed that based on topological and group theoretic considerations, disclinations rather than dislocations are to be considered as fundamental line defects in glasses and liquids. In the liquid state these defects are highly mobile and, according to their view, the glass transition occurs at the temperature at which the defects freeze in allowing the glass to become an elastic solid. Furthermore, disclination lines or loops are also considered as a microscopic model for the low-lying excitations (two level systems) characteristic of most glassy materials [6.125, 126]. Future studies on the behaviour of disclinations under high stress level will probably disclose new aspects also on the plasticity of metallic glasses.

Acknowledgements. I would like to express my sincere gratitude to Professor H.-J. Güntherodt for his constant interest and generous support for our experimental work in this field.

Particular acknowledgement is also made to T. Gabriel who skillfully prepared the samples and Dr. K. Agyeman and E. Armbruster who participated in much of the work described. I am also very much indepted to Dr. C.F. Hague for stimulating discussions and for carefully reading the manuscript.

Financial support from the Swiss National Science foundation and the "Kommission zur Förderung der wissenschaftlichen Forschung" is gratefully acknowledged.

References

6.1   G.Weiss, W.Arnold, K.Dransfeld, H.-J.Güntherodt: Sol. State Commun. **33**, 111 (1980)
6.2   H.S.Chen, J.T.Krause, E.Coleman: J. Non-Cryst. Sol. **18**, 157 (1975)
6.3   H.S.Chen, J.T.Krause, E.Coleman: Scripta Met. **9**, 787 (1975)
6.4   H.S.Chen, J.T.Krause: Scripta Met. **11**, 761 (1977)
6.5   L.A.Davis, C.-P.Chou, L.E.Tanner, R.Ray: Scripta Met. **10**, 937 (1976)
6.6   A.Kursumovic, M.G.Scott, E.Girt, R.W.Cahn: Scripta Met. **14**, 1303 (1980)
6.7   S.H.Wang, L.T.Kabacoff, D.E.Polk, B.C.Giessen: Met. Trans A **10A**, 1789 (1979)
6.8   A.Wolfenden: Met. Trans A **11A**, 1233 (1980)
6.9   B.S.Berry, W.C.Pritchet: IBM J. Res. Develop., July 1975, p. 334
6.10   B.S.Berry, W.C.Pritchet: J. Physique C-5, Suppl. no 10, **42**, C-5, 1111 (1981)
6.11   S.Jovanovic, C.S.Smith: J. Appl. Phys. **33**, 121 (1961)
6.12   M.F.Ashby, A.N.Nelson, R.M.A.Centamore: Scripta Met. **4**, 715 (1970)
6.13   T.Masumoto, R.Maddin: Acta Metall. **19**, 725 (1971)
6.14   H.J.Leamy, H.S.Chen, T.T.Wang: Metall. Trans. **3**, 699 (1972)
6.15   J.Logan, M.F.Ashby: Acta Metall. **22**, 1047 (1974)
6.16   C.-P.Chou, L.A.Davis, M.C.Narasimhan: Scripta Met. **11**, 417 (1977)
6.17   C.-P.Chou, L.A.Davis, R.Hasegawa: J. Appl. Phys. **50**, 3334 (1979)
6.18   J.J.Gilman: Phys. Today, p. 46 (May 1975)
6.19   B.Golding, B.G.Bagley, F.S.L.Hsu: Phys. Rev. Lett. **29**, 68 (1972)
6.20   M.D.Merz, R.P.Allen, S.D.Dahlgren: J. Appl. Phys. **45**, 4126 (1974)
6.21   I.N.Makarenko, A.M.Nikolaenko, S.M.Stishov: *Liquid Metals* 1976, 3rd Intern. Conf. on Liquid Metals, Bristol (1976) (Institute of Physics, Conference Series; No. 30)
6.22   M.Fenkner: Z. Metallkde. **61**, 10 (1970)
6.23   M.Barmatz, H.S.Chen: Phys. Rev. B **9**, 4073 (1974)
6.24   S.Kiss, G.Posgay, I.Z.Hrangozó, F.J.Kedves: Proc. of 7th Intern. Conf. on Internal Friction and Ultrasonic Attenuation in Solids, Lausanne (1981)
6.25   B.S.Berry: In *Metallic Glasses*, American Society for Metals (Metals Park, Ohio 1978)
6.26   H.S.Chen: J. Appl. Phys. **49**, 3289 (1978)
6.27   A.Kursumovic, M.G.Scott: Appl. Phys. Lett. **37**, 620 (1980)
6.28   U.Mizutani, T.Mizoguchi: J. Phys. F (Metal Phys.) **11**, 1385 (1981)
6.29   J.B.Suck, H.Rudin, H.-J.Güntherodt, H.Beck, J.Daubert, W.Gläser: J. Phys. C **13**, L167 (1980)
6.30   J.B.Suck, H.Rudin, H.-J.Güntherodt, D.Tomanek, H.Beck, C.Morkel, W.Gläser: J. Physique C-8, Suppl. no 8, **41**, C8–175 (1980)
6.31   D.Weaire, M.F.Ashby, J.Logan, M.J.Weins: Acta Metall. **19**, 779 (1971)
6.32   F.Cyrot-Lackmann: J. Physique C-8, Suppl. No. 8, **41**, C8–827 (1980)
6.33   W.Köster, H.Franz: Metall. Rev. **6**, 1 (1961)
6.34   K.Fukamichi, M.Kikuchi, S.Arakawa, T.Masumoto: Solid State Commun. **23**, 955 (1977)
6.35   K.Fukamichi, M.Kikuchi, T.Masumoto, M.Matsuura: Phys. Lett. **73**A, 436 (1979)
6.36   K.Fukamichi, M.Kikuchi, H.Hiroyoshi, T.Masumoto: In *Rapidly Quenched Metals III*, Vol. 2, ed. by B.Cantor (The Metals Society, London 1978) p. 117
6.37   K.Shirakawa, S.Ohnuma, M.Nose, T.Masumoto: IEEE Trans. MAG-**16**, 910 (1980)
6.38   T.Jagielinski, K.I.Arai, N.Tsuya, S.Ohnuma, T.Masumoto: IEEE Trans. MAG-**13**, 1553 (1977)
6.39   K.Fukamichi, H.M.Kimura, M.Kikuchi, T.Masumoto: IEEE Trans. MAG-**17**, 2701 (1981)
6.40   A.W.Simpson, W.G.Clements: IEEE Trans. MAG-**11**, 1338 (1975)
6.41   H.Fujimori, K.I.Arai, H.Shirae, H.Saito, T.Masumoto, N.Tsuya: Jpn. J. Appl. Phys. **15**, 705 (1976)
6.42   R.C.O'Handley: Solid State Commun. **21**, 1119 (1977)
6.43   R.C.O'Handley: Solid State Commun. **22**, 485 (1977)
6.44   K.I.Arai, N.Tsuya: J. Appl. Phys. **49**, 1718 (1978)
6.45   D.W.Forester, C.Vittoria, J.Schelleng, P.Lubitz: J. Appl. Phys. **49**, 1966 (1978)

6.46 R.C.O'Handley, C.-P.Chou: J. Appl. Phys. **49**, 1659 (1978)
6.47 R.C.O'Handley: Phys. Rev. B **18**, 930 (1978)
6.48 N.Tsuya, K.I.Arai: J. Appl. Phys. **50**, 1658 (1979)
6.49 R.C.O'Handley, M.C.Narasimhan, M.O.Sullivan: J. Appl. Phys. **50**, 1633 (1979)
6.50 H.Takagi, S.Tsunashima, S.Uchiyama: J. Appl. Phys. **50**, 1642 (1979)
6.51 T.Jagielinski: J. Appl. Phys. **50**, 7588 (1979)
6.52 K.Narita, J.Yamasaki, H.Fukunaga: J. Appl. Phys. **50**, 7591 (1979)
6.53 K.Twarowski, H.K.Lachowicz: Phys. Stat. Sol. (a) **53**, 599 (1979)
6.54 L.T.Baczewski, T.Jagielinski: IEEE Trans. MAG-17, 2692 (1981)
6.55 W.Dmowski, T.Jagielinski, H.Matyja: Proc. of Intern. Conf. on Metallic Glasses, Vol. 2, Science and Technology, Budapest (1980) p. 21
6.56 Cs.Kopasz, M.Stefan, J.Sulyok: As [Ref. 5.55, p. 75]
6.57 L.Potocky, R.Mlynek, E.Kisdi-Koszó, J.Takács, P.Samuely: As [Ref. 5.55, p. 101]
6.58 H.R.Hilzinger, H.Hillmann, A.Mager: Phys. Stat. Sol. **55**, 763 (1979)
6.59 H.R.Hilzinger, W.Kunz: J. Magn. and Magn. Mat. **15–18**, 1357 (1980)
6.60 R.C.O'Handley, L.I.Mendelsohn, A.E.Nesbitt: IEEE Trans. MAG-12, 942 (1976)
6.61 R.Hasegawa, R.C.O'Handley: J. Appl. Phys. **50**, 1551 (1979)
6.62 O.Kohmoto, K.Ohya, N. Yamaguchi, H.Fujishima, T.Ojima: J. Appl. Phys. **51**, 4342 (1980)
6.63 S.Ohnuma, T.Masumoto: J. Appl. Phys. **50**, 7597 (1979)
6.64 A.E.Clark: In *Ferromagnetic Materials*, Vol. 1, ed. by E.P.Wohlfarth (North-Holland, Amsterdam 1980) p. 531
6.65 H.Kronmüller, B.Gröger: J. Physique **42**, 1285 (1981)
6.66 H.Kronmüller: J. Physique Colloque C-8, Suppl. No. 8, C8–618 (1980)
6.67 R.Becker: Phys. Z. **33**, 905 (1932)
6.68 B.S.Berry, W.C.Pritchet: Phys. Rev. Lett. **34**, 1022 (1975)
6.69 B.S.Berry, W.C.Pritchet: J. Appl. Phys. **47**, 3295 (1976)
6.70 B.S.Berry, W.C.Pritchet: AIP Conf. Proc. **34**, 292 (1976)
6.71 K.I.Arai, N.Tsuya, M.Yamada, T.Masumoto: IEEE Trans. MAG-12, 936 (1976)
6.72 M.Kikuchi, K.Fukamichi, T.Masumoto, T.Jagielinski, K.I.Arai, N.Tsuya: Phys. Stat. Sol. (a) **48**, 175 (1978)
6.73 B.S.Berry, W.C.Pritchet: Sol. State Commun. **26**, 827 (1978)
6.74 G.Hausch, E.Török: Phys. Stat. Sol. (a) **50**, 159 (1978)
6.75 E.Török, G.Hausch: In *Rapidly Quenched Metals III*, Vol. 2, ed. by B.Cantor (The Metals Society, London 1978) p. 105
6.76 E.Török, G.Hausch: J. Magn. and Magn. Mat. **10**, 303 (1979)
6.77 M.A.Mitchell, J.R.Cullen, R.Abbundi, A.Clark, H.Savage: J. Appl. Phys. **50**, 1627 (1979)
6.78 M.Brouha, J.van der Borst: J. Appl. Phys. **50**, 7594 (1979)
6.79 G.Dietz, Th.Frechen, H.Strack: Proc. of Intern Conf. on Metallic Glasses, Vol. 2, Science and Technology, Budapest (1980) p. 15
6.80 T.Jagielinski, T.Walecki, W.Dmowski, T.Matya: as [Ref. 5.79, p. 49]
6.81 M.Kikuchi, K.Fukamichi, T.Masumoto: IEEE Trans. MAG-16, 913 (1980)
6.82 B.S.Berry, W.C.Pritchet: J. Appl. Phys. **52**, 1865 (1981)
6.83 L.T.Baczewski, T.Jagielinski: IEEE Trans. MAG-17, 2692 (1981)
6.84 J.R.Cullen, S.Rinaldi, G.V.Blessing: J. Appl. Phys. **49**, 1960 (1978)
6.85 K.Hathaway, M.Melamud, J.Cullen, G.Blessing: J. Appl. Phys. **50**, 1636 (1979)
6.86 M.Melamud, K.Hathaway, J. Cullen: Phys. Lett. **73A**, 58 (1979)
6.87 J.Cullen, H.Alperin, M.Melamud, K.Hathaway, J.Rhyne: J. Magn. and Magn. Mat. **15–18**, 593 (1980)
6.88 B.S.Berry, W.C.Pritchet: J. Appl. Phys. **50**, 1630 (1979)
6.89 A.Wenger, E.Török: J. Magn. and Magn. Mat. **13**, 283 (1979)
6.90 A.S.Nowick, B.S.Berry: *Anelastic Relaxation in Crystalline Solids* (Academic Press, New York 1972)
6.91 R.Maddin, T.Masumoto: Mater. Sci. Eng. **9**, 153 (1972)
6.92 H.Kimura, T.Murata, T.Masumoto: Sci. Rep. RITU A26, 270 (1977)
6.93 E.Woldt, H.Neuhäuser: J. Physique C-8, Suppl. No. 8, **41**, C-8, 846 (1980)

6.94   C.Zener: Phys. Rev. **52**, 230 (1937)
6.95   C.Zener: Phys. Rev. **53**, 90 (1938)
6.96   B.S.Berry, W.C.Pritchet: J. Appl. Phys. **44**, 3122 (1973)
6.97   T.Soshiroda, M.Koiwa, T.Masumoto: J. Non-Cryst. Sol. **22**, 173 (1976)
6.98   G.Hausch, E.Török: Proc. of 6th Int. Conf. on Internal Friction and Ultrasonic Attenuation in Solids, Tokyo (1977) p. 265
6.99   H.N.Yoon, A.Eisenberg: J. Non-Cryst. Sol. **29**, 357 (1978)
6.100 S.Tyagi, A.E.Lord: J. Non-Cryst. Sol. **30**, 273 (1979)
6.101 F.E.Luborsky: In *Ferromagnetic Materials*, Vol. 1, ed. by E.P.Wohlfahrt (North-Holland, Amsterdam 1980) p. 451
6.102 J.Perez, P.F.Gobin: Rev. Physique Appl. **12**, 819 (1977)
6.103 C.H.Bennet, P.Chaudhari, V.Moruzzi, P.Steinhardt: Phil. Mag. A**40**, 485 (1979)
6.104 W.Chambron, F.Lançon, A.Chamberod: J. Physique Lett. **43**, L-55 (1982)
6.105 B.S.Berry, W.C.Pritchet, C.C.Tsuei: Phys. Rev. Lett. **41**, 410 (1978)
6.106 B.S.Berry, W.C.Pritchet: In *Rapidly Quenched Metals III*, Vol. 2, ed. by B.Cantor (The Metals Society, London 1978) p. 21
6.107 A.J.Maeland, L.E.Tanner, G.Libowitz: J. Less-Common Met. **74**, 279 (1980)
6.108 F.H.M.Spit, J.W.Drijver, W.C.Turkenburg, S.Radelaar: J. Physique C-8, Suppl. No. 8, **41**, C8-890 (1980)
6.109 F.H.M.Spit, J.W.Drijver, S.Radelaar: Scripta Metall. **14**, 1071 (1980)
6.110 H.Peisel: In *Hydrogen in Metals I*, ed. by G.Alefeld, J.Völkl, Topics Appl. Phys., Vol. 28 (Springer, Berlin, Heidelberg, New York 1978) p. 53
6.111 H.U.Künzi, K.Agyeman: In *Internal Friction and Ultrasonic Attenuation in Solids*, ed. by C.C.Smith (Pergamon Press, London 1980) p. 371
6.112 H.U.Künzi, E.Armbruster, K.Agyeman: Proc. of Int. Conf. Metallic Glasses Vol. I, Science and Technology, ed. by C.Hargilai, I.Bakonyi, T.Kemeny, Budapest (1980) p. 107
6.113 B.S.Berry, W.C.Pritchet: Scripta Metall. **15**, 637 (1981)
6.114 B.S.Berry, W.C.Pritchet: Phys. Rev. B**24**, 2299 (1981)
6.115 H.U.Künzi, E.Armbruster, H.-J.Güntherodt: Proc. of Int. Conf. on Rapidly Quenched Metals RQ 4, Sendai 1981 (in press)
6.116 E.Wicke, H.Brodowsky: In *Hydrogen in Metals II*, ed. by G.Alefeld, J.Völkl, Topics Appl. Phys., Vol. 29, (Springer, Berlin, Heidelberg, New York 1978) p. 73
6.117 J.Völkl, G.Alefeld: In *Diffusion in Solids*, ed. by A.S.Nowick, J.J.Burton (Academic Press, New York 1975) p. 233; see also [Ref. 5.110, p. 321]
6.118 K.Agyeman, E.Armbruster, H.U.Künzi, A.DasGupta, H.-J.Güntherodt: J. Physique C-5, Suppl. No. 10, **42**, C5-535 (1981)
6.119 G.S.Cargill III: In *Liquid and Amorphous Metals*, ed. by E.Lüscher, H.Coufal (Sythoff and Noordhoff, Alphen aan den Rijn, The Netherlands 1980) p. 161
6.120 F.Spaepen: In *Rapidly Quenched Metals III*, ed. by B.Cantor (The Metals Society, London 1978) p. 253
6.121 P.Chaudhari, A.Levi, P.Steinhardt: Phys. Rev. Lett. **43**, 1517 (1979); see also Chap. 5
6.122 N.Rivier: Phil. Mag. A**40**, 859 (1979)
6.123 N.Rivier, D.M.Duffy: In *Numerical Methods in the Study of Critical Phenomena* (Springer Series in Synergetics) ed. by Vol. 9 (Springer, Berlin, Heidelberg, New York 1981)
6.124 N.Rivier, D.M.Duffy: J. Phys. C (Solid State Phys.) **15** (1982)
6.125 D.M.Duffy, N.Rivier: Physica **108**B, 1261 (1981)
6.126 N.Rivier, D.M.Duffy: J. Physique **43**, 293 (1982)

7. Vibrational Dynamics of Metallic Glasses Studied by Neutron Inelastic Scattering

J.-B. Suck and H. Rudin

With 16 Figures

Results from neutron inelastic scattering experiments made to investigate the atomic dynamics of binary metallic glasses are reviewed. Dynamical structure factors, frequency distributions and the dispersion of collective modes are discussed including the dynamics of gases in topologically disordered alloys. The characteristic features of the atomic dynamics of metallic glasses are pointed out and the consequences for their thermodynamics and other properties are shown.

7.1 Background

For the understanding of thermodynamics, transport and other properties of condensed systems on a microscopic level, knowledge of their atomic structure and dynamics on the basis of interatomic forces is required. Interatomic forces can be studied by inelastic scattering experiments if the energy change of the scattered particle can be directly related to the atomic dynamics of the system. The aim of such experiments is to correlate the observed energies (e.g., peaks in the measured spectra) with movements of atoms or groups of atoms and the forces between them.

Thermal, hot and cold neutrons are especially suited for such investigations because of their favourable properties. Their energies match those of vibrational and diffusive motions in gases, liquids and solids (meV range) and their wavelength is of the order of interatomic distances in condensed systems due to their large mass compared with other radiations like electrons or photons. Neutrons are neutral particles and they have a magnetic moment (this latter property will not be considered in this article). The scattering cross sections for thermal neutrons do not vary in a systematic manner from one nucleus to the next. For the investigation of atomic motions in condensed systems, the consequences of these favourable properties are: the energy shift in inelastic scattering processes are of the order of the energy of the incident neutron, i.e., the energy change is well resolved; the scattering event is sensitive to the atomic structure and it is possible to study single particle and collective dynamics separately depending on the scattering properties of the nuclei and momentum transfer; the dependence of atomic dynamics on momentum transfers (e.g., dispersion relations) can be investigated; due to the large penetration depth of thermal neutrons, bulk properties of condensed systems are studied and single scattering processes are normally dominant, which is important for the assignment of energy and momentum transfers to the dynamical process under investigation; the neutron is scattered at the nucleus and this interaction is characterized by a constant coupling parameter (scattering length instead of, e.g., an atomic form factor) in almost all interesting cases, which enables the effect of radiation on the observed spectra to be separated easily from the structural and dynamical information contained in them and no selection rules exist[1]. Finally, the nonsystematic change of the scattering cross sections allows, in favourable cases, the scattering of one element in polyatomic systems to be "marked" by replacing it by an isotope with different scattering properties.

The dynamics of crystals has been investigated successfully using neutron inelastic scattering techniques for 25 years now and there is no fundamental difficulty in applying this well-developed experimental technique to a new class of materials, the metallic glasses in our case, as long as the sample sizes are large enough to allow inelastic scattering experiments with the available neutron fluxes at the sample (roughly 10^5 to 10^7 n cm^{-2} s^{-1}). Nevertheless, in contrast to the investigation of the dynamics of crystals where most often only the first moment of the inelastic neutron spectrum (peak position) is used and relative intensities are measured, absolute intensities and the correct shape of the spectra have to be determined in inelastic neutron scattering experiments done on isotropic samples like polycrystals, glasses, liquids and gases to obtain the maximum information available. This requires demanding experiments and very careful data reduction including elaborate corrections.

The real problems, however, arise in the interpretation of the rather structureless distributions obtained. In the case of single crystals, the long-range order allows the theoretical description of the lattice dynamics to be reduced to that of the one atom (or the few atoms) in the primitive unit cell. In the experiment the movements of the atoms may be excited in definite directions relative to the coordinate system given by the unit cell and the measured excitations are sharp (δ-functions in the harmonic approximation). The assignment of measured energies and momentum transfers (dispersion) to specific atomic motions is therefore nearly always possible. For polycrystals the situation is already much less favourable. The theoretical description of their dynamics can still be reduced to that of the atoms in the unit cell, averaging the results obtained for a single crystal appropriately over all directions *if* one knows the dynamics of the corresponding single crystal in all directions, e.g., [7.1]. However, the results after this averaging procedure are of complex form and have to be compared with rather broad distributions measured in the experiment. For disordered systems like gases, liquids and topologically disordered solids, the situation is much more complicated. The lack of long-range order prevents the reduction of the many-body dynamics to that of one

---

1 With the exception that neutrons couple to transverse modes only via Umklapp processes.

or few atoms, like in the crystalline case, because dynamics and structure are intimately related in the measured quantities [7.2], i.e., the interatomic forces experienced by an atom depend on the actual local structure in its neighbourhood (quantitative disorder). An infinite number of ensemble averages of possible configurations would have to be calculated for an exact solution of the problem.

In fact, the broad and often structureless spectra measured in neutron inelastic scattering experiments on disordered isotropic samples can be interpreted quantitatively only by a comparison with the results obtained from complicated models for the dynamics of disordered systems or from computer simulations of the many-particle problem. However, even if such elaborate information is not available, some insight into the characteristic features of the dynamics of topologically disordered solids can be gained from a comparison with the spectra taken on the corresponding polycrystal[2]. Due to the lack of model calculations and computer simulations for the metallic glasses investigated so far, the results have mostly been compared with spectra of crystalline materials. The amount of information obtainable from neutron inelastic scattering experiments on disordered systems therefore depends on both the information principally accessible with this experimental method and the development of theoretical means for interpreting the experimental results.

In spite of these difficulties, neutron inelastic scattering experiments are still the best method available for studying the atomic dynamics of metallic glasses. In inelastic scattering experiments with electrons, phonons with larger wave vectors, q are difficult to measure because the cross section varies with $(1/q)^4$. Brillouin scattering experiments give information at very small momentum transfers, i.e., about long-range dynamical processes, which are adequately described by the dynamics of continua. This is independent of the nature (long-range order or not) of the atomic structure and no difference to results taken on crystalline materials were observed [7.3]. Infrared absorption and Raman scattering would give information about the spectrum of particle motions provided no selection rules prevented this. However, in this case the scattered intensity is very weak because the scattering volume is very small in samples with metallic conductivity and no results have been obtained so far [7.3]. Neutron inelastic scattering experiments provide the dynamical structure factor (scattering law) $S(Q, \omega)$ and the frequency distribution $G(\omega)$. $S(Q, \omega)$ contains information about the energy $\hbar\omega$ and the momentum transfer $\hbar Q$ of the interatomic motions. Of special interest in these investigations is the dispersion

---

2 A *corresponding* polycrystal we call a polycrystal with nearly the same density and similar chemical and short-range order as found in the disordered sample. (Otherwise too many assumptions have to be made in the comparison of the data taken on the disordered and the polycrystalline sample.) As the density change on crystallization is only 1–2 % for metallic glasses and the crystallization temperatures T_x are well below the melting point (so that no strong diffusion of atoms in the sample is possible during the crystallization process), polycrystals obtained by heating the glassy sample above T_x can be assumed to be "corresponding polycrystals", though they will often be mixtures of different crystalline phases.

and damping of propagating collective modes in metallic glasses. For practical applications, the frequency distribution $G(\omega)$ is perhaps even more important though it contains less information than $S(Q, \omega)$.

The investigation of the vibrational dynamics of metallic glasses using neutron inelastic scattering techniques has only started very recently, apart from an early experiment on $Co_{80}P_{20}$ [7.4] which was preliminary in character [7.5]. Though some of the characteristic vibrational properties of metallic glasses could be obtained from the few experiments over the last three years, many questions still have to be answered, no systematic investigations exist and the whole topic has to be regarded as still being in its infancy. One reason for this "retardation" compared with, e. g., the investigation of the atomic structure of glassy metals using neutron diffraction, is the fact that until recently it was difficult to obtain a sufficient amount of binary glassy alloys (10–50 g) necessary for neutron inelastic scattering experiments. Metallic glasses with more than two components are easier to produce; however, the interpretation of their vibrational spectra would be extremely difficult as the number of time-dependent pair correlations which have to be considered increases rapidly with the number of different elements in the sample. To start with the simplest case, therefore, the vibrational dynamics of two component metallic glasses have been investigated up to now.

Neutron inelastic scattering has been applied to more complicated metallic glasses only to study their magnetic properties [7.6–11]. Results from these experiments will be discussed in the context of magnetic properties of metallic glasses in Chap. 11.

7.2 Recent Computer Simulations and Model Calculations

As neutron inelastic scattering experiments and computer simulations and model calculations of the dynamics of disordered materials depend so intimately on one another (simulations and models giving some interpretation of the measured spectra, the experiment sometimes disproving the assumptions made in theoretical approaches[3]) we shall discuss some of the most recent results (without aiming for completeness) which appeared after the review given by *Hafner* [7.12].

What is required in this comparison between experimental and theoretical results is the dynamical structure factor which is determined in neutron inelastic scattering experiments. It is the Fourier transform of the time-dependent pair correlation function [7.13] which may be obtained from

---

3 One should keep in mind that even the most excellent agreement between a measured and calculated dynamical structure factor cannot prove the "correctness" of, e. g., the pair potential used in the calculation; however, it suggests that the atomic motions in the real system *may* be similar to those of the particles in the calculation.

computer simulations and some of the model calculations. For a monatomic isotropic system, one has with $Q = |Q|$ and $r = |r|$,

$$S(Q, \omega) = \frac{1}{2\pi} \int_{-\infty}^{\infty} dt \, e^{-i\omega t} \int dr \, e^{iQr} \, G(r, t). \tag{7.1}$$

Here $\hbar Q$ is the magnitude of the momentum transfer and $\hbar\omega$ the energy transfer, i.e., $S(Q, \omega)$ is a measure of the probability that in the dynamical processes in the sample a given momentum transfer $\hbar Q$ will be connected with an energy transfer $\hbar\omega$. In the classical limit, $G(r, t)$ can be interpreted as giving the degree of correlation between pairs of particles which are at a distance r from each other after a time t if one of them was at $r = 0$ at $t = 0$. It should be noted that the results of model calculations and computer simulations for the atomic dynamics of metallic glasses are often given as the one phonon part of the dynamical structure factor $S_1(Q, \omega)$ multiplied by ω^2, or as the momentum dependent frequency spectra $F(Q, \omega)$ [7.14–16] instead of $S(Q, \omega)$ itself. This representation plays up the structure in the inelastic spectra which is then difficult to compare quantitatively with experimental results because the one phonon part cannot be reliably separated from the measured dynamical structure factor.

For the general case of topologically disordered solids, collective excitations (often called "phonons") and frequency distributions were investigated in more detail [7.17–31] than the full dynamical structure factor [7.32, 33] and computer simulations of these systems are rare [7.34]. Regarding this last relation, the situation for metallic glasses is inverse: only one model calculation [7.35, 36] exists but several computer simulations with more or less realistic potentials have been made [7.37–44]. Of these, four have been published after the review of *Hafner* [7.12], though they are mentioned in the appendix.

Amorphous iron [7.41, 42] is not a metallic glass in the strict sense; nevertheless, being a one-component system it allows the influence of topological and quantitative disorder to be studied without the additional complication given by the chemical short-range order present in real metallic glasses. The amorphous structure was obtained by dense random packing of hard spheres (DRPHS) of 1450 atoms [7.45]. This initial configuration was then relaxed using the empirical effective pair potential of *Pak* and *Doyama* [7.46] which reproduces approximately the phonon dispersion curves and the phonon density of states of crystalline α-Fe. The parameters of the potential were determined from elastic data. As in all the simulations discussed here, harmonic interatomic forces were assumed. For the magnitude of the resulting interatomic forces, it is important that the density of amorphous iron is correctly achieved in the relaxation process.

Two of the common methods used in computer simulations of the dynamics of topologically disordered solids were compared: the equation of motion method and the recursion method [7.12]. In the former approach, the Green's functions of the dynamical problem are calculated by successive integration of the equation of motion of the vibrating atoms [7.47, 48], similar to what is done in conventional molecular dynamics calculations. In the latter method, the vibrational density of states $g(\omega)^4$ is obtained as an average over a number of local densities of states which are calculated from a sum over the appropriate Green's functions of a cluster of only a few atoms. The Green's functions are obtained from a continued fraction expansion by a recursive calculation of their coefficients [7.49, 50]. In both calculations the topological and the quantitative (force constant) disorder were taken into account.

For amorphous iron the results of both calculations agree reasonably well, though the vibrational DOS calculated with the equation of motion technique shows a rather unexpected flat tail at the high energy limit of the distributions and much less intensity at low energies than is found in the phonon DOS determined from the measured dispersion curves [7.51]. The vibrational DOS of amorphous iron exhibits two broad peaks with maxima near 20.7 and 32 meV. In the phonon DOS of crystalline iron, two of the van Hove singularities are found near 25 meV which can be correlated with the dispersion of transverse phonons in the crystal, and one is found near 35 meV which corresponds to longitudinal phonons. *Rehr* and *Alben* [7.39] concluded from their results that the inclusion of quantitative disorder completely destroys the structure in the vibrational DOS. From the survival of a two-peak structure in $G(\omega)$ of amorphous iron, the authors conclude that quantitative disorder leads to a complete smoothing of $G(\omega)$ only if the packing fraction is too low.

From the energies corresponding to the maxima in the momentum-dependent spectra, dispersion curves for longitudinal and transverse modes are derived. For longitudinal modes the position of the peaks can be assigned for momentum transfers as large as Q_p, the momentum transfer at the first peak of the static structure factor. Near Q_p a "roton-like" minimum of the dispersion curve is found. For transverse modes the dispersion can be specified for momentum transfers up to $Q_p/2$. For $Q > Q_p/2$ the distributions are so broad that no maxima can be defined. From the slope of the dispersion curves at low momentum transfers the velocities of transverse and longitudinal sound are obtained. Compared with crystalline iron, one finds a decrease in the sound velocity of transverse modes of 9 % and in the velocity of longitudinal modes of 16 %. This strong softening of the longitudinal modes, indicated also in the shift of the high energy peak in the vibrational DOS compared with the phonon

---

4 To distinguish between different kinds of frequency spectra, the following terminology will be used. *Phonon* density of states (crystals): $f(\omega)$ for one component crystals or partial DOS, $F(\omega)$ for crystals with several components. *Vibrational* density of states (topological disordered solids): $g(\omega)$ for one component systems or partial DOS, $G(\omega)$ for systems with several components. The frequency distributions determined in neutron inelastic scattering experiments differ from the DOS defined above because different elements contribute to the measured intensity with different weights $[(\sigma/M)\exp(-2W)]$ and approximations have been used in the corrections for coherence and multi-phonon contributions to the measured intensities. Where it seems necessary, these latter ones will be pointed to by a prime $[G'(\omega), g'(\omega)]$.

DOS of the crystal, is rather unexpected and the authors conclude that a central interatomic potential, as it was used in the calculation by Rehr and Alben, is inappropriate for atomic arrangements in which shear deformations are dominant.

The computer simulation of atomic motions in $Fe_{100-x}P_x$ glasses [7.43] with $x = 15.1$, 20.1, and 24.3 is the only theoretical investigation of the dynamics of a metal-metalloid glass done so far. The disordered matrix was obtained from DRPHS of about 1500–1600 hard spheres with two different diameters. Their ratio was determined from the average nearest-neighbour distances of Fe–Fe and Fe–P pairs in crystalline Fe_3P. This initial configuration was then relaxed using truncated Morse potentials for the Fe–Fe, Fe–P, and P–P interactions. The parameters for the potentials were determined from cohesive energy, compressibility and lattice constants of the crystalline materials. The density of the system was approached in the relaxation process within 5%. The vibrational DOS was calculated from about 200 Fe-atoms and 50 P-atoms in the central region of the spheric cluster using the recursion method. The total DOS and the Fe–Fe and Fe–P partial DOS were determined. They all showed a two-peak structure, of which the broad distribution between 0 and 40 meV is predominantly due to the acoustic modes of Fe atoms, and the narrower distribution between 40 and 60 meV is caused by optic modes due to the forces between Fe and P-atoms[5]. As iron atoms are most frequently nearest neighbours in this system, one would expect that the acoustic part of the vibrational DOS is similar to the vibrational DOS of amorphous iron [7.42]. However, no separation of transverse and longitudinal modes in the acoustic part of the vibrational DOS is observed and the authors attribute this to the complexity of the atomic arrangements in the $Fe_{100-x}P_x$ system. Compared with the results for crystalline Fe_3P, one finds an intensity increase at low energies in the vibrational DOS. Correspondingly, the first and second frequency moments of the vibrational DOS are lower by about 10–20%. The authors interpret these findings as being caused by the larger atomic volume in the glass compared with that in the crystal.

The Q-dependent spectra again show two maxima, the positions of which vary with momentum transfers. Thus, two dispersion branches are obtained, one for acoustic and one for optic modes. For both types of modes the dispersion of longitudinal and transverse vibrations could be determined for momentum transfers up to Q_p. Both optic branches show little dispersion, indicating the localized character of the Fe–P vibrations, and they are found in the same energy region. The acoustic branches show strong dispersion at $Q < Q_p/2$. For larger momentum transfers, transverse modes show little dispersion and the longitudinal vibrations show a "roton-like" minimum as it was found in other simulations before [7.39, 40, 42]. From the sound velocities

---

5 Strictly speaking, optic modes are difficult to define in disordered alloys without any unit cell. The expression will be used here for modes for which the equivalent modes in a crystal with two or more atoms in the unit cell would be optic modes.

obtained from the slope of the acoustic branches at low Q-values, a decrease in the sound velocity of longitudinal modes of 10% to 20% compared with the corresponding sound velocity in crystalline Fe_3P was found and there was no change in the velocity of the transverse modes. The authors argue that the larger atomic volume in the glass than in the crystal may effect the longitudinal density fluctuations more than the shear motions in transverse vibrations; however, this explanation is not fully convincing.

Of special interest to us is the investigation of the dynamics of the metal-metal glass $Cu_{57}Zr_{43}$ [7.44] because glassy $Cu_{46}Zr_{54}$ has been studied in a neutron inelastic scattering experiment. The model glass consisted of 874 Cu and 659 Zr atoms initially in an "as-poured loose random packed" structure which was relaxed under periodic boundary conditions using a modified Lennard-Jones 4–8 pair potential. Equal radii for both ions, but otherwise different parameters of the pair potentials between Cu–Cu, Cu–Zr, and Zr–Zr neighbours, were taken. These parameters were determined from the equilibrium next-nearest neighbour distances in Cu and Zr fcc crystals, the cohesive energies and the elastic constants [7.53].

The vibrational DOS was calculated using the recursion method. The resulting total and partial vibrational DOS are compared in Fig. 7.13 with the results of the neutron inelastic scattering experiment. The two partial vibrational DOS for Cu and Zr vibrations in the glass are nearly identical in shape and show no structure like the resulting vibrational DOS of $Cu_{57}Zr_{43}$. This can be understood from the ratios of the characteristic frequencies $\omega_{A-A} = (d_{A-A}/M_A)^{1/2}$, where d_{A-A} is the mean force constant between A–A neighbours and M_A is the mass of the atom A. As the same force constants were assumed between Cu–Cu and Zr–Zr neighbours, the characteristic frequencies are very similar: $\omega_{Cu-Cu} : \omega_{Cu-Zr} : \omega_{Zr-Zr} = 1:0.883:0.848$. In the case of $Fe_{100-x}P_x$ glasses, where the acoustic and optic modes were well separated in the vibrational DOS, the corresponding ratio of the characteristic frequencies was quite different: $1:1.738:0.638$.

From the Q-dependent spectra the dispersion of longitudinal and transverse modes was determined. In the longitudinal branch, again a "roton-like" minimum was found near Q_p while the spectra of the transverse vibrations for $Q > 0.6\,Q_p$ were already too broad to obtain a reliable dispersion curve. There may be a weak indication of optic modes at momentum transfers which almost correspond to the mean nearest-neighbour distance between Cu–Zr pairs, but no dispersion of these modes could be determined. From the slopes of the dispersions the velocity of longitudinal and transverse sound waves were obtained. They were used to calculate the elastic constants assuming elastic isotropy of the model system. In contrast to the results just discussed for the metal-metalloid glass $Fe_{100-x}P_x$, the bulk modulus which is determined by the velocity of longitudinal sound waves remains approximately unchanged compared with crystalline Cu and Zr, while the shear modulus is lowered by 40% to 50% indicating a strong softening of transverse modes in the metal-metal glass. Similar results have been reported in an early computer simulation [7.37] and in a recent model calculation of the elastic properties of metallic glasses [7.54, 55].

The contribution from diffuse Umklapp scattering processes to the one-phonon part of the dynamical structure factor has been analyzed very recently by analytical methods [7.31]. Under the assumption that the polarization vectors vary weakly from one atom to the next, not only in the long wavelength (small Q) limit but also for larger momentum transfers, the one-phonon part of the dynamical structure factor was split into two terms, one part containing the contributions from N (normal) processes, the second part those from diffuse Umklapp scattering. The structural term containing the Bragg scattering in the corresponding expression for Umklapp scattering in crystals is replaced by the continuous static structure factor of the metallic glass. Assuming linear dispersions, one for longitudinal and two for transverse acoustic vibrations, *Hafner* showed that an appreciable amount of the measured intensity for $Q > 0.6\, Q_p$ is due to diffuse Umklapp scattering. This contribution is most pronounced near the maxima of the static structure factor because 95% of the modes which are scattered via Umklapp processes in this Q-region are of a transverse type.

7.3 Experimental Techniques

In this section we shall briefly describe the experimental techniques which have been applied up to now in the investigation of the atomic dynamics of metallic glasses using neutron inelastic scattering.

In neutron inelastic scattering experiments, the intensity $I(k_0)$ of neutrons incident on the sample and $I(k)$, the intensity of the neutrons scattered under the scattering angle θ into the solid angle $\Delta\Omega$ (detector surface seen from the centre of the sample), are measured together with the wave vectors k_0 and k of the neutrons before and after the scattering, respectively. From the ratio of the measured intensities, a cross section $d^2\sigma/d\Omega dE$ is determined which gives the probability that a neutron is scattered with a given energy and momentum transfer from the moving atoms in the sample. The measured cross section contains contributions from two different types of scattering:

$$\frac{d^2\sigma}{d\Omega dE} = \frac{d^2\sigma^{\text{coh}}}{d\Omega dE} + \frac{d^2\sigma^{\text{inc}}}{d\Omega dE}. \tag{7.2}$$

In one case the neutron wave is scattered from many atoms with a fixed phase relation which leads to interference effects in the scattered wave. It is called *coherent* scattering and *collective* atomic motions (e.g., dispersion relations) are studied in this case. In the other case the phase relation between scattered waves are uncorrelated due to isotopic or spin disorder (or because the momentum transfer is so large that one approaches the incoherent approxima-tion). This is called *incoherent* scattering and *single* particle motions (e.g., frequency distributions) can be investigated in this case [7.56]. Separating out the neutron-nucleus interaction expressed by the scattering cross section σ, the double differential cross sections can be directly related to the static and dynamic properties of the scatterer, where collective and single particle motion have now been separated:

$$S(Q, \omega) = \frac{4\pi}{\sigma_{coh}} \frac{k_0}{k} \frac{d^2\sigma_{coh}}{d\Omega dE},$$ (7.3)

$$S_S(Q, \omega) = \frac{4\pi}{\sigma_{inc}} \frac{k_0}{k} \frac{d^2\sigma_{inc}}{d\Omega dE}.$$ (7.4)

Here $S_S(Q, \omega)$ is the self part of the dynamical structure factor, $\hbar Q = \hbar(k_0 - k)$, $\hbar\omega = \hbar^2(k_0^2 - k^2)/2m$ are the momentum and energy transfers, respectively, measured in the experiment and m is the mass of the neutron. For isotropic samples only the magnitude Q of the vector Q is relevant because all directions in the sample are equivalent. σ_{coh} and σ_{inc} are the cross sections for coherent and incoherent scattering from the bound atoms. They are nuclear properties. By an appropriate choice of substance and momentum transfer one can, therefore, study the collective and single particle motions of the atoms separately in neutron scattering experiments.

, These considerations apply to all neutron inelastic scattering experiments[6] and instruments differ only in the way k_0 and k are measured. Two main methods can be distinguished: the time-of-flight (TOF) method and the Bragg reflection from single crystals. Both methods were used in experiments done with metallic glasses.

In TOF experiments, pulses of neutrons with a selected wavelength (monochromatic neutrons) are sent upon the sample and part of the scattered neutrons are absorbed in a large number of detectors placed in fixed positions on a circle with a radius of 2–4 m around the centre of the sample (Fig. 7.1a). Thus, $S(Q, \omega)$ is obtained at many Q and ω values at the same time along paths of constant scattering angle in the dynamical region of the experiment (Fig. 7.1b). k_0 and k are determined from the velocity v_0 and v of the incident and scattered neutrons measured from their TOF over the known distances of the flight paths. The TOF method is especially suited for measurements of absolute intensities as is required in the investigation of isotropic samples like metallic glasses. This argument applies to both the experiment itself and to the data corrections which have to be done to obtain absolute intensities. This is because these corrections can be calculated more accurately (if at all) for "steady-state" experiments than for a strong variation of experimental parame-

---

6 We shall exclude spin-echo techniques and statistical pulsing of TOF spectrometers in the following discussion.

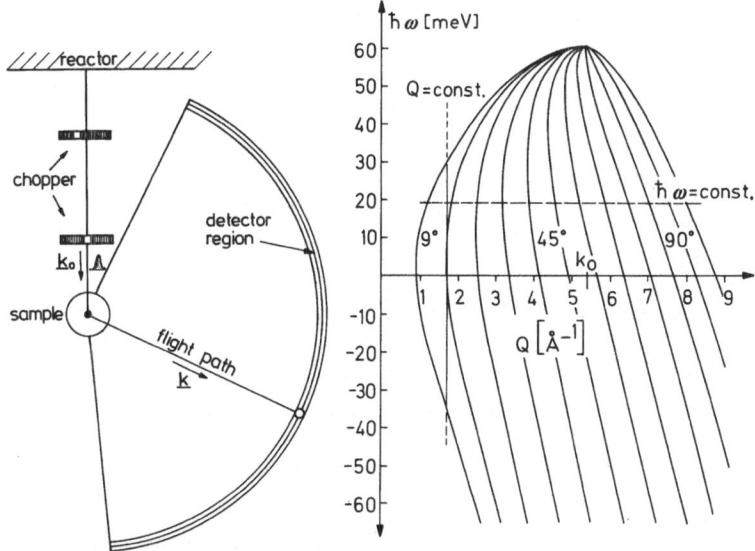

Fig. 7.1. (a) Time-of-flight spectrometer; k_0 is determined from the TOF between the two opened choppers and k from the TOF between the sample and the detector **(b)** Q-ω-range (dynamical range) of a neutron inelastic scattering experiment done with $k_0 = 5.4\,\text{Å}^{-1}$ in a range of scattering angles between 9 and 99 degrees. TOF spectra are measured along paths at constant θ. Values at constant Q have to be interpolated from these or they can be obtained directly using a TAS. Cuts at constant $\hbar\omega$ can be measured directly with both types of instruments. To reach large $\hbar\omega$ at low Q-values, experiments at small θ and high k_0 have to be done

ters. TOF spectrometers, therefore, have been used most often to determine the dynamical structure factor of metallic glasses. The disadvantage of the TOF method is the large amount of data treatment necessary after the experiment.

The Bragg reflection from single crystals is used in triple axis spectrometers (TAS) to determine k_0 and k from the known lattice plane distances d and the reflection angles θ_B ($2d \sin\theta_B = n2\pi/k$). A continuous beam of monochromatic neutrons is sent from the monochromator single crystal onto the sample. Part of the scattered neutrons fall on the analyzer single crystal and are reflected into the detector. The setting of the monochromator defines k_0. The position of the analyzer table relative to the incident beam gives the scattering angle θ, while k is given by the setting of the analyzer crystal (Fig. 7.2a). Each value of $S(Q, \omega)$, therefore, is obtained at a special setting of the spectrometer, i.e., $S(Q, \omega)$ is measured point by point and the instrumental parameters change between each of them. This makes absolute measurements of intensities, including the necessary quantitative corrections, very difficult, and relative intensities are usually measured on TAS. On the other hand the spectrometer has a low background, it is very flexible and one can chose the Q and ω value at which one wants to investigate the dynamical structure factor in advance. For

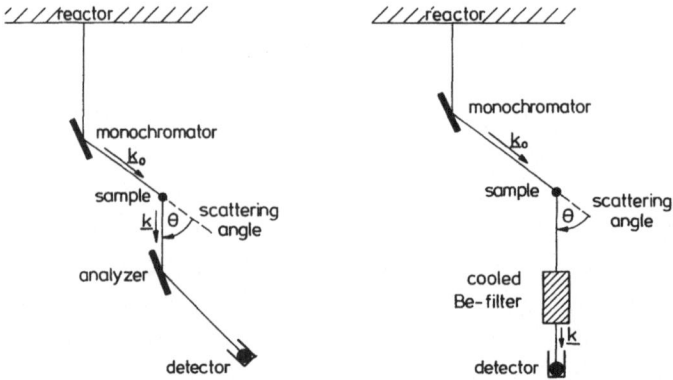

Fig. 7.2. (a) Triple axis spectrometer; k_0 is obtained from the setting of the monochromator single crystal and k from that of the analyzer crystal **(b)** Beryllium filter spectrometer; k_0 is determined from the setting of the monochromator crystal and k from the energy band transmitted by the cooled filter

example, cuts at constant momentum or constant energy transfers through $S(Q, \omega)$ can be obtained directly without the interpolations which are necessary after TOF experiments to get the same cuts through $S(Q, \omega)$ (Fig. 7.1b). TAS are therefore the obvious choice if the instrument has to be adapted to the inner coordinates of the sample (single crystal, anisotropic sample, magnetic field, etc.) and if the measurement of relative intensities is sufficient or if a few special cuts through the dynamical structure factor of an isotropic sample are needed. A TAS was therefore used in the search of the dispersion of collective excitations in glassy $Cu_{1-x}Zr_x$ alloys.

A second type of crystal spectrometer which was used to investigate glassy $Pd_{80}Si_{20}$ is a Beryllium filter spectrometer. The monochromator single crystal reflects a continuous monochromatic neutron beam onto the sample and some of the scattered neutrons are absorbed by the detector which can be moved to different scattering angles. This set up equals that of a diffractometer; however, the energy of the scattered neutron is analyzed by a Beryllium filter in front of the detector. This filter via Bragg reflection will scatter all neutrons with $E_n > 5.2$ meV (Bragg cutoff of Be) out of the beam and only neutrons with $E < 5.2$ meV can reach the detector. The monochromator setting defines k_0, the detector position relative to the incident beam gives the scattering angle θ and the filter defines k, which is assumed to be approximately constant because only a small energy band can pass the filter (Fig. 7.2b). This spectrometer always works in energy loss scattering because the energy of the scattered neutron has to be low to be able to pass the filter. The relative broad band (approximately 5 meV) transmitted by the filter considerably limits the resolution and the accessible Q-range of this instrument. However, the instrument has a very favourable intensity and this makes it attractive for the investigation of the atomic dynamics of metallic glasses if only a small amount of sample material is available.

7.4 The Dynamical Structure Factor

The dynamical structure factor contains the most detailed information one can get about the dynamics of monatomic samples. For polyatomic isotropic systems one has to determine partial dynamical structure factors to get the most detailed information available. For binary systems, which are the most interesting in this context, three time-dependent partial pair correlations have to be considered: AA, AB, and BB. Consequently, three partial dynamical structure factors S_{AA}, S_{AB}, and S_{BB} have to be determined:

$$\sigma S(Q,\omega) = 4\pi [c_A b_A^2 S_{AA}(Q,\omega) + 2\sqrt{c_A c_B} b_A b_B S_{AB}(Q,\omega) + c_B b_B^2 S_{BB}(Q,\omega)]. \quad (7.5)$$

Here σ is the effective scattering cross section of the binary system, $S(Q,\omega)$ is the dynamical structure factor measured in the neutron inelastic scattering experiment and b_A, b_B, and c_A, c_B are the scattering length and concentration of element A and B in the binary system, respectively. The three partial dynamical structure factors can be obtained by three measurements done on samples which are chemically the same but have different scattering properties. In neutron inelastic scattering experiments, this can be achieved by replacing one of the elements in the sample with the natural isotopic distribution by an isotope with sufficiently different scattering cross section. If this is done in two of the three experiments, one obtains three different spectra from chemically the same sample and can therefore separate out the partial dynamical structure factors and with these the dynamical properties of each of the two constituents. It is one of the advantages of thermal neutron scattering that this isotopic substitution can be applied due to the nonsystematic variation of scattering lengths for elements and their isotopes, which may even have opposite signs.

The partial dynamical structure factors S_{AA}, S_{AB}, and S_{BB} can be recaste in a different form [7.57, 58]. These new dynamical structure factors are then related to the fluctuations in the number density (NN) and the concentration (CC) with a corresponding cross term (NC).

In terms of these correlations, the measured dynamical structure factor is given as

$$\sigma S(Q,\omega) = 4\pi [\bar{b}^2 S_{NN}(Q,\omega) + 2\bar{b}\Delta b S_{NC}(Q,\omega) + (\Delta b)^2 S_{CC}(Q,\omega)]. \quad (7.6)$$

σ and $S(Q,\omega)$ have the same meanings as in (7.5), $\bar{b}=c_A b_A + c_B b_B$ is the mean value of the scattering lengths of the sample and $\Delta b = b_A - b_B$ is the difference between the scattering lengths. Choosing the scatterers and their concentrations so that $\Delta b = 0$ or $\bar{b} = 0$, one can, therefore, measure directly S_{NN} or S_{CC}, respectively. Up to now the possibility of measuring partial structure factors or S_{NN} and S_{CC} has only been used in structural investigations, especially also in those of metallic glasses (for a recent review see [7.59] and Chap. 3), but not for $S(Q,\omega)$ or $G(\omega)$ because the substitution of isotopes in the large samples necessary for neutron inelastic scattering experiments is very expensive and

Table 7.1. Relative weights of each of the partial structure factors given in row 1 according to (7.6) (columns 2–4) and (7.5) (columns 6–8), and the self part (column 5) of the measured dynamical structure factor for the metallic glasses given in column 1. The sample temperatures and the FWHM of the elastic peak in the TOF experiments are given in columns 9 and 10, respectively

$A_{1-x}B_x$	S_{NN}	S_{NC}	S_{CC}	S_S	S_{AA}	S_{AB}	S_{BB}	T [K]	ΔE_0 [meV]
$Cu_{46}Zr_{54}$	87	8	<1	4	33	32	32	296	3.0
$Ca_{70}Mg_{30}$	90	9	<1	<1	46	30	24	6, 296	2.7, 0.5
$Mg_{70}Zn_{30}$	92	5	<1	2	47	30	22	6, 296	3.0, 2.7

long measurements are necessary to obtain partial structure factors. Thus, all dynamical structure factors and frequency distributions discussed below will not be partial ones; however, we shall give the relative weights with which each of the three dynamical structure factors in (7.5, 6) enters the measured results for each of the metallic glasses investigated so far.

7.4.1 Dynamical Structure Factors of Binary Glassy Alloys

Up to now the dynamical structure factors have been determined for the binary metal-metal glasses $Cu_{46}Zr_{54}$ at 296 K [7.60, 61], $Ca_{70}Mg_{30}$ at 296 K [7.62] and at 6 K [7.36], and for $Mg_{70}Zn_{30}$ at 296 K [7.63, 64] and 6 K [7.65].

According to the weighting factors given in Table 7.1, about 90% of the measured dynamical structure factor is due to S_{NN}, the number density fluctuations, less than 10% stems from S_{NC} and less than 1% is due to S_{CC}, the concentration fluctuations. The contribution of the self part of the dynamical structure factor, see (7.2–4), to the measured intensity is smaller than 5%, i.e., predominantly collective atomic motions were studied in all cases. For metal-metalloid glasses, no dynamical structure factors have been determined up to now.

Though the dynamical structure factor of each glass differs from the others according to the different forces between the atoms in each of the glasses, the comparison of the dynamical structure factor of the glass with that of the same sample after crystallization reveals the same general results in all cases studied so far. We shall, therefore, quote only one typical example for each of the properties discussed below.

Some cuts at constant momentum transfers through the dynamical structure factor of $Ca_{70}Mg_{30}$ measured at 6 K [7.36] are shown in Fig. 7.3. As this was a low temperature measurement only the energy loss side of $S(Q, \omega)$ could be measured [7]. In the energy region displayed in Fig. 7.3, one finds a dynamical

---

7 It should be noted that the accuracy of this measurement was limited by the presence of a few percent H in the sample. Hydrogen is a very strong incoherent scatterer for thermal neutrons, but it contributes to the measured intensity strongest outside the energy region discussed here.

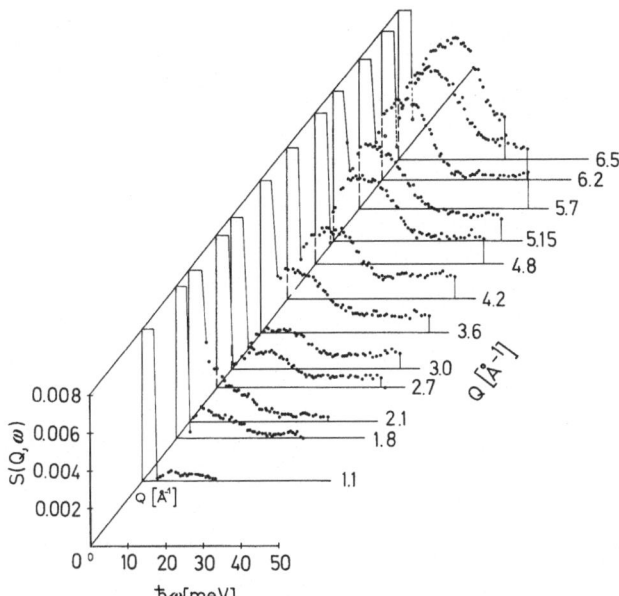

Fig. 7.3. Some cuts at $Q = \text{const}$ through the energy loss part of the dynamical structure factor of the metallic glass $Ca_{70}Mg_{30}$ measured at 6 K. Only the onset of the elastic peak is shown on the left hand side to augment the inelastic spectra

structure factor which is dominated by coherent scattering from the metal atoms, as can be seen from the distinct change of the *shape* of $S(Q,\omega)$ with momentum transfer. This reflects the interference scattering of the neutron from the moving atoms.

However, no dispersion of collective modes could be determined in this case. At momentum transfers larger than $6\,\text{Å}^{-1}$, the shape of the dynamical structure factor remains approximately the same and only the intensity of the inelastic spectra increases with increasing Q values. Here one is approaching the incoherent approximation which is valid for all scatterers at sufficiently high Q values. In this region of momentum transfers, mainly the frequency distribution is reflected in the inelastic part of $S(Q,\omega)$. This relatively smooth dynamical structure factor of the metallic glass has to be contrasted with that of the same sample after crystallization measured under identical experimental conditions[8]. In Fig. 7.4 the dynamical structure factor of the crystallized sample is displayed. It shows considerably more structure than that of the metallic glass (see, e.g., at $Q = 2.7\,\text{Å}^{-1}$, the sharp peak at $Q = 4.2\,\text{Å}^{-1}$, and all spectra for $Q > 6\,\text{Å}^{-1}$). However, in general, the two dynamical structure factors are very similar as far as the change in shape of $S(Q,\omega)$ with momentum transfer and the energy region covered by $S(Q,\omega)$ is concerned[9]. In fact, apart

---

8 In the case of $Ca_{70}Mg_{30}$, probably a mixture of $CaMg_2$ ($MgZn_2$ structure) and pure Ca (fcc structure) will be formed in the crystallization process [7.66].

9 The higher intensities in the dynamical structure factor of glassy $Ca_{70}Mg_{30}$ at energies above 25 meV and at momentum transfers above $2.7\,\text{Å}^{-1}$ are due to an incomplete correction for neutrons scattered from the hydrogen contamination.

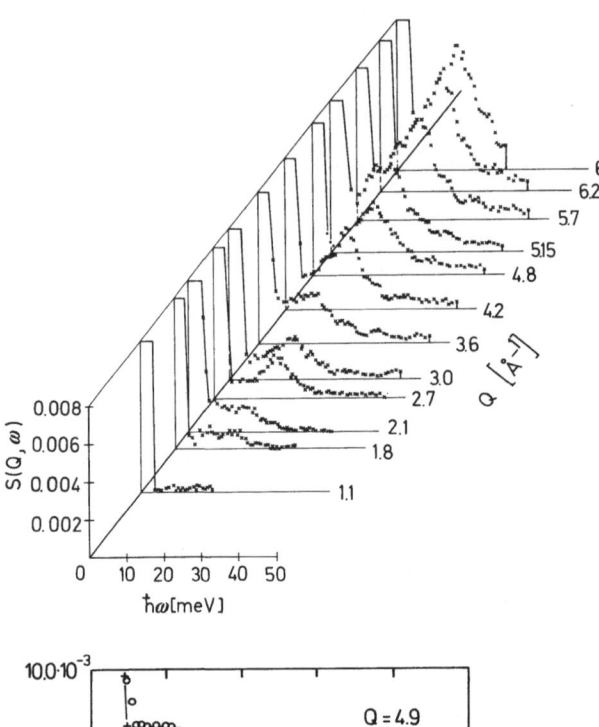

Fig. 7.4. Same as Fig. 7.3 for the same sample after crystallization measured at 6 K

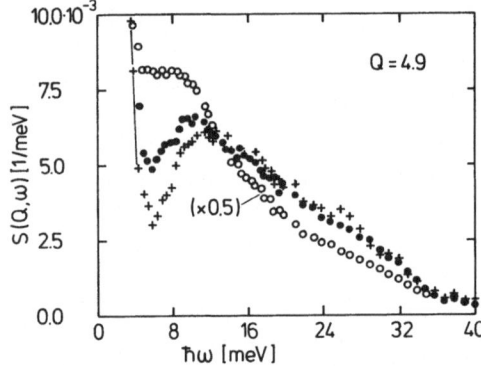

Fig. 7.5. Cuts at $Q = 4.9 \text{ Å}^{-1}$ through the energy loss part of the dynamical structure factors of the metallic glass $Mg_{70}Zn_{30}$ measured at 6 K (●) and at 273 K (○) and of the same sample after crystallization measured at 6 K (+). The intensity due to the low frequency modes is seen at energies below 12 meV in the spectra of the glass

from the more pronounced structure in the spectra of the polycrystal, one finds a striking similarity between the dynamical structure factor of the metallic glass and that of the crystallized sample for energies above 10 meV at all momentum transfers investigated so far ($1–7 \text{ Å}^{-1}$). This is shown in more detail in Fig. 7.5: the dynamical structure factor of the metallic glass $Mg_{70}Zn_{30}$ and the same sample after crystallization[10] measured at 6 K are very similar in shape and in intensity (absolutely measured, no normalization factor!) and only the structure near 16 and near 26 meV is slightly more pronounced in the case of the crystal. These results are representative for all metallic glasses investigated so

---

10 Most probably the crystallized sample consists, to a large degree, of $Mg_{51}Zn_{20}$ [7.67].

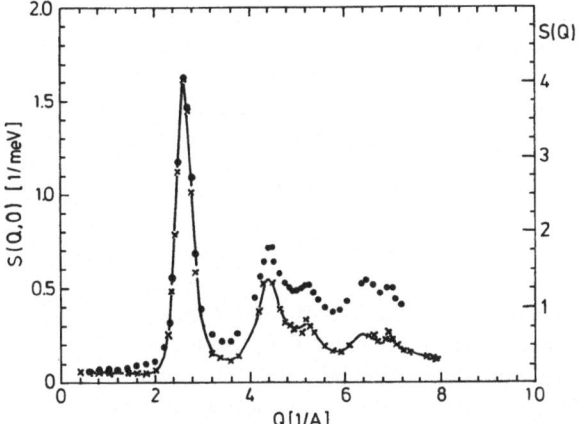

Fig. 7.6. A cut at $\hbar\omega = 0$ meV through the dynamical structure factor of the metallic glass $Mg_{70}Zn_{30}$ measured at 296 K (+, line of spline fit and left hand scale). The zeroth moment of $S(Q, \omega)$ of the same sample is shown for comparison ($\bullet$ and right hand scale). 8 % of the intensity is due to multiple scattering

far with the modification that the dynamical structure factors measured at room temperature show even less structure than those measured at low temperature because the low frequency modes, which we shall discuss below, and multi-"phonon" processes tend to smear out the remaining structure (see, e.g., the structure at 12 meV in Fig. 7.5 which is nearly lost in the data taken at 273 K).

From the dynamical structure factor the normalized frequency moments can be obtained:

$$\langle \omega^n \rangle = \int_{-\infty}^{\infty} \omega^n S(Q, \omega) d\omega \bigg/ \int_{-\infty}^{\infty} S(Q, \omega) d\omega. \tag{7.7}$$

The frequency moments for $n > 0$ are normalized to the zeroth moment, the static structure factor of coherent scatterers. The first and second moment are simple [7.68]. The fourth moment mainly contains an integral over the second derivative of an interatomic pair potential weighted with the static pair distribution function [7.69] and the sixth moment contains higher derivatives of the pair potential and static three particle correlation functions [7.70]. Up to now, only the zeroth moment could be successfully determined from all the dynamical structure factors measured so far (see, e.g., [7.64]) and the results agree rather well with those obtained from neutron and x-ray diffraction experiments. In the range of momentum transfers covered in the neutron inelastic scattering experiments, the structure of the zeroth moment is mainly determined by the structure of $S(Q, \omega)$ at $\omega = 0$, as shown in Fig. 7.6.

The measured dynamical structure factors have not yet been compared quantitatively with theoretical results. One reason for this is that the experimental results are not yet corrected for multiple scattering; another reason is that the theoretical results would have to be folded with the resolution function of the spectrometer before a quantitative comparison could successfully be made.

7.4.2 Low Energy Excitations

Figure 7.5 not only shows the similarity of the dynamical structure factor of the metallic glass to that of the same sample after crystallization for energies above 10–12 meV, but it also shows the difference between these two dynamical structure factors at energies below 10 meV. The low energy excitations which cause the intensity increase in the dynamical structure factor of the metallic glass in this $\hbar\omega$ region are characteristic of the atomic dynamics of metallic glasses and possibly all topologically disordered solids. They were seen for the first time in the investigation of $Cu_{46}Zr_{54}$ [7.60] and they disappear under identical experimental conditions as soon as the sample has been crystallized. Their dependence on momentum transfer was studied in several experiments [7.62, 64]. The difference in intensity between the dynamical structure factor of the metallic glass and the same sample after crystallization shows a characteristic dependence on momentum transfer. This can be seen in Fig. 7.7 where cuts through the dynamical structure factors of the metallic glass $Mg_{70}Zn_{30}$ and the same sample after crystallization at an energy transfer of $\hbar\omega = 5.3$ meV are displayed [7.64]. For momentum transfers below 0.6–$0.8\,Q_p$, the two dynamical structure factors measured at room temperature are essentially the same in this energy region. It should be mentioned, however, that the most recent investigation of $Mg_{70}Zn_{30}$ at 6 K indicates that a very small amount of extra intensity is also present for $Q < 0.6\,Q_p$ at the foot of the peak of elastically scattered neutrons. For larger Q-values, additional intensity is found in the dynamical structure factor of the glass. The difference between the two dynamical structure factors seems to be larger near the maxima of $S(Q)$ than in its minima. In the case of a crystal, this onset of scattering in the second Brillouin zone and this dependence of the measured intensity on momentum

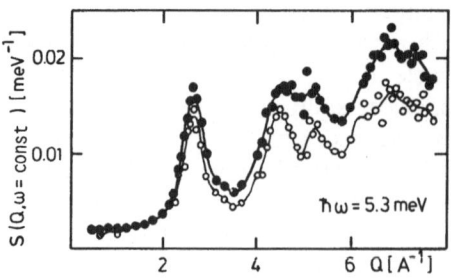

Fig. 7.7. Cuts at $\hbar\omega = 5.3$ meV through the dynamical structure factors of the metallic glass $Mg_{70}Zn_{30}$ (●) and that of the same sample after crystallization (○) measured at 296 K. The additional intensity in $S(Q,\omega)$ of the metallic glass is due to the low frequency modes

transfer would be characteristic of transverse modes. Analogous to this it was concluded that at least part of the observed additional intensity in $S(Q, \omega)$ of the glass is caused by modes of a predominantly transverse type as well.

The origin of these disorder characteristic atomic motions in the glass is not yet clear. Their low energy suggests that they take place in regions of reduced local density compared with the mean density of the glass. In an extreme case, it could be movements of atoms on the surface of very small voids in the rapidly quenched system which are removed in the crystallization process, as observed very recently [7.71]. Another source of the low energy excitations could be movements of atoms in lower density "defect" regions of the glass which have been investigated by *Egami* and his collaborators [7.72, 73]. These modes may, therefore, depend on the quenching conditions and on the thermal history of the sample. Neutron inelastic scattering techniques are being used at present to investigate this question. A third possibility will be discussed in context with the low energy part of the frequency distribution in Sect. 7.6.3.

From the arguments given above one could conclude that the low frequency modes may be localized, i.e., the excitations do not propagate through the glass. However, the Q dependence of the measured extra intensity suggests that at least part of the modes are not of a localized type.

7.4.3 Propagating Short Wavelength Collective Excitations

One of the most interesting subjects of the atomic dynamics of topologically disordered systems is the investigation of the dispersion and decay channels of propagating short wavelength collective excitations. Unfortunately, however, such experiments are extremely difficult. From the computer simulations reported in Sect. 7.2, and other considerations which will be discussed below, one should expect that the dispersion of propagating collective excitations could be detected – if at all – in the Q-region below and perhaps slightly above Q_p. The width of these excitations is smallest below $0.5\,Q_p$, so that this region of momentum transfers, which corresponds to the first Brillouin zone in a crystal, would be the most interesting to look at. However, incident neutrons with a velocity considerably higher than the sound velocity in the sample have to be used here to be able to excite the collective modes at these low Q-values (Fig. 7.1b). The high sound velocities found in crystals and glasses require very high incident energies (100–300 meV depending on the system under investigation) and the resolution of most of the existing neutron inelastic scattering instruments is insufficient for these investigations under such conditions. For these reasons the search for the dispersion of short wavelength collective excitations has concentrated on momentum transfers larger than $0.5\,Q_p$ where the line width of the excitations is already quite broad according to theoretical results.

The situation is different in the case of single crystals as long as the harmonic approximation is valid, i.e., the displacements of the atoms from their equilibrium positions are small compared to the interatomic distances. (This we shall assume throughout for glasses and crystals since all experiments were done below or at the Debye temperature of the system.) In crystals the propagating excitations (phonons) can be represented by a set of normal modes which are plane waves of infinite lifetime in the ideal case. This means that they are sharp excitations ($\Delta\hbar\omega/\hbar\omega \approx 1\%$ in real crystals) which can be measured relative to each reciprocal lattice point and which, therefore, do not have to be measured in the first few Brillouin zones (Fig. 7.1b).

In polycrystals the measured distributions are no longer sharp phonon peaks but rather broad maxima. This is caused by a superposition of phonons of the same dispersion branch at the same momentum transfer which have different energies in different directions of the single crystal, or even by a superposition of phonons of different dispersion branches at the same Q-value. As increasingly more reciprocal lattice points will be involved in the scattering process with increasing Q-values, it becomes very difficult to extract a dispersion relation from these maxima beyond the first few Debye-Scherrer peaks.

Finally, in glasses the Fourier transform $I(Q,t)$ of the density-density correlation function (7.1) will be governed by a superposition of normal modes of different frequencies and will therefore decay rapidly due to the phase mixing of these modes [7.33]. The collective excitations are therefore strongly damped and this damping increases with increasing Q and ω values [7.74]. For these reasons the peaks corresponding to the propagating short wavelength collective excitations are expected to be most clearly defined and therefore easiest to detect in the first Brillouin zone.

In fact, no such peaks could be detected in a first investigation made by *Holden* and collaborators [7.75] on several $Cu_{1-x}Zr_x$ glasses using a triple axis spectrometer, though in the extensive studies of $Cu_{60}Zr_{40}$ at 4 K and at 300 K and of $Cu_{45}Zr_{55}$ and $Cu_{27.5}Zr_{72.5}$ at 77 K, the results from the latter sample seem to show some extra intensity at a momentum transfer of $2\,\text{Å}^{-1}$ which might indicate a propagating collective excitation. Also in the study of $Cu_{46}Zr_{54}$ at 296 K [7.60] and $Ca_{70}Mg_{30}$ at 6 K [7.36] using a TOF spectrometer, no dispersion of collective excitations could be extracted in spite of strong coherence effects in the spectra measured at 6 K (Fig. 7.3). The first dispersion of propagating short wavelength collective excitations in a metallic glass was measured in $Mg_{70}Zn_{30}$ [7.63]. Peaks and shoulders were observed in the dynamical structure factor at several momentum transfers between $0.65\,Q_p$ and $0.87\,Q_p$.

Dispersions of longitudinal and transverse excitations had been found in a computer simulation of this glass using the equation of motion technique [7.40] and from an analytical calculation using a continued fraction up to the sixth moment of $S(Q,\omega)$ for the memory function [7.35] (see also [7.36]). However, the dispersion determined from the positions of the measured peaks in Q and ω did not coincide with any of the calculated dispersions, though the range of Einstein frequencies obtained in the computer simulation agreed reasonably well with the energy interval covered by the frequency distribution determined in the same neutron inelastic scattering experiment. This suggests that the effective pair potentials used in the computer simulation describe the distribution of interatomic forces in the system quite realistically.

The same experiment was repeated after crystallization of the sample and an equivalent dispersion of phonons was found in the same region of momentum transfers which could be interpreted as arising from transverse modes coupled to the neutron by Umklapp processes via the first reciprocal lattice point in this case. In analogy to this, the propagating short wavelength collective excitations found in the metallic glass were interpreted as being modes of predominantly transverse type coupling to the neutron by Umklapp scattering, the first peak of the structure factor acting as a "smeared out reciprocal lattice point". Thus, in spite of the absence of Debye-Scherrer peaks in the continuous, nearly liquid-like static structure factor, the short-range order in the metallic glass suffices to allow dynamical processes to take place which resemble those in crystals. Motivated by these experimental results, this problem was investigated in detail using analytical methods [7.31].

Taking a point of view which actually applies to crystals only, one could argue that the observed width of the maxima are made up of several contributions. One is due to the quantitative disorder, i.e., it results from all those interatomic forces which contribute to this special mode taken as an ensemble average over all existing atomic configurations. In the polycrystal this would correspond to the spread in phonon energies which one obtains in sampling over all directions in the crystal. This results in a broad dispersion band instead of a dispersion curve which one would find in a symmetry direction of an ideal single crystal. The second contribution is due to the topological disorder which causes a faster decay of the excitation as would be the case in the long-range ordered crystal[11]. A third contribution arises in Umklapp scattering processes because the infinite thin spherical shell, which represents the reciprocal lattice point of a single crystal in the case of a polycrystal, has a finite thickness in the case of the glass[12]. The width of the observed maxima has not been studied in detail yet; however, a rough comparison of the measured spectra taken on the metallic glass and the same sample after crystallization shows that the maxima in the spectra of the glass are not very much broader than those in the spectra of the crystal. This suggests that the major contribution to the width of the observed maxima is due to the quantitative disorder in the metallic glass.

Assigning also the energies of rather weak structures (maxima and shoulders) in the dynamic structure factor of $Mg_{70}Zn_{30}$ measured at 6 K [7.65], three different branches were obtained recently for momentum transfers between 1.7 and 6 Å^{-1}. As shown in Fig. 7.8, two of these branches show only

---

11 This contribution is also present to some extent in crystals with chemical or mass disorder.

12 Also in the case of real polycrystals, these shells have a finite thickness depending on the degree of imperfection of the crystal.

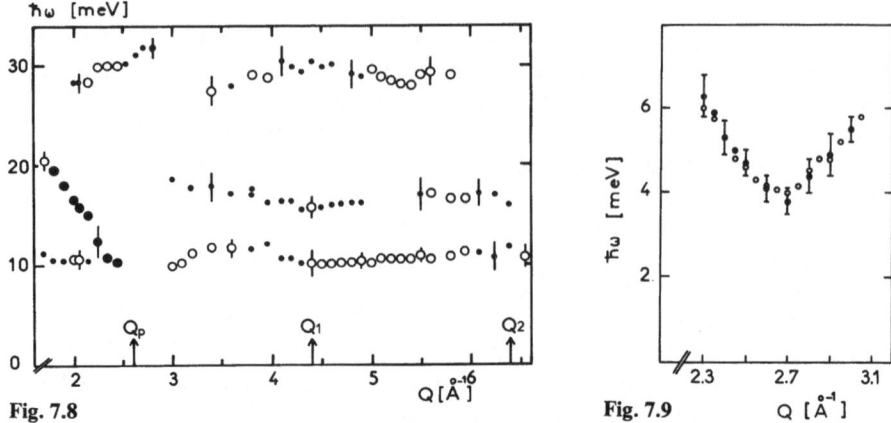

Fig. 7.8. Energies corresponding to peaks and shoulders in the dynamic structure factor of $Mg_{70}Zn_{30}$ measured at 6 K. Rather weak structures are included as points. The filled circles show the part where a change in the peak position with momentum transfer was observed. The error bars give the accuracy with which the peak positions could be assigned. Arrows on the Q scale mark the position of the maxima in the static structure factor (Fig. 7.6)

Fig. 7.9. Energies corresponding to the peaks in the dynamic structure factor of $Mg_{70}Zn_{30}$ measured at 296 K at momentum transfers around Q_p. Filled circles: sample as received. Open circles: same (still glassy) sample after heat treatment (3 h at 338 K)

little variation of the peak energy with momentum transfer and only the third shows the strong dispersion in the Q region discussed above. However, the peaks at higher energy ($\hbar\omega > 20$ meV) are rather weak. One reason for this is that the cross section in neutron inelastic scattering experiments varies like $1/\omega$, making high ω investigations more difficult. If the same arguments apply to metallic glasses, as have been used in context with insulating glasses [7.22, 77, 78], then one should expect that modes with higher energies are increasingly more localized than lower energy modes. Of the three branches, only the lowest showed an appreciable energy change on heating to 273 K.

Due to the limited resolution ΔE of these experiments done with $\Delta E_0 = 3$ meV, it could not be decided experimentally whether or not the dispersion of short wavelength collective excitations show a nonzero minimum near Q_p. A minimum of this kind was first observed in the case of rotons in superfluid ^{4}He [7.79]. For disordered systems it is expected from several theoretical investigations but for different reasons than in the case of rotons in ^{4}He (for a discussion of this question see [7.80 and 39])[13]. In a very recent experiment, however, done with a resolution of $\Delta E_0 = 0.19$ meV, the form of the dispersion of the lowest branch near Q_p could be determined. The energies corresponding to the peak positions in the dynamic structure factor of $Mg_{70}Zn_{30}$ measured at 296 K [7.76] at momentum transfers around Q_p are

---

13 It should be noted that a minimum of this kind was also found near the first peak of the structure factor in the case of a polycrystal [7.14].

shown in Fig. 7.9. The open circles demonstrate that the peak *positions* are rather uneffected by thermal treatment of the glass (in this case: 3 h at 338 K, i.e., 15 K below the crystallization temperature), though their absolute intensity decreases accompanied by an intensity transfer from the energy region around 5 meV to that of about 16 meV in $S(Q, \omega)$.

The most interesting study, however, would be an accurate measurement of the width of the peaks in $S(Q, \omega)$, because from these one could get detailed information about the damping of the propagating short wavelength collective excitations. Unfortunately, these experiments are extremely difficult as measurements with high precision have to be done at low Q values ($Q < Q_p/2$) and high incident energies.

7.5 Gases in Topologically Disordered Alloys

So far the dynamics of krypton in an amorphous alloy [7.81] and of hydrogen in a metallic glass [7.82] have been studied.

$Fe_{40}Zr_{60}$ with 0, 1, and 7% Kr was produced by sputtering FeZr in a krypton atmosphere with variable Kr pressure. Using a cold neutron TOF spectrometer with an incident energy of 5 meV, the generalized frequency distribution was determined at room temperature. In the measurements with the Kr doped samples, additional intensity was found with a maximum near 10 meV which extended as high as 20 meV. The amount of this extra intensity increased proportional to the Kr concentration. As the DOS of crystalline Kr ends at 6 meV [7.83], the extension of the additional intensity up to 20 meV shows that Kr is much more strongly bound in the amorphous metal than in a krypton crystal. This suggests that the Kr distribution in the amorphous alloy is either monatomic on interstitial sites or consists of very small clusters, rather than of bubbles or large clusters. If the interpretation of these experimental results is correct, then the interatomic forces between Kr and the metal atoms are about 10 times stronger than the interatomic forces in a krypton crystal. A possible application of this experimental finding could be the storage of radioactive Kr in amorphous alloys.

The vibrations of hydrogen in glassy and polycrystalline $CuTiH_{1.3}$ were studied at 78 K using a beryllium filter spectrometer. In the spectrum of the polycrystal, the hydrogen vibrations covered the energy region between 130 and 180 meV with a well-defined peak at 142 meV and a shoulder at 157 meV. In the spectrum taken on the metallic glass, the energies of the H vibrations covered a region between 60 meV and more than 200 meV with a maximum of this broad distribution near 142 meV, as in the case of the crystal (Fig. 7.10). The FWHM of this peak is about 75 meV, i.e., 4 times larger than the width of the energy distribution of hydrogen vibrations in the crystalline alloy. The authors conclude that the H atoms on the average occupy tetrahedral-type places in the metallic glass, though from vibrations of H atoms with energies

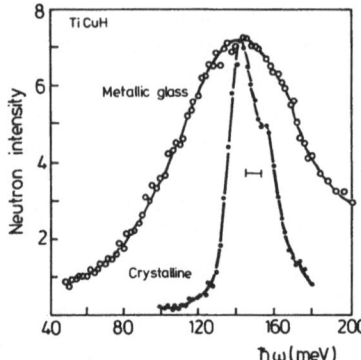

Fig. 7.10. Spectrum of hydrogen vibrations in glassy TiCuH$_{1.3}$ (O) and in polycrystalline TiCuH$_{0.93}$ (●). The energy resolution of the Be-filter experiment near the peak is indicated by a horizontal bar

below 100 meV in a crystal one would infer that the H atoms are preferably in octahedral positions.

The diffusion of hydrogen in metallic glasses is being studied at present using high resolution neutron inelastic scattering techniques.

7.6 Frequency Distributions

In order to determine the phonon density of states of a crystal, the relative number of phonons in the energy interval between ω and $\omega + \Delta\omega$, the one-phonon part of the double differential scattering cross section has to be separated from the measured intensity, i.e., from the peak of elastically scattered neutrons and from the intensity due to multiphonon processes in the sample. The intense elastic peak, represented in theory by a δ-function in ω for crystals and disordered solids, has a finite width in the experiment due to the convolution with the resolution function of the spectrometer near $\hbar\omega = 0$. In almost all cases it cannot be subtracted quantitatively, i.e., the measured intensity in the low energy region of the spectra, occupied by the elastic peak, has to be replaced by some reasonable model, for example, a Debye spectrum $[f(\omega) \sim \omega^2]$[14]. The multiphonon contributions can be subtracted by two equivalent methods which are both based on the multiphonon expansion of the double differential scattering cross section which applies to crystalline [7.84] and to topologically disordered monatomic coherent and incoherent scatterers [7.33] in the harmonic approximation.

For a monatomic *incoherently* scattering polycrystal with cubic structure and Bravais lattice, this phonon expansion of the inelastic part of $d^2\sigma/d\Omega dE$ is (e.g., [7.85])

$$\left.\frac{d^2\sigma_{inc}}{d\Omega dE}\right|_{inel} = \frac{\sigma_{inc}}{4\pi}\frac{k}{k_0}e^{-2W(Q)}\sum_{n=1}^{\infty}\frac{f_n(\omega)}{n!}\left(\frac{\hbar^2 Q^2}{2M}\right)^n \tag{7.8}$$

---

14 For the definitions of f, g, F, G, etc., see footnote 4.

with

$$f_n(\omega) = \int_{-\infty}^{\infty} f_{n-1}(\omega')f_1(\omega-\omega')d\omega' \qquad (n>1)$$

$$f_1(\omega) = \frac{f(\omega)}{\hbar\omega(1-e^{-\beta})}.$$

$$(7.9)$$

Here $\exp(-2W)$ is the Debye-Waller factor, $f(\omega)$ is the phonon DOS, $\beta = \hbar\omega/k_B T$ with k_B the Boltzmann constant and the other symbols are explained in context with (7.3, 4). The coefficients of this expansion in $(\hbar^2 Q^2/2M)$ are calculated by a convolution of the phonon DOS $[f_1(\omega-\omega')]$ with a multiphonon distribution $[f_{n-1}(\omega')]$, i.e., they only depend on ω, because only the intensity of the spectra is Q dependent. From (7.8) it follows that multiphonon contributions to the measured intensity are increasingly more important at higher temperatures [β small in (7.9)] and larger momentum transfers[15]. When the multiphonon contributions [terms with $n>1$ in (7.8)] are removed, the remaining one-phonon cross section is directly related to $f(\omega)$:

$$f(\omega) = \frac{4\pi}{\sigma_{inc}}\frac{k_0}{k}\hbar\omega(1-e^{-\beta})\frac{2M}{\hbar^2 Q^2}e^{2W(Q)}\frac{d^2\sigma_{inc}}{d\Omega dE}\bigg|_{1\ phon}. \qquad (7.10)$$

The two methods of removing the multiphonon contributions to the measured double differential cross section are the extrapolation method and the subtraction method.

The extrapolation method starts from (7.8) divided by Q^2. Plotting $\log[d^2\sigma_{inc}/d\Omega dE/(Q^2 k)]$ over Q^2 for each ω-value and extrapolating this function linearly towards $Q=0$, only the one-phonon term is retained. The extrapolated values, therefore, yield $f_1(\omega)$ and from this one gets $f(\omega)$ via (7.9) [7.86].

In the subtraction method the multiphonon terms are calculated according to (7.8) starting with the measured distribution in (7.10) (instead of the unknown one-phonon part) as a first approximation to $f(\omega)$. The calculated multiphonon contributions are subtracted from the measured double diffential scattering cross section and this procedure is repeated until self-consistency is achieved. This method has been used to determine the generalized frequency distributions of the binary glassy alloys discussed below.

For monatomic *coherently* scattering crystals with cubic structure and Bravais lattice, a phonon expansion equivalent to (7.8) can formally be given [7.84]. However, in this case the coefficients of the expansion are complicated functions of Q and ω because of the interference scattering of the neutrons. For

---

15 For this latter reason one risks large multiphonon corrections if one approaches the incoherent approximation at large Q-values to obtain $f'(\omega)$ in experiments done with coherent scatterers.

a polycrystal, an equation equivalent to (7.10) can be obtained from the first term of this phonon expansion [7.14]:

$$f(Q,\omega) = \frac{4\pi}{\sigma_{coh}} \frac{k_0}{k} \hbar\omega(1-e^{-\beta}) \frac{2M}{\hbar^2 Q^2} e^{2W(Q)} \frac{d^2\sigma_{coh}}{d\Omega dE}\bigg|_{1\ phon} . \tag{7.11}$$

In the expansion $f(Q,\omega)$ is given by

$$f(Q,\omega) = \left\langle \sum_j [\hat{Q}e^j(Q)]^2 \, \delta(\omega - \omega_j(Q)) \right\rangle . \tag{7.12}$$

Here e^j is the polarization vector of the displacement, j is the index of the dispersion band and $\hat{Q}$ is a unit vector in the direction of the momentum transfer. The angular bracket indicates that a mean value over all directions of the single crystal has to be taken to obtain the value for the polycrystal. This has been done numerically for the polycrystal [7.14]. For a topologically disordered system this average was evaluated formally for the first two terms of the phonon expansion and results for the one-phonon part of the dynamical structure factor for a very simple model were calculated [7.15, 16, 87].

To obtain the equivalent of the phonon DOS for coherent scatterers, $f(Q,\omega)$ has to be averaged over all polarization vectors and momentum transfers for each energy[16]. In experiments done on isotropic scatterers (polycrystals or topologically disordered systems), this can be achieved by taking the mean value weighted with $\sin\theta$ of the intensity over all accessible scattering angles between the smallest (θ_{min}) and the largest (θ_{max}) for each energy transfer [7.88, 89]:

$$f'(\omega) = \frac{4\pi}{\sigma_{coh}} \frac{k_0}{k} \hbar\omega(1-e^{-\beta}) \frac{8Mkk_0}{\hbar^2(Q_{max}^4 - Q_{min}^4)} \int_{\theta_{min}}^{\theta_{max}} e^{2W(Q)} \frac{d^2\sigma}{d\Omega dE}\bigg|_{1\ phon} \sin\theta d\theta , \tag{7.13}$$

that is, each energy is sampled in a large region of reciprocal space to average out the coherence effects measured in the spectra at each scattering angle.

Multiphonon contributions to the measured cross section have to be removed as described for incoherent scatterers. As higher-order terms of the multiphonon expansion are difficult to calculate exactly in the case of coherent scatterers, terms with $n > 1$ are usually approximated in the subtraction method

---

16 Roughly speaking the idea is that neutrons suffering the same energy transfer are scattered from a coherent scatterer into different directions with different intensities (in contrast to the incoherent case where the scattering is isotropic). However, if for each energy an average value of the intensity over all scattering angles (4π for a single crystal, $\pi/2$ for a polycrystal) is taken, then one should again obtain the phonon DOS, because on the average one cannot excite more phonons of one kind than there are in the phonon DOS, regardless of whether the neutrons are scattered incoherently or coherently. This means if $f(Q,\omega) = h(Q,\omega) \cdot f(\omega)$, then the mean value of $h(Q,\omega)$ will be $\bar{h}(Q,\omega) = 1$, if the averaging procedure was correctly applied.

by taking the equivalent phonon term from the phonon expansion of the double differential scattering cross section of incoherent scatterers, see (7.8). This is a reasonable approximation because scattering processes involving several phonons tend to be isotropic like incoherent scattering [7.85]. Thus, in the subtraction method, the average over all Q values is done first and independently of this the multiphonon correction is made afterwards. In the extrapolation method the mean value of the measured intensities over many Q-values for each energy transfer *and* the multiphonon correction have to be done in the same extrapolation procedure, i.e., extrapolation has to be done along a mean value over the structured function $\log[(d^2\sigma_{coh}/d\Omega dE)/(Q^2k)]$, the structure being due to interference scattering. This method has not yet been applied to metallic glasses. The equivalence of these two correction methods has been discussed in [7.1].

The deviation of the frequency distribution $f'(\omega)$ from the phonon DOS $f(\omega)$ can be estimated from the ratio of the volume of the Brillouin zone[17] to that of the reciprocal space covered in the experiment [7.88-91]. In experiments with sufficiently high incident energies (e.g., $E_0 > 20$ meV) and a large range of scattering angles [e.g., $(\theta_{max} - \theta_{min}) \geq 90$ degrees], the error due to the incompleteness of the averaging procedure is a few percent only.

For *polyatomic* systems, only a generalized phonon DOS $F(\omega)$ or frequency distribution $F'(\omega)$ in the case of crystals or a generalized vibrational DOS $G(\omega)$ or frequency distribution $G'(\omega)$ in the case of topologically disordered systems can be determined because the partial contributions $[f_i(\omega), f_i'(\omega), g_i(\omega), g_i'(\omega)]$ of each element i to the measured intensity is weighted according to the scattering properties of each element. For example, for the generalized frequency distribution of a coherently scattering polyatomic glass one has

$$G'(\omega) = \sum_{i=1}^{m} e^{2W_i(Q)} \frac{\sigma_{coh}^i}{M_i} g_i'(\omega) \Bigg/ \sum_{i=1}^{m} e^{2W_i(Q)} \frac{\sigma_{coh}^i}{M_i}, \tag{7.14}$$

where m is the number of different elements. The ratio of the Debye-Waller factors is often equal to 1 so that (7.14) mainly contains the partial vibrational DOS weighted with σ/M. It is this kind of generalized frequency distribution which was obtained for the coherently scattering binary glassy alloys using (7.13) and (for the subtractive correction) (7.8). It should be noted that $G'(\omega)$ determined in the experiment then implies another generalization because formulae given for polycrystals have been applied to the metallic glasses.

7.6.1 Frequency Spectrum of the Metal-Metalloid Glass $Pd_{80}Si_{20}$

The atomic dynamics of two metal-metalloid glasses have been studied so far using neutron inelastic scattering techniques [7.4, 92], of which only the results

---

17 For topologically disordered systems the volume of the Brillouin zone can be approximated by $\pi Q_p^3/6 \approx Q_p^3/2$.

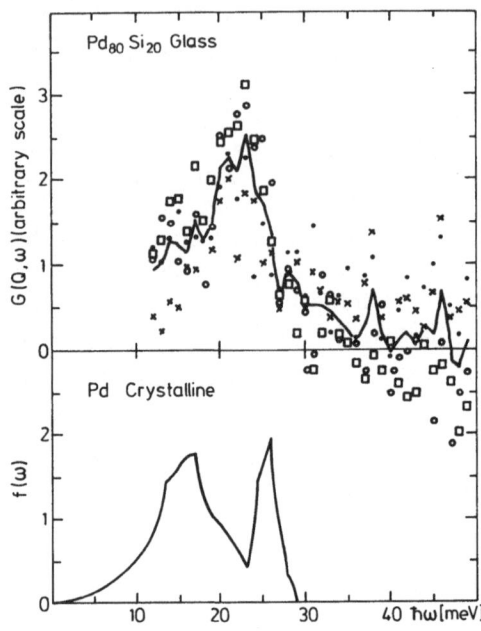

Fig. 7.11. Q-dependent frequency spectra of the metal-metalloid glass $Pd_{80}Si_{20}$ measured at 296 K using a Be-filter spectrometer. Four independent measurements and their average (line) are displayed. They are compared with the phonon DOS calculated from the dispersion curves measured in Pd single crystals (given below)

from the investigation of $Pd_{80}Si_{20}$ have been published [7.92]. The experiment was done at room temperature using a Be-filter spectrometer with a resolution of 2 meV. Energy transfers between 12 and 49 meV were scanned, thus the low energy region was excluded. From the measured intensities a Q dependent frequency distribution $G(Q, \omega)$ was determined for momentum transfers between 2.7 and 5 Å^{-1}. $Pd_{80}Si_{20}$ is a coherent scatterer for thermal neutrons ($\sigma_{inc} \ll \sigma_{coh}$), but the authors assume that the main features of the vibrational DOS is reflected in the spectra shown in Fig. 7.9. One finds a broad distribution with a maximum near 23 meV. The authors identified two further peaks at 28 and at 38 meV. The measured spectrum is compared with the phonon DOS of pure Pd which has two van Hove singularities [7.93] at 13.4 and 16.8 meV due to transverse phonons and one at 25.8 meV due to longitudinal phonons (Fig. 7.11). The maximum of the observed distribution is near this latter peak and the authors therefore assigned the measured spectrum to Pd–Pd longitudinal vibrations. However, this distribution extends so far down to lower energies that part of it also covers the region of transverse modes in pure Pd. To solve this problem, measurements with better statistical accuracy and a computer simulation like the one done for $Fe_{1-x}P_x$ [7.43] are needed.

7.6.2 Generalized Frequency Distribution of Binary Glassy Alloys

Up to now generalized frequency distributions of the type described in Sect. 7.6 have been determined for the metal-metal glasses $Mg_{70}Zn_{30}$ at room temperature [7.64] and at 6 K [7.65], $Ca_{70}Mg_{30}$ at 6 K and at 273 K [7.36] and $Cu_{46}Zr_{54}$ at room temperature [7.60].

According to Table 7.2, in all four cases one of the two elements predominates the one-phonon scattering by about a factor of two compared to the partner (see columns 2, 3 in Table 7.2), i.e., the ideal case in which each element contributes according to its concentration and where a true frequency distribution can be measured has not been investigated yet (see columns 4, 5 in Table 7.2).

Table 7.2. The relative weighting factors $W_x = (\sigma/M)_x \big/ \sum_A^B (\sigma/M)$ for the contributions of elements A and B (given in column 1) to the one-phonon cross section are given in columns 2, 3 (the unknown Debey-Waller factor is omitted). Columns 4, 5: relative one-phonon scattering "power" of elements A and B in the experiment due to their weighting factors W_x and concentrations C_x. Columns 6–9: sample temperatures, incident energies and angular ranges of the TOF experiments

$A_{1-x}B_x$	W_A	W_B	$C_A W_A$	$C_B W_B$	T [K]	E_0 [meV]	θ_{min}	θ_{max}
	[%]						[deg]	
$Mg_{70}Zn_{30}$	70	30	85	15	6, 296	50.4, 55.4	4.5	100
$Ca_{70}Mg_{30}$	33	67	54	46	6, 273	60.6	3	88
$Cu_{46}Zr_{54}$	64	36	60	40	296	60.8	9	97
$Pd_{80}Si_{20}$	36	64	69	31	296	Varied	–	–

The generalized frequency distributions of the binary glassy alloys have been compared with those of the same samples after crystallization. Such a comparison is reasonable as long as the density and the topological and chemical short-range order in the glassy and polycrystalline alloy are similar (see footnote 2)[18]. In Fig. 7.12, the generalized frequency distribution of the metal-metal glass $Mg_{70}Zn_{30}$ measured at room temperature is compared with that of the same sample after crystallization. In these samples, the Mg atoms contribute with 85 % to the one-phonon cross section. In fact, the two maxima in $F'(\omega)$ of the polycrystal at 17.9 and 26.7 meV are near the principal van Hove singularities in the phonon DOS of pure Mg which are at 16.7 and 27.5 meV [7.91, 94]. This suggests that the two maxima should be assigned to movements of Mg atoms. A similar assignment for the pronounced shoulder at 11.4 meV and the weaker shoulder at 30.4 meV cannot be done without assistance from detailed theoretical investigations. In the broad generalized frequency distribution $G'(\omega)$ of the metallic glass, the same structure with nearly the same relative weights is indicated as in $F'(\omega)$, though the peak near 17 meV seems to be shifted to slightly higher energies. Compared with $F'(\omega)$, additional intensity

---

18 This condition may not be sufficient in the case of Raman scattering and infrared absorption experiments because there, even identical frequency distributions might give different spectra if the selection rules are different for the polycrystal and the glass.

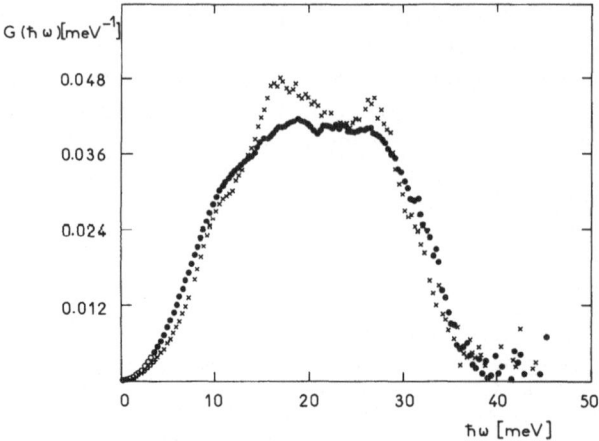

Fig. 7.12. Generalized frequency distribution of the metallic glass $Mg_{70}Zn_{30}$ (●) and of the same sample after crystallization (×) measured at 296 K. The first ten values at energies between 0 and 3.5 meV (○ and +) represent the Debye model fitted to the low energy end of the measured distributions

is found in $G'(\omega)$ at energies above 30 meV and below 10 meV. The former has been attributed to vibrations of atoms which are caught in a shorter distance to their neighbours in the rapid quenching process than that corresponding to the mean density of the glass, i.e., in regions of positive hydrostatic pressure [7.60]. The additional intensity below 10 meV is partly due to the low frequency modes found in the dynamical structure factor of metallic glasses (Fig. 7.5).

Investigation of the same system (but a different sample) at 6 K and at 273 K reproduced the results of the experiment done at room temperature within a few percent, except near 26 meV where about 7 % of the intensity was missing in $G'(\omega)$ *and* $F'(\omega)$ of the second experiment. As the accuracy of the two generalized frequency distributions is much better than the difference observed, this result may indicate that the details of the measured frequency distribution of the same glass could depend on the preparation conditions or the thermal history of the sample (the second sample was first cooled to 6 K and then measured at 273 K). Near 12 meV, the generalized frequency distribution of the glass at 6 K seems to be slightly harder than the one measured at 273 K, but this effect is smaller than the accuracy of the measurement. It is interesting to compare the $G'(\omega)$ determined in the neutron inelastic scattering experiment with the distribution of Einstein frequencies calculated in the computer simulation of the same system [7.40]. The statistical accuracy of this histogram only allows the identification of the positions of the two main peaks of the distribution at 18.3 and 28 meV, which are clearly separated. Their energy is about 1 meV higher than the maxima in $G'(\omega)$ and practically no intensity is found below 7 meV. In this comparison it should be kept in mind that the calculation of Einstein frequencies implies an extreme localization of the vibrations as only one atom moves while all the others are kept fixed. In contrast to what is observed in single crystals, the maxima of $G'(\omega)$ cannot be related to the extrema of the dispersion bands found in the computer simulations because these bands are already too broad at the Q-values corresponding to these extrema (e.g., $Q_p/2$).

The accuracy of the investigation of the metal-metal glass $Ca_{70}Mg_{30}$ at 6 K and 273 K was limited by the presence of a few percent hydrogen in the sample. However, the results do not fall outside the general characteristics of the results obtained for other binary glassy alloys. $G'(\omega)$ and $F'(\omega)$ both show one maximum at 16.4 meV, which is very sharp in $F'(\omega)$, but rather broad in the case of the glass. Two pronounced shoulders at 18.4 and 22.3 meV and a weaker shoulder at 10.9 meV in the generalized frequency distribution of the crystallized sample are still indicated in $G'(\omega)$ of the glass, but they are much weaker there. In this case, Ca and Mg vibrations are reflected to nearly equal parts in the measured distribution (see columns 4, 5 in Table 7.2) so that it is difficult to guess from the phonon DOS of the pure metals which maximum or shoulder can be correlated with the forces between a special pair of atoms in the sample.

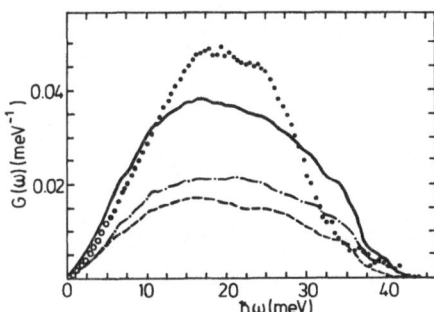

Fig. 7.13. The generalized frequency distribution of the metallic glass $Cu_{46}Zr_{54}$ (●) measured at room temperature is compared with the calculated total vibrational DOS of the model glass $Cu_{57}Zr_{43}$ (line) and the partial vibrational DOS of Cu atoms (chain curve) and Zr atoms (broken curve) in this system [7.44]. The first 9 values of the measured distribution (○) between 0 and 4.5 meV represent the $\omega^{3/2}$ approximation fitted to the low energy end of the measured distribution

An even smoother generalized frequency distribution was determined for a transition metal glass ($Cu_{46}Zr_{54}$) measured at room temperature[19]. Even in $F'(\omega)$ of the crystallized sample, a structure is only very weakly indicated near the maximum of the distribution and $G'(\omega)$ of the metallic glass has hardly any structure at all (Fig. 7.13). The region of the broad maximum extends from 15 to 24 meV approximately and $G'(\omega)$ and $F'(\omega)$ end near 36 and 34 meV, respectively. A hardening of $G'(\omega)$ by as much as 5 % compared with $F'(\omega)$ (not shown in Fig. 7.13) has not been observed in the other two cases where the generalized frequency distributions of the glass and the crystallized sample end at approximately the same energy.

The vibrational DOS has been calculated for metallic glass $Cu_{57}Zr_{43}$ using a modified Lennard-Jones potential and the recursion method [7.44] (Sect. 7.2).

---

19 A $G'(\omega)$ very similar to this one was determined for amorphous $Fe_{40}Zr_{60}$ at 296 K [7.81].

Though the concentrations of Cu and Zr are slightly different in the simulated system and in the glass studied experimentally, the calculated total and partial vibrational DOS and the measured generalized frequency distribution are compared in Fig. 7.13. The calculated $G(\omega)$ covers the energy region of the measured $G'(\omega)$ and also the energy region of the broad maximum is given correctly; however, no quantitative agreement could be obtained. The difference between the calculated and measured spectra cannot be accounted for by the difference in concentration of the two systems because the partial vibrational DOS of Cu and Zr in this glass are very similar [$g_{Cu}(\omega) \approx g_{Zr}(\omega)$]. The model could still be improved including the chemical short-range order and perhaps also by use of more realistic pair potentials though this will be difficult in the case of transition metals, especially Zr. The fact that the partial radial distribution functions and the density of $Cu_{57}Zr_{43}$ glass were well reproduced by the model system [7.53] suggest that these criteria are insufficient for a realistic model for the atomic dynamics of a metallic glass. But even without a quantitative agreement between the theoretical and experimental results, the simulation gives additional information which could only be obtained from the measurement, if partial vibrational DOS $g'(\omega)$ were measured. (In the case of neutron inelastic scattering, this could be done by isotopic substitution as described in Sect. 7.4, but no experiment of this kind has been performed yet.) According to the model, the lack of structure in $G'(\omega)$ is caused by the fact that the two $g_i(\omega)$ are nearly identical in shape. This is related to the fact that in this model the characteristic frequencies $\omega_{AA} = (d_{AA}/M_A)^{1/2}$ are nearly equal for all three atomic correlations (Cu–Cu, Zr–Zr, and Cu–Zr).

7.6.3 Low Energy Region of the Generalized Frequency Distribution of Binary Glassy Alloys

The most interesting part of the generalized frequency distributions of glasses is their low energy region because clear differences are found here between the distributions of the glass and the crystallized sample, and it is this part of $G(\omega)$ which determines predominantly the special thermodynamic properties of glasses at low temperatures.

In Fig. 7.14 the difference $\Delta G'(\omega) = G'(\omega) - F'(\omega)$ between the generalized frequency distributions of metallic glasses and of the corresponding crystallized sample are shown for $Mg_{70}Zn_{30}$ and $Cu_{46}Zr_{54}$. All four measurements were done at room temperature. For $Cu_{46}Zr_{54}$ this difference has a maximum near 5 meV and extends up to 7 meV; for $Mg_{70}Zn_{30}$ the corresponding values are 7 meV and 13 meV, respectively. Below 3.5 meV no values could be measured because of the finite resolution of the spectrometer. Clearly this energy region would be of greatest interest because the differences between the thermodynamic properties of glasses and crystals increase with decreasing temperatures and they become very large for $T < 1$ K (100 µeV). However, it is very difficult to get access to this energy region in neutron inelastic scattering experiments

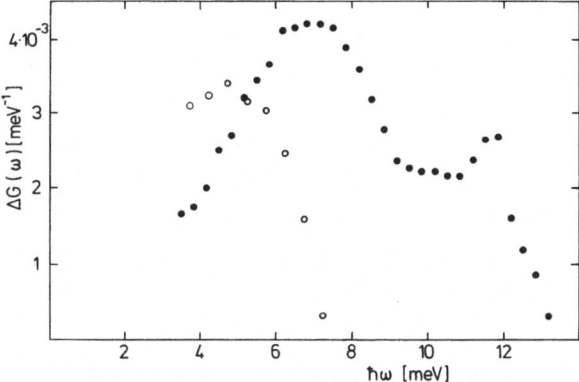

Fig. 7.14. Difference spectrum $\Delta G'(\omega) = G'(\omega) - F'(\omega)$ of the generalized frequency distributions of the metallic glasses $Mg_{70}Zn_{30}$ (●) and $Cu_{46}Zr_{54}$ (○) and of the same samples after crystallization for energies between 3.5 and 13 meV. All four measurements were done at room temperature. The first two values near 4 meV of the difference spectrum of $Cu_{46}Zr_{54}$ (○) fall into the energy region of the inserted model spectra ($\omega^{3/2}$ for the glass, ω^2 for the crystallized sample)

done with coherent scatterers. With incident energies of a few meV one could obtain the necessary resolution of 50–100 µeV near the elastic peak. However, experiments done with low incident energies do not provide a sufficiently large energy range and region of reciprocal space necessary to obtain a reliable frequency distribution.

The difference spectra shown in Fig. 7.14 are mainly due to the low frequency modes observed in the dynamical structure factors of the metallic glasses as described in Sect. 7.4.2. It is interesting to note that a spectrum of activation energies of the kind shown in Fig. 7.14 had to be assumed to explain the temperature dependence of the ultrasonic attenuation $[\alpha(T)]$ in PdSi [7.95, 96] and the term linear in ω and T $[\alpha(T) \sim \omega \cdot T]$ found in NiP and PdSi [7.97], which is different from what was observed in covalent glasses [7.96]. These results are very similar to those obtained for amorphous polymers like Se [7.98], which is possibly due to the fact that weak and nondirective forces are dominant in both disordered systems in contrast to the strong and directed forces in covalent glasses [7.98][20]. These weaker forces lead to lower energy barriers than have been found in covalent glasses. In fact, the results obtained in the ultrasonic attenuation experiments have been interpreted as being due to hopping processes over these low energy barriers [7.96]. Processes of this kind could, therefore, be the origin of low frequency modes found in the dynamical structure factor of glassy metals (see also the end of Sect. 7.4.2).

---

20 Also the coupling to the two-level system is weaker by about a factor of 5–10 in metallic glasses *and* amorphous polymers compared with covalent glasses [7.96].

Table 7.3. Second moment of the generalized frequency distribution (column 4) and the high temperature limit of the Debye temperature (column 5) for the glassy and polycrystalline alloys $Cu_{46}Zr_{54}$ and $Mg_{70}Zn_{30}$ given in columns 1 and 2 measured at the temperatures given in column 3

		T [K]	$(\frac{5}{3}\langle\omega^2\rangle)^{1/2}$ [meV]	$\theta_D(\infty)$ [K]
$Cu_{46}Zr_{54}$	glass	296	25.6	297
$Cu_{1-x}Zr_x$	cryst.	296	24.8	287
$Mg_{1-x}Zn_x$	cryst.	296	28.0	325
$Mg_{70}Zn_{30}$	glass	296	28.2	327
$Mg_{70}Zn_{30}$	glass	273	28.3	328
$Mg_{70}Zn_{30}$	glass	6	28.4	329
$Mg_{1-x}Zn_x$	cryst.	6	28.4	329

7.6.4 Moments and Thermodynamic Properties Calculated with the Generalized Frequency Distributions of Glassy Alloys

For the metallic glasses $Cu_{46}Zr_{54}$ [7.60] and $Mg_{70}Zn_{30}$ [7.64, 65], the moments of the generalized frequency distributions have been calculated:

$$\langle\omega_G^n\rangle = \int_0^\infty \omega^n G'(\omega)d\omega \bigg/ \int_0^\infty G'(\omega)d\omega . \tag{7.15}$$

From these the Debye "cutoff" frequencies [7.99] were determined. The second moment of the partial DOS $g_i(\omega)$ can be related to the mean value over the force constants between element i and all its neighbours [7.100]. In our case the second moments should, therefore, represent some mean force constant in the metallic glass and the same sample after crystallization. From Table 7.3 it follows that this mean force constant is nearly the same for the glassy and the polycrystalline $Mg_{1-x}Zn_x$ system. For the $Cu_{1-x}Zr_x$ sample the mean force in the glass is slightly stronger, i.e., the second moment is higher because the generalized frequency distribution is extended to higher energies and it is weighted with ω^2 in the integral. For this latter reason the second moment is practically insensitive to the choice of model in the low energy region of $G'(\omega)$. The second moment is also related to the high temperature value of the lattice part of the specific heat and with this to that of the Debye temperature (e.g., [7.101]). From the second moment it follows that the high temperature values of the Debye temperature are approximately the same in the glass and the corresponding crystal, in contrast to their low temperature values, as will be shown below.

The moments for $n=-2$ and $n=-1$ are related to the slope of the Debye-Waller coefficient $2W$ at temperatures above $\theta_D(\infty)/4$ and its value at $T=0$, respectively. In the case of $Cu_{46}Zr_{54}$ and $Mg_{70}Zn_{30}$, these moments cannot be obtained with good reliability because they are determined by the low energy part of the frequency distributions, which could not be measured but was replaced by a Debye model. However, even with these uncertainties it can be said that the low temperature limit and the slope of $2W(T)$ of the metallic glass are larger by a few percent compared with the crystallized sample.

The measured frequency distributions are not vibrational DOS. However, to compare the vibrational dynamics of the metallic glass with that of the crystallized sample, they have been used to calculate the temperature dependence of the "lattice" vibrational part of the specific heat

$$C_{\mathrm{VL}}(T) = 3R \int_{0}^{\beta_{\max}(T)} \frac{G'(\beta)\beta^2\, \mathrm{e}^{-\beta}}{(1-\mathrm{e}^{-\beta})^2}\, d\beta, \tag{7.16}$$

because the weighting factors given in columns 4, 5 of Table 7.2 are the same on the average for $G'(\omega)$ and $F'(\omega)$. In (7.16), $G'(\beta) = k_{\mathrm{B}} T G'(\omega)$ and R is the gas constant. The upper limit of the integral depends on the temperature for which $C_{\mathrm{VL}}(T)$ is calculated, i.e., $C_{\mathrm{VL}}(T)$ "scans" $G'(\beta)$ times a weighting function with increasing temperatures over the energy region of the frequency distribution. Due to its tailing towards higher energies, the weighting function has the effect of including parts of the generalized frequency distribution which are about a factor of 3 higher in energy than the temperature for which $C_{\mathrm{VL}}(T)$ is calculated (e.g., [7.102]). This means that $C_{\mathrm{VL}}(T)$ can still be reliably calculated at temperatures which are lower than the low frequency end of the *measured* distribution.

As an example, the temperature dependence of the vibrational part of the specific heat of the transition metal glass $\mathrm{Cu}_{46}\mathrm{Zr}_{54}$ and the same sample after crystallization, both measured at room temperature, are shown in Fig. 7.15. Temperatures above 60 K have not been included because at temperatures above 70 K to 100 K, the vibrational part of the specific heat of the metallic glass and the same sample after crystallization are essentially the same. This is due to the fact that the generalized frequency distributions of the glassy and the polycrystalline alloy are not very different for energies above 10 meV and that the specific heat is fairly insensitive to the details of the vibrational DOS at higher energies. For temperatures below 40 K the vibrational part of the specific heat of the metallic glass is larger than that of the crystallized sample. In Fig. 7.15, two sets of data are given for C_{VL} of the glass to show the effect on the specific heat of two different models for the low energy region of the generalized frequency distribution of the glass. The low energy part of $F'(\omega)$ could in all cases be bridged by a Debye spectrum which could be fitted to the low energy end of the measured distribution without any problems. This was not so in the case of metallic glasses [7.60, 64], though it was normally done for reasons of consistency between $G'(\omega)$ and $F'(\omega)$. In the case of $\mathrm{Cu}_{46}\mathrm{Zr}_{54}$ a better fit to the measured distribution could be obtained using an $\omega^{3/2}$ law. Thus, the difference between the open (Debye spectrum) and the filled circles ($\omega^{3/2}$ approximation)

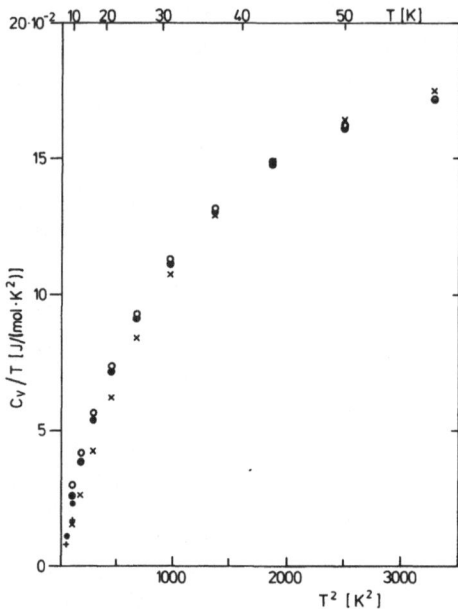

Fig. 7.15. Temperature dependence of the vibrational ("lattice") part of the specific heats calculated from the generalized frequency distributions of the metallic glass $Cu_{46}Zr_{54}$ (●, ○) and of the same sample after crystallization (×). The two sets of data given for the glass show the influence of the models inserted to bridge the low energy region of the frequency distributions: $\omega^{3/2}$ (●) and ω^2 (○). At 5 and 10 K, the measured specific heat of the metallic glass $Cu_{57}Zr_{43}$ (·) and of the same sample after after crystallization (+) are shown [7.103]

in Fig. 7.15 shows the importance of the choice of the model for the low energy part of $G'(\omega)$ for the vibrational part of the specific heat. The difference is most important at lowest temperatures.

There are no measured values of the specific heat of $Cu_{46}Zr_{54}$ to our knowledge which would allow a reasonable comparison between the values obtained by direct measurements and those calculated from the generalized frequency distributions determined in neutron inelastic scattering experiments. In addition, the specific heat is usually measured in a temperature region below 10 K which corresponds to an energy region where $F'(\omega)$ and especially $G'(\omega)$ are difficult to determine. But the specific heat of glassy and crystallized $Cu_{57}Zr_{43}$ was measured and they are shown in Fig. 7.15 [7.103]. The linear part of the measured values has been fully attributed to the electronic specific heat and has been subtracted already. Keeping in mind all the assumptions which have to be made in this comparison [slightly different Cu and Zr concentrations, $F'(\omega)$, $G'(\omega)$ measured at 296 K, model for low energy part of $F'(\omega)$ and $G'(\omega)$, etc.], the agreement between the calculated and measured value for the crystallized sample at 10 K is satisfactory – or accidental. For the measured and calculated specific heat of the metallic glass the agreement is much worse, but in this case the choice of model for the low energy region of $G'(\omega)$ is still an open question, while in the case of the crystallized sample the Debye spectrum is a reasonable approximation.

From the vibrational part of the specific heat the temperature dependence of the Debye temperature $\theta_D(T)$ was calculated using the Debye model for the temperature dependence of the specific heat. As an example, the values

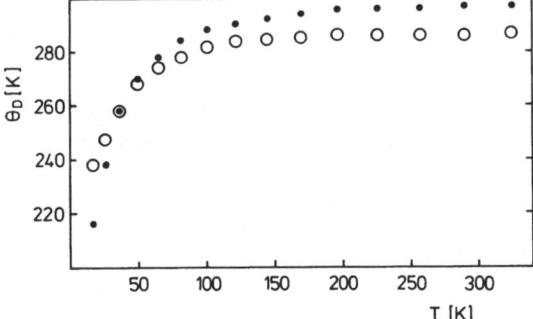

Fig. 7.16. Temperature dependence of the Debye temperatures of the metallic glass $Cu_{46}Zr_{54}$ (●) and of the same sample after crystallization (○) calculated from the values of the specific heat displayed in Fig. 7.15 (●, ×)

obtained for the metallic glass $Cu_{46}Zr_{54}$ and of the same sample after crystallization are shown in Fig. 7.16. In the calculation, the values of the specific heat obtained with the $\omega^{3/2}$ approximation in the low energy region of the measured $G'(\omega)$ were used. The Debye temperature of the crystallized sample is essentially constant for temperatures above 130 K, while θ_D of the glass approaches the high temperature value given in column 5 of Table 7.3 only near 200 K. From these temperatures towards lower temperatures, $\theta_D(T)$ of both samples decreases. Below 70 K the Debye temperature of the glass decreases much more strongly than θ_D of the crystallized sample and the crossing point of the two curves is at 40 K. For the metallic glass $Cu_{57}Zr_{43}$, a value of approximately 200 K was determined from low temperature specific heat measurements [7.103]. Similar results as for $Cu_{46}Zr_{54}$ were obtained for $Mg_{70}Zn_{30}$ and the same sample after crystallization measured at 6 K and at 296 K with the difference that the Debye temperatures of the glassy and the polycrystalline alloy are nearly the same at high T in the case of $Mg_{70}Zn_{30}$. These results explain why the Debye temperatures $\theta_D(0)$ of metallic glasses obtained from low temperature specific heat measurements are considerably lower than those determined for the crystallized sample, while at high temperatures the situation is sometimes inverse due to the higher intensities found at the high energy side of the generalized frequency distribution of the metallic glass (see Fig. 7.12 between 30 and 35 meV).

The difference between the specific heats and, accordingly, between the Debye temperatures of the glassy and the crystallized alloy is due to the low energy modes found in the dynamical structure factor of the glass which lead to the additional intensity in the generalized frequency distribution of the metallic glass. Due to the fact that the inelastic spectra could not be separated from the intense elastic peak at energies below 3.5 meV in the results from the few neutron inelastic scattering experiments done so far, one cannot decide from the existing results how much of the excess specific heat observed in metallic glasses at the lowest temperatures (e.g., [7.104]) is due to these characteristic modes and whether they contribute to the additional part of the T^3 term of C_{VL} only.

7.7 Concluding Remarks

We have tried to show that some insight into the atomic dynamics of glassy metals and especially of binary glassy alloys has already been gained from neutron inelastic scattering experiments, although these investigations have only been started very recently.

In the comparison with crystalline materials the dynamical structure factor which contains the most detailed information about the atomic dynamics obtainable in neutron inelastic scattering experiments can be characterized mainly by three properties:

1) Its dependence on momentum transfer predominantly reflects the lack of long-range order in the metallic glass as does the zeroth moment of $S(Q, \omega)$ (and with this the static structure factor).

2) Its dependence on energy for energies *above* 10 meV reflects the strong similarity of the atomic dynamics and of the mean interatomic forces in the metallic glass and the same sample after crystallization in all cases investigated so far. This means on the $\hbar\omega$-scale, $S(Q, \omega)$ can be regarded as a "smeared out" version of the dynamical structure factor of the corresponding crystal. The same characterization applies in this energy region to the generalized frequency distributions and most likely to the vibrational DOS as well. As the long-range order, which defines the crystal, is absent in the glass, this shows that the atomic dynamics is dominated in this energy region by the short-range order which in the mean must be similar in the glass and the sample after crystallization, though this latter sample will, in general, consist of several crystalline phases. Glasses are obtained by quenching from the melt, where all atoms of the system are present at the same time [21] and the onset of the future structure can, therefore, already be formed in the liquid before the quenching process [7.105]. Metallic glasses are, therefore, most easily obtained with concentrations *between* composition regions where crystals are formed, e.g., in deep eutectica [7.106]. The formation of several crystalline phases in the crystallization of glassy metals therefore is nearly a necessary consequence. This situation may be different if amorphous metals are obtained by the successive deposition of atoms as is done in sputtering or vapour deposition techniques where elementary amorphous metals and alloys near a crystalline phase can be obtained. In these cases the frequency distribution of the crystallized sample may show considerably more structure than that of the amorphous metal [7.81].

3) The characteristic difference in the atomic dynamics of glassy and polycrystalline alloys appears in the dynamical structure factor and the vibrational density of states at energies *below* 10 meV. In this region of energies additional intensity is found in the dynamical structure factor of the metallic

---

21 In contrast to sputtered samples where the atoms are added to the amorphous sample one after the other.

glass which disappears on crystallization. For the momentum transfers investigated so far ($Q \leq 7$ Å^{-1}), the Q-dependence of the inelastic part of $S(Q, \omega)$ in this energy region is largely given by the structure of $S(Q, 0)$. The origin of the low energy modes observed in this energy region has not yet been understood. We assume that the observed intensity in this energy region is due to a superposition of probably localized modes with propagating collective modes, especially of transverse modes which are softer in the glass than in the crystal [7.104]. Near the first (and perhaps also the second) maximum of the static structure factor, where the energies of the dispersion bands of propagating collective excitations decrease, also longitudinal modes will contribute to the measured intensity in this energy region, but it has not been possible to separate them from the transverse modes up to now. The localized modes could be due to the hopping of atoms over barriers with a broad distribution of activation energies[22]. These barriers could possibly be related to the atomic configurations which are involved in the tunneling process observed at temperatures below 1–4 K.

These low energy modes contribute to the vibrational DOS at energies below 10 meV. This has mainly two consequences:

1) The vibrational DOS is higher in this energy region than that of the crystallized sample and this leads to thermodynamic properties which are different from those of crystals at temperatures below 70 K to 100 K approximately, like the enhanced specific heat or the lower Debye temperatures.

2) Down to the lowest energies investigated so far, the vibrational DOS does not show the Debye spectrum observed in the case of the crystallized samples for energies up to 7 meV.

The observation of the dispersion of collective short wavelength excitations in metallic glasses demonstrates that at least part of the atomic vibrations in the glass are not localized, and the possibility for Umklapp processes in these topologically disordered systems demonstrates their well developed short-range order.

The comparison of results obtained from neutron inelastic scattering experiments with those from theoretical calculations is still difficult at present. In most cases the glassy metals studied in experiments were different from those investigated theoretically as the availability of reasonable pair potentials does not necessarily imply the suitability of the system for neutron inelastic scattering experiments. In nearly all cases theoretical results have not been presented as dynamical structure factors but rather as momentum dependent "one phonon" spectra which cannot be obtained from neutron inelastic scattering experiments without further assumptions. Also the experimental results have still to be improved until a quantitative comparison can rule out assumptions and approximations made in the model for the atomic dynamics

---

22 If this picture of the excitations were correct then we would expect friction peaks to be observed between 40 and 100 K approximately, which deviate from simple Debye peaks due to the broad distribution of activation energies.

of metallic glasses. The dispersion of longitudinal and transverse propagating collective excitations in glassy alloys could not be separated experimentally as clearly as was possible in the theoretical investigation of amorphous iron [7.41, 42] (see also Sect. 7.2). High energy modes with hardly any dispersion were observed in $Mg_{70}Zn_{30}$ [7.65]. However, whether these modes can be labeled as "optic modes" in a system without an elementary unit cell is not clear to us. This was done in a theoretical investigation of FeP [7.43] and $Mg_{70}Zn_{30}$ [7.35], because the modes found in the same energy region in $S(Q, \omega)$ of the corresponding crystal are optic modes. In nearly all of the theoretical investigations the low energy modes, which play the dominant role in the experimental results, are not or are insufficiently reproduced. This could be due to the limited size of the glassy cluster investigated in the computer, or it could be due to a deficiency of the model system which is not reflected in the pair correlation function usually well reproduced by the theoretical results.

The dynamical structure factor of the glassy alloys investigated so far displays strong interference effects up to momentum transfers of 6–7 Å^{-1}. Only at higher momentum transfers is the incoherent approximation approached, though the static structure factor still oscillates out to much higher Q-values. In calculations of the temperature dependence of the electrical resistivity of glassy metals at temperatures below $\theta_D/2$ [7.107, 108] and of the Eliashberg function of superconducting metallic glasses [7.109], the incoherent approximation of the dynamical structure factor was used in the Baym formula [7.110]. In view of the strong coherence effects found in the dynamical structure factor of the glassy alloys at momentum transfers below 2 k_F, this approximation seems somewhat questionable.

We have not compared the results obtained for the atomic dynamics of glassy metals with those of neutron inelastic scattering experiments done on covalent glasses. The reason for this is that the atomic dynamics of glasses, as we have shown, is less strongly determined by the lack of long-range order (which is common to both systems) than by the short-range order in the glass. And this is very different in these two types of glasses. In covalent glasses the interatomic forces are strong and highly oriented due to the localization of the electrons in these insulating glasses. The resulting structure, therefore, is an open network structure which includes an appreciable number of holes, depending on the production conditions [7.111]. This results in a density change on crystallization which may amount to as much as 20% in these systems. In glassy metals, on the contrary, the interatomic forces are weak and they are much less oriented than those of covalent glasses due to the screening produced by the nearly free conduction electrons. The density change on crystallization is a few percent only, showing that holes do not play such an important role in the structure of these glasses. For these reasons the results obtained for the atomic dynamics of metallic glasses are similar to those obtained in the case of insulating glasses in very general terms, but they differ considerably in the details. Under these circumstances, a comparison could not be done without many restrictions. However, we refer the interested reader to some of the investigations of the atomic dynamics of insulating glasses and amorphous systems done with neutron inelastic scattering techniques [7.74, 112–121].

Acknowledgements. It is a pleasure to acknowledge helpful discussions with H. Beck, J. M. Carpenter, P. A. Egelstaff, F. Gompf, J. Hafner, W. Reichardt, H. Rietschel, and T. Springer. G. Hallam and F. Gompf kindly supplied information on their experimental results before publication. We are especially indebted to W. Reichardt for his perusal of the manuscript which was patiently typed by Mrs. M. Müller and Mrs. E. Thun.

References

7.1   J.-B. Suck: KFK report 2231 (1975)
7.2   J. D. Axe: "Inelastic Coherent Neutron Scattering in Amorphous Solids" in *Physics of Structurally Disordered Solids (1976)*, ed. by S. S. Mitra (Plenum Press, New York 1977) pp. 507–524
7.3   A. P. Malozemoff: "Brillouin Light Scattering from Metallic Glasses", in *Glassy Metals I*, ed. by H.-J. Güntherodt, H. Beck, Topics Appl. Phys., Vol. 46 (Springer, Berlin, Heidelberg, New York 1981) Chap. 5, pp. 79–91
7.4   S. C. Moss, D. L. Price, J. M. Carpenter, D. Pan, D. Turnbull: Bull. Am. Phys. Soc. **19**, 321 (1974)
7.5   D. Weaire, P. C. Taylor: "Vibrational Properties of Amorphous Solids", in *Dynamical Properties of Solids*, Vol. 4, ed. by G. K. Horton, A. A. Maradudin (North-Holland, Amsterdam, New York, Oxford 1980) Chap. 1, pp. 1–61
7.6   H. A. Mook, N. Wakabayashi, D. Pan: Phys. Rev. Lett. **34**, 1029–1033 (1975)
7.7   J. D. Axe, G. Shirane, T. Mizoguchi, K. Yamauchi: Phys. Rev. B**15**, 2763–2770 (1977)
7.8   H. A. Mook, C. C. Tsuei: Phys. Rev. B**16**, 2184–2190 (1977)
7.9   J. A. Tarvin, G. Shirane, R. J. Birgeneau, H. S. Chen: Phys. Rev. B**17**, 241–248 (1978)
7.10 R. J. Birgeneau, J. A. Tarvin, G. Shirane, E. M. Gyorgy, R. C. Sherwood, H. S. Chen, C. L. Chien: Phys. Rev. B**18**, 2192–2195 (1978)
7.11 J. J. Rhyne, J. W. Lynn, F. E. Luborsky, J. L. Walter: J. Appl. Phys. **50**, 1583–1585 (1979)
7.12 J. Hafner: "Theory of the Structure, Stability, and Dynamics of Simple-Metal Glasses", in [Ref. 7.3, pp. 93–140]
7.13 L. van Hove: Phys. Rev. **95**, 249–262 (1954)
7.14 F. W. de Wette, A. Raman: Phys. Rev. **176**, 784–790 (1968)
7.15 J. M. Carpenter, C. A. Pelizzari: Phys. Rev. B**12**, 2391–2396 (1975)
7.16 J. M. Carpenter, C. A. Pelizzari: Phys. Rev. B**12**, 2397–2401 (1975)
7.17 P. Fulde, H. Wagner: Phys. Rev. Lett. **27**, 1280–1282 (1971)
7.18 S. Takeno, M. Goda: Prog. Theor. Phys. **45**, 331–352 (1971)
7.19 S. Takeno, M. Goda: Prog. Theor. Phys. **47**, 790–806 (1972)
7.20 M. Goda, S. Takeno: In *Phonon Scattering in Solids*, ed. by L. J. Challis, V. W. Rampton, A. F. G. Wyatt (Plenum Press, New York, London 1976) pp. 126–128
7.21 P. Dean: Rev. Mod. Phys. **44**, 127–168 (1972) and references therein
7.22 H. Böttger: Phys. Stat. Sol. (b) **51**, 139–142 (1972)
7.23 H. Böttger: Phys. Stat. Sol. (b) **59**, 517–523 (1973)
7.24 H. Böttger: Phys. Stat. Sol. (b) **62**, 9–42 (1974) and references therein
7.25 D. Weaire, R. Alben: J. Phys. C (Solid State Phys.) **7**, L 189–191 (1974)
7.26 M. F. Thorpe: In [Ref. 7.2, pp. 623–663] and references therein

7.27   M.Goda: J. Phys. C: Solid State Phys. **10**, 1121–1131 (1977)
7.28   Y.P.Joshi: Phys. Stat. Sol. (b) **95**, 317–324 (1979)
7.29   P.Kleinert, R.Leihkauf: Phys. Stat. Sol. (b) **97**, 491–499 (1980)
7.30   C.G.Montgomery: J. Low Temp. Phys. **39**, 13–20 (1980)
7.31   J.Hafner: J. Phys. C (Solid State Phys.) **14**, L 287–291 (1981)
7.32   J.Hubbard, J.L.Beeby: J. Phys. C (Solid State Phys.) **2**, 556–571 (1969)
7.33   K.Kim, M.Nelkin: Phys. Rev. B**7**, 2762–2771 (1973)
7.34   A.Rahman, M.J.Mandell, P.C.McTague: J. Chem. Phys. **64**, 1564–1568 (1976)
7.35   D.Tomanek: Diplomarbeit, Universität Basel (1979)
7.36   J.-B.Suck, H.Rudin, H.-J.Güntherodt, D.Tomanek, H.Beck, C.Morkel, W.Gläser: J. Phys. (Paris) C**8**, 175–178 (1980)
7.37   D.Weaire, M.F.Ashby, J.Logan, M.J.Weins: Acta Met. **19**, 779–788 (1971)
7.38   L.v.Heimendahl, M.F.Thorpe: J. Phys. F (Metal Phys.) **5**, L 87–91 (1975)
7.39   J.J.Rehr, R.Alben: Phys. Rev. B**16**, 2400–2407 (1977)
7.40   L.v.Heimendahl: J. Phys. F (Metal Phys.) **9**, 161–169 (1979)
7.41   R.Yamamoto, T.Mihara, K.Haga, M.Doyama: Solid State Commun. **36**, 377–379 (1980)
7.42   R.Yamamoto, K.Haga, T.Mihara, M.Doyama: J. Phys. F (Metal Phys.) **10**, 1389–1399 (1980)
7.43   Y.Ishi, T.Fugiwara: J. Phys. F (Metal Phys.) **10**, 2125–2136 (1980)
7.44   S.Kobayashi, S.Takeuchi: J. Phys. C (Solid State Phys.) **13**, L 969–974 (1980)
7.45   R.Yamamoto, K.Haga, H.Shibuta, M.Doyama: J. Phys. F (Metal Phys.) **8**, L 179–182 (1978)
7.46   H.M.Pak, M.Doyama: J. Fac. Engng., Tokyo Univ. B**30**, 111 (1969)
7.47   M.F.Thorpe, R.Alben: J. Phys. C (Solid State Phys.) **9**, 2555–2567 (1976)
7.48   L.v.Heimendahl: Phys. Stat. Sol. (b) **86**, 549–556 (1978)
7.49   R.Haydock, V.Heine, M.J.Kelly: J. Phys. C (Solid State Phys.) **5**, 2845–2858 (1972)
7.50   R.Haydock, V.Heine, M.J.Kelly: J. Phys. C (Solid State Phys.) **8**, 2591–2605 (1975)
7.51   B.N.Brockhouse, H.E.Abou-Helal, E.D.Hallman: Solid State Commun. **5**, 211–216 (1967)
7.52   T.Fujiwara, Y.Ishii: J. Phys. F (Metal Phys.) **10**, 1901–1911 (1980)
7.53   S.Kobayashi, K.Maeda, S.Takeuchi: J. Phys. Soc. Japan **48**, 1147–1152 (1980)
7.54   F.Cyrot-Lackmann: In J. Phys. (Paris) C**8**, 827–830 (1980)
7.55   F.Cyrot-Lackmann: Phys. Rev. B**22**, 2744–2748 (1980)
7.56   W.Marshal, S.W.Lovesey: *Theory of Thermal Neutron Scattering* (University Press, Oxford 1971)
7.57   A.B.Bhatia, D.E.Thornton: Phys. Rev. B**2**, 3004–3012 (1970)
7.58   A.B.Bhatia, D.E.Thornton: Phys. Rev. B**4**, 2325–2328 (1971)
7.59   P.Chieux, H.Ruppersberg: In J. Phys. (Paris) C**8**, 145–152 (1980)
7.60   J.-B.Suck, H.Rudin, H.-J.Güntherodt, H.Beck, J.Daubert, W.Gläser: J. Phys. C (Solid State Phys.) **13**, L 167–172 (1980)
7.61   J.-B.Suck, H.Rudin, H.-J.Güntherodt, H.Beck, J.Daubert, W.Gläser: In *Liquid and Amorphous Metals*, ed. by E.Lüscher, H.Coufal (Sijthoff & Noordhoff, Alphen 1980) pp. 649–652
7.62   J.-B.Suck, H.Rudin, H.-J.Güntherodt, H.Beck: J. Phys. F (Metal Phys.) **11**, 1375–1383 (1981)
7.63   J.-B.Suck, H.Rudin, H.-J.Güntherodt, H.Beck: J. Phys. C (Solid State Phys.) **13**, L 1045–1051 (1980)
7.64   J.-B.Suck, H.Rudin, H.-J.Güntherodt, H.Beck: J. Phys. C (Solid State Phys.) **14**, 2305–2317 (1981)
7.65   J.-B.Suck, H.Rudin, H.-J.Güntherodt, H.Beck: In *Proc. of the Fourth Intern. Conf. on Rapidly Quenched Metals*, ed. by T.Masumoto, K.Suzuki (Sendai, Japan 1982) p. 407
7.66   M.Hansen, K.Anderko: *Constitution of Binary Alloys*, 2nd ed. (McGraw-Hill, New York 1958)
7.67   I.Higashi, N.Shiotani, M.Uda, T.Mizoguchi, H.Katoh: J. Solid State Chem. **36**, 225–233 (1981)
7.68   G.Placzek: Phys. Rev. **86**, 377–388 (1952)

7.69   P.G.deGennes: Physica **25**, 825–839 (1959)
7.70   D.Forster, P.C.Martin, S.Yip: Phys. Rev. **170**, 155–159 (1968)
7.71   E.Cartier, F.Heinrich, H.-J.Güntherodt: Phys. Lett. **81**A, 393–396 (1981)
7.72   T.Egami, K.Maeda, V.Vitek: Phil. Mag. A**41**, 883–891 (1980)
7.73   T.Egami, K.Maeda, D.Srolovitz, V.Vitek: J. Phys. (Paris) C**8**, 272–275 (1980)
7.74   A.J.Leadbetter: In *International Conference on Phonon Scattering in Solids*, ed. by H.G.Albany (CEN Saclay 1972) pp. 338–352 and references therein
7.75   T.M.Holden, J.S.Dugdale, G.C.Hallam, D.Pavuna: J. Phys. F (Metal Phys.) **11**, 1737–1748 (1981)
7.76   J.-B.Suck, H.Rudin, H.-J.Güntherodt, H.Beck: Phys. Rev. Lett. **50**, 49–52 (1983)
7.77   R.J.Bell, N.F.Bird, P.Dean: J. Phys. C (Solid State Phys.) **1**, 299–303 (1968)
7.78   R.J.Bell, P.Dean, D.C.Hibbins-Butler: J. Phys. C (Solid State Phys.) **3**, 2111–2118 (1970)
7.79   R.A.Cowley, A.D.B.Woods: Can. J. Phys. **49**, 177–200 (1971)
7.80   R.Alben, D.Weaire, J.E.Smith, Jr., M.H.Brodsky: Phys. Rev. B**11**, 2271–2296 (1975)
7.81   F.Gompf, H.J.Schmidt, K.B.Renker: Proc. of the Intern. Conf. on Phonon Scattering, Bloomington (1981) to appear in J. Physique C (1982)
7.82   J.J.Rush, J.M.Rowe, A.J.Maeland: J. Phys. F (Metal Phys.) **10**, L 283–285 (1980)
7.83   J.Skalyo, Y.Erdoh, G.Schirane: Phys. Rev. B**9**, 1797–1803 (1974)
7.84   A.Sjölander: Ark. Fys. **14**, 315–371 (1958)
7.85   V.F.Turchin: *Slow Neutrons* (Sivan, Jerusalem 1965)
7.86   P.A.Egelstaff, P.Schofield: Nucl. Sci. Eng. **12**, 260–270 (1962)
7.87   J.M.Carpenter: J. Chem. Phys. **46**, 465–468 (1967)
7.88   M.M.Bredov, B.A.Kotov, N.M.Okuneva, V.S.Oskotskii, A.L.Shakh-Budagov: Fiz. Tverd. Tela **9**, 214–218 (1967) [English transl.: Sov. Phys. – Solid State]
7.89   V.S.Oskotskii: Fiz. Tverd. Tela **9**, 420–422 (1967) [English transl.: Sov. Phys. – Solid State]
7.90   N.Breuer: Z. Physik **271**, 289–293 (1974)
7.91   H.Eschrig, L.vanLoyen, P.Ziesche: Phys. Stat. Sol. (b) **66**, 587–593 (1974)
7.92   C.G.Windsor, H.Kheyrandish, M.C.Narasimhan: Phys. Lett. **70**A, 485–488 (1979)
7.93   L.vanHove: Phys. Rev. **89**, 1189–1193 (1953)
7.94   R.Pynn, G.L.Squires: Proc. R. Soc. A**326**, 347–360 (1972)
7.95   M.Dutoit: Phys. Lett. **50**A, 221–223 (1974)
7.96   G.Bellessa: J. Phys. (Paris), C**8**, 723–730 (1980)
7.97   P.Doussineau, A.Levelut, G.Bellessa, O.Bethoux: J. Phys. Lett. (Paris) **38**, L 483–487 (1977)
7.98   J.Y.Duquesne, G.Bellessa: J. Phys. C (Solid State Phys.) **13**, L 215–219 (1980)
7.99   T.H.K.Barron, W.T.Berg, J.A.Morrison: Proc. R. Soc. A**242**, 478–492 (1957)
7.100   W.Reichardt: Private communication
7.101   R.Becker: *Theory of Heat*, 2nd ed. (Springer, Berlin, Heidelberg, New York 1967)
7.102   R.B.Stephens: Phys. Rev. B**8**, 2896–2905 (1973)
7.103   T.Mizoguchi, S.vonMolnar, G.S.Cargill III, T.Kudo, N.Shiotani, H.Sekizawa: In *Amorphous Magnetism II*, ed. by R.A.Levy, R.Hasegawa (Plenum Press, New York 1977) pp. 513–520
7.104   B.Golding, B.G.Bagley, F.S.L.Hsu: Phys. Rev. Lett. **29**, 69–70 (1972)
7.105   M.Sakata, N.Cowlam, H.A.Davies: J. Phys. F (Metal Phys.) **11**, L 157–162 (1981)
7.106   P.Duwez: ASM Trans. Quart. **60**, 605–633 (1967)
7.107   K.Froböse, J.Jäckle: J. Phys. F (Metal Phys.) **7**, 2331–2348 (1977)
7.108   J.Jäckle, K.Froböse: J. Phys. F (Metal Phys.) **9**, 967–986 (1979)
7.109   J.Jäckle, K.Froböse: J. Phys. F (Metal Phys.) **10**, 471–476 (1980)
7.110   G.Baym: Phys. Rev. **135**A, 1691–1692 (1964)
7.111   S.A.Brawer: Phys. Rev. Lett. **46**, 778–781 (1981)
7.112   P.A.Egelstaff: In *Physics of Non-Crystalline Solids*, ed. by J.A.Prins (North-Holland, Amsterdam 1965) pp. 127–151
7.113   B.A.Kotov, N.M.Okuneva, A.R.Regel, A.L.Shakh-Budagov: Fiz. Tverd. Tela **9**, 955–957 (1967) [English transl.: Sov. Phys. – Solid State]
7.114   A.Axmann, W.Gissler, A.Kollmar, T.Springer: Disc. Farad. Soc. **50**, 74–81 (1970)

7.115 A.J.Leadbetter, A.C.Wright: J. Non-Cryst. Solids **3**, 239–254 (1970)

7.116 A.J.Leadbetter, D.Litchinsky: Disc. Farad. Soc. **50**, 62–73 (1970)

7.117 A.J.Leadbetter, M.W.Springfellow: In *Neutron Inelastic Scattering* (IAEA, Vienna 1972) pp. 501–514

7.118 J.D.Axe, D.T.Keating, G.S.Cargill III, R.Alben: AIP Conf. Proc. **20**, 279–283 (1974)

7.119 O.L.Kukhto, V.I.Mikhailov, R.P.Ozerov, S.P.Solov'ev: Fiz. Tverd. Tela **18**, 991–994 (1976) [English transl.: Sov. Phys. – Solid State]

7.120 A.J.Leadbetter, P.M.Smith, P.Seyfert: Phil. Mag. **33**, 441–456 (1976)

7.121 F.Gompf: J. Phys. Chem. Solids **42**, 539–544 (1981)

8. Laser Quenching

M. von Allmen

With 10 Figures

Glassy metals have been made by mechanical quenching methods for almost a quarter of a century. While the equipment has been constantly improved and refined since Duwez' first shock tube, the basic working principle has always been the same: Fast cooling of the melt is achieved by establishing sudden mechanical contact with a good thermal conductor. The different quenching methods are usually compared on the basis of the average cooling rates achieved; cooling rates ranging up to 10^6 or 10^7 K/s are expected to arise with the best mechanical methods (incidentally, we will show below that the cooling rate is not a very well-defined concept for characterizing melt quenching). The heat flux out of the melt during quenching is determined by two limiting effects, namely by the final thickness of the melt being quenched (usually ranging between 10 and 100 μm) and second, by the thermal conductivity of the interface between the melt and the heat sink, which tends to contain poor conductors such as oxides, trapped gas, etc.

The use of laser pulses as a means of quenching materials into a glassy state was first attempted only a few years ago and exploitation of the method has barely started. In laser quenching heat is produced in a solid sample by very short irradiation. A surface layer of a width of typically one micrometer or less is heated, melted and subsequently cooled by conduction into the bulk of the specimen. Owing to the small width of the molten zone as well as to the absence of a poorly conducting interface, the heat flux during cooling after laser irradiation exceeds that achievable by mechanical quenching by orders of magnitude. Relevant cooling rates will be shown to reach 10^{10} K/s under typical conditions. Not surprisingly, the faster cooling is found to result in a drastically extended range of materials that can be obtained and studied in the glassy state. Incidentally, similar effects should be expected from short electron (or other light particle) beam pulses, provided the penetration depth can be made small enough.

In what follows, we first give an overview of experimental results obtained to date. In the second part we analyse the criteria of glass formation from the point of view of laser quenching.

8.1 Experimental Results

8.1.1 General Remarks

The application of laser pulses to produce metallic glasses, as stated before, is rather recent, and the experimental situation is preliminary at best. Thus, the aim in writing a review at the present time must be to illustrate the potential of the new method, rather than to give a comprehensive account.

Experimental work published so far, scarce as it is, tends to fall into two categories. In the first group of investigations, short pulses have been used with the goal of extending present ranges of glass forming ability to new materials or material combinations, so far mainly to Si-metal systems. In the second group, typically directed towards achieving surface hardening of machine alloys, mostly continous laser beams in a fast-scanning mode have been adopted. While there is considerable technological interest in the latter application, its results have not been shown to be qualitatively different from those of splat cooling. We shall, therefore, focus mainly on the first group in this article.

8.1.2 Short Laser Pulses: Alloys

The interest in using laser pulses in material technology is several years old. Q-switch laser pulses have been used by numerous authors to recrystallize the surface of ion-implanted single-crystal Si or Ge wafers, or to form ohmic contacts by alloying thin deposited metal layers to Si substrates (see, e.g., [8.1]). The latter process involves melting of the metal layer as well as part of the semiconductor, followed by intermixing in the liquid and formation of silicides upon recrystallisation. The alloyed layers usually exhibit a rather complicated microstructure due to the nonequilibrium nature of the solidification process. Cellular growth patterns with a cell size down to a few tens of nm have been found as well as apparently amorphous zones. Quantitative investigations in such structures are difficult, because the composition may vary widely from point to point.

A more controlled technique to investigate the properties of laser alloyed structures was developped in [8.2] and has since been applied to a number of material systems. In these experiments several alternating layers of metal and Si are vapor-deposited onto inert substrates, e.g. onto sapphire. The thickness of the layers is adjusted so as to yield the desired average film composition. In order to guarantee complete intermixing of the elements during the lifetime of the liquid, individual layers should be no more than about 15 nm thick if ns pulses are used to melt the structure; total film thickness may be a few 100 nm. The laser pulse energy required is of the order of $1 \, \text{J cm}^{-2}$, depending mainly on the reflectivity of the multilayer.

The reflectivity of metals, as a rule, decreases with temperature. This has the consequence that the choice of pulse energy is critical and that "hot spots" in

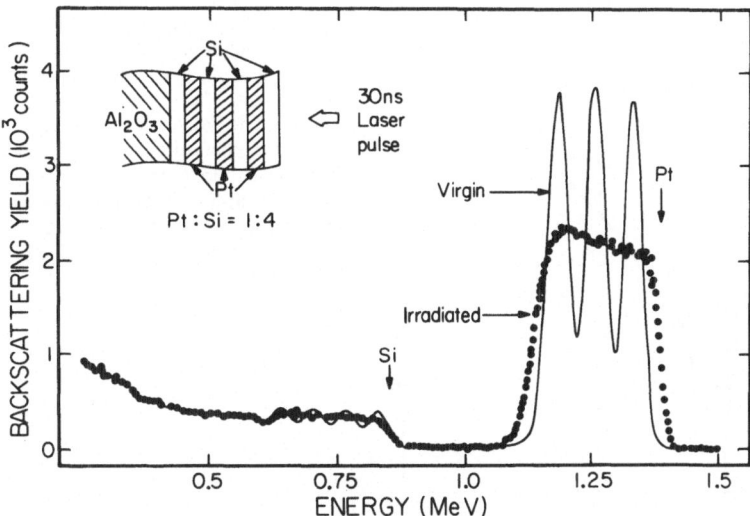

Fig. 8.1. 1.5 MeV ^{4}He$^+$ backscattering spectra of a Pt–Si sample with 20 at. % Si, deposited on a Sapphire substrate. The thin solid line shows the multilayered virgin, the dots represent the spectrum after laser-induced mixing and quenching

the laser intensity distribution tend to be amplified. The problem arises mainly with laser wavelengths in the infrared, where cold metals have high reflectivities. In our experiments, the surface layer was chosen to be Si, which not only serves the purpose of an antireflection coating, but also has the property of stabilizing the light absorption process, since its reflectivity increases upon melting [8.3].

The described procedure is exemplified in Fig. 8.1, which shows ^{4}He$^+$ backscattering spectra of a Pt–Si film (20 at. % Si) before and after irradiation with a 30 ns pulse from a Nd-glass laser ($\lambda = 1.06\,\mu$m). Such lasers deliver typically pulse energies of about two tenths of a Joule, enough to transform a spot of 3–5 mm diameter. More extended areas can be transformed by applying several partially overlapping shots. Experimental results obtained with this method in various systems are summarized below.

a) Au–Si System

Au and Si are well known to form a metallic glass by splat cooling at compositions near the eutectic (17 at. % Si). In fact, the Au–Si glasses were the first to be ever obtained [8.4]. Although the technological interest in this system is limited, it may serve as a model case for many features of laser quenching. Amorphous films were prepared by the above method with compositions ranging from 9 to 91 at. % Si (compositions outside this range were not tried). Figure 8.2 shows the Au–Si phase diagram [8.5] with the

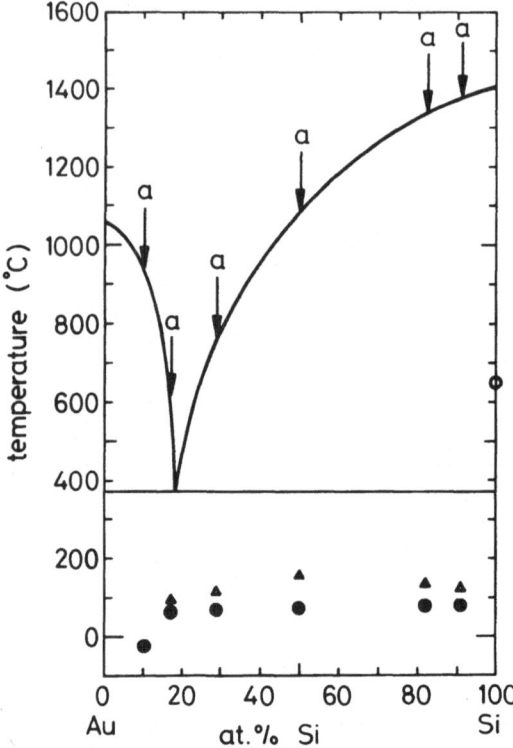

Fig. 8.2. Au–Si phase diagram [8.5] with composition of investigated films indicated by arrows; all films were amorphous ("a") after laser irradiation. The dots indicate the crystallization temperature (measured under heating at 3 K/min), the triangles mark the appearance of equilibrium phases; a metastable silicide exists in the range between dots and triangles

composition of investigated films marked by arrows. The thermal stability of the amorphous phases was determined by monitoring the electrical resistivity of the films during heating in a furnace at a rate of 3 °C/min. The crystallization temperature was taken as that temperature at which an abrupt change in film resistivity occurred. The structure of the films was investigated by x-ray diffraction at various stages of thermal decomposition. The crystallization temperature at various compositions is indicated in Fig. 8.2 by dots. It can be seen that, while the crystallization temperature increases monotonically with the Si content, its slope is very different in different regions of the phase diagram. Pure amorphous Si (prepared by ion-implantation or vapor deposition) is known to crystallize around 650 °C. The presence of only 9 at. % Au decreases the crystallization temperature to about 80 °C. Another dramatic decrease is observed near the Au-rich end. The glasses with 9 at. % Si had to be kept in liquid nitrogen during and after irradiation.

The compositional range of the Au–Si glasses reported here greatly exceeds previously established limits for glass formation by melt quenching. Glass-forming ability is thought to be related to melting point depression [8.6] which is largest at the eutectic point. The most Si-rich glasses shown in Fig. 8.2 have melting point depressions (relative to the weighted average of the melting

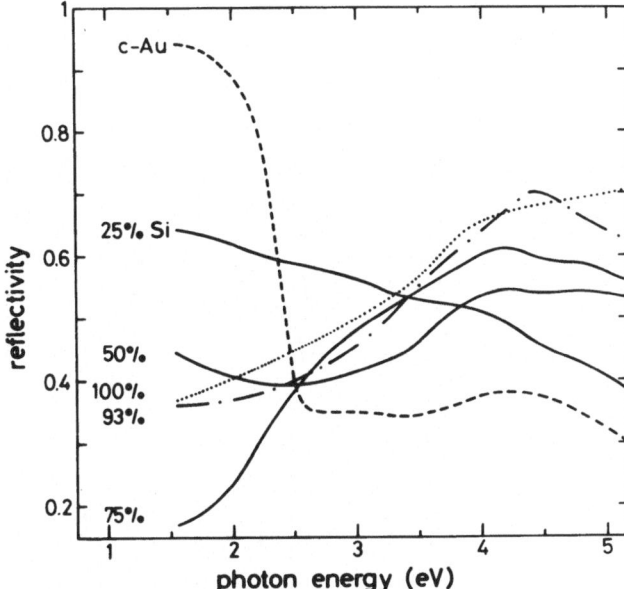

Fig. 8.3. Optical reflection spectra of the glassy Au–Si films. Also included are spectra of evaporated amorphous Si (···) and pure crystalline Au (---)

points of the pure elements) of only about 2%. This may be compared to the 20% depression which is generally found to be necessary for glass formation by splat-cooling [8.7]. It may be added here that neither of the pure constituents can be obtained in a glassy state by irradiation with Q-switch pulses at the present wavelength. However, the formation of pure semiconducting amorphous Si by laser quenching has been reported (Sect. 8.1.3).

Careful annealing of the amorphous Au–Si films resulted in nucleation and growth of a metastable silicide. The silicide is different from previously described metastable gold-silicides [8.8] and has a hexagonal structure with a c/a ratio of 1.24; its exact stoichiometry has not yet been worked out. The temperature at which the silicide decays into Au and Si is around 150°C, as indicated by the triangles in Fig. 8.2. It may be added that irradiation of the same samples with longer pulses (300 μs instead of 30 ns) leads directly to formation of the same metastable silicide [8.2].

The availability of glassy phases covering most of the phase diagram invites comparative studies of glass properties as a function of composition. As an illustration, Fig. 8.3 shows optical reflection spectra of the Au–Si glasses; also included are the spectra of pure Au and pure amorphous Si (evaporated). Analysis of the optical and electrical properties of the glassy films reveals a tremendous variability of the material properties achievable: Below 30 at. % Si the films are purely metallic alloys, as the Si atom itself is in a metallic state.

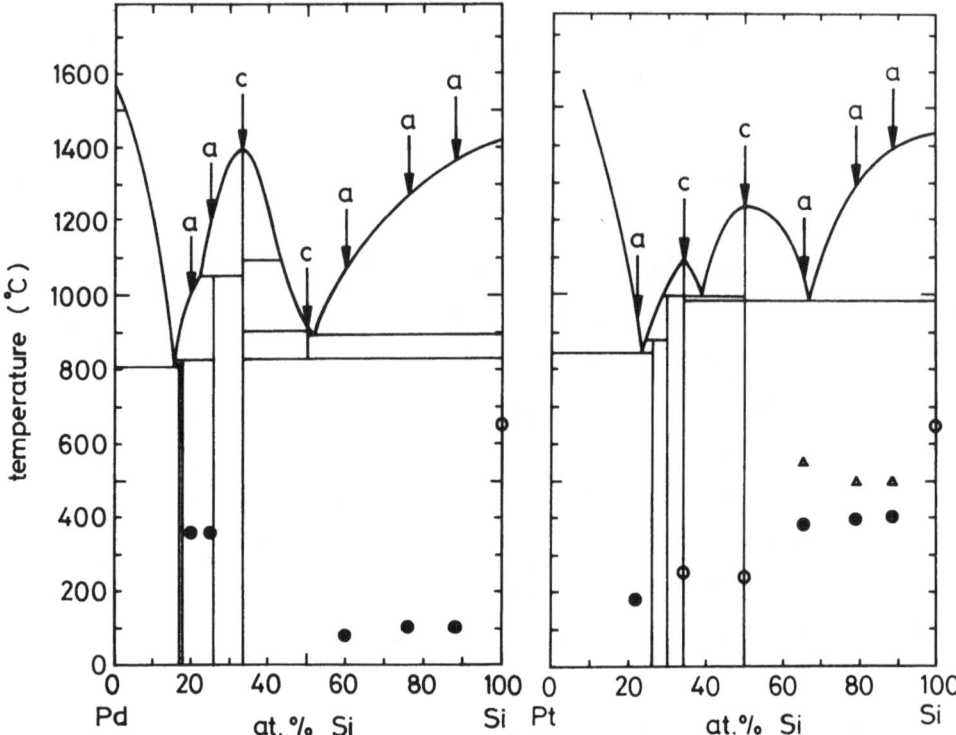

Fig. 8.4. Phase diagrams of Pd–Si [8.5, 10] and Pt–Si [8.11] systems. Arrows marked "c" indicate films that were crystalline after laser irradiation; open circles give the crystallization temperatures of such films, subsequently amorphized by ion-implantation. The other symbols have the same meaning as in Fig. 8.2

Between 30 and 85 at. % Si the Si is semiconducting and Au is metallic; the films have properties somewhere intermediate between metallic and semiconducting (remember, however, that the glassy films are atomically homogenous materials, not just mixtures of Au and Si). Around 85 at. % Si the Au undergoes an Andersen transition and turns insulating; correspondingly, the films with more than 85 % at. % Si behave as heavily doped amorphous semiconductors [8.9].

b) Pd–Si and Pt–Si Systems

From a thermodynamical point of view, the main difference between Au–Si and the systems Pd–Si and Pt–Si in the presence of stable intermediate phases in the latter. The phase diagrams of the Pd–Si [8.5, 10] and Pt–Si [8.11] systems are shown in Fig. 8.4. Pd–Si is known as a glass former from splat cooling (at the metal-rich eutectic, 18 at. % Si), whereas Pt–Si is not.

Laser quenching yields metallic glasses over most of the phase diagrams in both systems [8.12]. Compositions of investigated films are marked by arrows in Fig. 8.4. Except for $Pd_{50}Si_{50}$, all compositions not coinciding with one of the congruently melting phases led to glass formation (indicated by an "a" in the figure), while the others were found to be crystalline ("c") after irradiation. Why laser quenching fails to yield a glass for certain compositions is not clear a priori; the reason may be either an insufficient cooling rate, or a low crystallization temperature of the glass. A hint to this question is provided by the measured crystallization temperatures. In Fig. 8.4 the crystallization temperatures of laser quenched amorphous films, measured by the method mentioned before, are indicated by closed circles; values indicated by open circles were obtained from films amorphized by implantation of Xe ions after laser irradiation and are included for comparison. The relative stability of the implanted amorphous films suggests that it is probably nucleation from the melt (during quenching) which prevents these films from becoming amorphous by laser irradiation.

The crystallization temperatures as a function of composition reveal quite different trends, notably in the Si-rich half of the two phase diagrams: the Pt–Si glasses are stable up to 400 °C and then transform into the metastable silicides Pt_2Si_3 or Pt_4Si_9 discovered recently [8.13]. In contrast, the Si-rich Pd–Si glasses all transform at temperatures as low as 100 °C, due to growth of the equilibrium PdSi phase. (It cannot be excluded that nuclei of this phase, undetectable by x-ray diffraction, were already present in the as-irradiated films. Nevertheless, the low-temperature growth of PdSi is in contrast to the findings of [8.10], according to which PdSi is not stable below 824 °C). There is also no obvious correlation between the crystallization temperatures in the Si-rich and the metal-rich part of the same phase diagram. This may suggest that the thermal stability of a binary glass is primarily determined by the nucleation and growth behavior of that stable or metastable crystalline phase which is closest to its composition.

Glassy phases offer the possibility of studying the mutual influence of different atomic species in close contact at compositions outside those of stoichiometric compounds. Here laser quenching is able to extend the range of investigation considerably. The tendency of the metal-silicon glasses to become semiconducting above a certain Si content, which is also observed in the Pd–Si and Pt–Si systems, again illustrates how strongly electronic properties can be made to vary within the same binary system. A convenient method of investigating such effects is electron spectroscopy of glassy phases of different composition. As an example, Fig. 8.5 shows UPS spectra of the electron distribution up to the Fermi level, measured at two different photon energies in laser quenched Pd–Si glasses [8.14]. The peak position can be seen to shift by several eV as a function of composition. Similar observations were made for Pd and Si core levels. These results are preliminary and further work is required in order to draw quantitative conclusions. Incidentally, the same study also confirmed the equivalence of mechanically quenched glasses with laser quenched ones of the same composition: spectra obtained from near-eutectic Pd–Si glasses prepared by both methods were indistinguishable.

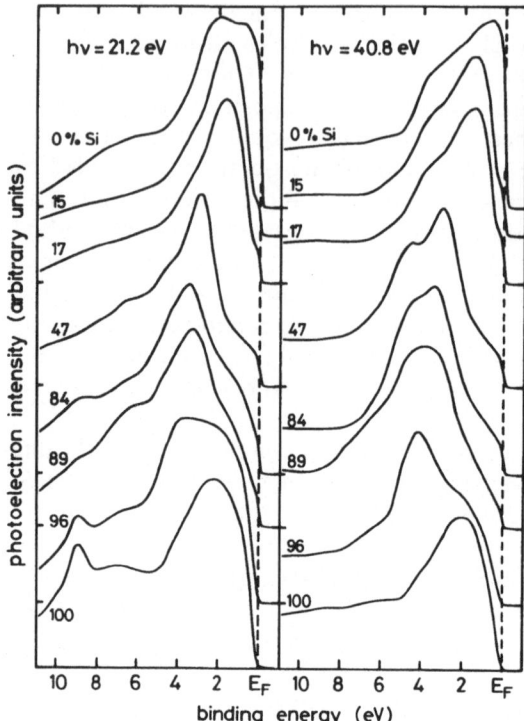

Fig. 8.5. UPS spectra of glassy Pd–Si films with compositions between 15 and 96 at. % Pd obtained at two different photon energies. Also included are spectra for pure Pd and Si (crystalline)

c) V–Si and Nb–Si Systems

In contrast to the systems considered above, V–Si and Nb–Si are pronounced high-temperature systems. As a glance at the phase diagrams [8.5, 11, 15] in Fig. 8.6 shows, both must be considered very unlikely glass formers from the criterion of melting point depression. Nevertheless, a series of glasses has been obtained by laser quenching in both systems [8.16]. Compositions that led to glass formation are marked in Fig. 8.6 by an "a", those which didn't by a "c". It is not clear at present what limits glass forming ability in these systems. Certainly there is no correlation with melting point depression, which is even negative for some of the glassy films. In some cases (e. g., the Si-rich Nb–Si films) the crystallinity of the as-irradiated films may be due to low-temperature nucleation of equilibrium phases. It is noteworthy that many of the amorphous films contained crystalline precipitates just detectable by x-ray diffraction. This suggests that these two systems are somewhat at the limit of the glass forming power of laser quenching in the ns regime. The dots in Fig. 8.6 indicate the temperature where crystallization of amorphous films first sets in. Note that some of the glasses are stable up to 600 °C, in spite of the probable presence of quenched-in crystalline nuclei. Some of the amorphous films in Fig. 8.6 form yet unidentified metastable crystals upon annealing. Metastable crystals seem

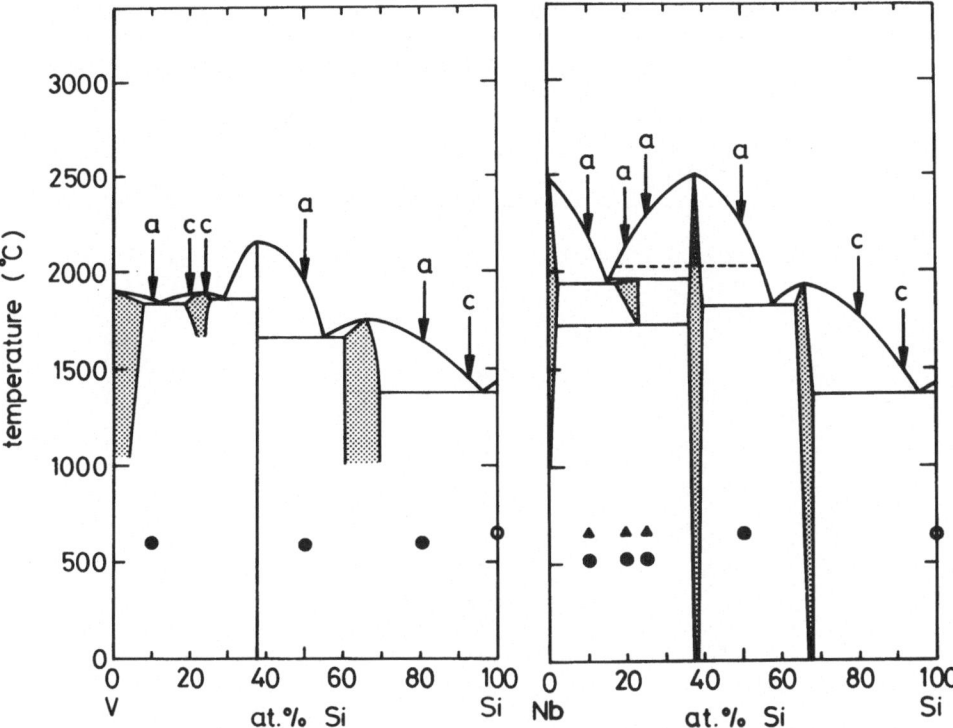

Fig. 8.6. Phase diagrams of Nb–Si and V–Si sytems [8.5]. The symbols have the same meaning as in Fig. 8.2, 4

to coexist with equilibrium phases up to temperatures around 1000 °C; the triangles in the figure indicate the temperature where the latter begin to appear.

As a summary of data on glass forming ranges and glass stability obtained by ns-pulse laser quenching in Si-metal systems, Table 8.1 lists the composition, the crystallisation temperature and the nature of the first nucleating crystalline phase (metastable or equilibrium) of some of the phases so far investigated.

d) Other Systems

The extended glass-forming power of laser quenching is, of course, not limited to Si-metal systems, but exploration of other systems has barely started. A series of glassy films of Fe–B with compositions between 5 and 24 at. % B have been produced by *Lin* and *Spaepen* using 30 ps laser pulses [8.17]. Glassy films of binary transition metal systems including Au–Ti, Co–Ti and Cr–Ti have recently been obtained by the same method, as used for the Si-metal glasses [8.18]. Metastable crystalline superconducting Au–Te films produced by laser quenching were described by *Stritzker* [8.19]. *Picraux* and *Follstaedt* [8.20] reported amorphous Al(Ni) alloy formation by irradiation with 65 ns electron beam pulses. Other systems are certain to follow.

Table 8.1. Composition, crystallization temperature and nature of first nucleating crystalline phase (e: equilibrium, m: metastable) of some laser-quenched amorphous films. The asterisk indicates the presence of very small-grained precipitates in the as-irradiated film (faint, broad lines in x-ray spectrum)

Metal	Composition [at. % Si]	Crystallization temperature [°C]	First phase
Au	9	− 25	m
	18	60	m
	50	70	m
	91	81	m
Pd	20	365	e
	25	365	e
	60	70	e
	76	100	e
	88	100	e
Pt	20	>170	e
	64	390	m
	80	400	m
	90	400	m
V	10*	600	e
	50	600	e
	80*	600	e
Nb	10*	500	e + m
	25*	500	e + m
	50*	600	?

8.1.3 Short Laser Pulses: Pure Elements

Except for some nonmetals like B or Se, none of the pure elements has ever been obtained in a glassy state by splat cooling. The reason is that crystal nucleation and growth is fastest in pure melts where compositional constraints at the crystal-melt interface are absent. Therefore, the cooling rates required to bypass crystallization in pure elements are larger than in alloy melts.

In a few cases the formation of glassy phases from pure metallic melts by short laser pulses has been reported. In these investigations, bulk specimens rather than thin deposited films were irradiated and amorphous zones formed at the surface of the crystalline material. Tsu et al. [8.17] used frequency-quadrupled Nd-laser pulses ($\lambda = 0.265\,\mu\text{m}$) of 10 ns duration incident on single-crystal Si to obtain amorphous zones a few tens of nm thick. Radiation of twice that wavelength ($\lambda = 0.53\,\mu\text{m}$) did not result in glass formation because the larger absorption length resulted in a lower cooling rate. The tendency for glass formation was also found to depend on the crystallographic direction of the irradiated Si wafer. This may be understood on the basis of the well-known fact that the growth velocity in Si is about 25 times higher for a $\langle 100 \rangle$ surface than for a $\langle 111 \rangle$ surface [8.22]. The formation of glassy Si was also achieved by *Liu* et al. [8.23] by using 30 ps pulses of both 0.53 and 0.266 μm radiation. The thickness of the amorphous zone was estimated to be about 30 nm and pulse energy was found to be very critical. Kinetic and morphological aspects of the amorphization of Si by short laser pulses were also discussed by *Cullis* et al. [8.24] and by *Rozgonyi* et al. [8.25]. Finally, the occurrence of an "essentially amorphous" region was even reported in pure Al [8.26] after irradiation by a 15 ns ruby laser pulse ($\lambda = 0.69$ μm). The layer was found to contain small grains or crystalline AlN, probably originating from adsorbed nitrogen at the surface of the specimen before irradiation. These few results, as preliminary as they may be, certainly indicate that some long-held beliefs about the impossibility of quenching pure metals into a glassy state have to be revised in view of the potential of short-pulse laser quenching.

8.1.4 Continuous Beams in a Scanning Mode

The obvious advantage of using continuous, rather than pulsed laser radiation is that glassy phases can be obtained in a continous process with potentially high throughput. This may be achieved in two ways: either the continuous laser beam is directed onto a sample mounted on a moving or rotatable fixture [8.27, 28] or, alternatively, the beam is scanned with the aid of moving mirrors across a fixed specimen. High beam intensity is required if interesting cooling rates are to be achieved. Absorbed intensities of the order of $100 \, \text{kW cm}^{-2}$ (corresponding to about $1 \, \text{MW cm}^{-2}$ of incident intensity for a CO_2 laser) and scan speeds of m/s are required in order to achieve cooling rates of 10^6 K/s in a bulk metal specimen. Suitable devices for this purpose include noble-gas ion lasers (the radiation of which, due to their limited output power, must be sharply focussed) or fast-flow CO_2 lasers delivering up to several kW of continous or quasi-continous output power (pulses in the ms range may be considered equivalent to continous radiation as far as melt quenching is concerned). Cooling rates significantly exceeding those of the best splat cooling methods do not seem to be realizable with present equipment.

While the glass forming potential of scanned laser beams should be roughly the same as that of the state-of-the-art mechanical quenching methods, the former offer technological and practical advantages such as spatial resolution, versatility and direct applicability to preformed structural metal parts.

An example of the structures obtainable from quenching by a scanned CO_2-laser beam, Fig. 8.7 shows a cross-sectional micrograph of a glassy layer of $Ni_{60}Nb_{40}$ on top of a steel substrate (St 52) [8.29]. The Ni–Nb alloy was applied to the steel base as a powder and subsequently melted and quenched by the scanned beam. Intermittent pulses of ms duration, rather than a continous beam, were used in this particular case in order to reduce the heat load on the specimen. Some of the problems yet to be solved in this technique include nucleation in the quenched layer due to a change in melt composition by

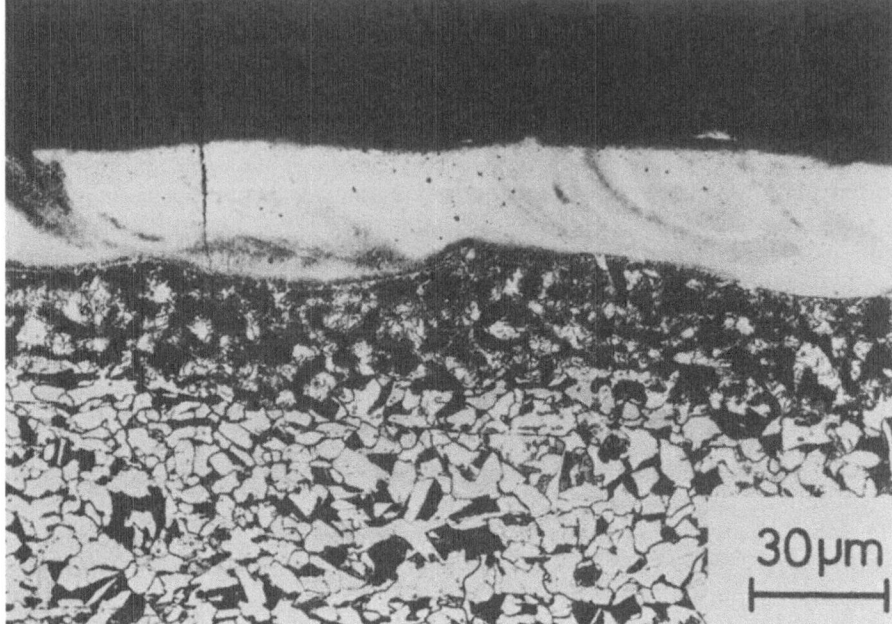

Fig. 8.7. Cross-sectional micrograph of an amorphous $Ni_{60}Nb_{40}$ layer on a steel (St 52) substrate produced by an intermittently pulsed CO_2 laser beam scanned across the surface [8.29]

admixtures from the substrate, as well as material cracking caused by different thermal expansion coefficients of the layer and the substrate (Fig. 8.7).

After this survey of current experimental results, let us now take a look at some fundamental aspects of the laser quenching process itself.

8.2 Analysis of the Laser Melting and Quenching Process

8.2.1 Kinetic Conditions for Glass Formation

The objective of melt quenching is to bypass crystal nucleation and growth. Therefore, the conditions for glass formation are to be derived from nucleation and growth theory. The growth velocity of a crystal into undercooled melt, according to classical theory, is essentially given by

$$u(T) = u_0 e^{-Q/kT} \{1 - \exp[-s_m(T_m - T)/kT]\}. \tag{8.1}$$

Here u_0 is a constant which depends on the detailed structure of the crystal-melt boundary, s_m is the entropy of melting per particle and Q

is the activation energy for molecular rearrangement at the boundary (assumed here to be a simple thermally activated process). Q may be estimated from the activation energy for viscous flow in the undercooled melt (see, e.g., [8.30]). The growth velocity increases rapidly with increasing undercooling just below the melting point but, due to the presence of a thermally activated process, it approaches zero at large undercooling. Thus, there is a maximum velocity at which the crystal can grow.

The rate of crystal nucleation, if regarded as a function of temperature, behaves qualitatively similarly to the growth velocity in that there too is an optimum temperature range, determined by a compromise between chemical driving force and particle mobility. However, the nucleation rate depends explicitly also on time. If a melt is suddenly undercooled, then a finite time-lag is observed before measurable nucleation sets in. The time-lag t_n may be interpreted as the time required to establish an equilibrium population of clusters corresponding to the new temperature [8.31]. The rate of homogenous nucleation may be expressed in a somewhat simplified form as

$$I(T,t) = I_\infty(T)\left[1 + 2\sum_{n=0}^{\infty}(-1)^n \exp(-n^2 t/t_n)\right],$$
$$I_\infty(T) = I_0\, e^{-Q/kT}\exp[-\sigma^3/N^2 s_m^2(T_m - T)^2 kT].$$

$$(8.2)$$

Here I_∞ is the steady-state nucleation rate, σ is the crystal-melt interface energy, N is the particle density and I_0 is a constant. Note the very strong dependence of I_∞ on undercooling; unlike growth, homogenous nucleation sets in only when the undercooling exceeds a certain finite amount, even in the steady-state case. The time-lag should be minimum in the temperature range where I_∞ is at its maximum; a lower limit to the minimum time-lag is given by the ratio of the number of atoms in a critical nucleus to the molecular rearrangement frequency. This yields values of the minimum time-lag of the order of ns for metallic melts. At large undercooling the nucleation rate approaches zero, in a similar way as the growth velocity. At the same time the time-lag, which is expected to scale approximately as $\exp(-Q/kT)$, becomes very large [8.32].

Once the nucleation rate and the growth velocity are known, one may formulate a condition for glass formation by requiring that the crystallized volume fraction X, given for small X by

$$X(T,t) \sim I(T,t)u^3(T)t^4,$$

$$(8.3)$$

stays below some arbitrary limit, e.g., 10^{-6} [8.33].

What (8.1, 2) do not allow for is the fact that the compositions of the crystal and the melt may differ. It is clear that a change in composition must slow down the nucleation and growth process, the more so the larger the difference in composition. Growth slows down further with time in such a case, as one species in the melt is depleted. Eventually, nucleation of a different compond with composition closer to that of the melt may be required for further crystallization [8.34].

If applied to laser quenching, a few additional remarks on the predictions of nucleation and growth theory seem appropriate. The usual theory assumes that the volume of the undercooled melt is large compared to that of a critical nucleus and that it is uniformly undercooled. However, in the presence of the large thermal gradients characteristic of laser quenching only a narrow layer of melt is effectively undercooled at any given time; its thickness is clearly of the order (assuming one-dimensional heat flow)

$$\Delta z \sim \Delta T/(\partial T/\partial z), \qquad (8.4)$$

where ΔT is the undercooling required for rapid crystallization and z is the coordinate perpendicular to the material surface. If Δz comes in the order of the critical nucleus size, as may be the case with short pulses, the nucleation rate should be substantially reduced [8.35]. The lifetime of a molten phase for ns pulse irradiation will be shown below to be of the order of 10^{-7} s; the time during which a given portion of the melt is undercooled,

$$\Delta t \sim \Delta T/(\partial T/\partial t) \qquad (8.5)$$

is likely to be an order of magnitude smaller and approaches the range of the expected minimum nucleation time-lag.

In order to see how the above conditions work out in laser quenching, we must next consider the characteristics of the heat flow caused by absorption of a laser pulse.

8.2.2 Heat Flow

Lasers provide powerful pulses of collimated and nearly monochromatic radiation at wavelengths covering the near-ultraviolet, visible and near-infrared spectrum. In metals, the absorption lengths for all wavelengths of interest are of the order of 10 nm and the radiation is converted into heat on a timescale of ps or less. This means that the absorbed power densities can be made very high and that heating remains sharply localized. Let us illustrate this regime with an example. Consider a 10 ns laser pulse of 1 J cm^{-2} (easily obtainable from a commercial device), 10 % of which may effectively absorbed in a metal specimen. The volumetric heat production rate created near the surface is then of the order of 10^{13} W cm^{-3}. During the pulse duration τ the heat spreads over a depth $\delta \cong 2\sqrt{\kappa\tau}$, where κ is the thermal diffusivity of the material. For the 10 ns pulse this is several 100 nm; the average enthalpy within this layer is a few kJ cm^{-3}, enough to melt at least part of the material. The temperature gradient created is, roughly speaking, of the order of $(10^3 \, \text{K}/10^{-5} \, \text{cm}) \sim 10^8 \, \text{K/cm}$. This corresponds to a heat flux $f = -K(\partial T/\partial z)$ of the order of 10^8 W cm^{-2} (K is the thermal conductivity and z the coordinate perpendicular to the specimen surface) leaving the melt just after the pulse has ended. At this instant the cooling rate is approximately $f/\varrho c_p \delta \sim 10^{12}$ K/s, where ϱ and c_p denote the density and specific heat of the material, respectively. These simple arguments also illustrate the importance of a short radiation absorption length: if the latter is of the order of, or exceeds, the diffusion length δ (as is the case, e.g., in Si irradiated by short visible or infrared pulses) then heat is created over a larger depth, independent of pulse duration, and the cooling rate is accordingly reduced.

A larger absorbed power density during melting thus means a larger cooling rate after the pulse ends. However, there are obvious limits to the maximum heating rates one can apply to a specimen in practice. The thickness of the molten layer decreases with increasing radiation intensity since the temperature profile inside the material becomes steeper. The practical limit is given by material evaporation which occurs if the surface is heated too far above the boiling point. Liquid metals, owing to their relatively large surface tension, can be significantly superheated without developing volume boiling on a ns time scale; however, comparable superheating may also occur in the solid phase. For estimating purposes, the temperature distribution at the end of the pulse may be represented as

$$T(z, t) = [4F(1 - R)/\varrho c_p \delta]\, \text{ierfc}(z/\delta), \tag{8.6}$$

where F is the fluence [J cm^{-2}] of the laser pulse and R is the reflectivity of the specimen (ierfc denotes the first integral of the error function). Here we have assumed one-dimensional heat flow (which is well satisfied as long as δ is small compared to the beam diameter) and we have neglected the latent heat of melting of the material.

The shortest and most intense laser pulses with durations down to a few ps are obtained from mode-locked lasers. Application of such pulses to metallic glass formation is tempting, although the thickness of the melt is reduced to a few tens of nm, probably too thin for most potential applications. Most experiments to date have used pulses from Q-switched lasers which are in the range of 10–100 ns, and on which the above estimate was based. Such pulses yield molten layers between 0.1 and 1 μm thick. Thicker melts – perhaps 10 μm or more – may be produced by pulses in the ms regime [8.36] or, alternatively, by continous beams scanned at high velocity across the specimen surface [8.27]. The "pulse-duration" here corresponds to the dwell time of the beam, given by the ratio of the beam diameter to the scan speed. The cooling rates achievable are accordingly reduced and should be comparable with those of the fastest mechanical quenching methods, as mentioned before.

This leads us to the central question, namely, how the cooling rate (or whatever measures the "glass forming power" of a quenching process) scales with experimental parameters such as the laser pulse duration or the material data. A realistic answer to these questions requires a little more than the rough estimates produced above. What one has do to is to combine the equations governing heat flow with those describing the molecular rearrangment kinetics.

8.2.3 A Model of Laser Quenching

The interplay of heat flow and molecular kinetics becomes evident if we write the heat flow equation in the following generalized form [8.34]:

$$(\partial H/\partial t) = \partial/\partial z(K\partial T/\partial z) + A(z,t) + (\partial L/\partial t). \tag{8.7}$$

Here $H(T)$ is the enthalpy per unit volume of the material, A is the volumetric rate of heat production by light absorption, and L is the latent heat content of the material which varies between 0 and L_m, the heat of melting. The rate $(\partial L/\partial t)$ depends on temperature as well as on time and is determined by nucleation and growth kinetics. The solution to (8.7) clearly requires numerical treatment – not even the simple heat flow equation alone can be solved analytically under realistic assumptions.

A computer code was designed to solve (8.7) with allowance for the dynamical processes during melting and solidification. In short, the program is based on the following model.

i) Radiation is absorbed in a thin surface layer and converted into heat instantaneously. This is a valid approximation for irradiation of metals.

ii) The material is divided into discrete volume elements which are being heated by the laser and exchange heat according to (8.7). Volume elements absorb or liberate latent heat at a rate that depends on local temperature as well as on time. The quantity $(\partial L/\partial t)$ is determined by a rate function, a generalized and simplified combination of (8.1, 2). The temperature-dependent part of the rate function is schematized in Fig. 8.8; positive values lead to liberation, negative ones to absorption of latent heat. There is admittedly some arbitrariness in the parameters of this curve; however, it turns out that its detailed shape (apart from the width and height) has little influence on the results of the calculation. The time-dependent part of the rate function allows latent heat to be absorbed or liberated only after a specified time has elapsed since the volume element was first superheated or undercooled. This accounts for condition (8.5). Condition (8.4) is ignored.

iii) To simplify the presentation of the numerical results, a liquid-crystal boundary is arbitrarily defined at the position of that volume element which has absorbed or liberated 50 % of the latent heat of melting. This boundary does not have a physical meaning in all cases, as shown in the following.

A more detailed account of the calculation will be published elsewhere [8.34]. As an illustration, we present some results for the case of a 30 ns laser pulse of 10 MW cm^{-2} of absorbed fluence, incident on a crystalline bulk metal speci-

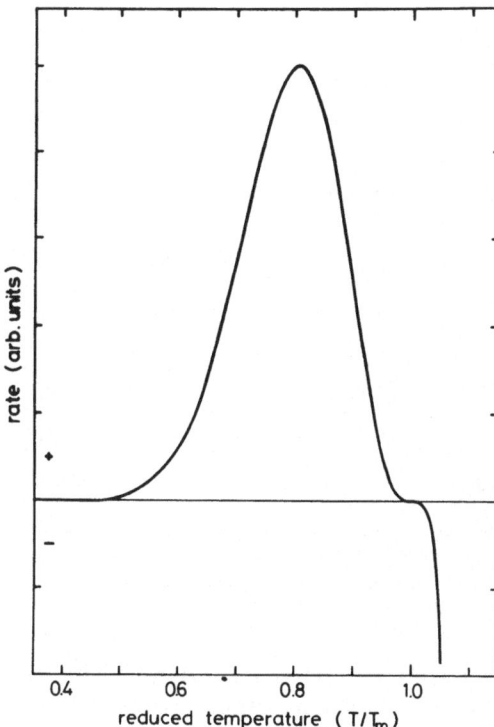

Fig. 8.8. Temperature-dependent part of the rate function for melting and crystallization used in the calculation shown in Fig. 11.9, 10

men. Typical values for the thermal data were chosen and kept constant during the process in order to facilitate interpretation of the results. Figure 8.9 shows the position of the liquid-solid boundary as a function of time. The three curves shown differ only by a scale factor in the crystallization rate; this simulates three different material compositions: for curve a the peak rate (assumed to occur at reduced temperature $T/T_m = 0.81$ in all cases) is $3 \times 10^9 \, \text{s}^{-1}$ for curve b it is $6 \times 10^6 \, \text{s}^{-1}$, and for curve c is $3 \times 10^6 \, \text{s}^{-1}$.

The largest rate (curve a) results in crystal growth at moderate undercooling ($T/T_m = 0.98$) and a nearly constant velocity of 3.2 m/s which is limited by heat flow rather than by the intrinsic growth kinetics. Curves of this kind are well-known from laser annealing studies [8.37]; only in this case does the crystal boundary as defined above, have a clear physical meaning: it represents the interface at which latent heat is being liberated. Curve b presents a case close to the limit of crystal growth: After an initial period of slow growth (limited by the rate function), the undercooling grows larger and the boundary accelerates; eventually it moves at a velocity close to maximum ($T/T_m = 0.85$). During the last stage the whole remaining melt is strongly undercooled. The boundary tends to lose its meaning, since the crystallization process is of a volume nature. Curve c is just beyond the limit for crystal growth: the undercooling at the boundary exceeds the critical value ($T/T_m = 0.81$) after a few tens of ns and the

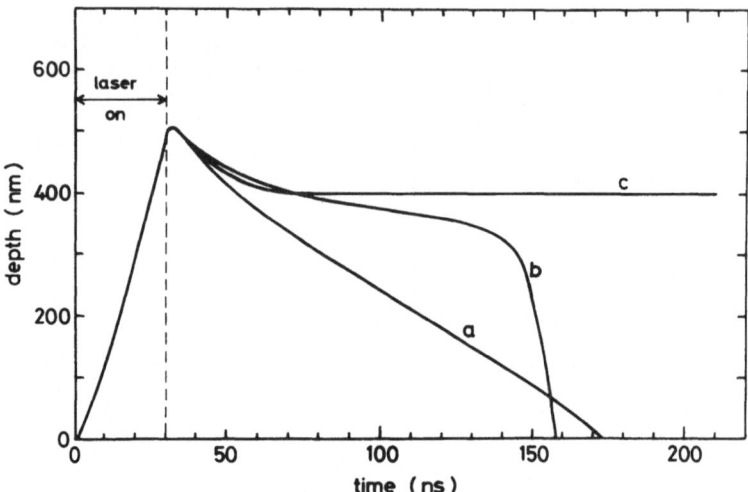

Fig. 8.9. Calculated position of the crystal boundary as a function of time in a bulk metal sample, irradiated by a 30 ns laser pulse. Curves *a–c* are obtained with different scale factors for the rate function of Fig. 8.8. Curve *a, b* correspond to crystal growth, Curve *c* to glass formation

"boundary" stops; the liquid cools down without a phase transition. The latent heat essentially remains stored in the structure, which may thus be called a glass (of course, the model accounts only for the energy content, not for the structure of the glass).

Figure 8.10 shows the surface temperature as a function of time for the cases of Figure 8.9. The numbers in the Figure indicate the instantanous value of the cooling rate at the boundary (which may slightly differ from that at the surface). In all three cases the cooling rate is of the order of 10^{12} K/s just after the pulse has ended. However, this cooling rate is of little relevance since it occurs in the superheated melt. The cooling rate first decreases by heat flow and then, in addition, by the liberation of latent heat. It stays close to zero until crystallization is completed in cases a and b. In case c liberation of latent heat is suppressed. The cooling rate stabilizes around 2×10^9 K/s until the temperature has dropped well below the range where the rate function is appreciably different from zero: the glass is stable. The calculation shows, however, that even in this case, some ten percent of the latent heat has been emitted which may be interpreted as partial nucleation having taken place. Only with a ten times smaller scale factor for the rate function is nucleation found to be completely suppressed. The cooling rate then remains around 10^{10} K/s until solidification is completed. Incidentally, the enormous range of cooling rates shown in the figure (notably all pertaining to the same laser pulse!) should caution against the use of cooling rates as a global concept in characterizing quenching processes. The only relevant cooling rates are those in the temperature interval where the rate of nucleation and growth is significant.

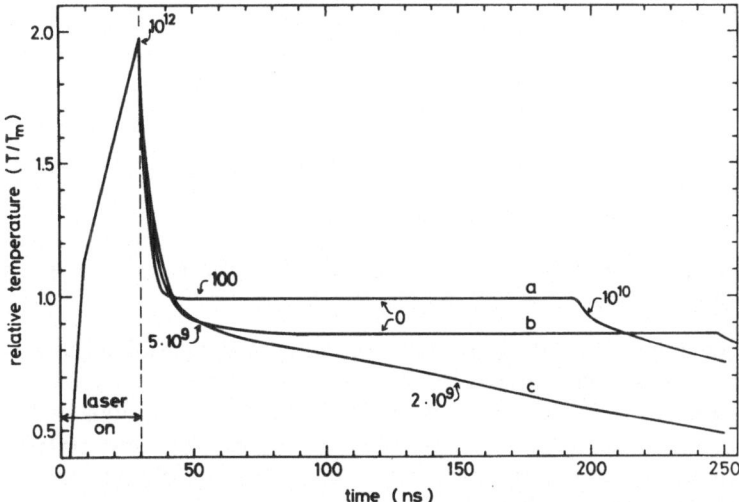

Fig. 8.10. Surface temperature as a function of time for the cases of Fig. 8.9. Inserted numbers indicate the instantaneous cooling rate at the crystal boundary

8.2.4 Discussion

The sample calculation presented above emphasizes the influence of the melt properties. For example the experimental glass-forming range in the Pd–Si and Pt–Si systems is readily understood if the different scale factors in cases a–c above are thought of as due to melt composition: The rate of crystallization events at a given undercooling should be largest if the melt composition coincides with that of a crystal of large driving force (such as a congruently melting phase). Away from that composition the rate should decrease, making glass formation easier. The criterion based on melting point depression mentioned earlier also has its correspondence in the rate function approach. For glass formation to be possible, the rate function must not only be small enough in its absolute value, but also in its width on the temperature scale.

Further conclusions regarding the influence of various experimental parameters can be drawn from the model. For example, it turns out that material (b) becomes amorphous if irradiated with a ten times shorter pulse of the same energy. Pulse energy also matters: A 30 ns pulse of twice the intensity used in Figures 8.8 and 9 melts a layer 1000 nm instead of 500 nm thick; cooling is slowed down sufficiently even for material (c) to crystallize. These examples demonstrate how critical the choice of parameters can be and how little can be concluded from roughly estimated cooling rates, as often used in the literature.

The model calculation presented above relates known or measurable quantities – the laser parameters, the thermal material data, as well as the degree of crystallinity after irradiation – to the intrinsic crystallization kinetics of the material which, in general, is poorly known. This suggests that laser quenching and its analysis should be able to provide new insight into the properties of strongly undercooled melts.

8.3 Conclusions

The main properties of laser quenching as a method for metallic glass formation can be summarized as follows:

i) Laser quenching with short pulses offers cooling rates exceeding those of splat cooling by several orders of magnitude. Concomitant with the increase in cooling rate is a decrease in the thickness of the obtainable amorphous phase. For quenching with ns pulses the thickness is typically a few 100 nm. Irradiation with ms pulses or scanned beams yields thicknesses of up to several μm, at correspondingly reduced cooling rates.

ii) The increased cooling rates result in an unprecedented choice of materials that can be prepared in the glassy state. New insights into the mechanisms that ultimately limit glass formation and glass stability in metallic systems can be gained. Comparative studies of glass structure and glass properties as a function of composition over extended ranges in binary or multiple systems are made possible. Work in these areas has barely started.

iii) The freedom in the choice of glass composition should make it possible to create specific metastable crystals, not available by other means, by growth from an amorphous phase of the desired composition.

iv) Laser quenching can be applied to samples in unusual environments – e.g., immersed in liquid nitrogen, mounted in a UHV chamber, or even moved in an assembly line. Further, it can be kept sharply localized, enabling one to create amorphous patterns in a crystalline matrix or to form a glassy skin covering a crystalline structure.

v) It should be kept in mind that the choice of laser parameters can be quite critical. Proper sample design and careful choice of laser wavelength, pulse energy and pulse duration is necessary. The relations between crystallization kinetics, cooling rate and laser parameters certainly warrant further study.

Part of this list admittedly reads more like a program than like a review, emphasizing again the preliminary nature of the subject. It is hoped, however, that the potential of the new technique will stimulate work on a broader basis, in order that in a few years a somewhat more extensive survey on laser quenching will have to be written.

Acknowledgements. The author is indebted to numerous colleques who have contributed to this work in one way or another, and to Prof. H. P. Weber for his interest and support. Financial support by the Swiss Commission for the Encouragment of Scientific Research is gratefully acknowledged.

References

8.1   J.M.Poate, J.W.Mayer (eds.): "*Laser-Annealing of Semi-Conductors*" (Academic Press, New York 1982)
8.2   M.von Allmen, S.S.Lau, M.Mäenpää, B.Y.Tsaur: Appl. Phys. Lett. **36**, 205–207 (1980)
8.3   M.von Allmen: In [Ref. 8.1, pp. 43–74]
8.4   W.Klement, R.H.Willens, P.Duwez: Nature **187**, 869 (1960)
8.5   W.G.Moffat: "The Handbook of Binary Phase Diagrams" (General Electric Co., Schenectady, NY 1979) and supplements
8.6   I.W.Donald, H.A.Davies: J. Non-Cryst. Solids **30**, 77–85 (1978)
8.7   H.J.VindNielsen: J. Non-Cryst. Solids **33**, 285–289 (1979)
8.8   A.K.Green, E.Bauer: J. Appl. Phys. **47**, 1284–1291 (1970)
      C.Suryanarayana, T.R.Anatharaman: Mat. Sci. Engr. **13**, 73–81 (1979)
8.9   E.Huber, M.von Allmen: Phys. Rev. B (to appear)
8.10 H.Langer, E.Wachtel: Z. Metallkunde **72**, 769–775 (1981)
8.11 M.Hansen: *Constitution of Binary Alloys*, 2nd ed. (McGraw-Hill, New York 1958)
      F.A.Shunk, Second Supplement (McGraw-Hill, New York 1969)
8.12 M.von Allmen, S.S.Lau, M.Mäenpää, B.Y.Tsaur: Appl. Phys. Lett. **37**, 84–86 (1980)
8.13 B.Y.Tsaur, Z.L.Liau, J.W.Mayer: Phys. Lett. **71**A, 270 (1979)
8.14 P.Oelhafen, R.Lapka, J.Krieg, M.von Allmen, K.Affolter (to be published); see Chap. 9
8.15 J.L.Jorda, J.Muller, J.Less: Common Metals **84**, 39–48, (1982)
8.16 K.Affolter, M.von Allmen, H.P.Weber, M.Wittmer: J. Noncryst. Solids (to appear)
8.17 C-J.Lin, F.Spaepen: Appl. Phys. Lett. **42**, 721–724 (1982)
8.18 K.Affolter, M.von Allmen: Unpublished (1983)
8.19 B.Stritzker: In *Laser and Electron-Beam Interaction with Solids*, ed. by B.R.Appleton, G.K.Celler (North-Holland, Amsterdam 1982) pp. 363–375
8.20 S.T.Picraux, D.M.Follstaedt: Pres. at MRS meeting, Boston (November 1982)
8.21 R.Tsu, R.T.Hodgson, T.Y.Tan, J.E.Baglin: Phys. Rev. Lett. **42**, 1356–1358 (1979)
8.22 S.S.Lau, W.F.van der Weg: In *Thin Films*, ed. by J.M.Poate, K.N.Tu, J.W.Mayer (Wiley, New York 1978) pp. 433–480
8.23 P.L.Liu, R.Yen, N.Bloembergen, R.T.Hodgson: Appl. Phys. Lett. **34**, 864–866 (1979)
8.24 A.G.Cullis, H.C.Webber, N.G.Chew: In *Laser and Electron-Beam Interaction with Solids*, ed. by B.R.Appleton, G.K.Celler (North-Holland, Amsterdam 1982) pp. 131–140
8.25 G.A.Rozgonyi, H.Baumgart, F.Phillip, R.Uebbing, H.Oppholzer: In *Laser and Electron-Beam Interaction with Solids*, ed. by B.R.Appleton, G.K.Celler (North-Holland, Amsterdam 1982) pp. 177–182
8.26 P.Mazzoldi, G.DellaMea, G.Battaglin, A.Miotello, M.Servidori, D.Bacci, E.Jannitti: Phys. Rev. Lett. **44**, 88–91 (1980)
8.27 E.M.Breinan, B.H.Kear, C.M.Banas: Phys. Today 44–50 (November 1976)
8.28 R.Becker, G.Sepold, P.L.Ryder: Scripta Metallurgica **14**, 1283–1285 (1980)
8.29 R.Becker, W.Jüptner, G.Sepold: Private communication (1982)
8.30 K.A.Jackson: In *Treatise on Solid State Chemistry*, ed. by N.B.Hannay (Plenum Press, New York 1975) pp. 233–282
8.31 D.Kashiev: Surf. Sci. **14**, 209–220 (1969)
8.32 J.Köster, U.Herold: In *Glassy Metals I*, ed. by H.J.Güntherodt, H.Beck, Topics Appl. Phys., Vol. **46** (Springer, Berlin, Heidelberg, New York 1981) pp. 225–259
8.33 D.R.Uhlmann: J. Non-Cryst. Solids **7**, 337–348 (1972)
8.34 M.von Allmen: *Physics of Laser Effects on Materials* (Academic Press, New York) in preparation
8.35 D.Turnbull: Contemp. Phys. **10**, 473–483 (1969)
8.36 See, e.g., W.A.Elliot, F.P.Gagliano, G.Krauss: Metallurgical Trans. **4**, 2031–2037 (1973)
8.37 J.C.Wang, R.F.Wood, P.P.Pronko: Appl. Phys. Lett. **33**, 455–458 (1978)

9. Electron Spectroscopy on Metallic Glasses

P. Oelhafen

With 36 Figures

This chapter will deal with electron spectroscopy measurements on metallic glasses. The most important aim of these experiments is to get detailed information about the electronic structure of the solid to be studied. Since many physical properties like electronic specific heat, magnetic susceptibility, ferromagnetism, superconductivity and thermodynamic data are directly related to the electronic structure of a solid, the importance of electron spectroscopy measurements is evident.

9.1 Overview

Beside the usual questions which can be investigated on crystalline alloys, additional interesting issues related to the electronic structure of glassy alloys arise: (i) Do the alloys which are good glass formers (alloys which can be obtained in the amorphous phase by rapid quenching from the melt by *low cooling rates* and which have *high crystallization temperatures*) have any common electronic properties? (ii) Are there any correlation between parameters which define the glassy state (e.g., glass and crystallization temperature, lowest required cooling rate, reduced glass temperature $T_{rg} = T_g/T_m$ where T_g and T_m are the glass and melting temperatures, respectively) and electronic properties? The former question is related to the model of *Nagel* and *Tauc* [9.1] which treats the glass as a nearly-free-electron metal. They found that there will be increased stability against crystallization when the Fermi level E_F is located at a minimum in the density of states.

Therefore, the metallic glasses offer the opportunity to study new and interesting problems besides the standard questions to be solved in investigations on crystalline alloys. In addition, they make it possible to study alloys in continuous ranges of concentrations and at concentrations at which no crystalline phase exists.

The significance of electron spectroscopy in studying the electronic structure of alloys and the progress in preparing new metallic glasses has stimulated many workers to perform measurements in this field. A summary of the papers published so far is given in Table 9.1. Besides the *alloys* studied by electron spectroscopy, the different kind of *experiments* and *excitation energies* are listed

Table 9.1. Electron spectroscopy measurements on metallic glasses. Measurements on crystalline samples are included if especially performed for comparison with measurements on glassy samples

First author/ Reference		Alloy	Valence band photo-electron spectroscopy excitation energy hv [eV]	Core level XPS	AES	Other
Nagel	[9.2]	$Pd_{77.5}Cu_6Si_{16.5}$	21.2	+		
		c-$Pd_{77.5}Cu_6Si_{16.5}$	21.2	+		
Nagel	[9.3]	$Ni_{50}Nb_{50}$	21.2	+		
		$Ni_{60}Nb_{40}$	21.2	+		
Shen	[9.4]	Pd–Cu–Si				EELS
		$(Pd_{80}Ni_{20})_{80}P_{20}$			+	EELS
		$(Pd_{60}Ni_{40})_{80}P_{20}$			+	EELS
		$(Pd_{20}Ni_{80})_{80}P_{20}$			+	EELS
Amamou	[9.5]	$Cu_{60}Zr_{40}$	21.2, 40.8, 1253.6			
Riley	[9.6]	$Pd_{81}Si_{19}$	21.2, 1486.6 (M)	+	+	
Oelhafen	[9.7]	$Pd_{85}Si_{15}$	21.2			
		$Pd_{82}Si_{18}$	21.2			
		$Pd_{81}Si_{19}$	21.2, 40.8, 1253.6			
		$Pd_{79}Si_{21}$	21.2			
Oelhafen	[9.8]	$Pd_{35}Zr_{65}$	40.8			
		$Pd_{30}Zr_{70}$	40.8			
		$Pd_{25}Zr_{75}$	40.8			
		$Cu_{60}Zr_{40}$	21.2			
		$Cu_{40}Zr_{60}$	21.2			
		$Cu_{30}Zr_{70}$	21.2			
Amamou	[9.9]	$Co_{78}P_{14}B_8$	21.2, 40.8, 1486.6	+		
		c-Co_3B	21.2, 40.8, 1486.6			
Waclawski	[9.10]	$Pd_{85}Si_{15}$	21.2			
		$Pd_{80}Si_{20}$	21.2			
		$Pd_{75}Si_{25}$	21.2			
		$Pd_{80}Cu_3Si_{17}$	21.2			
		$Pd_{80}Cu_6Si_{14}$	21.2			
Cartier	[9.11]	$Pd_{80}Si_{20}$	1486.6 (M)			
		$Fe_{80}B_{20}$	1486.6 (M)	+		
		$Fe_{40}Ni_{40}P_{14}B_6$	1486.6 (M)	+		
		$Fe_{32}Ni_{36}Cr_{14}P_{12}B_6$	1486.6 (M)	+		
		c-$Fe_{32}Ni_{36}Cr_{14}P_{12}B_6$	1486.6(M)	+		
		$Cu_{60}Zr_{40}$	1486.6 (M)			
Matsuura	[9.12]	$Fe_{88}B_{12}$	1486.6	+		
		$Fe_{86}B_{14}$	1486.6	+		
		$Fe_{84}B_{16}$	1486.6	+		
		$Fe_{80}B_{20}$	1486.6	+		
		$Fe_{75}B_{25}$	1486.6	+		
Amamou	[9.13]	$Fe_{80}B_{20}$	21.2, 1486.6			
		$Fe_{79}P_{13}B_8$	21.2, 1486.6			
		c-$Fe_3P_{0.1}B_{0.9}$	21.2, 1486.6			
		c-Fe_3P	21.2, 1486.6			

Table 9.1 (continued)

First author/ Reference		Alloy	Valence band photo-electron spectroscopy excitation energy hv [eV]	Core level XPS	AES	Other
Amamou	[9.14]	$Ni_{25}Zr_{75}$	21.2, 1486.6 (M)			
		c-Ni Zr$_2$	1486.6 (M)			
		$Co_{30}Zr_{70}$	40.8			
Oelhafen	[9.15]	$Cu_{40}Zr_{60}$	21.2			
		$Ni_{24}Zr_{76}$	21.2			
		$Co_{22}Zr_{78}$	21.2			
		$Fe_{24}Zr_{76}$	21.2			
		$Pt_{21}Zr_{79}$	21.2			
		$Pd_{25}Zr_{75}$	21.2			
		$Ni_{37}Zr_{67}$	21.2			
		$Cu_{40}Ti_{60}$	21.2			
		$Ni_{60}Nb_{40}$	21.2			
		$Ni_{63}Ta_{37}$	21.2			
Amamou	[9.16]	$Mo_{48}Ru_{32}B_{20}$	40.8, 1486.6			
		$Mo_{40}Ru_{40}P_{20}$	40.8, 1486.6			
Oelhafen	[9.17]	$Cu_{60}Zr_{40}$		+		
		$Ni_{37}Zr_{63}$		+		
		$Co_{40}Zr_{60}$		+		
		$Fe_{24}Zr_{76}$		+		
		$Pd_{35}Zr_{65}$		+		
		$Rh_{25}Zr_{75}$		+		
		$Pt_{21}Zr_{79}$		+		
		$Cu_{40}Ti_{60}$		+		
		$Ni_{60}Nb_{40}$		+		
		$Ni_{63}Ta_{37}$		+		
Amamou	[9.18]	$Ni_{78}P_{14}B_8$	1486.6			
		c-Ni$_3$B	1486.6			
		c-Ni$_3$P	1486.6			
Güntherodt	[9.19]	$Cu_{60}Zr_{40}$	40.8			
		c-Cu$_3$Zr$_2$	40.8			
Güntherodt	[9.20]	$Rh_{25}Zr_{75}$	21.2			
		$Co_{33}Gd_{67}$	40.8			
Petõ	[9.21]	$Fe_{84}B_{16}$	10.2			
		c-Fe$_{84}$B$_{16}$	10.2			
Bevolo	[9.22]	$Fe_{82}B_{14}Be_4$			+	
		$Fe_{82}B_{13}Be_5$			+	
Mizoguchi	[9.23]	$Cu_{70}Ti_{30}$	21.2			
		$Cu_{60}Ti_{40}$	21.2			
		$Cu_{50}Ti_{50}$	21.2			
		$Cu_{40}Ti_{60}$	21.2			
		$Cu_{35}Ti_{65}$	21.2			
Colavita	[9.24]	$Fe_{80}B_{20}$				EELS
Oelhafen	[9.25]	$(Pd_{30}Zr_{70})_{80}H_{20}$	21.2	+	+	

Table 9.1 (continued)

First author/ Reference		Alloy	Valence band photo-electron spectroscopy excitation energy hv [eV]	Core level XPS	AES	Other
Honda	[9.26]	$Ni_{82}B_{18}$	1253.6			
		$Ni_{67}B_{33}$	1253.6			
Spit	[9.27]	$(Ni_{64}Zr_{36})_{100-x}H_x$			+	
DasGupta	[9.28]	$Nb_{55}Rh_{45}$	16.8, 21.2, 40.8, 1253.6	+		
		$Nb_{55}Ir_{45}$	40.8	+		
		$Ta_{55}Rh_{45}$	40.8	+		
		$Ta_{55}Ir_{45}$	40.8	+		
Nagel	[9.69]	$Ca_{70}Al_{30}$	21.2			SXS
		$Ca_{50}Al_{50}$	21.2			

in the following columns. As can be seen from the second column, most of the alloys fall into two groups: the transition-transition metal[1] alloys (T–T group) and alloys containing (at least) one transition metal and a normal metal (nontransition or rare-earth metal) or metalloid (T–N group). From all the studies given in Table 9.1, a fairly precise picture of the electronic structure of metallic glasses emerges.

Even though electron spectroscopy measurements can yield information on both surface and bulk properties of a solid (the escape depth of the electrons to be detected in these experiments is between 5 and 30 Å, depending on their energies), in most of the experiments done so far the *bulk* properties have been studied. However, there are many interesting phenomena related to the surface properties of glassy alloys, like corrosion resistance, segregation, catalysis and wear, for which electron spectroscopy could make important contributions. Therefore, an increasing effort in studying the *surface properties* of metallic glasses in the future is to be expected.

As mentioned above, this chapter deals with electron spectroscopy measurements performed on *metallic glasses*, i.e., the samples have been obtained by rapid quenching from the melt, either by the splat cooling or melt spinning techniques. For comparison, a few measurements on crystalline counterparts are included as well. Not considered are measurements on amorphous films obtained by evaporating or sputtering.

It is evident that other kinds of spectroscopy such as optical absorption, x-ray emission spectroscopy, bremsstrahl isochromate spectroscopy and soft x-ray appearance potential spectroscopy also yield information about the electronic structure of solids. In many cases the results of these experiments are complementary to those obtained by electron spectroscopy.

---

1 In this context, the noble metals are considered as belonging to the transition metals.

However, to date only a few experiments have been performed by these other techniques on metallic glasses. Most of these measurements will be mentioned in the following chapters as far as they are related to the electron spectroscopy measurements.

In Sect. 9.2 we shall discuss the experimental techniques in studying metallic glasses such as photoelectron spectroscopy, Auger electron spectroscopy and electron energy loss spectroscopy. In Sect. 9.3 a summary of the results obtained on glassy transition metal alloys (T–T group) will be given. The other large group, glassy alloys containing a transition metal and a normal metal or metalloid (T–N group) will be presented in Sect. 9.4, and in Sect. 9.5, measurements on other groups of metallic glasses like hydrogenated glassy alloys and normal metal glasses (N–N group) will be discussed.

9.2 Experimental Techniques

9.2.1 Background

In an electron spectroscopy experiment the electrons emitted from a solid are detected in an electron energy analyzer. The electrons are emitted as a consequence of the excitations in the solid (or the electrons in the solid) by bombarding the sample surface either by photons or particles. The photons used in a *photoelectron spectroscopy* (PES) experiment may vary from the ultraviolet range (*ultraviolet photoelectron spectroscopy*, UPS) with excitation energies from the photoelectric threshold to about 50 eV, to the soft x-ray range covered, e.g., by synchrotron radiation, up to the x-ray range with excitation energies of typically 1.5 keV (*x-ray photoelectron spectroscopy*, XPS). In *Auger electron spectroscopy* (AES), any excitation which is able to create a hole in an inner electron shell may be used such as photons, electrons or ions. In an *electron energy loss spectroscopy* experiment (EELS), the incoming particles are electrons and the resulting energy loss due to the scattering at the sample surface is measured. A brief summary of the different experimental techniques is given in Table 9.2.

As mentioned before, in all these experiments the detected electrons emitted from the samples originate from a surface layer with a thickness of the order of 5–30 Å. This escape depth depends mainly on the energy of the emitted electron and far less on the sample material itself [9.29]. It turns out that electrons with energies lower than 10 eV and greater than 1 keV are usually emitted from a depth of approximately 8–10 monolayers. In the intermediate energy range, near 100 eV, the escape depth has a minimum and can be as low as 2 monolayers. The energy dependence of the escape depth of the electrons, and its dimension, has several consequences for the electron spectroscopy experiments: (i) the experimental results can reflect surface or bulk properties of the sample depending on the electron energy range used in the experiment; (ii)

Table 9.2. Experimental techniques of electron spectroscopy

Method	Excitations by	Excitation energy range [eV]	Information obtained
Photoelectron spectroscopy PES	*Photons*		
Ultraviolet photoelectron spectroscopy UPS	e.g., resonance lamps	$\leqq 50$	Valence band spectroscopy (electron states densities)
Soft x-ray photoelectron spectroscopy SXPS	Synchrotron radiation	$20 \ldots 500$	Valence band spectroscopy (electron states densities)
X-ray photoelectron spectroscopy XPS	e.g. x-rays ($Al\ K_{\alpha}$, $Mg\ K_{\alpha}$)	1.5×10^3	Core electron spectroscopy (binding energy shifts, surface composition, local states densities E_F)
Auger electron spectroscopy AES	Electrons, photons, ions	$5 \times 10^2 \ldots 5 \times 10^3$	Surface composition in some cases: local valence band states densities
Electron energy loss spectroscopy EELS	Electrons	$1 \ldots 10^5$	Interband transitions, plasmon energies, adsorbate vibration modes, plasmon dispersion

surface contamination like an oxide layer can change the experimental results to a large extent. Therefore, special care has to be taken in *cleaning* the sample surface in ultrahigh vacuum before doing the experiments. The usual techniques in cleaning alloy surfaces are either *ion etching* or *mechanical cleaning*.

An important problem when using ion etching in alloy studies is the *preferential sputtering* of one alloy constituent which changes the surface composition. Information about the changes in composition during sputtering of the surface may be obtained by XPS or AES and in some cases by UPS (Sect. 9.3).

A second problem in sputter-cleaning of solid surfaces is the structural disorder caused by the impinging high energy ions. In studying metallic glasses, however, this effect is not expected to have an apparent influence on the electronic structure of the alloys.

Because of the problems connected with sputter cleaning of surfaces, mechanical cleaning procedures are often used in electron spectroscopy experiments. The tools used are of different designs but the common goal is to remove the contamination by grinding the sample surface.

The information which may be obtained by UPS, XPS, AES, and EELS will be discussed in the following section. For a more detailed introduction to the different electron spectroscopy techniques, the reader is referred to the specialized literature [9.29–31].

9.2.2 Photoelectron Spectroscopy (PES)

Photoelectron spectroscopy is the most widely used technique in studying metallic glasses by electron spectroscopy (Table 9.1). In these experiments an electron in the solid with initial energy E_i is brought up to a final energy E_f by absorbing an incoming photon of energy hv. If the energy level E_f is above the vacuum level E_∞, the electron has the possibility of being emitted from the solid into the vacuum. The electron energy analyzer measures the energy E_f of the photoelectron with respect to the Fermi level E_F of the solid. Since energy conservation requires $E_f = E_i + hv$, the energy level E_i from which the photoelectron originates is known.

Depending on the photon energy used in the experiment, either valence bands or core levels can be studied. In *UPS experiments*, the excitation energies are of the order of 20 eV to 40 eV, and except for a few exceptions, no core levels can be ionized. Therefore, these experiments provide mainly valence band spectra. In an amorphous solid, the probability for absorbing a photon by a valence electron is proportional to the optical matrix element M_{if}^2 and the density of states at E_i and E_f. For sufficiently high photon energy hv, M_{if}^2 does not show a great hv dependence and, in general, $D(E_f)$ does not show variations with energy as big as $D(E_i)$. Therefore, the main information obtained from UPS experiments is about the occupied electron state densities of the alloys and the photoelectron valence band spectra reflects more or less the density of states $D(E)$ function below the Fermi level. Since the valence electrons of all the alloy constituents contribute to the spectra, photoelectron spectroscopy does not provide local information about the valence band structure at the different sites of alloy constituents. However, by varying the alloy composition one can, in many cases, assign features of special interest to an element in the alloy.

The experimental *energy resolution* in UPS experiments is of the order of 0.1 eV to 0.2 eV, which gives from this point of view a clear advantage over the x-ray photoelectron valence band spectroscopy, with a resolution of about 1 eV, using nonmonochromatized Mg $K_\alpha(hv = 1253.6 \text{ eV})$ or Al $K_\alpha(hv = 1486.6 \text{ eV})$ radiation. The resolution for the latter excitation may be improved by using an x-ray monochromator and values of 0.5–0.6 eV may be obtained with commercially available monochromators. However, in this case, the intensity is about 3 orders of magnitude smaller than that in UPS.

In XPS *experiments*, the excitation energy is high enough to ionize inner electron shells as well and, therefore, information about the core electrons may be obtained in addition to the valence band spectra. Important information is the *binding energy* E_B (always measured with respect to the Fermi level E_F) of a core electron related to an atom in the alloy. The *binding energy shift* is the binding energy difference $\Delta E_B = E_B^{alloy} - E_B^{element}$, where E_B^{alloy} and $E_B^{element}$ are the binding energies of a core electron related to an atom in the alloy and in the corresponding pure metal. E_B is also referred to as *chemical shift*, since the physical origin of the binding energy shift is related to the charge transferred on alloying, or forming a chemical bond between two dissimilar atoms, and the altered screening conditions of the positive nuclear charge. In fact, the binding energy shifts, of the order of 1 eV, may be determined with high accuracy by directly comparing the core level spectra of the alloy and the corresponding pure metals. However, the interpretation of the core level binding energy shifts in terms of a simple charge transfer model is mostly not applicable, since in many cases (Sect. 9.3), the binding energy shifts ΔE_B of both alloy constituents have the same sign. A change of the Fermi level itself on alloying could be one possible reason for this behavior.

Further important information from x-ray photoelectron core level spectroscopy may be obtained from an analysis of the *core line shapes*. Due to many-electron processes, the photoelectron spectrum of a core level shows some broadening on the low energy side, resulting in an asymmetry of the core level peak [9.32, 33]. The origin of the tailing at the low energy side are energy losses related to electron-hole excitations near E_F. Since the probability for these excitations is related to the local density of states at E_F of the corresponding atom, the asymmetry α of the core level peak becomes a local probe for the density of states at E_F. So far, this effect has only been used for qualitative interpretation of the experimental data. However, the increasing success in understanding the asymmetry α will make it possible to use this kind of study more quantitatively [9.34].

Another result which may be obtained from x-ray photoelectron core electron spectroscopy is the *elemental composition* of the sample surface within the escape depth of the photoelectrons, since the binding energies of the core electrons are specific for each element. The absolute composition of the alloy surface can be determined from theoretical photoionization cross sections within an accuracy of 20 % or better, depending on the elements to be analyzed. The *change* in surface composition (e.g., during sputtering) can be monitored with a much higher accuracy. And using the energy dependence of the escape depth of the photoelectrons, information about the concentration *depth profile* can be obtained by measuring the intensities of core electrons with different binding energies of the same element.

9.2.3 Auger Electron Spectroscopy (AES)

In general, three different electron states (XYZ) are involved in Auger electron spectroscopy. The Auger electron emission process requires an empty electron state in shell X of an atom, created by an incoming electron, ion or photon of sufficient energy. As a consequence, an electron from shell Y combines with the hole in shell X and the energy difference involved in this recombination process is transferred to an electron in shell Z which can be emitted from the solid as an Auger electron.

The first information which may be obtained from AES measurements is an *elemental analysis* of the surface composition of the alloy. The varying escape depths of the emitted electrons may be used again to study different depths of the sample surface region.

In some cases, information about the *valence band structure* may be obtained from Auger transitions in which one or two *valence electrons* are involved (XYV or XVV transitions). Depending on the hole-hole interaction U_{eff} of the two final state holes as compared with the valence band width W, the Auger transition is either determined by localized states, when $U_{eff} \gtrsim W$, or by band-like states, when $U_{eff} \ll W$. In the latter case, the Auger line of a XVV transition represents a weighted self-convolution of the occupied local valence state density. As a consequence the line shape will change if the atom is placed in a different chemical environment. Though studies of this kind can provide important information about the local valence band structure, only a few measurements of this kind have been performed so far in alloy studies. Similar information may be obtained from x-ray emission spectroscopy (XES) and a few experiments in the field of metallic glasses have been performed. These measurements will be mentioned in Sect. 9.3.

9.2.4 Electron Energy Loss Spectroscopy (EELS)

In EELS, the energy losses of the impinging electrons due to the interaction with the solid are measured. From the experimental point of view, three different kind of EELS experiments are usually performed: (i) EELS experiments in reflection geometry with a resolution of the order of 0.5 eV, usually performed with the AES electron gun and spectrometer. The electron primary energies range typically from 10^2 eV to 10^3 eV; (ii) high resolution EELS experiments in reflection geometry using a monochromatized primary electron beam with a primary energy of a few eV; a resolution of the order of 10 meV is typically achieved; (iii) EELS experiment in transmission geometry with high primary energies of the order of 10^4 eV to 10^5 eV.

The first type of experiments (i) provide information about single particle excitations like *interband transitions*, comparable to an optical absorption experiment. In addition, *collective excitations*, and surface and bulk plasmons can be examined. The high resolution EELS (ii) is often used in adsorption studies, since information about *vibration modes* of adsorbates may be obtained. The high energy EELS (iii) experiment is mostly used to study the dispersion properties of plasmon excitations. So far, only the first kind of EELS has been applied for studying metallic glasses [9.4, 24].

9.3 Glassy Transition Metal Alloys (T–T Alloys)

9.3.1 Experimental Results

In this section the experimental results obtained from binary glassy T–T alloys are summarized. The alloys studied so far consist of one early and one late transition metal. The assignment early and late refers to the position of the alloy constituents in the periodic table and can therefore be characterized by the d electron number n_d, $n_d \leq 5$, and $n_d > 5$ for the early and late transition metal, respectively. This alloy family is especially well suited for electron spectroscopy since the d bands of both alloy constituents are easy to detect and to distinguish in most cases due to the high states density in the d bands. As a consequence, the bonding mechanism, i.e., the d band binding energy shifts, and the splitting of the alloy valence bands, can be directly observed in the valence band spectra.

a) Valence Band Spectroscopy

This behavior is clearly demonstrated in Fig. 9.1 [9.8] with the UPS valence band spectra of three different glassy Pd–Zr alloys and those for pure polycrystalline metals Pd and Zr. The spectra of the pure transition metals are essentially determined by their d-band state densities with maxima near the Fermi edge E_F (at binding energy $E_B = 0$). The alloy spectra do not simply consist of a superposition of the spectra related to the pure metals but show a distinct *splitting in two peaks*. The concentration variation studies of the alloy have led to the conclusion that the peak with the higher binding energy is mainly determined by Pd $4d$ states, whereas the peak with the lower binding energy contains mainly Zr $4d$ states. From this behavior, it becomes clear that the center of gravity of the Pd $4d$ band is *shifted to a higher binding energy* in the alloys, compared with pure Pd, by as much as $2\,\mathrm{eV}$.

A second example is given in Fig. 9.2 [9.8]. Again, three different glassy alloys of Cu–Zr are compared with the UPS valence band spectra of pure Cu and Zr. The d band derived from the late transition metal is shifted again to

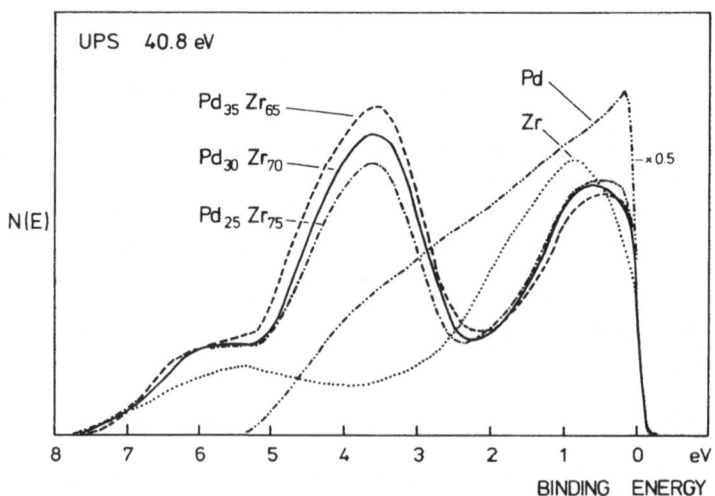

Fig. 9.1. UPS ($hv = 40.8\,\mathrm{eV}$) spectra of three glassy Pd–Zr alloy compositions, polycrystalline Pd and Zr [9.8]

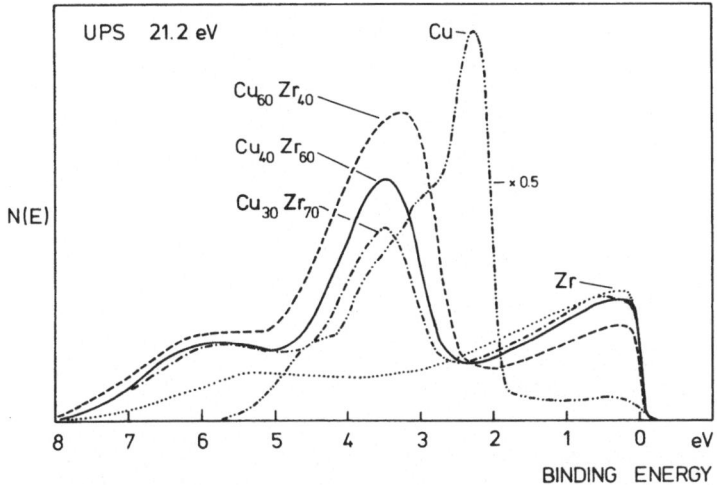

Fig. 9.2. UPS ($hv = 21.2$ eV) spectra of three Cu–Zr samples, polycrystalline Cu and Zr [9.8]

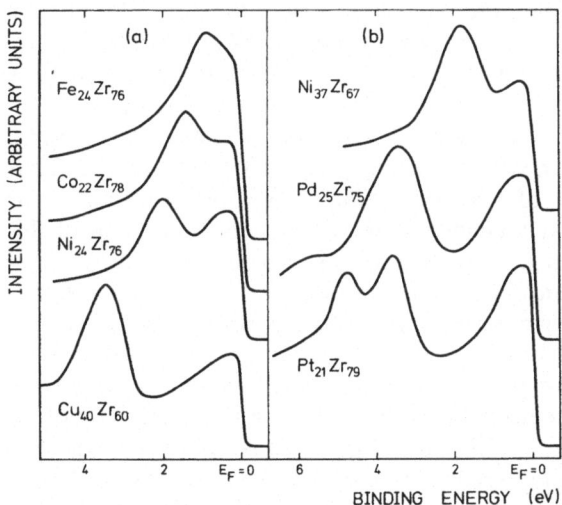

Fig. 9.3a, b. UPS spectra ($hv = 21.2$ eV) of glassy Zr alloys. (**a**) Zr with a *late* transition metal of the *same series*. (**b**) Zr with elements of the *same group* [9.15]

higher binding energies. Measured at the high energy edge of the Cu $3d$ band, the shift is of the order of 0.5 eV. A comparison of the UPS valence band spectra of different glassy Zr alloys is shown in Fig. 9.3 [9.15]. Alloys with $3d$ transition metals (Fe, Co, Ni, and Cu) are shown on the left side, and alloys with transition metals from the same group in the periodic table are compared on the right side (Ni, Pd, and Pt). Again, from an alloy concentration variation and a core level line shape analysis (which will be discussed later), the experiments indicate that the peaks at the high binding energy side of the valence band spectra are related to the d-states of the late transition metal. The

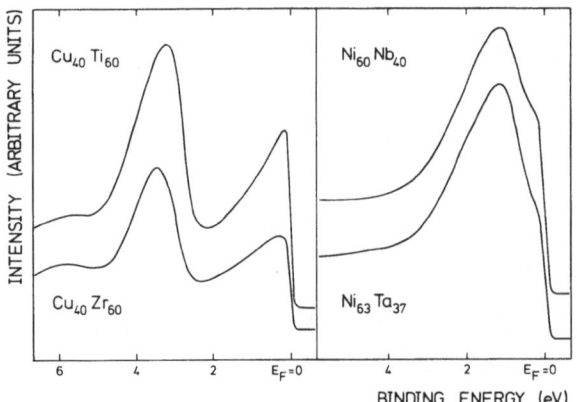

Fig. 9.4. Comparison of UPS spectra ($hv = 21.2$ eV) of glassy Cu and Ni alloys with an *early* transition metal of the *same group* [9.15]

sequence of UPS valence bands of the alloys with the $3d$ elements clearly shows a decrease in the $3d$-band peak binding energies with decreasing valence difference Δn (or group number difference in the periodic table). However, even in the case of $Fe_{24}Zr_{76}$, the valence band maximum (which is related to the Fe $3d$ states) is shifted towards higher binding energies since the pure $3d$ transition metals shown here have their valence band[2] maxima near E_F. The UPS valence band spectra of the alloys with Ni, Pd, and Pt show that the d band splitting increases as the atomic number of the late transition metal increases.

In Fig. 9.4 [9.15], the UPS valence band spectrum of Cu–Ti is compared with Cu–Zr, and that of Ni–Nb with Ni–Ta, i.e., the early transition metal has been replaced by an element of the same group in the periodic table. The comparison clearly shows that a change of the early transition metal of this kind has no significant influence on the Cu and Ni d band positions.

In Fig. 9.5 [9.14], the valence band of glassy $Ni_{25}Zr_{75}$ measured by XPS is shown (full line). Due to the lower resolution in this experiment, the splitting of the valence band into two peaks (Fig. 9.3) is not directly visible. However, the position of the Ni d band is roughly the same as in the UPS measurements. In addition, due to different photoelectric excitation cross sections, the ratio of the peak intensities related to the Ni $3d$ and Zr $4d$ states is different in the two experiments. In Fig. 9.5 the alloy spectrum is reconstructed from the (shifted) valence bands of the pure elements. From this superposition, the *narrowing* of the Ni $3d$ band upon alloying becomes evident.

An example of a UPS study of the valence band spectra of the glassy Cu–Ti alloys over a wide range of composition, and of the pure crystalline metals, is given in Fig. 9.6 [9.23]. The two peaks in the alloy spectra related to the Cu and Ti $3d$ band clearly show less structure than the corresponding peaks in pure metals. In addition, the Cu d band shows a weak concentration dependence on the peak position. A similar shift of increasing binding energy with decreasing

---

2 In this context, the expression valence band always refers to the occupied part of the conduction band.

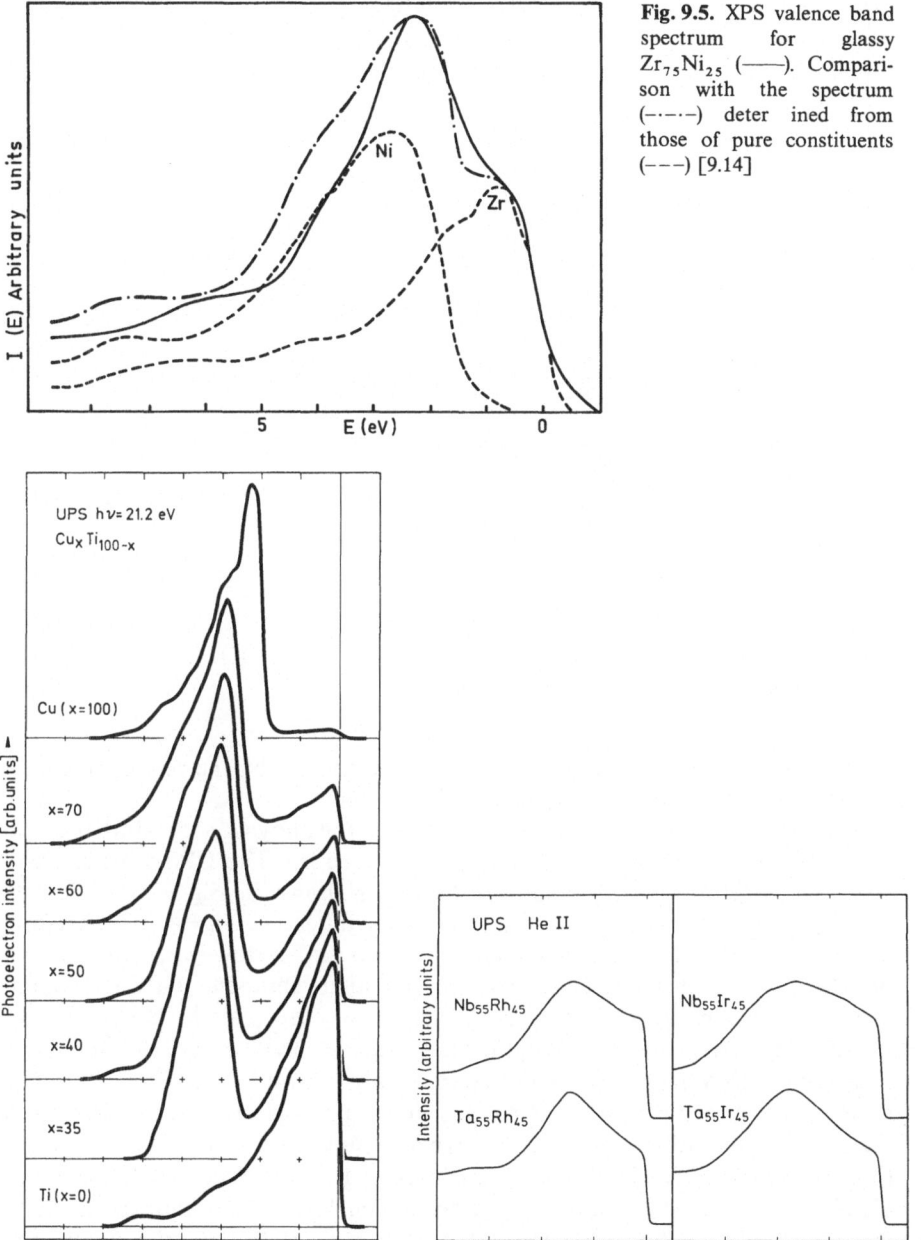

Fig. 9.5. XPS valence band spectrum for glassy $Zr_{75}Ni_{25}$ (——). Comparison with the spectrum (–·–·–) deter ined from those of pure constituents (–––) [9.14]

Fig. 9.6. Valence band spectra ($hv = 21.2$ eV) of glassy Cu–Ti alloys and polycrystalline Cu and Ti [9.23]

Fig. 9.7. Valence band spectra ($hv = 40.8$ eV) of glassy $Nb_{55}Ir_{45}$, $Ta_{55}Ir_{45}$, $Nb_{55}Rh_{45}$, and $Ta_{55}Rh_{45}$ alloys [8.28]

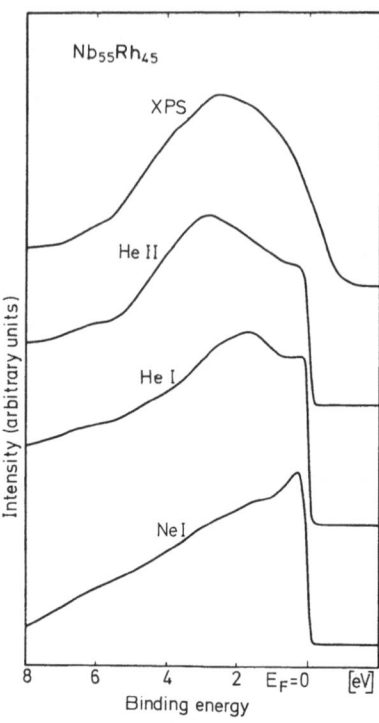

Fig. 9.8. Valence band photoelectron spectra of the metallic glass $Nb_{55}Rh_{45}$ measured with different excitation energies: XPS ($h\nu = 1253.6$ eV), He II (40.8 eV), He I (21.2 eV), and Ne I (16.8 eV) [8.28]

content of the late transition metal has been observed by the author in glassy Pd–Zr, Cu–Zr, Rh–Zr, Ni–Zr, and Fe–Zr alloys.

As mentioned above, the d-band splitting of the alloy valence band into two peaks depends strongly on the valence difference Δn. The UPS valence band spectra of a group with a common d band is shown in Fig. 9.7 [9.28]. For Nb–Ir, Ta–Ir, Nb–Rh, and Ta–Rh, the valence difference $\Delta n = 4$ and therefore, as seen in Fe–Zr, the d band is not split into two peaks. However, similar to the $3d$ alloys, the d states of the late transition metal are shifted to a higher binding energy. As a consequence, the local states density at F_F at the late transition metal site is strongly decreased in the alloy. Valence band spectra of $Nb_{55}Rh_{45}$, measured with different excitation energies, are shown in Fig. 9.8 [9.28]. The spectra clearly demonstrate the change in the photoelectron excitation cross section for the different d bands when going from the XPS spectrum ($h\nu = 1253.6$ eV) to the UPS Ne I spectrum ($h\nu = 16.8$ eV). Whereas the XPS and He II spectra are quite similar (note the much lower resolution in the XPS measurements), the lower photon energies emphasizes the electron states near E_F. The photoelectron spectra measured with higher photon energies ($h\nu > 40$ eV) are believed to be a better representation of the density of states than those of lower excitation energies because the latter are dominated by strongly energy-dependent matrix elements for optical transition and final states effects.

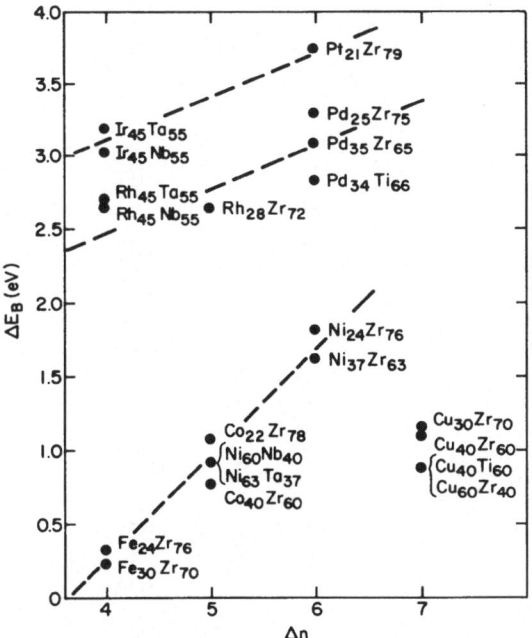

Fig. 9.9. The binding energy shifts ΔE_B of the d-band maximum of the late transition metals on alloying, plotted as a function of the valence difference (or group number difference) Δn. The dashed lines indicate that the experimental results fall into three groups: alloys with $3d$, $4d$ or $5d$ metals as the late transition metal

A summary of the d-band maximum binding energy shifts ΔE_B on alloying is given in Fig. 9.9. The values obtained from UPS measurements are plotted as a function of the valence difference. Three different groups have to be distinguished: alloys with $3d$, $4d$, and $5d$ elements as the late transition metal. Obviously, for all groups of the transition metal alloys studied here there is an *increase* in the *binding energy shift ΔE_B with increasing group number difference Δn*. Oelhafen et al. [9.15] have correlated the increase in the binding energy shifts with an increase in the (exothermal) alloy heat of formation [9.35] and in turn with an increase in the glass-forming ability. The latter observation has been confirmed by measurements of the critical cooling rates for obtaining the amorphous phase for Zr alloys with Fe, Co, Ni, and Cu [9.36]. Since recent measurements on glassy Zr alloys with Mn, Cr, and V ($\Delta n = 3$, 2, and 1, respectively) have been reported [9.37], an extension of the correlation between d-band shifts versus critical cooling temperature could be very interesting. The d-band shifts for these alloys are expected to be very small [9.38].

b) Core Electron Spectroscopy

The core electron binding energy shifts with respect to the pure alloy constituents of the most intense core lines and the changes in the line asymmetry α are given in Table 9.3 [9.17, 28]. Note that apart from some exceptions, *the binding energy shifts ΔE_B are relatively small* (below 0.5 eV). This indicates that only a small charge transfer occurs on alloying. This is in agreement with band structure calculations for ordered alloys [9.38] where it

Table 9.3. Core level binding energy shifts ΔE_B for glassy binary transition metal alloys with respect to the pure constituents. The asymmetry change $\Delta\alpha$ of the core lines on alloying are indicated by $+$ and $-$ signs for an increase and a decrease of the asymmetry, respectively [9.17, 28]

Alloy A–B	Core level A			Core level B		
	Level	ΔE_B [eV]	$\Delta\alpha$	Level	ΔE_B [eV]	$\Delta\alpha$
$Cu_{60}Zr_{40}$	Cu $2p_{3/2}$	0.33	$+$	Zr $3d_{5/2}$	0.20	$+$
$Ni_{37}Zr_{63}$	Ni $2p_{3/2}$	0.55	$-$	Zr $3d_{5/2}$	0.05	$+$
$Co_{40}Zr_{60}$	Co $2p_{3/2}$	0	$-$	Zr $3d_{5/2}$	0	$+$
$Fe_{24}Zr_{76}$	Fe $2p_{3/2}$	0.05	$-$	Zr $3d_{5/2}$	-0.10	$+$
$Pd_{35}Zr_{65}$	Pd $3d_{5/2}$	1.51	$-$	Zr $3d_{5/2}$	0.20	$+$
$Rh_{25}Zr_{75}$	Rh $3d_{5/2}$	0.17	$-$	Zr $3d_{5/2}$	0.14	$+$
$Pt_{21}Zr_{79}$	Pt $4f_{7/2}$	0.88	$-$	Zr $3d_{5/2}$	0.05	$+$
$Cu_{40}Ti_{60}$	Cu $2p_{3/2}$	0.30	$+$	Ti $2p_{3/2}$	0	$+$
$Ni_{60}Nb_{40}$	Ni $2p_{3/2}$	0.37	$-$	Nb $3d_{5/2}$	0.50	$-$
$Ni_{63}Ta_{37}$	Ni $2p_{3/2}$	0.48	$-$	Ta $4f_{7/2}$	0.80	$-$
$Rh_{45}Nb_{55}$	Rh $3d_{5/2}$	0.13	$-$	Nb $3d_{5/2}$	0.77	$-$
$Rh_{45}Ta_{55}$	Rh $3d_{5/2}$	0.27	$-$	Ta $4f_{7/2}$	0.75	$-$
$Ir_{45}Nb_{55}$	Ir $4f_{7/2}$	-0.27	$-$	Nb $3d_{5/2}$	0.86	$-$
$Ir_{45}Ta_{55}$	Ir $4f_{7/2}$	-0.17	$-$	Ta $4f_{7/2}$	0.83	$-$

was found that the number of d electrons remains essentially constant on alloying. This is also valid for the alloys with large d-band energy shifts such as Pd–Zr: the shift of the Pd d band to a higher binding energy does not imply a filled d band (containing essentially $10d$ electrons) of the late transition metal in the alloy. The *changes in asymmetry* α of the core lines of the elements A and B on alloying are included in Table 9.3. The $+$ and $-$ signs represent an increase and decrease in asymmetry, respectively. As mentioned in Sect. 9.2.2, the asymmetry α provides a local probe for the state density at the Fermi level. Except in the alloys with a noble metal, α is strongly decreasing for the late transition metal in all the studied alloys. This indicates a strong decrease in the d-electron state density at E_F at the T_L site. The increase of α at the Cu site indicates an increase in the local states density at E_F in these alloys. The asymmetry of the early transition metal core lines increases on alloying (except for Nb and Ta alloys), indicating an increase in the density of states at E_F at the Zr and Ti site. This increase is because in pure Zr and Ti, E_F is near a minimum of the density of states [9.39]. In Nb and Ta, E_F is near a maximum of the density of states [9.40] and therefore a decrease of α is observed on alloying.

9.3.2 Discussion

Band structure calculations, in general, make important contributions to the understanding of electron spectroscopy data. However, no calculations for structurally disordered alloys have been performed so far. As a consequence, comparison of the experimental data with theoretical results is restricted to

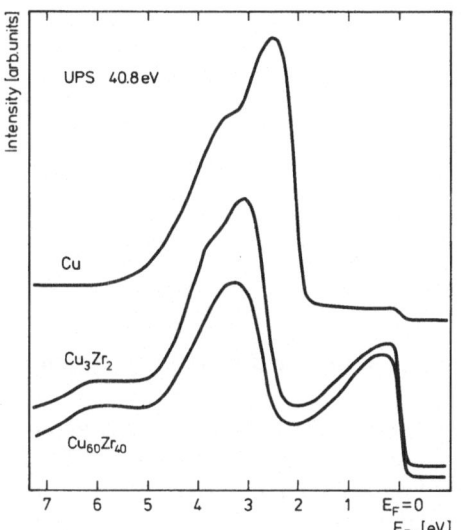

Fig. 9.10. Photoelectron valence band spectra of pure Cu, the crystalline compound Cu_3Zr_2 and glassy alloy $Cu_{60}Zr_{40}$ [9.19]

band structure calculations for *alloy clusters* (ordered or disordered) and *ordered alloys*. For the case of ordered alloys, the question arises as to how far the calculations are relevant to the experimental data obtained from amorphous alloys. Structural studies of different kinds, such as EXAFS [9.41], diffraction methods [9.42], and NMR [9.43] measurements, have clearly shown that a *structural* and *chemical* short-range order comparable to crystalline phases is present in the glassy phase. From this point of view, band structure calculations for the ordered alloys in the appropriate structure are expected to provide a good approximation to the electronic structure of the glassy alloy. In addition, the comparison of valence band spectra of both the glassy and crystalline phases indicate that the gross features of the electronic structure are the same. Therefore, band structure calculations for *ordered* alloys in an appropriate close-packed (e.g., fcc-like) model structure seem to be relevant to a large extent for the interpretation of measurements obtained on amorphous samples.

In Fig. 9.10, a comparison of the UPS spectra for the metallic glass $Cu_{60}Zr_{40}$ with the corresponding result for the crystalline compound Cu_3Zr_2 [9.19] shows that the d-band splitting and d-band binding energy shift is not a specific property of the glassy alloys but is also found in the crystalline phase. We find essentially the same Cu d-band peak positions in the crystalline and glassy states. However, the shape of the d-band is changed. Qualitatively, the same behavior has been observed in Pd–Si alloys [9.2]. In the crystalline compound Cu_3Zr_2, the Cu d-band exhibits covalent splitting which is typical of the pure Cu d-band spectrum, whereas in the glassy state the Cu d-band becomes more like a Gaussian. A second example is given in Fig. 9.11 where the valence bands of glassy $Ni_{25}Zr_{75}$ and crystalline Zr_2Ni are compared. Though

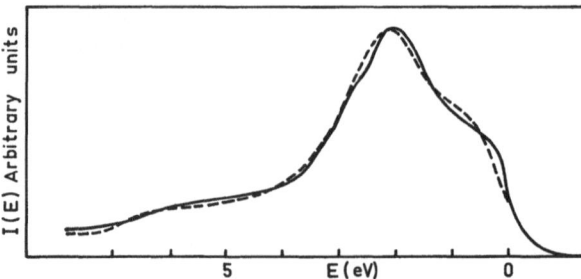

Fig. 9.11. X-ray photoelectron valence band spectra ($hv = 1486.6\,\text{eV}$) of glassy $Zr_{75}Ni_{25}$ (——) and crystalline Zr_2Ni (– – –) [8.14]

the two samples do not have the same composition, the *d*-band properties for the two alloys are essentially the same.

The strong similarity of the *d*-band positions in the crystalline and glassy state has an important consequence. Since the alloy heats of formation ΔH are mainly determined by the *d*-band properties, ΔH for the glassy state is only slightly different from the one for the crystalline state. This fact is supported by measurements of heats of crystallization which are small compared to the alloy heat of formation. This means that the glassy state lies energetically very close to the crystalline state.

Figures 9.12–17 show a comparison of UPS valence bands measured on glassy Zr alloys close to the 1 : 3 stoichiometry with calculated states densities [9.44]. The calculations are performed by the self-consistent ASW (augmented spherical wave) method [9.45] for the fcc-like $AuCu_3$ type symmetry. The calculated state densities shown here refer to the theoretical equilibrium lattice separation obtained by energy minimization. The only input to these calculations are the atomic numbers of the constituents and the crystal structure. The site decomposed densities of states for both alloy constituents are included in addition to the total density of states. From the sequence (Figs. 9.12–17), it is obvious that the UPS valence bands are very closely related to the theoretical density of states for the Cu_3Au structure.

The main results of the comparison of the experimental and theoretical results are summarized in the following points. (i) The measured *d-band peak position*, essentially related to the *d* states of the late transition metal, is in agreement with the calculated values within 0.2 eV. (In the case of $Co_{22}Zr_{78}$, the deviation is explainable by the slightly different alloy composition.) (ii) The calculations clearly show the *decrease of state densities* on the late transition metal site at E_F (with the exception of $CuZr_3$) and therefore, they explain the observed decrease in the core line asymmetry on alloying for the late transition metal. (iii) The electronic properties *at E_F* are *dominated by Zr* (mainly *d*) electron states. Even in $FeZr_3$ with an appreciable contribution from the Fe 3*d* states at E_F, the amount arising from the three Zr atoms dominates. (iv) The calculations show a *splitting of the Zr band itself*. This is clearly visible for $PdZr_3$, $RhZr_3$, and $CuZr_3$. This splitting into two peaks, which is not present in pure Zr, has been experimentally confirmed by x-ray emission spectroscopy by

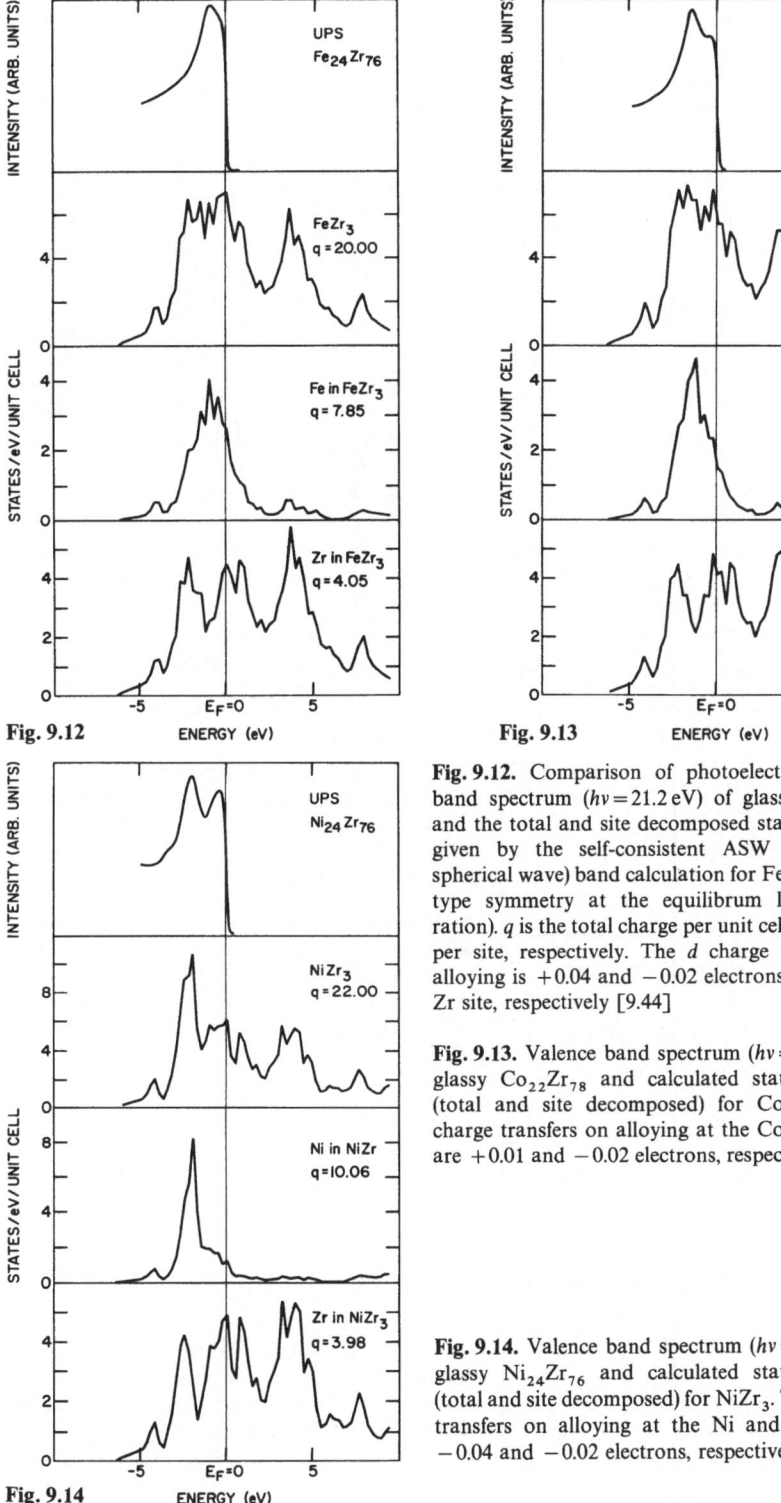

Fig. 9.12. Comparison of photoelectron valence band spectrum ($h\nu = 21.2$ eV) of glassy $Fe_{24}Zr_{76}$ and the total and site decomposed states densities given by the self-consistent ASW (augmented spherical wave) band calculation for $FeZr_3$ ($AuCu_3$ type symmetry at the equilibrium lattice separation). q is the total charge per unit cell and charge per site, respectively. The d charge transfers on alloying is $+0.04$ and -0.02 electrons per Fe and Zr site, respectively [9.44]

Fig. 9.13. Valence band spectrum ($h\nu = 21.2$ eV) of glassy $Co_{22}Zr_{78}$ and calculated states densities (total and site decomposed) for $CoZr_3$. The d charge transfers on alloying at the Co and Zr site are $+0.01$ and -0.02 electrons, respectively [9.44]

Fig. 9.14. Valence band spectrum ($h\nu = 21.2$ eV) of glassy $Ni_{24}Zr_{76}$ and calculated states densities (total and site decomposed) for $NiZr_3$. The d charge transfers on alloying at the Ni and Zr site are -0.04 and -0.02 electrons, respectively [9.44]

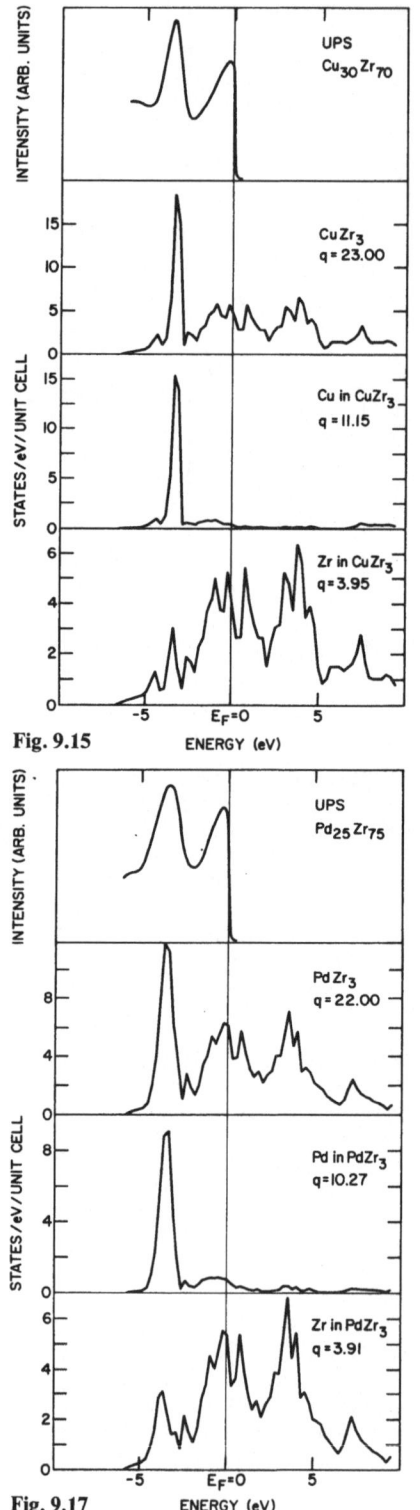

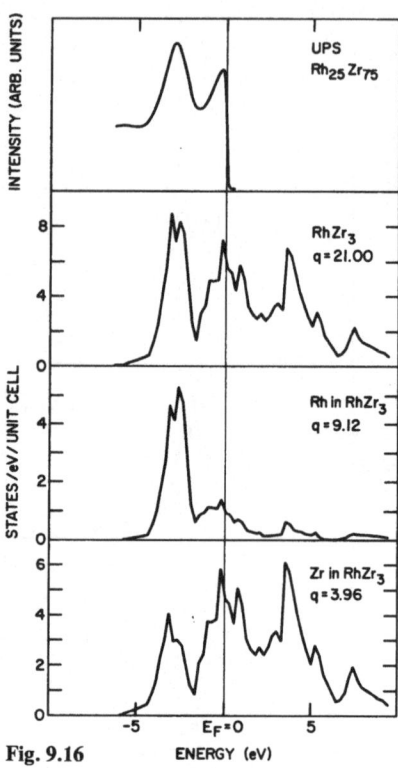

Fig. 9.15. Valence band spectrum ($h\nu = 21.2\,\text{eV}$) of glassy $Cu_{30}Zr_{70}$ and calculated states densities (total and site decomposed). The d charge transfers on alloying at the Cu and Zr site are -0.12 and 0.01 electrons, respectively [9.44]

Fig. 9.16. Valence band spectrum ($h\nu = 21.2\,\text{eV}$) of glassy $Rh_{25}Zr_{75}$ and calculated states densities (total and site decomposed). The d charge transfers on alloying at the Rh and Zr site are $+0.12$ and -0.07 electrons, respectively [9.44]

Fig. 9.17. Valence band spectrum ($h\nu = 21.2\,\text{eV}$) of glassy $Pd_{25}Zr_{75}$ and calculated states densities (total and site decomposed). The d charge transfers on alloying at the Pd and Zr site are -0.04 and -0.07 electrons, respectively [9.44]

Table 9.4. Superconducting transition temperature T_c of Zr-based glassy transition metal alloys

Glassy alloy	T_c [K]
$Fe_{30}Zr_{70}$	1.9[a]
$Co_{30}Zr_{70}$	3.3[b]
$Ni_{30}Zr_{70}$	2.9[b]
$Cu_{30}Zr_{70}$	2.8[c]
$Rh_{23}Zr_{77}$	4.1[d]
$Pd_{30}Zr_{70}$	2.7[e]

[a] [9.37].   [b] [9.54].   [c] [9.50].   [d] [9.56].   [e] [9.52].

Hague et al. [9.46] for $Pd_{30}Zr_{70}$, and by means of AES for Pd–Zr and Rh–Zr alloys by *Oelhafen* et al. [9.47]. (v) The photoionization cross section σ of the mainly *d*-band electron states of the late transition metal relative to that of Zr and for He I ($h\nu = 21.2$ eV) excitation is *decreasing* appreciably within the 3*d* series from Fe to Cu: in $Fe_{24}Zr_{76}$ the UPS spectrum is essentially determined by the Fe 3*d* states with almost no contribution from the Zr states. The 3*d* contribution relative to that of Zr 4*d* decreases in $Co_{22}Zr_{78}$ and $Ni_{24}Zr_{76}$ and in $Cu_{30}Zr_{70}$ the excitation cross section for Cu 3*d* and Zr 4*d* states are about equal. The fact of clearly defined contributions to the total density of states of both alloy constituents (as, e.g., in Cu–Zr, Pd–Zr, Ni–Zr alloys) makes it possible to obtain information about relative photoionization cross sections, not only for the different elements in a series at a given excitation energy, but also for a given alloy at *different excitation energies*.

The *superconductivity* in Zr-based metallic glasses with Cu, Ni, Co, Fe, Mn, Cr, V, Pd, and Rh has been discussed by *Tennhover* and *Johnson* [9.37] (see also [Ref. 9.48, Chap. 9]). The T_c values for some Zr-based glasses are given in Table 9.4. *Tennhover* and *Johnson* related the increase of T_c from $Cu_{30}Zr_{70}$ over $Ni_{30}Zr_{70}$ to $Co_{30}Zr_{70}$ to the increase in the density of states at the Fermi level $D(E_F)$. They found from a *Varma* and *Dynes* [9.49] type of analysis that, much like the case of crystalline superconductors, T_c is simply related to $D(E_F)$ and that T_c increases as $D(E_F)$ increases. The increase in $D(E_F)$ for the sequence Zr–(Cu, Ni, Co, Fe) could be established by calculating $D(E_F)$ from measured H_{c2} (T) values (the perpendicular upper critical field) and normal state resistivities. By using the photoelectron valence band spectra, the $D(E_F)$ behaviour was explained qualitatively in terms of the different binding energies of the *d* band related to the late transition metal. This picture has been confirmed by density-of-states calculations (Figs. 9.12–15); whereas the Zr 4*d* contribution at E_F is roughly constant for Zr alloys with Cu, Ni, Co, and Fe, the *d*-states density contribution from the late transition metal at E_F increases in going from $Cu_{30}Zr_{70}$ to $Co_{30}Zr_{70}$ and $Fe_{30}Zr_{70}$. *Tennhover* and *Johnson* [9.37] pointed out that the trend in T_c is broken for glassy Zr alloys with elements left of Co. They explained this behavior by the occurrence of spin fluctuations and

Table 9.5. Density of states at the Fermi level for Zr compounds. The values are obtained from self-consistent ASW band structure calculations for ordered compounds in the Cu_3Au and CuAu type symmetry at the theoretical lattice separation [9.38, 44]

Alloy	Density of states at E_F [States/eV·Atom]	
	Zr site	Total
Co_3Zr	0.40	1.17
Co Zr	1.23	1.15
Co Zr_3	1.43	1.50
Ni_3Zr	0.05	0.10
Ni Zr	1.22	0.94
Ni Zr_3	1.62	1.50
Cu_3Zr	2.26	1.07
Cu Zr	1.62	1.03
Cu Zr_3	1.43	1.23
Pd_3Zr	0.40	0.20
Pd Zr	1.35	0.88
Pd Zr_3	1.82	1.56

the formation of localized magnetic moments at the $3d$ element. Both of these effects act as pair-breaking mechanisms and therefore depress T_c.

The *concentration dependence of the T_c* values of Zr-based metallic glasses is another interesting observation and which is probably related to the states density at E_F. The *increase* in T_c with *increasing* Zr content is a common feature of the Cu–Zr [9.50], Ni–Zr [9.51], Pd–Zr [9.52], and Be–Zr [9.53] glasses[3]. The increase in the state density at E_F with increasing Zr content has been established by susceptibility measurements on Ni–Zr [9.51], Cu–Zr [9.50], and Pd–Zr [9.52], or measurements of the upper critical field $H_{c2}(T)$ [9.53]. In addition, the increase in the state densities at E_F in the above alloys is confirmed by band structure calculations of the ordered counterparts [9.38]. The state densities $D(E_F)$ at the Zr site and the average state density per atom for three different compositions of the alloys (in the CuAu and Cu_3Au structure) are given in Table 9.5. Though we do not claim that the density of states calculations for the ordered phase yield exact $D(E_F)$ values for the *amorphous* phase, we believe that the values given in Table 9.5 show the general trend of the concentration dependence of $D(E_F)$. Except for the Cu–Zr alloy, a distinct increase in $D(E_F)$ has always been found with increasing Zr content. In Cu_3Zr and CuZr, the Fermi level is located at very narrow peaks with a high density of states related to Zr $4d$ states. It is obvious that in the amorphous phase, these peaks are washed out and therefore the state density at E_F will be reduced. Nevertheless, the total density of states per atom is clearly higher in $CuZr_3$ than in CuZr.

---

3 The same observation has been made on amorphous Fe–Zr and Co–Zr films [9.54].

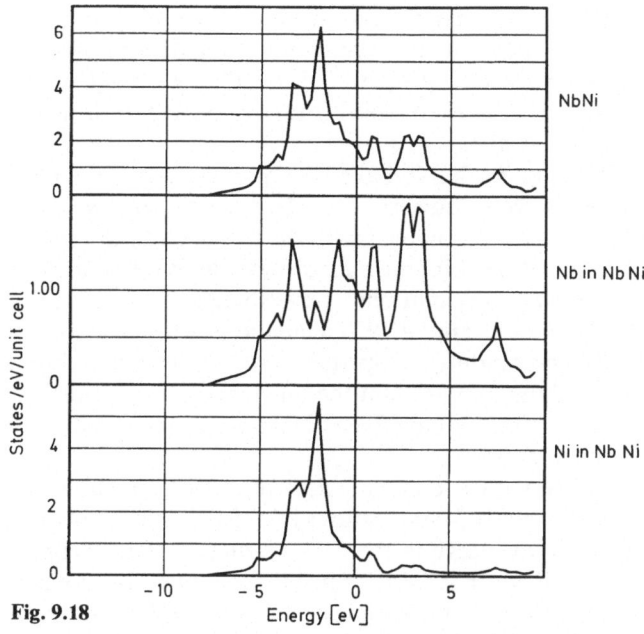

Fig. 9.18

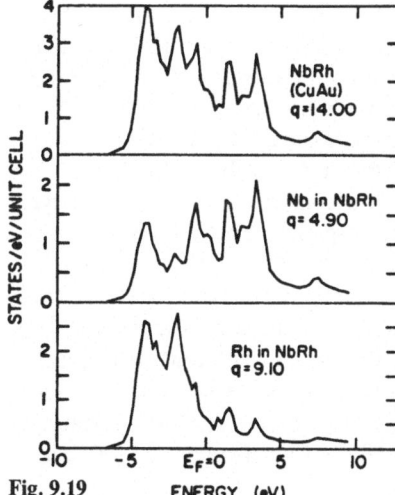

Fig. 9.19

Fig. 9.18. Total and site decomposed states densities given by the self-consistent ASW band calculation for NbNi in the CuAu type symmetry at the equilibrium lattice separation. The charges at the Nb and Ni sites are 4.98 and 10.02 electrons, respectively. The d charge transfers on alloying are -0.05 electrons on both the Nb and Ni site [9.38]

Fig. 9.19. Total and site decomposed states densities from ASW band calculations for NbRh in the CuAu type symmetry at the equilibrium lattice separation. q is the total charge per unit cell and the charge per site, respectively. The d charge transfers on alloying are -0.12 and $+0.07$ electrons at the Nb and Rh site, respectively [9.28]

To make a comparison with the experimental data obtained from $Ni_{60}Nb_{40}$ given in Fig. 9.4 (see also [9.3]), the ASW density-of-states calculation for the CuAu type symmetry is shown in Fig. 9.18 [9.38]. The UPS results obtained from glassy Ni–Nb alloys have been interpreted by *Nagel* et al. [9.3] in terms of the Nagel and Tauc model [9.1] for the valence band structure of a glassy alloy with a minimum in the density of states at E_F. To some extent, the DOS calculation shown in Fig. 9.18 supports the conclusion of *Nagel* et al. [9.3]. However, the position of the Fermi level in a minimum of the density of states (or at a low density) is not a general property of metallic glasses. For the Zr-alloys shown in Figs. 9.12–17, the Fermi level is rather located in or near a *maximum* of the density of states at the Zr site.

Density-of-states calculations for NiNb have been performed for the CsCl and NaCl type symmetry as well. However, the best agreement with the experimental Ni *d*-band position was obtained with the CuAu structure shown in Fig. 9.18 [9.38], since even bigger binding energies for the Ni *d*-band were obtained for the CsCl and NaCl calculations. Nevertheless, there is still a discrepancy between the UPS data and ASW calculations which is not yet understood: the UPS Ni *d*-band binding energy (measured at the maximum) is lower by about 0.5 eV after taking into account the difference in composition[4]. Similar deviations of the order of 0.3 eV have been observed in CuTi, CuZr, and PdZr alloys in the 1:1 stoichiometry.

A calculation for an alloy with a common *d*-band, NbRh, is shown in Fig. 9.19. Most of the experimentially found features (Fig. 9.8) can be explained by band structure calculations for NbRh in the CuAu structure. The occupied portion of the calculated *d*-band width is 5.4 eV, while the measured band width is 5.3 eV. The computed Rh *d*-band complex consists of two peaks located at 4.1 eV and 1.9 eV, while the measured Rh *d*-band complex is at 2.8 eV. The Nb *d*-band has two peaks located at 0.7 eV and at the bottom of the valence band. Comparison of the calculated site decomposed state densities at E_F with the state densities of the pure elements at E_F explains the observed change in core level line shape asymmetries (Table 9.3): 0.6 states/eV-atom for Rh in NbRh compared with 1.4 state/eV-atom for a fcc Rh metal and 1.15 state/eV-atom for Nb in NbRh, compared with 1.5 state/eV-atom for a bcc Nb metal. The calculated charge transfer from the Nb to Rh atoms on alloying is relatively small (0.1 electron per atom). The almost vanishing charge transfer is consistent with the small core level binding energy shift at the Rh site (Table 9.3).

Cluster calculations provide another approach to studying the electronic structure of alloys. In contrast to the ASW calculation, the cluster-type calculations make it possible to choose a great variety of different coordination numbers and any degree of chemical disorder. From this point of view, information about the *short-range order* in metallic glasses should be obtained by comparing cluster-type band calculations with electron spectroscopy measurements.

Cluster calculations have been performed for Cu–Zr by *Delley* et al. [9.58] and *Jaswal* et al. [9.59], for Pd–Zr by *Hague* et al. [9.46], and for Pd_cZr_{1-c}, Fe_cZr_{1-c}, Co_cZr_{1-c}, Ni_cZr_{1-c}, and Cu_cZr_{1-c} with $c \sim 0.3$ by *Fairley* et al. [9.60]. The latter calculations have been performed to study the significance of both the *positional* and *chemical order* in the Zr-based glasses. The results and

---

4 According to other measurements on glassy alloys [9.57], a Ni *d*-band shift proportional to $1-c$ was assumed, c being the Ni alloy concentration.

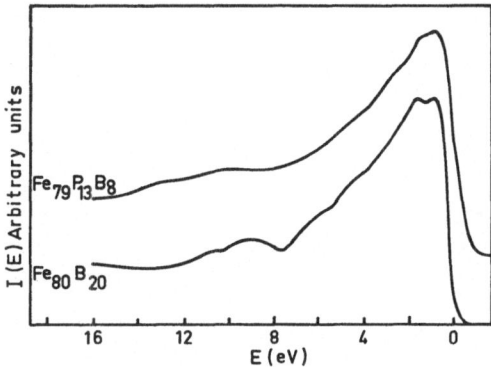

the comparison with experimental data (photoelectron spectroscopy and x-ray emission spectroscopy) suggest that the *chemical order* is the more important effect as far as the electronic states are concerned.

9.4 Glassy Alloys Containing Transition Metals and Normal Metals or Metalloids (T–N Alloys)

9.4.1 Alloys of 3d Transition Metals and Metalloids

Besides the glassy transition-transition metal alloys, most electron spectroscopy measurements have been performed in this class of alloys. The Fe, Co, and Ni alloys with B and P are of special interest since many glasses of this kind reveal interesting magnetic properties. In this section the results obtained from measurements on the Fe, Co, and Ni alloys are summarized.

a) Glassy Fe-Alloys

XPS valence band spectra of the glassy alloys $Fe_{80}B_{20}$ and $Fe_{79}P_{13}B_8$ are shown in Fig. 9.20 [9.13]. For comparison, the spectra of crystalline compounds Fe_3P and $Fe_3B_{0.9}P_{0.1}$ (assigned as "Fe_3B") and of pure Fe are shown in Fig. 9.21 [9.13]. The comparison of the compounds and glassy alloys show only minor differences in the valence band structure between the two phases. A strong similarity was found between the spectra of glassy $Fe_{80}B_{20}$ and the crystal $Fe_3B_{0.9}P_{0.1}$ ("$Fe_{80}B_{20}$"), and $Fe_{79}P_{13}B_8$ is related to Fe_3P in the same way. For example, these alloys reveal about the same *increase* in the Fe d-band width (FWHM) compared to pure Fe. A similar increase in the d-band width has been observed in crystalline FeB and Fe_2B compounds by *Joyner* et al. [9.61].

The total valence *band width* of the alloys shown in Figs. 9.20, 21 is strongly increased with respect to that of pure Fe: a width of about 17 eV was found in

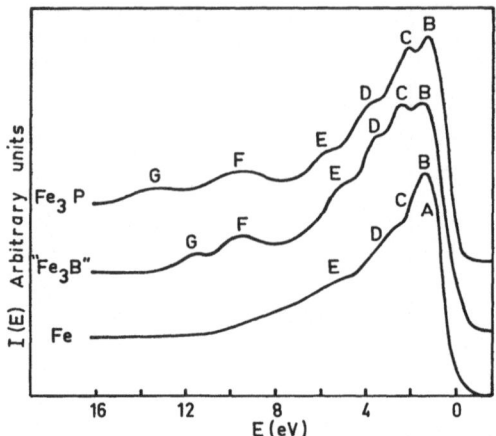

Fig. 9.21. XPS valence band spectra obtained with the Al K_α radiation for crystalline Fe, $Fe_3B_{0.9}P_{0.1}$ ("Fe$_3$B") and Fe_3P [9.13]

Fe_3P (and $Fe_{79}P_{13}B_8$) and 13 eV in Fe_3B (and $Fe_{80}B_{20}$). The structures F and G in Fig. 9.21 (which are also visible in the spectra of the corresponding glasses, Fig. 9.20) have been attributed to *metalloid s states*, and the structures E, D (and eventually C) to the mixing of the metalloid p and the Fe d states [9.13]. The *core levels* of Fe and the metalloids (Fe 3s, 3p; P 2p and B 1s) show a shift towards a lower binding energy and therefore a simple interpretation in terms of a charge transfer is obviously not applicable. From the shape of the Fermi edge the authors [9.13] concluded that, in contrast to the crystalline phases, E_F is located in *a high density-of-states* region and therefore they did not find any support for the Tauc and Nagel model. The high state density at E_F is confirmed by the calculations shown in Figs. 9.22, 23 and those performed by *Fujiwara* [9.62] on clusters containing 1500 atoms of the compositions $Fe_{84.9}P_{15.1}$ and $Fe_{84.9}B_{15.1}$. The results shown in Figs. 9.22, 23 have been obtained from a self-consistent ASW calculation for Fe_3B and Fe_3P in the Cu_3Au crystal structure [9.38]. Again, the close-packed Cu_3Au structure (fcc-like) is a good approximation both to the real crystal structure in the compounds and to short-range order in the glassy alloys since many features of the valence band spectra can be explained by these calculations. They confirm the origin of the structures at the bottom of the valence bands which are related to metalloid s states hybridized with Fe s and p states, and show a band width of about 12 and 15 eV for Fe_3B and Fe_3P, respectively. The structures observed in the Fe d-band of the compounds are probably related to the distinct maxima found in the calculated density of states in Fe_3B and Fe_3P. However, a conclusive interpretation is difficult since the Fermi level was not determined accurately in the valence band spectra of Figs. 9.20, 21.

An XPS study of glassy $Fe_{100-x}B_x$ alloys ($x = 12, 14, 16, 20, 25$) has been published by *Matsuura* et al. [9.12]. Similar to the results of *Amamou* and *Krill* [9.13], they found, compared to pure Fe, an additional peak in the valence band spectra of the glasses near 10 eV binding energy. They concluded that

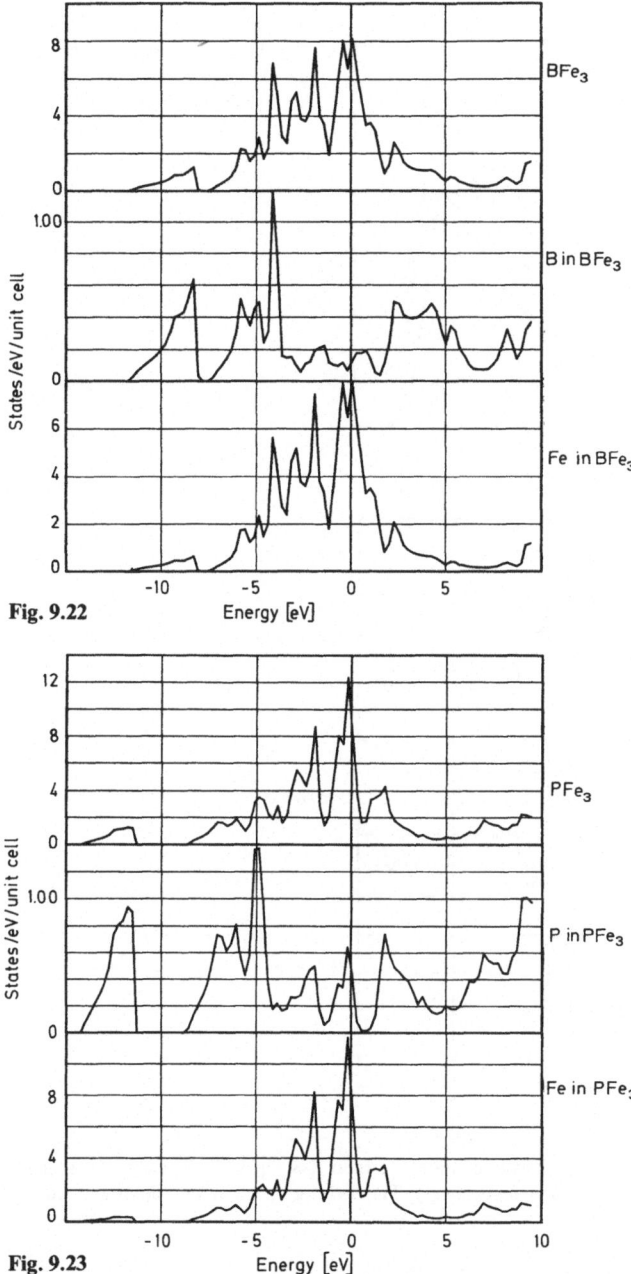

Fig. 9.22

Fig. 9.23

Fig. 9.22. Total and site decomposed density of states obtained from self-consistent ASW band calculation for BFe$_3$ in the AuCu$_3$ type symmetry at the theoretical equilibrium lattice separation. The charges at the B and Fe sites are 2.80 and 8.07 electrons, respectively. The d charge transfer on alloying at the Fe site is +0.08 electrons [9.38]

Fig. 9.23. Total and site decomposed density of states for PFe$_3$ (as Fig. 9.22). The charges at P and Fe sites are 5.27 and 7.91 electrons, respectively. The d charge transfer on alloying at the Fe site is +0.06 electrons [9.38]

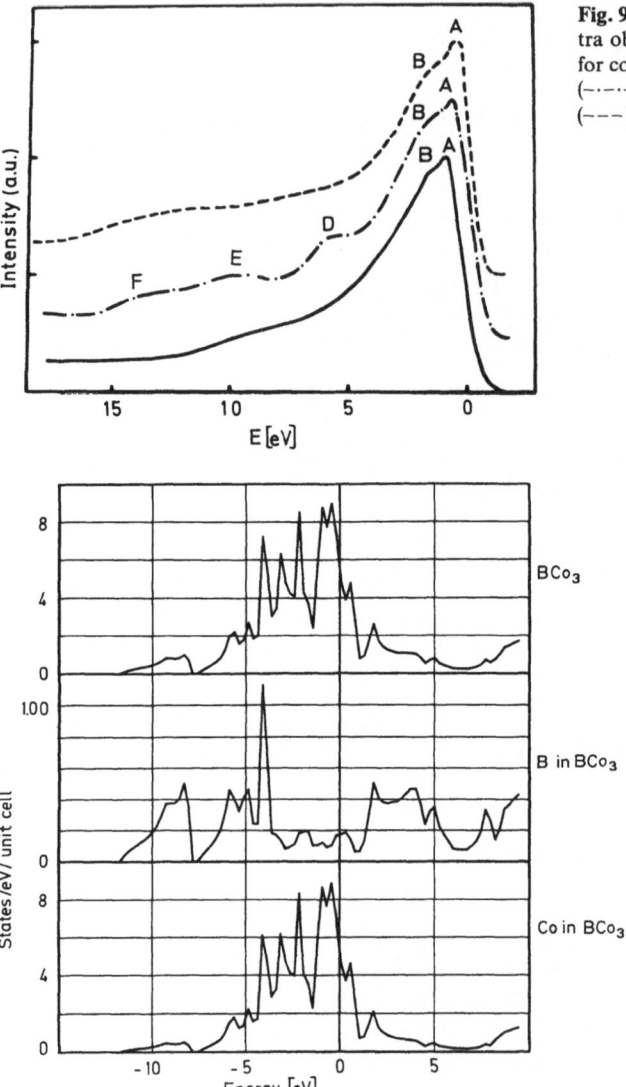

Fig. 9.24. XPS valence band spectra obtained with Al K_α radiation for cobalt (——), crystalline Co_3B (–·–·–) and glassy $Co_{78}P_{14}B_8$ (– – –) [9.9]

Fig. 9.25. Total and site decomposed density of states for BCo_3 (as Fig. 9.22). The charges at the B and Co sites are 2.70 and 9.10 electrons, respectively. The d charge transfer on alloying at the Co site is 0.07 electrons [9.38]

bonding states between B 2s and Fe 3d electrons are the most plausible explanation for this feature.

A high resolution XPS spectrum of $Fe_{80}B_{20}$ has been published by *Cartier* et al. [9.11]. However, the valence band has only been shown to a binding energy of about 7 eV and therefore the B 2s bonding states are not visible.

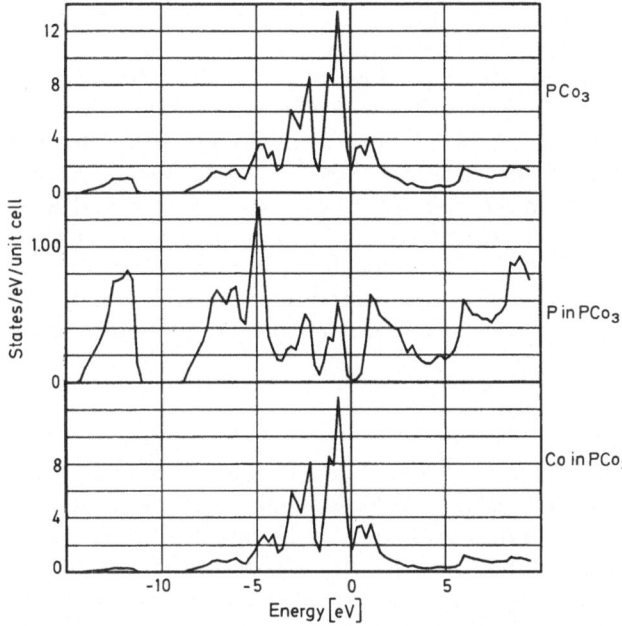

Fig. 9.26. Total and site decomposed density of states for PCo_3 (as Fig. 9.22). The charges at the P and Co sites are 5.11 and 8.96 electrons, respectively. The d charge transfer at the Co site on alloying is $+0.07$ electrons [9.38]

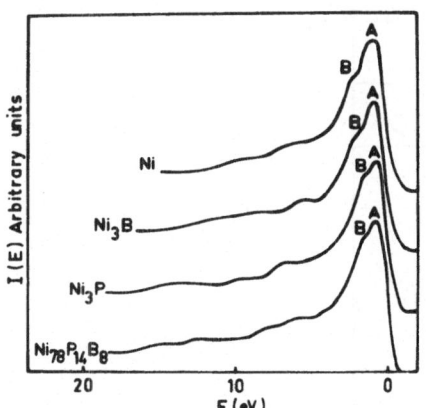

Fig. 27. XPS valence band spectra obtained with Al K_α radiation in crystalline and amorphous nickel phosphoborides [9.18]

b) Glassy Co-Alloys

The XPS valence band spectra for glassy $Co_{78}P_{14}B_8$, Co_3B and pure Co are shown in Fig. 9.24 (dotted, broken, and full line, respectively) [9.9]. The calculated (ASW) density of states for Co_3B and Co_3P in the Cu_3Au structure are presented in Figs. 9.25, 26. The experimental results (Fig. 9.24) obtained on the glassy $Co_{78}P_{14}B_8$ and Co_3B are qualitatively comparable to those obtained on the corresponding Fe alloy. For the band width, a value of 15 eV was obtained which is close to the calculated band width for Co_3P. The experimentally observed hybridization of metalloid s with Co s and p states is again confirmed by the density-of-states calculations. The Co *core level* binding energy shifts measured in Co_3B and the glassy alloy $Co_{78}P_{14}B_8$ was 0 ± 0.2 and -0.5 ± 0.2 eV, respectively. As in the case of the Fe-glasses, *Amamou* and *Krill* [9.9] concluded from their measurements that E_F is located in a *high density-of-states region* in $Co_{78}P_{14}B_8$. The calculation shown in Fig. 9.26 for the Co_3P compound does not confirm this observation, probably due to the different alloy composition. No binary Co–P glasses have been reported so far near the $3:1$ stoichiometry

c) Glassy Ni-Alloys

The experimental results for the glassy alloy $Ni_{78}P_{14}B_8$, the compounds Ni_3P, Ni_3B and pure Ni are shown in Fig. 9.27. Again, for comparison the calculated ASW density of states are shown in Figs. 9.28, 29 for Ni_3B and Ni_3P, respectively. The XPS valence band spectra of Fig. 9.27 show that the d-band width (FWHM) is the same in the four samples. Differences can be observed only in the lower part of the valence bands.

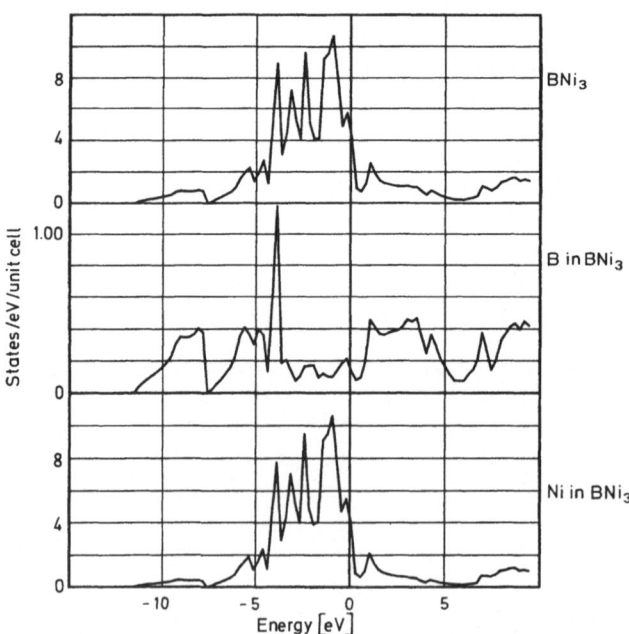

Fig. 9.28. Total and site decomposed density of states for BNi_3 (as Fig. 9.22). The charges at the B and Ni sites are 2.59 and 10.14 electrons, respectively. The d charge transfer on alloying at the Ni site is $+0.02$ electrons [9.38]

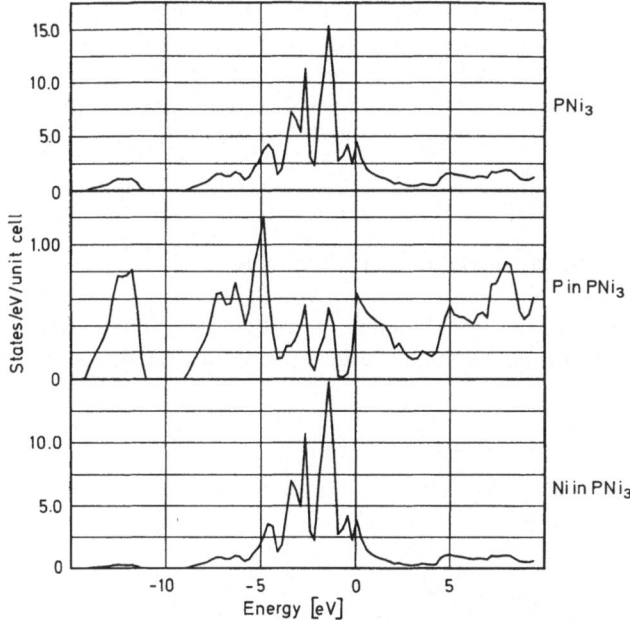

Fig. 9.29. Total and site decomposed density of states for PNi_3 (as Fig. 9.22). The charges at the P and Ni sites are 4.99 and 10.00 electrons, respectively. The d charge transfer on alloying at the Ni site is $+0.07$ electrons [9.38]

The experimental results obtained from glassy Fe, Co, and Ni alloys clearly show that in Fe–P–B alloys, the d-band is *strongly modified* compared to that of pure Fe; the Co–P–B alloys reveal only *minor differences* in the Co d-band and the differences in the Ni–P–B alloys from Ni are almost negligible. However, the same bond model is applicable for the alloy formation of all the Fe, Co, and Ni alloys with B and P: the experimental results clearly demonstrate, in agreement with band structure calculations [9.38, 60, 61] that an *ionic bond model*, with an appreciable charge transfer from metalloid to transition metal atom, is not applicable, but rather a *covalent bond* between the $3d$ metal atom and the metalloid. The P $3s$ or B $2s$) states hybridize with Fe s and p states at the bottom of the valence band (at about $10\,eV$) and, in addition, bonding states between P $3p$ (or B $2p$) and transition metal $3d$ states are formed.

An interesting application of AES as a local probe for the electronic valence band structure has been reported by *Bevolo* et al. [9.22]. They studied the Be KVV Auger transition in the ternary glasses $Fe_{82}B_{18-x}Be_x$ which show an anomalous peak in the average magnetic moment per Fe atom between $x=4$ and 5. The authors have found a $14\,eV$ shift of the Be KVV Auger transition at the same concentration x as the magnetic anomaly occurs. From their measurements, including the Be $1s$ binding energy, they concluded that the energy shift of Be KVV Auger transition on going from $x=4$ to $x=5$ must be due to a *change in the valence local density of states* of Be. They pointed out that because of the highly local nature of the Auger process, the large 14 eV energy shift should only occur if the nearest-neighbor environment of all of the Be atoms is drastically altered when x goes from 4 to 5. The main point in their work is the fact that the average magnetic moment per Fe atom shows an anomaly at the same composition.

9.4.2 Glassy Alloys of 4d Transition Metals and Metalloids

A group of glassy ternary 4d transition metal alloys with B, P, and Si have been studied extensively due to their interesting mechanical properties, hardness and superconductivity.

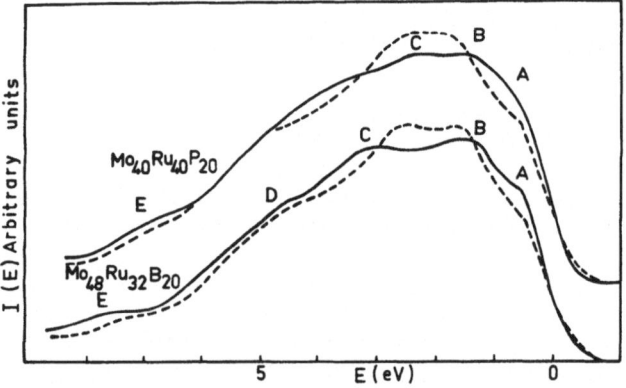

Fig. 9.30. Photoelectron valence band spectra obtained with the Al K_α radiation for glassy MoRuB and MoRuP ($-$). The dashed lines are valence band spectra for Mo–Ru alloys generated from those of pure Mo and Ru, as described in the text [9.16]

In glassy Mo–Ru alloys with B or P, a decrease in T_c due to the presence of the metalloid has been observed and attributed to a smaller density of states at the Fermi level $D(E_F)$ in the ternary alloys [9.63]. The valence bands of glassy $Mo_{48}Ru_{32}B_{20}$ and $Mo_{40}Ru_{40}P_{20}$ have been studied by photoelectron spectroscopy by *Amamou* and *Johnson* [9.16]. The XPS valence bands are shown in Fig. 9.30. From their measurements, the authors have concluded that the valence bands of the glassy alloys are very similar to that of a binary Mo–Ru alloy. For $Mo_{48}Ru_{32}B_{20}$, a total width of the valence band of 10 eV and a FWHM of 5.4 eV was reported [9.16]. As shown in Fig. 9.30, structure in the spectrum (assigned A–D) is clearly observed in the alloy with B and is less distinctly visible in the P alloy. The valence band spectra of the glassy alloys are compared with hypothetical $Mo_{1-c}Ru_c$ alloys (dashed curves in Fig. 9.30). The latter were obtained by a superposition of the valence band spectra of pure Mo and Ru, weighted according to the relative core level to valence band intensities in the glassy alloys and the pure metals, respectively. Figure 9.30 clearly shows the close similarity of the valence bands of the hypothetical $Mo_{1-c}Ru_c$ and the glassy alloys.

From the measurements it was concluded [9.16] that the Fermi edge is sharper for the glassy alloys than for the hypothetical compound. In $Mo_{48}Ru_{32}B_{20}$ the same structures, somewhat shifted, have been found whereas in $Mo_{40}Ru_{40}P_{20}$, these structures are washed out. This observation has been attributed to a stronger *disordering effect* in the P based alloy: Similar behavior has been observed in the glassy Fe–B and Fe–P alloys, respectively. From the photoemission results and NMR measurements [9.64], the authors concluded that the electronic structure of the investigated alloys can be interpreted within a covalent bonding model where the upper part of the valence band is related to *d* electrons. The metalloid states are either spread out over the valence band, or more likely, at the bottom of this valence band since the density of states of the metalloid *p* electrons is negligible at the Fermi level. From UPS results there is evidence that at the Fermi level, the density of states (mainly due to *d* electrons) is intermediate between that of Mo and Ru [9.16]. This is confirmed by susceptibility data [9.16, 65].

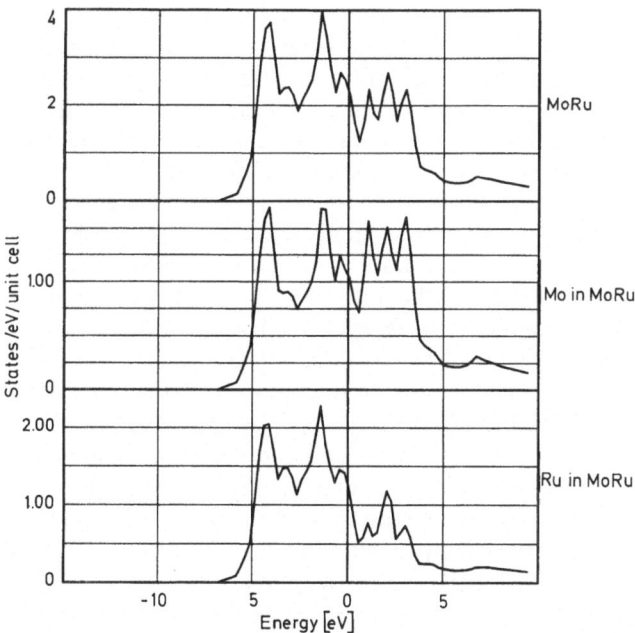

Fig. 9.31. Total and site decomposed density of states obtained from self-consistent ASW band calculation for MoRu in the CuAu type symmetry at the theoretical lattice separation. The charges at the Mo and Ru sites are 5.94 and 8.06 electrons, respectively. The *d* charge transfers on alloying at the Mo and Ru sites are -0.04 and $+0.08$ electrons, respectively [9.38]

In Fig. 9.31, the ASW density of states calculation for MoRu (in the CuAu) structure is shown [9.38]. The total and site decomposed state densities clearly show that the two alloy constituents form a *common d band*. The contributions from the two sites are *very similar* with respect to the bandwidth, band shape and states densities at E_F, which can be explained by the close positions of the two metals in the periodic table (valence difference $\Delta n = 2$) to the left and right side of technetium. Therefore, a similar contribution to the superconducting properties is expected from both Mo and Ru in a ternary alloy Mo–Ru–B (or P). It has been pointed out by *Amamou* and *Johnson* [9.16] that the decrease in T_c due to the presence of B and P in the glassy alloys with respect to crystalline Mo–Ru alloys does not seem to be attributable to a variation in the d state density at the Fermi level. However, a conclusive interpretation of the Ru concentration dependence of T_c in the Mo–Ru–B (P) alloys can hardly be given from valence band spectra of the ternary alloys alone. More information about the state densities at E_F could be obtained from a valence band and core level line shape comparison of the ternary glassy alloy and the binary crystalline Mo–Ru alloys with the pure elements.

9.4.3 Glassy Pd-Alloys with Si

The first study of photoelectron spectroscopy on a metallic glass was performed by *Nagel* et al. [9.2] on a $Pd_{77.5}Cu_6Si_{16.5}$ glass, which is a very easy glass former. From their UPS valence band spectra and core level XPS measurements on the glass and the crystallized sample, they concluded that (i) the electronic structure of the glass and the crystal are very similar, (ii) no evidence was found for strong chemical bonds (a Pd 3d core level shift of $+0.25$ eV was measured in the alloy with respect to pure Pd), and (iii) the UPS valence band spectrum of the $Pd_{77.5}Cu_6Si_{16.5}$ glass was interpreted as evidence for the nearly-free-electron model with the Fermi level in a minimum of the density of states for good glass formers. However, the authors have pointed out that the decrease in the density of states at E_F in the glass with respect to pure Pd is mainly due to the fact that the Fermi level is shifted above the Pd d band and therefore, it has been difficult to deduce the contribution of structural effects.

Riley et al. [9.6] have studied the glassy $Pd_{81}Si_{19}$ alloy by valence band and core level XPS, UPS, and AES. The XPS valence band spectrum is shown in Fig. 9.32. Two features are clearly visible by comparing the $Pd_{81}Si_{19}$ glass with the pure Pd spectrum: (i) the change in the Pd d band on alloying with Si, resulting in a strong decrease in the density of states at E_F, and (ii) an additional peak (E) at a binding energy of about 10 eV in the glass.

The concentration dependence of the glassy Pd–Si alloys have been studied by *Oelhafen* et al. [9.7] and *Waclawski* and *Boudreaux* [9.10]. The valence band spectra in the vicinity of E_F are shown in Fig. 9.33 for three different compositions. The UPS spectra are normalized to the same d band peak intensity. No measurable change was observed in the intensity at E_F. From

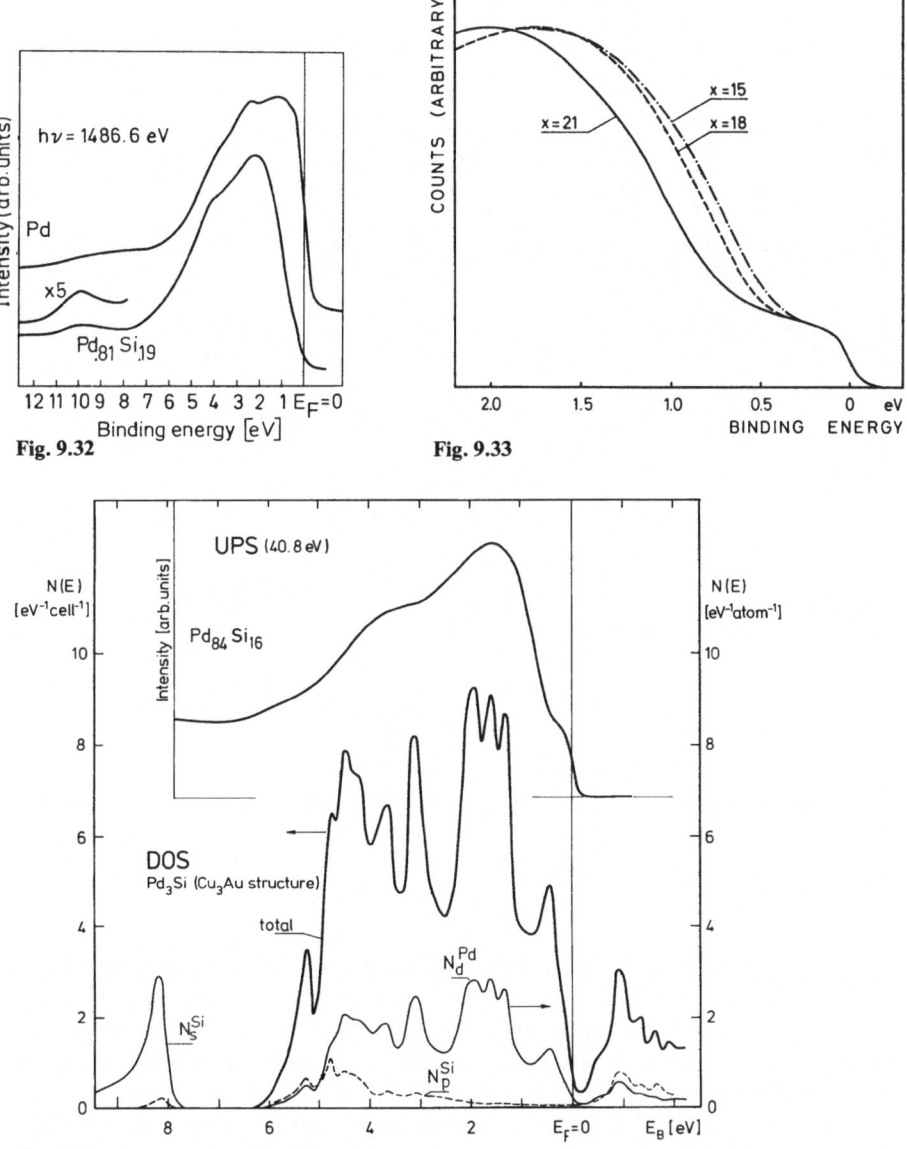

Fig. 9.32. XPS ($h\nu = 1486.6$ eV) valence band spectra of glassy $Pd_{81}Si_{19}$ and Pd [9.6]

Fig. 9.33. UPS (21.2 eV) spectra of three $Pd_{100-x}Si_x$ samples with different Si content. UPS ($h\nu = 21.2$ eV) spectra of three $Pd_{100-x}Si_x$ glasses near the Fermi level [9.7]

Fig. 9.34. The calculated total and partial density of states of Pd_3Si in the Cu_3Au structure and the UPS spectrum of glassy $Pd_{84}Si_{16}$ [9.19]

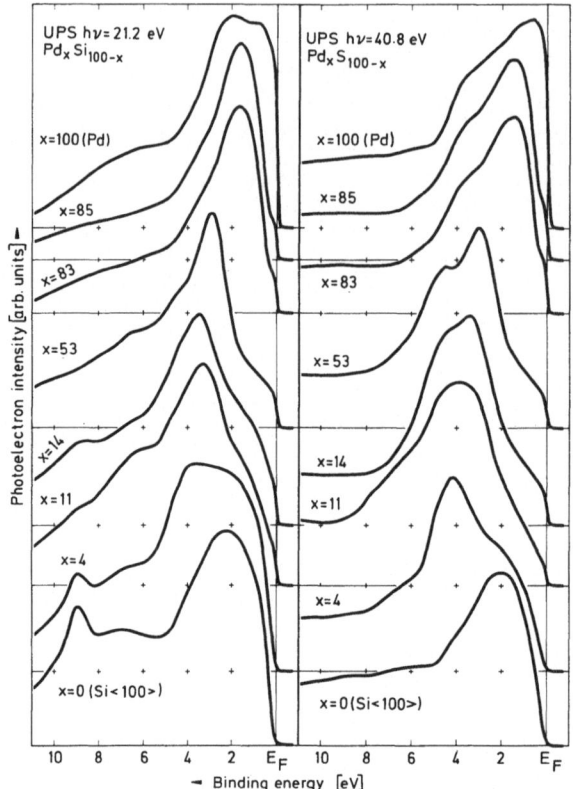

Fig. 9.35. UPS valence band spectra of glassy Pd_xSi_{100-x} and polycrystalline sputter-cleaned Pd and Si, measured with He I ($hv = 21.2$ eV) and He II (40.8 eV) excitation. The Pd–Si alloys have been prepared by laser glazing. The compositions of the sample surfaces have been determined from XPS core level intensities [9.67]

UPS measurements at different excitation energies [9.7], it was concluded that the Pd d states provide an appreciable contribution to the state density at E_F. This conclusion has been confirmed by ASW band structure calculations for Pd_3Si in the Cu_3Au structure by *Güntherodt* et al. [9.19] which are shown in Fig. 9.34. Though the calculation is performed for a higher Si content, the Pd d contribution at E_F is of the order of 0.4 states/eV-unit cell, compared with 0.05 states/eV-unit cell from Si p electron states. In alloys with a lower Si content, E_F is expected to be shifted to a lower energy (with respect to the valence band) and the contribution from the Pd site would even be increased. The density-of-states calculation confirms the local density-of-states picture derived from AES and PES valence band spectroscopy by *Riley* et al. [9.6]. The calculation shows an atomic-like Si s peak at 9 eV which was observed in the glassy $Pd_{81}Si_{19}$ near 10 eV. The general shape of the Si $3p$ contribution (Fig. 9.34) with a main contribution at a binding energy of about 5 eV is also in agreement with the experimentally determined local density of states at the Si site, except for the behavior at E_F where the calculation yields a deep minimum.

Laser glazing is a very promising technique for preparing metallic glasses [9.25, 66], especially in connection with electron spectroscopy measurements and in situ preparation work. Compared to other techniques (splat cooling, melt spinning), the higher cooling rates in laser glazing (of the order of 10^{10} K/s) makes it possible to considerably extend the range of concentration in which a glassy alloy can be obtained [9.25, 66]. The first UPS/XPS, AES measurements on Pd_xSi_{100-x} ($4 \leqq x \leqq 85$) in the glassy state prepared by laser glazing revealed the following features [9.25]: (i) at concentrations at which the glassy phase can be obtained by splat cooling ($78 \leqq x \leqq 85$), the spectra obtained from samples prepared by laser glazing and splat cooling are identical; (ii) the Pd d-band peak binding energy increases with decreasing Pd content monotonically up to a binding energy of 4.3 eV in Pd_4Si_{96} (the valence band spectra of a variety of Pd–Si alloys are shown in Fig. 9.35); (iii) the Pd $3d$ core level binding energy shift shows a maximum of about 2.1 eV at roughly $x \sim 20$; (iv) a transition from the glassy metallic alloy to a glassy semiconductor occurs at $x \sim 10$.

9.5 Other Groups of Metallic Glasses

There are other important groups of metallic glasses not discussed so far: the glassy alloys containing a rare earth and a transition metal (or normal metal), the R–T group, and the metallic glasses containing only normal metals – the N–N group. Almost no electron spectroscopy data have yet been published for these alloys. The main reason is the experimental difficulties in obtaining atomically clean, oxygen-free surfaces on the highly reactive samples.

An example of the valence band of glassy R–T alloys is shown in Fig. 9.36 [9.20]. The UPS spectrum of $Co_{33}Gd_{67}$ was obtained after prolonged sputter cleaning of the sample surface. Nevertheless, a small oxygen contribution at a binding energy of 6 eV was still visible. The valence band spectrum of $Co_{33}Gd_{67}$ shows that, with respect to the d band behavior, similar effects occur in R–T alloys as in the case of the T–T group: the Co d band maximum shifts in the glass with respect to pure Co by about 1 eV to higher binding energies. The distinct peak at a binding energy of 8.3 eV arises from the Gd $4f$ electron states. In pure Gd the $4f$ peak position is located at 8.0 eV [9.68].

The first UPS/XPS measurements on glassy Ca–Al alloys were performed by *Nagel* et al. [9.69]. The results obtained from these measurements and the comparison with a self-consistent ASW band structure calculation are interesting in two respects: (i) the valence band splits into two parts, separated by a gap of a few tenths of eV: one contains essentially Al $3s$ states and the other Al $3p$ states; (ii) the density of states decreases towards E_F and E_F is close to a minimum in the density of states. The first effect was unexpected since the

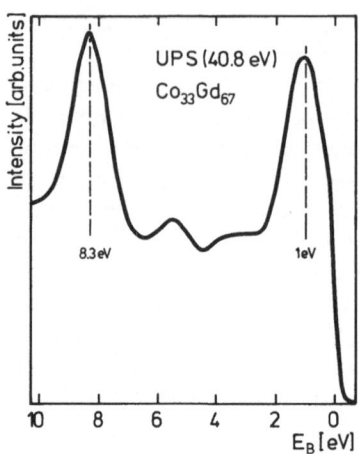

Fig. 9.36. UPS valence band spectrum of glassy $Co_{33}Gd_{67}$ [9.20]

splitting of the valence band into two parts upon alloying two simple free electron metals had never been observed. The latter effect (ii) supports the electronic model for glass formation as proposed by *Nagel* and *Tauc* [9.1].

a) Hydrogenated Metallic Glasses

UPS/XPS/AES measurements on the glassy hydride $(Pd_{30}Zr_{70})_{100-x}H_x$ $(x \sim 20)$ have been performed by *Oelhafen* et al. [9.25]. Although the hydrogen content was kept low, striking changes in the alloy surface properties have been observed after hydrogenation. First, the hydride shows a H-induced Pd segregation at the sample surface and second, the relative sputter yield for Zr and Pd from the glassy hydride is changed appreciably by the presence of H. Whereas the first effect could play a crucial role for the catalytic activity of transition metal alloy catalysts, the latter one is important for quantitative alloy surface studies like SIMS (secondary ion mass spectroscopy) [9.70] or photoelectron spectroscopy [9.31].

The hydrogen sorption kinetics of the metallic glass $Ni_{64}Zr_{36}$ has been studied by *Spit* et al. [9.27]. The surface state of the samples has been characterized by means of Rutherford backscattering, AES and XPS. Interesting correlations between the surface oxide layer and the hydrogen sorption kinetics were found and discussed in terms of Ni surface segregation in small clusters and cracks in the oxide layer.

9.6 Conclusion

The electronic structure of metallic glasses reveals one important feature: the valence bands show distinct bonding effects which are directly visible in the valence band spectra. In the case of transition metal glasses (T–T alloys), the covalent bonding results in a valence band splitting and in d-band shifts with respect to the Fermi level *and* to inner electron levels. The d-band splitting and the related bonding effects are clearly visible in alloys with constituents that have large valence differences and large alloy heats of formation, and which represent the alloys with a high glass forming ability. In the case of transition metal-metalloid alloys (T–N alloys), the transition metal and metalloids form covalent bonds by hybridization between energetically low-lying metalloid s and higher-lying p electron states with transition metal s, p, and d electron states, respectively. These observations clearly demonstrate the inadequacy of a ionic bond model in T–N metallic glasses.

An important question is the behavior of the density of states near the Fermi level, which is related to the stability model for metallic glasses of *Nagel* and *Tauc* [9.1]. In addition, the electronic properties at E_F account for the superconducting properties of many metallic glasses. The valence band spectra of glassy Pd–Cu–Si and Nb–Ni alloys have been interpreted in terms of the *Nagel-Tauc* model with a minimum in the states density at E_F [9.2, 3]. These alloys, in fact, have a much lower density of states at E_F compared with the pure majority component. However, in many other cases, photoelectron spectroscopy measurements have shown that E_F could be located in a maximum of the density of states as well, as in Zr rich glassy alloys with Fe, Co, Ni, Cu, Rh, Pd and in glassy Fe–B, Fe–P–B, and Co–P–B alloys. Density-of-states calculations have confirmed this observation.

Though the position of the Fermi level in a deep minimum of the density of states is indeed a stabilizing factor for an amorphous (or crystalline) phase, the electron spectroscopy measurements have shown that this is not a necessary condition for the occurrence of a glassy phase.

Acknowledgements. I would like to thank Professor H.-J. Güntherodt, Professor J. Kübler, Professor K. H. Bennemann, Professor S. R. Nagel, and Dr. A. R. Williams for valuable suggestions and stimulating discussions. I am very grateful to Dr. V. L. Moruzzi for having disclosed to me the results of the band structure calculations prior to publication. I would also like to thank the IBM Research Center, Yorktown Heights, for its hospitality while this work was performed. I am indebted to Claire Oelhafen, Dr. M. P. Vecchi, and Dr. J. L. Freeouf for preparing and careful reading of the manuscript. Financial support from the Swiss National Science Foundation, the "Kommission zur Förderung der wissenschaftlichen Forschung", the "Eidgenössische Stiftung zur Förderung Schweizerischer Volkswirtschaft", and the "Fonds für Lehre und Forschung" is gratefully acknowledged.

References

9.1   S. R. Nagel, J. Tauc: Phys. Rev. Lett. **35**, 380 (1975)
9.2   S. R. Nagel, G. B. Fisher, J. Tauc, B. G. Bagley: Phys. Rev. B **13**, 3284 (1976)
9.3   S. R. Nagel, J. Tauc, B. C. Giessen: Solid State Commun. **22**, 471 (1977)
9.4   L. Y. L. Shen, H. S. Chen, R. C. Dynes, J. P. Garno: J. Phys. Chem. Solids **39**, 33 (1978)
9.5   A. Amamou, G. Krill: Solid State Commun. **28**, 957 (1978)
9.6   J. D. Riley, L. Ley, J. Azoulay, K. Terakura: Phys. Rev. B **20**, 776 (1979)

322   P. Oelhafen

9.7   P.Oelhafen, M.Liard, H.-J.Güntherodt, K.Berresheim, H.D.Polaschegg: Solid State Commun. **30**, 641 (1979)
9.8   P.Oelhafen, E.Hauser, H.-J.Güntherodt, K.H.Bennemann: Phys. Rev. Lett. **43**, 1134 (1979)
9.9   A.Amamou, G.Krill: Solid State Commun. **31**, 971 (1979)
9.10 B.J.Waclawski, D.S.Boudreaux: Solid State Commun. **33**, 589 (1980)
9.11 E.Cartier, Y.Baer, M.Liard, H.-J.Güntherodt: J. Phys. F (Metal Phys.) **10**, L21 (1980)
9.12 M.Matsuura, T.Nomoto, F.Itoh, K.Suzuki: Solid State Commun. **33**, 895 (1980)
9.13 A.Amamou, G.Krill: Solid State Commun. **33**, 1087 (1980)
9.14 A.Amamou: Solid State Commun. **33**, 1029 (1980)
9.15 P.Oelhafen, E.Hauser, H.-J.Güntherodt: Solid State Commun. **35**, 1017 (1980)
9.16 A.Amamou, W.L.Johnson: Solid State Commun. **35**, 765 (1980)
9.17 P.Oelhafen, E.Hauser, H.-J.Güntherodt: In *Inner Shell and X-Ray Physics of Atoms and Solids*, eds. by D.J.Fabian, H.Kleinpoppen, L.M.Watson (Plenum Press, New York 1981) p. 575
9.18 A.Amamou, D.Aliaga-Guerra, P.Panissod, G.Krill, R.Kuentzler: J. Phys. Paris C 8, **41**, 396 (1980)
9.19 H.-J.Güntherodt, P.Oelhafen, R.Lapka, H.U.Künzi, G.Indlekofer, J.Krieg, T.Laubscher, H.Rudin, U.Gubler, F.Rösel, K.P.Ackermann, B.Delley, M.Fischer, F.Greuter, E.Hauser, M.Liard, M.Müller, J.Kübler, K.H.Bennemann, C.F.Hague: J. Physique C8, **41**, 381 (1980)
9.20 J.-J.Güntherodt, P.Oelhafen, E.Hauser, F.Greuter, R.Lapka, F.Rosel, R.Jacobs, J.d'Albuquerque e Castro, J.Kübler, C.F.Hague, K.H.Bennemann, R.H.Fairlie, W.M.Temmerman, B.L.Gyorffy: Inst. Phys. Conf. Ser. No. 55, Chap. 13, p. 619 (1981)
9.21 G.Petõ, J.Kanski: Solid State Commun. **38**, 377 (1981)
9.22 A.J.Bevolo, C.S.Severin, C.W.Chen: Phys. Rev. Lett. **47**, 733 (1981)
9.23 T.Mizoguchi, U.Gubler, P.Oelhafen, H.-J.Güntherodt, N.Akutsu, N.Watanabe: Proc. 4th Intern. Conf. on Rapidly Quenched Metals, eds. by T.Masumoto and K.Suzuki (1982) p. 1307
9.24 E.Colavita, M.De Gescenzi, L.Papagno, R.Scarmozzino, G.Chiarello, L.S.Caputi, R.Rosei: Proc. 4th Intern. Conf. on Rapidly Quenched Metals, ed. by T.Masumoto and K.Suzuki (1982) p. 1271
9.25 P.Oelhafen, R.Lapka, U.Gubler, J.Krieg, A.DasGupta, H.-J.Güntherodt, T.Mizoguchi, C.Hague, J.Kübler, S.R.Nagel: Proc. 4th Intern. Conf. on Rapidly Quenched Metals, ed. by T.Masumoto and K.Suzuki (1982) p. 1259
9.26 T.Honda, F.Itoh, K.Suzuki: Proc. 4th Intern. Conf. on Rapidly Quenched Metals, ed. by T.Masumoto and K.Suzuki (1982) p. 1303
9.27 F.Spit, K.Blok, E.Hendriks, G.Winkels, W.Turkenburg, J.W.Drijver, S.Radelaar: Proc. 4th Intern. Conf. on Rapidly Quenched Metals ed. by T.Masumoto and K.Suzuki (1982) p. 1635
9.28 A.DasGupta, P.Oelhafen, U.Gubler, R.Lapka, H.-J.Güntherodt, V.L.Moruzzi, A.R.Williams: Phys. Rev. B**25**, 2160 (1982)
9.29 V.V.Nemoshkalenco, V.G.Aleshin: *Electron Spectroscopy of Crystals* (Plenum Press, New York 1979)
9.30 H.Ibach (ed.): *Electron Spectroscopy of Surface Analysis*, Topics Current Phys., Vol. 4 (Springer, Berlin, Heidelberg, New York 1977)
9.31 M.Cardona, L.Ley (eds.): *Photoemission in Solids I and II*, Topics Appl. Phys., Vols. 26 and 27 (Springer, Berlin, Heidelberg, New York 1978)
9.32 N.J.Shevchik: Phys. Rev. Lett. **42**, 846 (1974)
9.33 N.J.Shevchik, D.Bloch: J. Phys. F (Metal Phys.) **7**, 543 (1977)
9.34 J.C.W.Folmer, D.K.G. de Boer: Solid State Commun. **38**, 1135 (1981)
9.35 P.Oelhafen: J. Phys. F (Metal Phys.) **11**, L41 (1981)
9.36 Y.Nishi, T.Morohoshi, M.Kawakami, K.Suzuki, T.Masumoto: Proc. 4th Intern. Conf. on Rapidly Quenched Metals, ed. by T.Masumoto and K.Suzuki (1982) p. 111
9.37 M.Tennhover, W.L.Johnson: Physica **108**B, 1221 (1981)
9.38 V.L.Moruzzi, C.D.Gelatt, Jr., A.R.Williams: Private communication
9.39 V.L.Moruzzi, J.F.Janak, A.R.Williams: In *Calculated Electronic Properties of Metals* (Pergamon Press, New York, 1978)

9.40 L.F.Mattheis: Phys. Rev. B1, 373 (1970)
9.41 D.Raoux, J.F.Sadoc, P.Lagarde, A.Sadoc, A.Fontaine: J. Physique C8, 41, 207 (1980)
9.42 P.Chieux, H.Ruppersberg: J. Physique C8, 41, 145 (1980)
9.43 P.Panissod, D.AliagaGuerra, A.Amamou, J.Durand: Phys. Rev. Lett. 44, 1465 (1980)
9.44 V.L.Moruzzi, P.Oelhafen, A.R.Williams, R.Lapka, H.-J.Güntherodt: Phys. Rev. B27, 2049 (1983)
9.45 A.R.Williams, J.Kübler, C.D.Gelatt, Jr.: Phys. Rev. B19, 6094 (1979)
9.46 C.F.Hague, R.H.Fairlie, W.M.Temmerman, B.L.Gyorffy, P.Oelhafen, H.-J.Güntherodt: J. Phys. F (Metal Phys.) 11, L95 (1981)
9.47 P.Oelhafen, F.Greuter, H.-J.Güntherodt, C.Hague, V.L.Moruzzi, A.R.Williams: To be published
9.48 H.J.Güntherodt, M.Beck (eds.): Glassy Metals I, Topics Appl. Phys., Vol. 46 (Springer, Berlin, Heidelberg, New York 1981)
9.49 C.M.Varma, R.C.Dynes: In Superconductivity in d- and f-band Metals, ed. by D.H.Douglass (Plenum, New York 1976)
9.50 Z.Altounian, TuGuo-hua, J.O.Strom-Olsen: Solid State Commun. 40, 221 (1981)
9.51 E.Babić, R.Ristić, M.Miljak, M.G.Scott, G.Gregan: Solid State Commun. 39, 139 (1981)
9.52 G.R.Gruzalski, J.A.Gerber, D.J.Sellmyer: Phys. Rev. B19, 3469 (1979)
9.53 R.Hasegawa, L.E.Tanner: Phys. Rev. B16, 3925 (1977)
9.54 H.J.Güntherodt: Private communication
9.55 O.Rapp, B.Lindberg, H.S.Chen, K.V.Rao: J. Less Common Metals 62, 221 (1978)
9.56 K.Togano, K.Tachikawa: Phys. Lett. 54A, 205, (1975)
9.57 J.Kübler, K.H.Bennemann, R.Lapka, F.Rösel, P.Oelhafen, H.-J.Güntherodt: Phys. Rev. B23, 5176 (1981)
9.58 B.Delley, D.E.Ellis, A.J.Freeman: J. Physique C8, 41, 437 (1980)
9.59 S.S.Jaswal, W.Y.Ching, D.J.Sellmyer, P.Edwardson: To be published
9.60 R.H.Fairlie, W.M.Temmerman, B.L.Gyorffy: To be published
9.61 D.J.Joyner, O.Johnson, D.M.Hercules, D.W.Bullett, J.H.Weaver: Phys. Rev. B24, 3132 (1981)
9.62 T.Fujiwara: In Fourth Intern. Conf. on Rapidly Quenched Metals, ed. by T.Masumoto and K.Suzuki (1982) p. 1267
9.63 W.L.Johnson: J. Physique, C8, 41, 731 (1980)
9.64 D.Aliaga Guerra, J.Durand, W.L.Johnson, P.Panissod: Solid State Commun. 31, 487 (1979)
9.65 W.L.Johnson, S.J.Poon, J.Durand, P.Duwez: Phys. Rev. B18, 206 (1978)
9.66 M.von Allmen, S.S.Lau, M.Mäepää, B.Y.Tsaur: Appl. Phys. Lett. 37, 84 (1980)
9.67 P.Oelhafen, R.Lapka, H.-J.Güntherodt, M.von Allmen: To be published
9.68 Y.Baer, G.Busch: J. Electron Spectroscopy and Related Phenomena 5, 611 (1974)
9.69 S.R.Nagel, U.Gubler, C.Hague, J.Krieg, R.Lapka, P.Oelhafen, H.-J.Güntherodt, J.Evers, A.Weiss, V.Moruzzi, A.Williams: Phys. Rev. Lett. 49, 575 (1982)
9.70 A.Benninghoven, C.A.Evans, Jr., R.A.Powell, R.Shimizu, H.A.Storms (eds.): Secondary Ion Mass Spectrometry SIMS II, Springer Ser. Chem. Phys., Vol. 9 (Springer, Berlin, Heidelberg, New York 1979)
       A.Benninghoven, J.Giber, J.László, M.Riedel, H.W.Werner (eds.): Secondary Ion Mass Spectrometry SIMS III, Springer Ser. Chem. Phys., Vol. 19 (Springer, Berlin, Heidelberg, New York 1982)

10. Low Temperature Electron Transport in Metallic Glasses

R. Harris and J. O. Strom-Olsen

With 7 Figures

Our aim in this chapter is to provide an account of the electron transport properties of metallic glasses below about 50 K. The principal focus will be the negative ln T dependence of resistivity found in many metallic glasses, usually below 15 K. Although such behavior is reminiscent of the Kondo effect, it was recognized from the outset [10.1–3] that there were many features of its appearance in metallic glasses which could not be explained by a straightforward application of the Kondo model. Specifically, the effect was generally not sensitive to the presence of magnetic order nor to the application of a magnetic field, quite unlike the Kondo effect in crystalline alloys. It therefore seemed reasonable that the amorphous structure played a significant role. Many authors have attempted to define this role and their ideas centre about two main models: one preserving the magnetic basis of the effect and invoking the structure only indirectly, the other insisting that the effect is entirely and directly structural in origin. To this moment we do not believe that the debate has been settled as to which (if either!) of these two approaches is correct; we shall attempt to present the evidence for both in as clear a way as possible and leave the reader to decide for himself. We shall discuss other contributions to the low temperature resistivity and shall also extend our review to cover thermopower, Hall effect and magnetoresistance. Our treatment of these last three will be necessarily brief, partly because the available material is so slight and partly because they show few characteristics unique to low temperature.

10.1 The Electrical Resistivity

10.1.1 General Behavior

As outlined above, the electrical resistivity of many metallic glasses shows a negative logarithmic dependence on temperature below, typically, 15 K. At higher temperatures, say above half the Debye temperature, the resistivity may have either a positive or a negative temperature coefficient, α, as reviewed by *Cote* and *Meisel* [10.4], but there appears to be no particular correlation between the existence of a logarithmic region and the sign of α.

For example, Fig. 10.1 shows data for an alloy system showing logarithmic regions with both positive and negative α. Figure 10.2, a counter example,

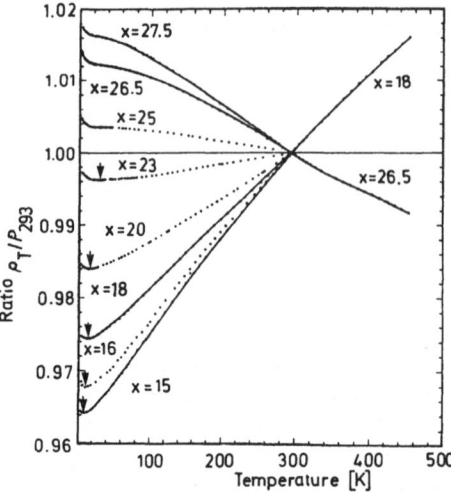

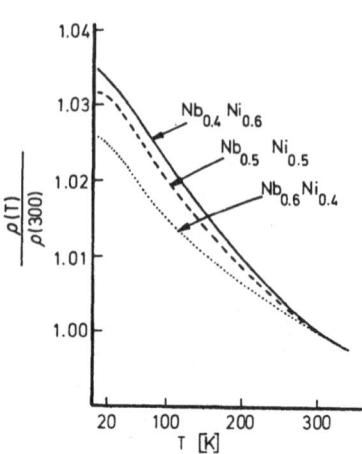

Fig. 10.1. The temperature dependence of the resistivity ratio ϱ_T/ϱ_{293} in amorphous $(Ni_{0.5}Pd_{0.5})_{1-x}P_x$ from [10.5]. The existence of a logarithmic region in the same ternary alloy system, but at slightly different concentrations, is clearly displayed in [10.2]

Fig. 10.2. The temperature dependence of the resistivity ratio ϱ_T/ϱ_{300} in three amorphous alloys of NbNi, from [10.29]

shows negative α with no logarithmic behavior. Examples of no logarithm with positive α are quite hard to find – an ironic result, since such behavior represents the norm for crystalline metals. We give one example in Fig. 10.3.

Although there is no hard and fast rule, generally metallic glasses based on 3-*d* transition metals and certain rare-earth systems [10.1–3, 5, 8–28] show behavior resembling that shown in Fig. 10.1, whereas those glasses based on nontransition metals [10.17, 29–32] tend to resemble the behavior shown in Fig. 10.2. Otherwise, there seems to be no systematic way to characterize the mass of data in the literature: a survey was given by *Cochrane* [10.33] in 1978 and some more recent contributions can be found in topical conference proceedings such as those of the 1980 Liquid and Amorphous Metals Conference [10.34].[1]

If we now focus more closely on those systems showing a ln *T* region, we find that, in fact, this is only an approximate description: at low temperatures the resistivity saturates and the data can usually be better approximated by an

---

1 Most samples have been made by rapid quenching from the melt, either by the piston and anvil technique or by so-called "melt spinning". A few were made either by sputtering, by evaporation onto a cold substrate or by chemical deposition. So far as we can determine, there is to date no evidence for any difference in the general characteristics of the resistivity of samples made by different techniques, although there are probably important differences in other physical quantities. In this review, therefore, we shall no longer concern ourselves with the technique of manufacture.

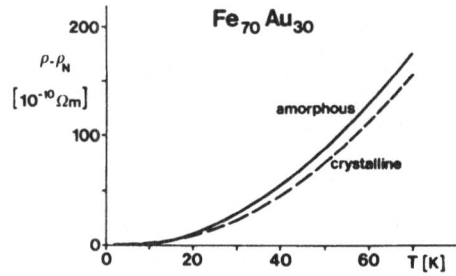

Fig. 10.3. The temperature dependence of the resistivity of $Fe_{70}Au_{30}$ in both the amorphous and crystalline states [10.7]

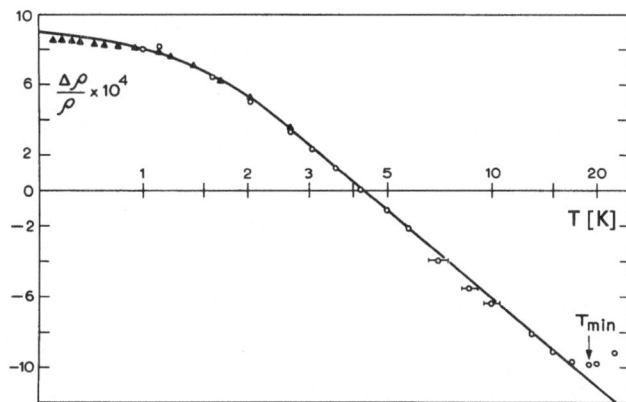

Fig. 10.4. The low temperature resistivity of Metglas 2826 ($Fe_{40}Ni_{40}P_{14}B_6$) [10.18]. The solid curve is a fit to $-\ln(T^2 + \Delta^2)$ with $\Delta = 0.6\,K$

expression of the form $\ln(T^2 + T_0^2)^{1/2}$, with T_0 typically of order 1 K [10.18]. The use of such an expression – despite its implications for the models which will be discussed below – is heuristic: other expressions undoubtedly can be found which represent the data well. A good example of the saturation and the above behavior is shown in Fig. 10.4; there are many other examples in the literature. Samples which also have positive α at high temperature must have a resistance minimum; the temperature of this minimum is typically 20 K but some systems, most notably those containing Cr and Mn [10.8], have minima at much higher temperatures, in some cases above room temperature.

It is important to recognize that in contrast to statements in the literature [10.35], the effect is not particularly small. It appears to be small only because the background resistivity is so high. The data in Fig. 10.5 show that the change in resistivity compares with that for around 0.2 at. %. Mn in Zn, which is an entirely typical Kondo alloy [10.36].

The observation of $\ln T$ behavior by itself, of course, would attract little attention since, as mentioned, it merely resembles the Kondo effect, well established in crystalline alloys containing magnetic impurities [10.37, 38]. But what distinguishes metallic glasses is the fact that the effect is observed, inter alia, in systems which order ferromagnetically. In *crystalline* alloys, ferromagnetic ordering eliminates the $\ln T$ dependence [10.39] since it places the

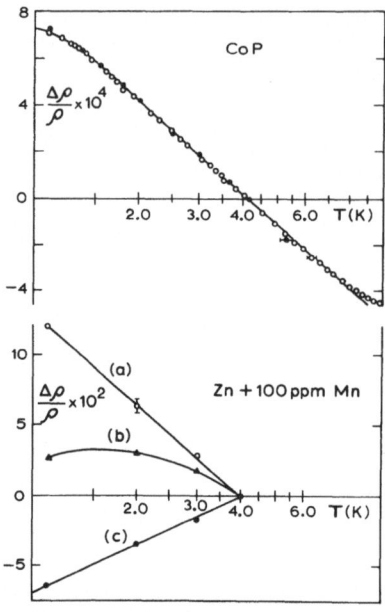

Fig. 10.5. *Upper curve*: the temperature dependence of the resistivity of amorphous $Co_{76}P_{24}$ in zero field (○○○) and in 45 kOe (4.5 T) (●●●) [10.18]. The solid curve is a fit to $-\ln(T^2 + \varDelta^2)$ with $\varDelta = 0.8$ K

Lower curve: the temperature dependence of the resistivity of Zn + 100 ppm Mn in zero field (○○○), 10 kOe (1 T) (△△△) and 45 kOe (4.5 T) (●●●) [10.18]. For a precise value of the concentration dependence of the ln T region in Zn Mn, see [10.27]

magnetic moments in a large internal field and so removes the local spin degrees of freedom necessary for Kondo scattering. The persistence of the effect in ferromagnetic metallic glasses, then, in some way reflects the amorphous structure.

10.1.2 Magnetic Models for the ln *T* Behavior

Since the ln T behavior existed in amorphous ferromagnets and since the collective spin degrees of freedom in a ferromagnet are magnons, the first models [10.40–42] tried to invoke magnon scattering, though without marked success. An anomalously large density of excitations at zero frequency and wave number was required and since this is essentially the same as requiring free spins, such models merely restated the Kondo problem without specifying the origin of the free spins.

The first attempt to provide independent and systematic evidence for the possible presence of free spins was given by *Sharon* and *Tsuei* [10.43] in a study of the Mössbauer effect in alloys of FePdP. They found that the Mössbauer spectrum they observed could not be described by a unique internal field, but instead required a distribution of internal fields $P(H)$. In fitting their results, they found that although their distribution was peaked at some high field H_0, it also had a long "tail" below H_0 which in fact extended all the way to zero field. Hence, they had apparently predicted the existence of a number of magnetic moments experiencing no internal field which could thus give rise to orthodox Kondo scattering.

Although this approach provided a plausible framework within which to understand the low temperature resistivity data, it is open to criticism. To begin with, the value of $P(H)$ deduced at fields well below H_0 is very unreliable since it comes from regions in the Mössbauer spectrum where the six lines overlap. By contrast, high field values of $P(H)$ are much better defined. In fact, later data by *Raj* et al. [10.44] on $Fe_{80}B_{20}$ were successfully fitted without a long tail to $P(H)$. The strongest inference one can therefore draw is that the existence of free spins is *not inconsistent* with the Mössbauer data. Even then, as already mentioned, typical data (Fig. 10.5) requires of order 0.2 at. % spins lying within fields of perhaps ± 1 kOe, placing a severe constraint on the magnitude of the tail of the fitted hyperfine field distribution.

A more fundamental criticism of models relying on a magnetic interaction (there being other models which do not depend upon the existence of free spins [10.45, 46]) is derived from the experimental observation [10.13, 18] that the existence and even the magnitude of the ln T region does not generally depend upon the existence of ferromagnetic order. Indeed, alloys such as $Fe_{80}B_{20}$, $Co_{80}P_{20}$, and $Ni_{76}P_{24}$ all show *identical* effects even though the first two are ferromagnetic and the last diamagnetic, or at the most, weakly paramagnetic. A further damaging observation [10.20, 22] is that in almost all systems, the ln T slope is totally independent of applied magnetic field[2] up to about 60 kOe, in strong contrast to a simple crystalline Kondo system such as ZnMn (Fig. 10.5). Such a result can be made consistent with a Kondo explanation, but only if a distribution of internal fields exists which is essentially independent of the field out to 60 kOe: such a distribution seems, at first sight, to be most unlikely.

Recently, however, *Grest* and *Nagel* [10.35] have proposed a model within which such a distribution occurs in a plausible manner. Their novel suggestion is that the glass formers (P, B, C, Si or even Al) present in many metallic glasses can provide a large antiferromagnetic superexchange component to the coupling between the spins. Using a Monte Carlo simulation calculation, they show that reasonable superexchange parameters, together with suitable direct and RKKY exchange, can produce an appropriate distribution of the effective field. They also show that their model can explain qualitatively the variation with Mn and Cr concentration of the temperature of the minimum of resistivity in FeMn B alloys [10.47].

The model, however, offers no independent evidence for the existence of superexchange and, indeed, cannot explain the ln T terms observed in Y Ni, SmCo and other metal-metal systems since the superexchange is inextricably linked to the presence of a glass former. Thus, no existing versions of the magnetic interaction model are free from criticism. Perhaps, since all such models are empirical in nature, this is not surprising. Nevertheless, any future versions must encompass equally systems which are either ferromagnetic or diamagnetic and which contain or do not contain glass formers.

---

2 Magnetic field dependence of the ln T dependence has been observed in some systems [10.27], particularly in those containing rare-earth components [10.21, 26].

10.1.3 Scattering from the Structure: Two Level Systems

The empirical nature of the magnetic models is shared by the alternative model proposed by *Cochrane* et al. [10.13]. These authors rejected any magnetic mechanism, preferring instead to link the effect directly to the structure of the amorphous state. In this way they hoped to circumvent the problems of the magnetic models. However, as will become evident, their model has its own special difficulties: in the final analysis these are, perhaps, no less severe than those described above. The model uses the so-called "two-level-systems" (TLS) as a source of electron scattering. As described in detail by *Black* [10.48], TLS provide a consistent explanation for many other anomalous low temperature properties. In the simplest picture (and computer simulation suggests [10.49] that such a picture is quite reasonable), the TLS are internal vibrational degrees of freedom which result from atoms free to tunnel between two alternative positions of local equilibrium. Such double wells result directly from the disordered structure and disappear on crystallization. Although the internal degrees of freedom provide scattering channels for the electron analogous to those provided by spin degrees of freedom, it is not immediately apparent that such channels give rise to a $\ln T$ resistivity. For this reason, and also because no survey parallel to the exhaustive surveys of the orthodox Kondo effect exists, we find it necessary to give a rather detailed discussion of the problem.

We start by writing the Hamiltonian for a TLS [10.48] as

$$H_{TLS} = \frac{1}{2}\begin{pmatrix} \Delta & -\Delta_0 \\ -\Delta_0 & \Delta \end{pmatrix}, \tag{10.1}$$

using basis states corresponding to the atom being either in one well or in the other. Δ is the energy difference between the bottoms of the two wells. Δ_0 is given by $\Delta_0 = \hbar\omega_0 \exp(-\lambda)$, with ω_0 the zero-point frequency. λ is related to the barrier height V via $\lambda = dh^{-1}\sqrt{2mV}$, with d the separation of the wells and m the mass of the tunneling atom. It will be convenient, later, to write this Hamiltonian in diagonalized pseudo-spin form

$$H_{TLS} = EI_z, \tag{10.2}$$

where E, the splitting between the two levels of the TLS, is given by $E^2 = \Delta^2 + \Delta_0^2$, and I_z is a spin operator having eigenvalues $\pm 1/2$ which correspond to the upper and lower levels.

The simplest treatment of the electrons is to assume that they occupy plane-wave states, as discussed below, but in the original paper, *Cochrane* et al. [10.13] realized that a TLS – plane-wave interaction could not give a logarithmic anomaly of the required form because such an interaction does not have the same *mathematical* structure as the usual Kondo interaction. The physical reason is that there is no analogue of the spin degree of freedom for the electrons: such an analogue – a degree of freedom conjugate to, but independent of, the internal degree of freedom of the TLS – has no meaning within a plane-wave model. *Cochrane* et al. therefore postulated that the electrons – perhaps corresponding to *d*-like wave functions localized around the atom – could distinguish between the two positions of the tunneling atom, so gaining an appropriate pseudo-spin label. The interaction Hamiltonian for one TLS thus becomes

$$H' = V_c \sum_{kk'} [(a^+_{k_+} a'_{k_+} - a^+_{k_-} a'_{k_-}) I_z + e^{-\lambda} (a^+_{k_+} a_{k'_-} I_+ + a^+_{k_-} a_{k'_+} I_-)] , \tag{10.3}$$

where + and − are the pseudo-spin labels and the pseudo-spin operators I_z and I_+ refer to the TLS. V_c is the potential scattering matrix element for processes without pseudo-spin flip, and $V_c \exp(-\lambda)$ is the matrix element for processes with pseudo-spin flip in which the scattering event simultaneously causes tunneling and a change of electron pseudo-spin label.[3]

In direct analogy with the usual (magnetic) Kondo effect, the resistivity caused by one TLS per unit volume then becomes [10.37, 38, 51]

$$\delta\varrho \simeq \frac{M}{\hbar e^2 Nn} (NV_c)^3 e^{-2\lambda} \ln(k_B^2 T^2 + E^2)/D^2 , \tag{10.4}$$

where N is the density of electron states per atom at the Fermi energy, n is the number of electrons per unit volume and D is the electron band width. Thus, the sign of the resistivity contribution corresponds to the sign of V_c: an attractive Coulomb interaction giving a resistivity increasing with decreasing temperature. The temperature dependence of the contribution is in accord with the data (Sect. 10.1.1) if E is of order 1 K.

To estimate the resistivity due to all the TLS in a sample, Cochrane et al. introduced a factor f as the fraction of all atoms which contribute to the resistivity so that $\varrho = N_0 f \delta\varrho$ where N_0 is the total number of atoms per unit volume. f was defined in terms of those TLS which tunnel fast enough that they can contribute within a reasonable measurement time $(E > E_{min})$ and yet which have barriers sufficiently wide that electron wave functions corresponding to the two positions are essentially orthogonal $(\lambda > \lambda_{min}$ or $E < E_{max})$. Thus, if the number of TLS per unit energy and per unit volume is $\bar{P}$, then $f = (E_{max} - E_{min})\bar{P}$. The original estimate that f be "a few percent" is questionable in view of the measured values of $\bar{P}$ which are between 10^{21} and 10^{22} eV^{-1} cm^{-3} and plausible values of $(E_{max} - E_{min})$ which might be perhaps 10 K; thus, as argued by *Black* et al. [10.52], the model in its original form becomes unrealistic.

---

3 The relationship of V_c with the matrix elements $V_\perp$ and $V_\parallel$ given by *Black* [10.48] will be given below (see p. 332). However, following *Zawadowski* [10.50], a matrix element of the form $V_c \exp(-\lambda)$ is readily seen to correspond to scattering processes involving tunneling. If fluctuations in the electron density can cause changes in the barrier through which the TLS can tunnel, then the changes in the tunneling frequency become $V_c \partial(\Delta_0)/\partial V$ where V, as before, is the barrier height. This expression is readily seen to be proportional to $V_c \exp(-\lambda)$.

A related difficulty is that in compounding the contributions from TLS with different splittings, the simple form $\ln(k_B^2 T^2 + E^2)$ with $E \simeq 1$ K is lost. An integral of the form

$$\int_{E_{min}}^{E_{max}} dE\bar{P} \ln(k_B^2 T^2 + E^2) \tag{10.5}$$

is required, and such an integral does not behave logarithmically with temperature unless $(E_{max} - E_{min})$ is very small. The resulting problems have been noted by several workers [10.25]. Perhaps a more fundamental problem is that the model in its original form cannot explain the absence of resistivity anomalies in alloys such as CuZr [10.17]. A qualitative argument can be advanced that the magnitude of the matrix element V_c would be largest for 3-d metals where the electron wave functions are most closely localized around the two positions of the TLS, but this argument has little predictive ability. We shall return to this point later.

Despite these objections, however, further work by *Zawadowski* and co-workers [10.53–55] suggests that the model may still be satisfactory. These workers started with a Hamiltonian suggested by *Kondo* [10.56, 57], which at first sight does not contain any pseudo-spin labels for the electron states. Kondo's Hamiltonian was originally written down with respect to the eigenstates of the TLS, but it is more convenient to write it as

$$H = H_{TLS} + \sum_k E_k a_k^+ a_k + \sum_{kk'} a_k^+ V_{kk'}^i a_{k'} \sigma^i, \quad i = x, y, z, \tag{10.6}$$

where the Pauli matrices σ^i refer to the two possible *positions* of the TLS. Thus, the matrix element $V_{kk'}^z$ is related to the V_c of *Cochrane* et al. [10.13] and $V_{kk'}^x$ and $V_{kk'}^y$ are related to $V_c \exp(-\lambda)$. The matrix elements $V_\parallel$ and $V_\perp$ of *Black* [10.46, 50] should not be confused with V^z and V^x. They arise from the matrix element V^z after a transformation to a representation diagonal in the TLS eigenstates.

In the so-called "commutative model" where the matrix elements have no explicit momentum dependence (or, equivalently, all have identical dependences on momentum), it can be shown that there are no terms in the resistivity proportional to $\ln(T/D)$ or $[\ln(T/D)]^2$ in either 3rd or 4th order perturbation theory [10.58]. Results showing small 4th order $[\ln(T/D)]^2$ terms [10.54, 59] have been shown to be incorrect.

A more realistic treatment of the Hamiltonian [10.50, 56b, 57] leads to explicit momentum dependences of the form

$$\begin{aligned} V_{k'k}^z &\simeq \tfrac{1}{2}i[(k-k')d]V_c, \\ V_{k'k}^x &\simeq [(k-k')d]^2 V_c e^{-\lambda}, \end{aligned} \tag{10.7}$$

where d is the separation of the two positions of the TLS so that V^z and V^x do not commute in momentum space:

$$V^x_{k'k}V^z_{kk''} - V^z_{k'k}V^x_{kk''} \neq 0. \tag{10.8}$$

It is conventional to take $V^y = 0$ because this simplifies the mathematics without changing the character of the Hamiltonian. In such a case, *Kondo* [10.56b] showed that use of the leading logarithmic approximation leads to a resistivity with a fourth-order logarithmic term proportional to $[\ln(T/D)]^2$. He concluded, using arguments much the same as those presented earlier for Cochrane et al.'s model, that such terms would be unobservable.

Zawadowski et al.'s main contribution [10.50, 53, 55] was to recognize that further progress could be made by writing the momentum dependence of the matrix elements in a different manner. Instead of using a free electron model for the electron wave functions, as did *Kondo* [10.56b] and *Black* et al. [10.52], they considered an expansion

$$V^i_{kk'} = \sum_{\alpha\beta} f^*_\alpha(k') V^i_{\alpha\beta} f_\beta(k), \tag{10.9}$$

where the f_α are a complete orthogonal set of functions. In a simplified picture where only 2 functions f_1 and f_2 are retained, *Zawadowski* [10.53] shows that the scaled Hamiltonian has exactly the form of the Kondo (spin) Hamiltonian for antiferromagnetic (and anisotropic) coupling.

One way to see this is to identify in the original Hamiltonian terms corresponding to the nonspin-flip and spin-flip parts of the conventional Kondo spin Hamiltonian. In other words, it is useful to show that the label α can serve as a pseudo-spin label in the spirit of the original analysis of *Cochrane* et al. [10.13]. It is convenient to choose tight-binding-like sets of electron wave functions of the form $f_\alpha(k) = e^{ikr}\chi(r - R_\alpha)$, $\alpha = 1, 2$, where R_1 and R_2 are the two possible positions of the tunneling atom and $\chi(r - R_1)$ and $\chi(r - R_2)$ are the orthogonal wave functions localized about these positions. In general, none of the coefficients $V^i_{\alpha\beta}$ will vanish, but by taking suitable linear combinations of the $\chi(r - R_\alpha)$ functions it will always be possible to reduce $V^z_{\alpha\beta}$ to diagonal form. (Intuitively, the choice of functions localized about R_1 and R_2 should ensure that $V^z_{\alpha\beta}$ is already diagonal.) Thus, V^z will contain a part transforming as $\sigma^z_{\alpha'\beta'}$. The functions which diagonalize $V^z_{\alpha\beta}$ will not, in general, diagonalize $V^x_{\alpha\beta}$ but the Hamiltonian will, nevertheless, contain terms $V^x_{\alpha'\beta'} \sim \sigma^x_{\alpha'\beta'}$ as stated by *Zawadowski* [10.50, 53].

Reverting to a spin-operator notation which is more natural for the spin Kondo situation, and defining I_z and $I_\pm$ for the TLS and S_z and $S_\pm$ (with α' and β' as basis states) for the electrons, the Hamiltonian thus contains terms proportional to both $V^z I_z S_z$ and $V^x(I_+ S_- + I_- S_+)$. Clearly, this is the usual Kondo Hamiltonian, where the electron pseudo-spin label refers to the states α' and β' localized on the alternate positions of the tunneling atom, much as anticipated by *Cochrane* et al. [10.13].

Thus, the scaling equations associated with the traditional Kondo problem [10.60] will lead to a rescaling of the interaction $V^x(I_+S_- + I_-S_+)$ towards a critical value given by $NV^x \sim 1/8$, but will leave V^z substantially unaffected. The critical value occurs at a (Kondo) temperature T_k given by [10.50, 55]

$$k_B T_k \simeq D \left(\frac{V^x}{4V^z} \right)^{1/4NV^z} N(V^x V^z)^{1/2},$$ (10.10)

below which the electron gas condenses around the TLS, neutralizing the dynamical aspects of·the scattering and thus, by analogy with earlier work [10.61], leading to a resistivity proportional to $\ln[k_B^2(T^2 + T_k^2)/D^2]$. Further, in a later paper, Vladar and Zawadowski [10.55] discussed the same simplified model to higher order in the scaling equations and showed that the splitting E is also renormalized. If, for a given TLS, $E(T_k) < k_B T_k$, then $E \rightarrow 0$ (or at least, $\Delta_0 \rightarrow 0$ for $\Delta = 0$) as $T \rightarrow 0$. This has the effect of defining a number of TLS, say $\bar{P} k_B T_k$, all of which contribute resistivities of the form given above and for which the maximum contribution to ϱ from the unitarity limit is

$$\varrho_{max} \simeq \frac{4}{\pi \hbar} \frac{m}{ne^2} v \frac{\bar{P} k_B T_k}{N},$$ (10.11)

where v is the volume per atom of the material.

These results clearly have an important bearing on the ability of the model to explain experimental data. Vladar and Zawadowski [10.55] demonstrated that values of T_k around 1 K are quite consistent with existing data, e.g., $T_k \sim 1$ K for $D \sim 10$ eV, $NV^x \sim 10^{-3}$ and $NV^z \sim 0.25$, and, therefore, that the unitarity limit contribution to the resistivity can be of order 10^{-1} μΩ cm. This result comes from the effective elimination of the $\exp^{(-2\lambda)}$ factor (or the S^2) in previous expressions for the resistivity. It is as if an electron – TLS "bound state" masks all fluctuations of electron density associated with the tunneling motion, so that the only relevant potential matrix element is V_c.

It also becomes possible to understand, in a natural way, how small values of V_c can lead to the absence of observable resistivity anomalies in some alloys. The expression for T_k depends critically on the exponent $1/NV^z$ so that a small value of $V^z(\sim V_c)$ leads directly to a small T_k. It is at least possible that such a situation may sometimes occur, as discussed earlier, in alloys not containing 3d elements so that such alloys will not show the resistivity anomaly.

These considerations go some way towards answering the criticisms of the tunneling model in its original form. However, only further theoretical and experimental studies can settle the question: particularly useful would be ultrasonic measurements on samples with well-defined resistivity anomalies. If these anomalies could be associated with corresponding changes in ultrasonic parameters, the model would be tested in a much more severe manner. In this way, together with renewed study of the magnetic interactions, there is some hope that a consistent description of the low temperature anomaly will soon emerge.

10.1.4 Discussion

The previous two sections have assembled the evidence for and against the two contrasting mechanisms for the logarithmic resistivity anomaly: in neither case can definite conclusions be drawn. There is, of course, the possibility that both mechanisms exist and even that both contribute simultaneously in the same alloys. In fact, there are certainly systems for which the tunneling state mechanism cannot be the sole contributor: these are the systems which show a field dependence to the low temperature resistivity. A good example of this is found in the work of *Kastner* et al. [10.27] who carried out a high-resolution study of a diamagnetic $Pd_{80}Si_{20}$ alloy which showed the characteristic resistivity anomaly below 5 K. They found (Fig. 10.6) that the anomaly had a small field dependence which, however, became largely saturated by 50 kOe, leaving a much larger field insensitive anomaly. The field sensitive part (caused by about 1–2 ppm Fe [10.62]) remained after crystallization, even to its magnitude and suppression by 50 kOe, while the field insensitive part disappeared. Less direct evidence comes from the measurement of *Asomoza* and co-workers on rare-earth nickel amorphous alloys [10.21, 26]. Here, there are clear magnetic effects superimposed upon an anomalous term similar to that observed earlier [10.23]. This term remains unchanged in fields of about 20 kOe and appears to be of nonmagnetic origin.

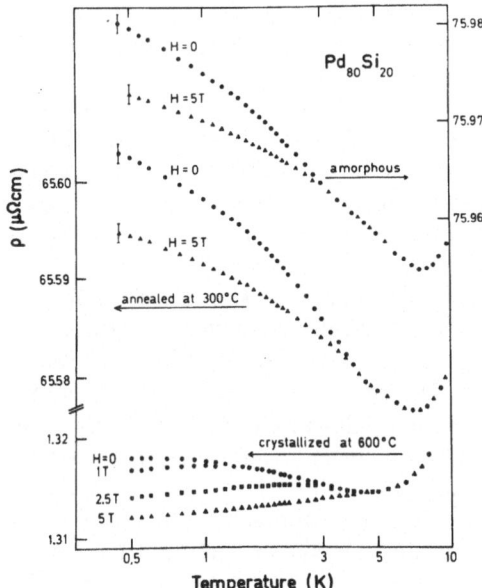

Fig. 10.6. The temperature dependence of the resistivity of $Pd_{80}Si_{20}$ in zero and up to 50 kOe (5 T) [10.27]

In principle, the appropriate description of the resistivity anomaly – be it magnetic, structural, both or neither in origin – will permit the identification of the "normal" behavior of amorphous metallic resistivity at low temperature. Such behavior has been characterized theoretically by a T^2 law arising from a modified Ziman model [10.4] but has been unambiguously supported by experimental evidence in only a few cases [10.7]. Since examples of positive $d\varrho/dT$ behavior *without* a low temperature anomaly seem hard to find (Sect. 10.2.1), much of the literature refers to data for which a logarithmic term has been subtracted off. Such a subtraction, at a practical level, often seems to indicate other than quadratic behavior [10.63] and, evidently, must remain adhoc until the anomalous contribution is better understood. The evidence for the validity of the Ziman model at low temperatures thus remains cloudy.

10.2 The Thermoelectric Power

By contrast with the plethora of data avilable for electrical resistance, the thermopower of metallic glasses has been studied relatively little. This is remarkable, because, although in principle it is more difficult to interpret than resistivity, thermopower is the more sensitive quantity since it depends upon the energy derivative of the scattering. Until recently not only had there been few measurements of thermopower but of these, few were systematic – most published articles before 1982 [10.10, 64–72] presenting measurements on one alloy only or else on a heterogeneous collection of different alloys. In the last twelve months however, a number of significant additions to the literature have appeared [10.73–78] and the situation is improving monthly. Nonetheless few of these studies focus on the low-temperature behavior, so our comments will not just be limited to that region.

   The first measurements were made on Ni–Pt–P alloys by *Sinha* [10.10] who reported a positive thermopower of about $2\,\mu$V/K at room temperature, linear in temperature but which did not extrapolate through the origin. This implies that the thermopower had to deviate from linearity at low temperatures since the third law of thermodynamics requires $S \rightarrow 0$ as $T \rightarrow 0$. Similar results on non-magnetic alloys have been reported by other authors [10.65, 66–78], and Fig. 10.7 shows some representative data. The characteristic "knee" at 50 K is clearly shown: it can also be seen in other published material, though not always so clearly, and its presence suggests that the low temperature thermopower is interesting in its own right. The thermopower of magnetic alloys [10.66–70] is typically quite different, being nonlinear and of variable sign and magnitude. Since a clear description is lacking even for non-magnetic systems, we shall say no more about magnetic alloys.

   As with high temperature resistivity, the main approach to the understanding of high-temperature thermopower in metallic glasses has been the Ziman model for liquid metals. The thermopower, or to give it its full name, the

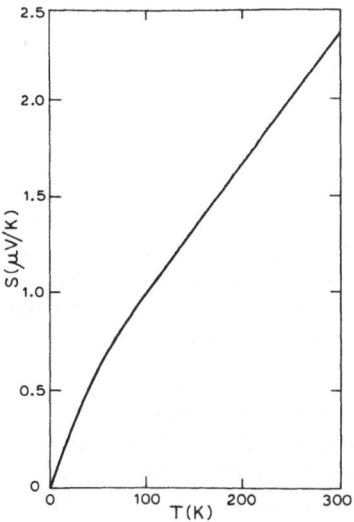

Fig. 10.7. The temperature dependence of the absolute thermoelectric power of amorphous $Cu_{30}Zr_{70}$ [10.73] (see also [10.71])

absolute thermoelectric power S, measures the electric field set up when electrons diffuse down a fixed thermal gradient. The magnitude of this diffusion is ultimately limited by electron scattering and the basic expression is given by [10.79, 80] for nearly free electrons as

$$S = -\frac{\pi^2}{3}\frac{k_B^2}{|e|E_F}\xi T,\tag{10.12}$$

where

$$\xi = \frac{3}{2} - \frac{\partial \ln[1/\tau(E)]}{\partial \ln E}\bigg|_{E=E_F}.$$

$\tau(E)$ is the energy dependent relaxation time and the other parameters take their usual meanings. When ξ is constant, S is linear in T, and since this is usually the case at high temperatures, the thermopower is often presented in terms of ξ.

In simple liquid metals, the scattering is most usefully expressed in terms of the structure factor $s(Q)$ and the scattering potential matrix element $V(Q)$ so that the parameter ξ becomes [10.81]

$$\xi = 3 - 2q - \tfrac{1}{2}r,$$

where

$$q = |V(2k_F)|^2 s(2k_F)/\langle|V^2|s\rangle$$
$$r = \langle k_F \partial|V|^2/\partial ks\rangle/\langle|V^2|s\rangle\tag{10.13}$$

and averages of the form $\langle x \rangle$ are defined by

$$\langle x \rangle = \int_0^1 x(q) 4(q/2k_F)^3 d(q/2k_F), \tag{10.14}$$

where k_F is the Fermi momentum.

When applied to liquid metals at high temperatures, these expressions use the static structure factor. However, at low temperatures, as exhaustively reviewed by *Cote* and *Meisel* [10.4] for resistivity, $s(Q)$ should be generalized to an average over the dynamical structure factor. The thermopower expressions can clearly be generalized in exactly the same way.

With alloys, complications arise because the integral over the structure factor must be replaced by a series of integrals over the partial structure factors for various components. In liquids this has led to the Faber-Ziman theory [10.81] which has met with some success in liquid metal alloys. Only in the last twelve months has a serious attempt been made to extend this to glasses. Previously *Sinha* [10.10] and later *Nagel* [10.65] attempted to use a simple averaged structure factor. Unfortunately this approach led *Carini* et al. [10.70] to predict a simple correlation between S and $d\varrho/dT$, an assertion in contradiction to the behavior of [10.75]. The deficiencies of this approach led *Baibich* et al. [10.75] and *Altounian* et al. [10.76] to use the Faber-Ziman theory with, in general, reasonable success.

Another correlation established in simple liquids [10.81] is a relation between the pressure dependence of the resistivity and the thermopower. The scanty material available on pressure dependence of the resistivity in amorphous alloys [10.82–84] prevents any convincing conclusions being drawn at this moment.

Since the gross features of thermopower are not yet fully understood, it may seem premature to comment on the "knee". However a few conclusions may be drawn. First, the short electron mean-free path precludes any explanation based on phonon drag [10.80]. Second, scattering from TLS cannot per se be the origin, since the Nordheim-Gorter rule [10.80] states that the contribution to S of any particular scattering mechanism is weighted by its contribution to ϱ, which for the TLS is of order 10^{-3}. One possible cause is the temperature dependence of the various partial structure factors entering the expression for S. However, a calculation along these lines for MgZn alloys [10.85] gives no indication of a knee.

A more recent and interesting suggestion [10.71, 86] is that the "knee" is caused by changes in the electron mass enhancement caused by phonons. Data on Zr-rich Cu–Zr and Ni–Zr, where the electron phonon coupling is particularly strong, appear to support this idea [10.87] and some further support comes from the analysis of *Altounian* and *Strom-Olsen* [10.88] in superconductivity in these, and other related, alloys.

10.3 Hall Effect and Magnetoresistance

Existing published data, in these areas [10.18, 20, 25, 26, 36, 89–94], is really too insubstantial to allow meaningful analysis. The majority of Hall effect studies have been carried out on magnetic systems which thus makes it difficult to extract characteristic behavior due to the amorphous structure per se. So far as can be determined there are no features yet observed which are unique to low temperatures. A similar scanty situation obtains for magnetoresistance. Those alloys for which the ln T resistivity is independent of field show, as a necessary corollary, a temperature independent magnetoresistance [10.18], which in most cases in quite unremarkable. The only unusual feature appears to be a small positive magnetoresistance associated with the magnetization in certain soft magnetic alloys [10.20, 89] for which microscopic explanation has yet been given [10.95]. Once again there appear to be no features unique to low temperatures.

10.4 Conclusions

We must conclude this survey by noting that our understanding of low-temperature electron transport in metallic glasses is still very unsatisfactory. No unified account of even so simple a property as resistivity exists, much less an account embracing all aspects of electron transport. We believe that the main reason for this state of affairs is the relative scarcity of systematic experimental studies on simple glasses. Even a cursory examination of the literature shows that the majority of glasses studied have been multicomponent magnetic systems (usually based on Fe, and of interest primarily for technological reasons) which, from the standpoint of fundamental interpretation, are just about the worst possible.

What is required is a number of different transport measurements, such as resistivity, thermopower, or pressure dependence, on a set of samples of different concentration within a given binary alloy system, which should itself be as simple as possible. Such data, especially when coupled with the related information obtainable from ultrasonic measurements [10.48] and from super-conductivity [10.88] would provide a proper testing ground for established ideas, such as the Ziman model [10.4], whose range of validity could then be correctly evaluated. Once this were done, it would become feasible to assess new and controversial ideas, such as the structural origin of the resistivity anomaly, or the role of the electron-phonon interaction in the temperature dependence of the thermo-electric power.

Acknowledgements. Conversations with many of our colleagues, especially those at McGill, have provided much of our insight into the topics treated in this review. Special thanks are also due to A. Zawadowski and C. L. Foiles, whose specific contributions have greatly improved our presentation. M. N. Baibich provided us with much valuable unpublished data. Our work was supported by the NSERC of Canada and the FCAC program of the Province of Quebec.

References

10.1   S.C.H.Lin: J. Appl. Phys. **40**, 2173 (1969)
10.2   P.Maitrepierre: J. Appl. Phys. **41**, 498 (1970)
10.3   C.C.Tsuei, R.Hasegawa: Solid State Commun. **7**, 1581 (1969)
10.4   P.J.Cote, L.V.Meisel: In *Glassy Metals I*, ed. by H-J.Güntherodt, H.Beck, Topics Appl. Phys. Vol. 46 (Springer, Berlin, Heidelberg, New York 1981) p. 141
10.5   B.Y.Boucher: J. Now-Cryst. Solids **7**, 277 (1972)
10.6   M.N.Baibich: Unpublished data
10.7   G.Bergmann, P.Marquardt: Phys. Rev. B**17**, 1355 (1978)
10.8   R.Hasegawa, C.C.Tsuei: Phys. Rev. B**2**, 1631 (1970)
10.9   R.Hasegawa, C.C.Tsuei: Phys. Rev. B**3**, 214 (1971)
10.10 A.K.Sinha: Phys. Rev. B**1**, 4541 (1970)
10.11 A.K.Sinha: J. Appl. Phys. **42**, 5184 (1971)
10.12 V.K.C.Liang, C.C.Tsuei: Solid State Commun. **9**, 579 (1971)
10.13 R.W.Cochrane, R.Harris, J.O.Strom-Olsen, M.J.Zuckermann: Phys. Rev. Lett. **35**, 676 (1975)
10.14 P.J.Cote: Solid State Commun. **18**, 1311 (1976)
10.15 J.Logan, M.Yung: J. Non-Cryst. Solids **21**, 151 (1976)
10.16 M.N.Baibich, R.W.Cochrane, W.B.Muir, J.O.Strom-Olsen: In *Amorphous Magnetism II*, ed. by R.A.Levy, R.Hasegawa (Plenum Press, New York 1977) p. 297
10.17 F.R.Szofran, G.R.Gruzalski, J.W.Weymouth, D.J.Sellmyer, B.C.Giessen: Phys. Rev. B**14**, 2160 (1976)
10.18 R.W.Cochrane, J.O.Strom-Olsen: J. Phys. F**7**, 1799 (1977)
10.19 O.Rapp, S.M.Bhagat, Ch. Johannesson: Solid State Commun. **21**, 83 (1977)
10.20 R.W.Cochrane, J.O.Strom-Olsen: Physica **86–88**B, 779 (1977)
10.21 R.Asomoza, A.Fert, I.A.Campbell, R.Meyer: J. Phys. F**7**, L 327 (1977)
10.22 S.J.Poon, J. Durand, M.Yung: Solid State Commun. **22**, 475 (1977)
10.23 R.W.Cochrane, J.O.Strom-Olsen, Gwyn Williams, A.Lienard, J.P.Reboillat: J. Appl. Phys. **49**, 1677 (1978)
10.24 E.Babic, Z.Marohic, F.Hajdu, M.Tegze, I.Vincze: Solid State Commun. **27**, 441 (1978)
10.25 A.M.Stewart, W.A.Phillips: Philos. Mag. B**33**, 1 (1978)
10.26 R.Asomoza, I.A.Campbell, A.Fert, A.Lienard, J.P.Rebouillat: J. Phys. F**9**, 349 (1979)
10.27 J.Kastner, H.-J.Schink, E.F.Wassermann: Solid State Commun. **33**, 527 (1980)
10.28 R.W.Cochrane, F.T.Hedgcock, B.J.Kastner, W.B.Muir: J. Phys. (Paris) **39**, C6–939 (1978)
10.29 S.R.Nagel, J.Vassiliou, P.M.Horn, B.C.Giessen: Phys. Rev. B**17**, 462 (1978)
10.30 H.H.Buschow, N.M.Beekmans: Phys. Rev. B**19**, 3843 (1979)
10.31 G.R.Gruzalski, J.A.Gerber, D.J.Sellmyer: Phys. Rev. B**19**, 3469 (1979)
10.32 J.Hafner, E.Gratz, H.-J.Güntherodt: J. Phys. (Paris) **41**, C8–512 (1980)
10.33 R.W.Cochrane: J. Phys. (Paris) **39**, C6–1540 (1978)
10.34 Proc. of the Fourth Intern. Conf. on Liquid and Amorphous Metals (LAM 4), J. Phys. (Paris) **41**, C8 (1980)
10.35 G.S.Grest, S.R.Nagel: Phys. Rev. B**19**, 3571 (1978)
10.36 J.Kastner, E.F.Wassermann: J. Low Temp. Phys. **29**, 411 (1977)
10.37 J.Kondo: In *Solid State Physics*, Vol. 23, ed. by F.Seitz, D.Turnbull, H. Ehrenreich (Academic Press, New York 1969) p. 184
10.38 A.Heeger: In *Solid State Physics*, Vol. 23, ed. by F. Seitz, D. Turnball, H. Ehrenreich (Academic Press, New York 1969) p. 284
10.39 P.Monod: Phys. Rev. Lett. **19**, 1113 (1967)
10.40 R.Hasegawa: Phys. Lett. **36**A, 207 (1971)
10.41 A.Madhukar, R.Hasegawa: Solid State Commun. **14**, 61 (1974)
10.42 R.N.Silver, T.C.McGill: Phys. Rev. B**9**, 272 (1974)
10.43 T.E.Sharon, C.C.Tsuei: Phys. Rev. B**5**, 1047 (1972)

10.44 K.Raj, A.Amamou, J.Durand, J.I.Budnick, R.Hasegawa: In *Amorphous Magnetism II*, ed. by R.A.Levy, R.Hawegawa (Plenum Press, New York 1977) p. 207
10.45 T.Kaneoyoshi: Phys. Status Solidi (b) **66**, K1 (1974)
10.46 M.A.Continentino, N.Rivier: J. Phys. F**8**, 1187 (1978)
10.47 H.Gudmundsson, H.U.Åstrom, D.New, K.V.Rao, H.S.Chen: J. Phys. (Paris) **39**, C6–943 (1978)
10.48 J.L.Black: In *Glassy Metals I*, ed. by H.-J.Güntherodt, H.Beck Topics Appl. Phys., Vol. **46** (Springer, Berlin, Heidelberg, New York 1981) p. 167
10.49 M.Banville, R.Harris: Phys. Rev. Lett. **44**, 1136 (1980)
       R.Harris, L.J.Lewis: Phys. Rev. B**25**, 4997 (1982)
10.50 A.Zawadowski: In *Trends in Physics* 1981, ed. by I.A.Dorobantu (European Physical Society, Petit-Lancyl Geneva, Switzerland)
10.51 K.Matho, M.T.Beal-Monod: Phys. Rev. B**5**, 1899 (1972)
10.52 J.L.Black, B.L.Gyorffy, J.Jackle: Philos. Mag. **40**, 331 (1979)
10.53 A.Zawadowski: Phys. Rev. Lett. **45**, 211 (1980)
10.54 A.Zawadowski, K.Vladar: Solid State Commun. **35**, 217 (1980)
10.55 K.Vladar, A.Zawadowski: Solid State Commun. **41**, 649 (1982)
10.56 J.Kondo: Physica **84**B, 40 (1976)
       J.Kondo: Physica **84**B, 207 (1976)
10.57 T.Matsubara (ed.): *The Structure and Properties of Matter*, Springer Ser. Solid-State Sci., Vol. 28 (Springer, Berlin, Heidelberg, New York 1982)
10.58 J.L.Black, K.Vladar, A.Zawadowski: Phys. Rev. B**26**, 1559 (1982)
10.59 J.L.Black, B.L.Gyorffy: Phys. Rev. Lett. **41**, 1595 (1978)
10.60 J.Solyom: J. Phys. F**4**, 2269 (1974)
10.61 D.R.Hamann: Phys. Rev. **158**, 570 (1967)
10.62 E.F.Wassermann: Private communication
10.63 A.Mogro-Campero: Private communication
10.64 D.Korn, W.Murer: Z. Phys. B**27**, 309 (1977)
10.65 S.R.Nagel: Phys. Rev. Lett. **41**, 990 (1978)
10.66 S.N.Teoh, W.Teoh, S.Arajs, C.A.Moyer: Phys. Rev. B**18**, 2666 (1978)
10.67 M.N.Baibich, W.B.Muir, G.Belanger, J.Destry, H.S.Elzinga, P.A.Schroeder: Phys. Lett. **73**A, 328 (1979)
10.68 P.J.Cote, L.V.Meisel: Phys. Rev. B**20**, 3030 (1979)
10.69 S.Basak, S.R.Nagel, B.C.Giessen: Phys. Rev. B**21**, 4049 (1980)
10.70 J.P.Carini, S.Basak, S.R.Nagel: J. Phys. (Paris) **41**, C8–463 (1980)
10.71 B.L.Gallagher: J. Phys. F**11**, L207 (1981)
10.72 R.W.Cochrane, J.Destry, J.Brebner, M.N.Baibich, W.B.Muir: Physica **107**B, 131 (1981)
10.73 M.N.Baibich, W.B.Muir, Z.Altounian, Tu Guo-hua: Private communication
10.74 B.L.Gallagher, D.Greig: J. Phys. F**12**, 1721 (1982)
10.75 M.N.Baibich, W.B.Muir, Z.Altounian, Tu Guo-hua: Phys. Rev. B**27**, 619 (1983)
10.76 Z.Altounian, C.L.Foiles, W.B.Muir, J.O.Strom-Olsen: Phys. Rev. B**27**, 1955 (1983)
10.77 M.N.Baibich, W.B.Muir, Z.Altounian, Tu Guo-hua: Phys. Rev. B**26**, 2963 (1982)
10.78 T.Matsuda, U.Mizutani: J. Phys. F**12**, 1877 (1982)
10.79 N.F.Mott, H.Jones: *Theory of the Properties of Metals and Alloys* (Dover, New York 1958)
10.80 R.D.Barnard: *Thermoelectricity in Metals and Alloys* (Taylor and Francis, London 1972)
10.81 T.E.Faber: *An Introduction to the Theory of Liquid Metals* (Cambridge University Press, Cambridge 1972)
10.82 D.Lazarus: Solid State Commun. **32**, 175 (1979)
10.83 R.W.Cochrane, J.O.Strom-Olsen, J.-P.Rebouillat, A.Blanchard: Solid State Commun. **35**, 199 (1980)
10.84 D.Greig, M.A.Howson: Solid State Commun. **42**, 729 (1982)
10.85 R.Harris, B.G.Mulimani: Phys. Rev. B**27**, 1382 (1983)
10.86 J.Jackle: J. Phys. F**10**, L43 (1980)
10.87 A.B.Kaiser: J. Phys. F**12**, L223 (1982) and private communication

10.88 Z. Altounian, J.O. Strom-Olsen: Phys. Rev. B**27**, 4149 (1983)

10.89 M.R. Bennett, J.G. Wright: Phys. Lett. **38**A, 419 (1972)

10.90 Z. Marohnic, E. Babic, D. Pavuna: Phys. Lett. **63**A, 348 (1977)

10.91 G. Bergmann: Phys. Rev. B**15**, 1514 (1977)

10.92 G. Bergmann, P. Marquardt: Phys. Rev. B**18**, 326 (1978)

10.93 R. Malmhall, K.V. Rao, G. Backstrom, S.M. Bhagat: Physica **86–88**B, 796 (1977); Solid State Commun. **19**, 193 (1976)

10.94 R.W. Cochrane, J. Destry, M. Trudeau: Phys. Rev. B**27**, 5955 (1983)

10.95 E.W. Lee: Physica **86–88**B, 781 (1977)

11. Magnetic Properties of Metallic Glasses

J. Durand

With 10 Figures

This chapter is restricted to fundamental magnetic properties of liquid-quenched amorphous alloys. Emphasis is placed on the interplay between the atomic structure and the magnetic properties. An introductory section presents the aim of the chapter together with some preliminary remarks about the possible influence of the sample size and of the fabrication technique on the magnetic properties of the amorphous materials. The first part analyses those magnetic measurements which can yield some structural information on metallic glasses (MG's). Some knowledge about the average symmetry of the first atomic shell around a reference atom is obtained through local measurements of the electric-field gradient and through bulk magnetic studies of the crystal field effects in amorphous alloys containing rare-earth (RE) elements. Medium-range fluctuations are evidenced by both bulk magnetic properties and hyperfine field distribution measurements in ferromagnetic MG's.

The second part discusses the influence of structural disorder on magnetic properties of transition-metal (TM) base MG's. The models (localized, itinerant) commonly used to describe TM magnetism are schematically presented along with the main physical ingredients on which structural disorder can play a role. As for MG's which do not exhibit a long-range, homogeneous magnetic order, three points are given special attention, namely, the inhomogeneous character of the appearance of magnetism, the RKKY interaction in MG's and the reentrant magnetism behaviour. As for ferromagnetic MG's based on TM, we summarize the data on zero-temperature properties and then, the data at finite temperature, namely, the magnetic excitations, the Invar properties and the critical phenomena.

The third part deals with RE base MG's, analysing first the magnetic properties of MG's containing S state RE ions (Gd, Eu^{2+}). Then, the case of non-S state RE ions, especially those with intermediate valence, is shortly discussed. Finally, preliminary data on uranium based MG's and on TM based MG's containing RE additives are listed. Brief remarks conclude the chapter.

11.1 Background

The historical development, during the last two decades, of the research effort on magnetic properties of metallic glasses (MG's) seems to have been spurred by three major incentives. Early studies were concerned mainly with purely scientific questions raised by amorphous magnetism [11.1]. Later on, interest in this field was renewed by the potential for application of these novel magnetic materials. More recently, magnetic properties were recognized as a powerful tool for systematic investigations of the atomic-scale structure of amorphous alloys. While studies of application-oriented properties tend to constitute an autonomous field of research, the investigation of the basic magnetic properties of MG's appears to be naturally connected with the effort toward a better knowledge of the atomic structure as can be achieved through magnetic measurements. Before any attempt to discriminate in magnetic properties of MG's between alloying effects and manifestations of long-range disorder, the determination of short-range order (SRO) is a prerequisite.

Even restricted to basic magnetism and to structure-oriented studies, the bibliography on magnetic properties of amorphous alloys has become overwhelming [11.2]. On the other hand, magnetic properties of amorphous alloys have been the subject of numerous reviews in the recent past, either from a general point of view [11.3–6] or with emphasis on MG's containing a transition metal (T) [11.7–11], on magnetic thin films [11.12], on amorphous alloys containing rare earth metal (RE) [11.13, 14], or on hyperfine fields [11.15, 16]. Thus, the present review will not attempt to be exhaustive. Instead it will focus on some recent magnetic studies which are believed to be of particular significance in the relationship with both the atomic and electronic structure of MG's.

By following a widely-accepted convention, metallic glasses refer to those amorphous alloys which are produced by liquid-quenching. Priority will be given in this review to magnetic properties of metallic glasses according to the scope of this book. Results on amorphous alloys obtained by other techniques such as deposition or vapor-quenching will be analyzed, however, when complementary to data on metallic glasses or for an illustrative comparison. The technique of preparation will then be specified in the text. This might be not a simple matter of taxonomy. A fundamental question can be raised, indeed, as to what extent all those amorphous alloys are amorphous in the same sense. More precisely, are some magnetic properties a signature of the amorphous character of the materials, or to what extent are they related to the techniques of preparation or to the size of the samples? The pertinence of this question is illustrated by some recent work along this line.

The magnetic properties of amorphous particles of $Fe_{75}Si_{15}B_{10}$ produced by spark erosion exemplify the influence of both the sample size and the fabrication technique [11.17]. The Curie temperature (T_c) and the saturation magnetization at 4.2 K for the largest particles (20–30 μm) whose diameters are similar to the thickness of the liquid-quenched ribbon, are significantly lower (by 5 and 10%, respectively) than the corresponding values for the ribbon. These quantities continue to decrease with decreasing particle diameter. The moment is reduced by 24% in the smallest particles (0.5–5.0 μm) with an accompanying decrease of 89 K in T_c. These effects were attributed to the quenching rate being faster for the particles compared to the ribbons; in addition, the quenching rate would increase for decreasing particle diameter. The influence of the preparation technique becomes more drastic in amorphous Fe-base alloys over the concentration range for the disappearance of magnetic order, as illustrated by the case of amorphous YFe_2. A dc sputtered amorphous YFe_2 is reported to undergo a true paramagnetic to spin-glass transition at 58 K, followed at lower temperature (around 20 K) by a cluster-glass "pseudo-transition" [11.18]. Other amorphous YFe_2 samples produced by different groups with the same technique [11.19, 20] or by coevaporation [11.21] seem to exhibit a magnetic behaviour which is basically the same. In contrast, liquid-quenched YFe_2 is ferromagnetic with a T_c of about 270 K [11.22], while the crystalline compound has a T_c of 548 K [11.23]. Here again, sputtered or evaporated samples seem to be produced with a faster effective quenching rate than the melt-spun samples, which results in increased disorder and lowered magnetic-ordering temperatures. Systematic studies of the influence of the size and of the preparation technique on the magnetic properties of amorphous alloys have remained scarce to data. However, it can be conjectured that effects such as those occurring in the above examples are expected every time the magnetic properties of a given crystalline alloy or compound are largely affected by structural disorder (e.g., magnetic alloys in the critical concentration range for the disappearance of magnetism, especially those containing Fe, Laves phases, see [11.13], Heusler alloys [11.24], etc. ...). Thus, magnetic properties must be discussed with explicit reference to the fabrication technique. In many cases, where structural disorder alters the values of T_c and of magnetic moment (e.g., Fe_3B, Co_3B) very slightly, no significant differences are observable between liquid and vapor-quenched samples, as verified for amorphous FeB alloys [11.25].

These preliminary remarks can usefully stress the point that the structural disorder is not always an unambiguous characterization of the amorphicity. Sometimes magnetic properties can be sensitive to various types of SRO prevailing in amorphous alloys more than to the lack of periodicity. The need then exists for detailed atomic-scale characterization, in addition to measurements of bulk and local magnetic properties. This chapter will first review the structural information yielded by local and bulk magnetic measurements themselves. A following section will summarize the most salient features of the magnetic properties of MG's containing T. The magnetic behaviour of RE base MG's will be analyzed in another section. We shall then conclude with some final remarks.

11.2 Magnetic Properties of Metallic Glasses as Probes of Their Atomic-Scale Structure

The knowledge of average atomic arrangement around each atomic species in amorphous alloys has been considerably improved in the recent years. More accurate determination of the coordination numbers and of fluctuations around these average numbers, along with more precise evaluation of interatomic distances from a given reference atom and of fluctuations on these average distances, were obtained through selective techniques such as EXAFS. As well as this, partial pair distribution functions were determined for an increasing number of amorphous alloys through sophisticated diffraction techniques using isotopic substitutions, combined x-ray and neutron spectroscopy, etc. .... On the other hand, experimental evidence for atomic correlations over medium-range scale was provided in some MG's through small-angle x-ray or neutron diffraction techniques or through high-resolution microscopy. However, concerning the SRO, information yielded by diffraction or EXAFS techniques is basically radial in nature and information about angular atomic distributions can be derived only indirectly from these experiments. Magnetic measurements may allow, in favorable cases, the average local symmetry around a reference atom in an amorphous alloy to be determined directly. On the other hand, concerning the medium-range atomic correlations, bulk and local magnetic studies can give information over a range which is complementary to that accessible by small-angle diffraction techniques. First we will review the results of recent studies of the local symmetry. Then we will summarize the information about the medium-range scale [11.26].

11.2.1 Local Symmetry in Amorphous Alloys

The local symmetry around a given atomic species in amorphous alloys can be experimentally determined in two main ways. One is to study the interaction between the nuclear quadrupole moment Q (for a nuclear spin $I > 1/2$) and the electric field gradients (EFG) created by the electric charges distributed around the nucleus. This can be done by local techniques such as NMR or Mössbauer spectroscopy. Another way is to study the interaction between a nonspherical electronic charge distribution and the electrostatic field originating from the neighbouring electric charge distribution. For example, RE ions with non-spherical $4f$ electronic shells (non-S ions) can be studied to probe the symmetry of their atomic environment through the so-called "crystal field" (CF) effects. These CF effects are reflected even through bulk properties of the alloy.

a) Local Studies of EFG [11.27]

By assuming that the higher-order terms are negligible, the EFG are usually expressed as a tensor whose eigenvalues in a principal axes system are $|V_{zz}| > |V_{yy}| > V_{xx}|$. In metallic systems, the trace vanishes. We are thus left with two independent components, namely, V_{zz} and the asymmetry parameter $\eta = |V_{xx} - V_{yy}|/|V_{zz}|$. V_{zz} is readily determined in non-magnetic alloys through Mössbauer or NMR spectroscopy, since the values of ΔE or of v_Q, the quadrupole frequency, respectively, are proportional to V_{zz}. In most cases, the experimental determination of η is a difficult task. For amorphous alloys, where there is a distribution of the quadrupole components, an independent evaluation of both $P(V_{zz})$ and $P(\eta)$ can seldom be achieved. But, at least, the nature of the average local symmetry can be clearly characterized. For a local spherical symmetry, V_{zz} is zero. Thus, $v_Q(\Delta E)$ is a measure of the departure from spherical symmetry. For an axial symmetry, $V_{xx} = V_{yy}$. Thus, $\eta(0 \leqq \eta \leqq 1)$ is a measure of the departure from axial symmetry. First, we review the recent experimental studies of EFG in amorphous alloys with emphasis on NMR data (Mössbauer results are reviewed in Chap. 4). Then, we will compare the data to model predictions.

Many experimental observations of quadrupole interactions in MG's (mainly through Mössbauer spectroscopy) were reported, before these results were first analyzed in terms of atomic-scale structure [11.28]. However, systematic investigations of EFG in MG's with respect to their significance for structural information are rather recent. From these experimental studies, it has already become clear that the nature of the local symmetry along with the distributions of the parameters V_{zz} and η for the different atomic species vary considerably from an amorphous alloy to another one. In most cases investigated so far, the local symmetry is rather well defined with narrow distributions for V_{zz} and η, which are not compatible with the larger disorder predicted by a dense-random packing model of hard spheres (DRPHS) [11.29]. However, this latter approximation was found to be adequate for a description of the EFG experimentally observed by the perturbed angular correlation technique on amorphous Ga films prepared by evaporation [11.30]. Similarly, analysis of ^{155}Gd spectra in sputtered amorphous GdNi alloys was performed by using analytical expressions for $P(V_{zz})$ and $P(\eta)$ deduced for amorphous solids with random ionic coordination [11.31]. But, such broad distributions of the EFG parameters are not necessarily the rule for sputtered amorphous alloys since ^{11}B NMR spectrum in sputtered Mo_2B exhibits the quadrupolar structure characteristic of an axial symmetry ($\eta = 0$) with a mean value of the quadropular frequency close to that measured for the tetragonal Mo_2B compound. The disorder for this amorphous alloy can be expressed by the rms half-width σ of a Gaussian distribution of v_Q; the ratio $\sigma/\bar{v}_Q$ was determined to be 0.2–0.3 [11.29]. Recent NMR investigations of EFG in MG's have shown that the average local symmetry around the constituents of the MG's tend to reproduce to some extent the symmetry prevailing in the crystalline counterparts [11.29]. Thus, the local symmetry around Ga in a-La_3Ga is spheric on average ($V_{zz} = 0$), as in the metastable crystalline compound. Similarly, the average symmetry around Al in a-La_3Al is axial ($\eta = 0$), as in the hexagonal crystalline compound [11.32]. In other cases such as $Eu_{80}Au_{20}$, for example, local symmetry was studied through ^{151}Eu Mössbauer spectroscopy and was found to be well defined around Eu (with a positive sign for V_{zz} without significant distribution on V_{zz} and with a value for η close to zero) [11.33], while no crystalline compound of same composition was reported to exist (Fig. 11.1). The high degree of SRO observed for this latter alloy might be related to the fact that its composition is close to that of a deep eutectic in the equilibrium phase diagram. It has recently been noted, indeed, that the ^{11}B NMR spectrum

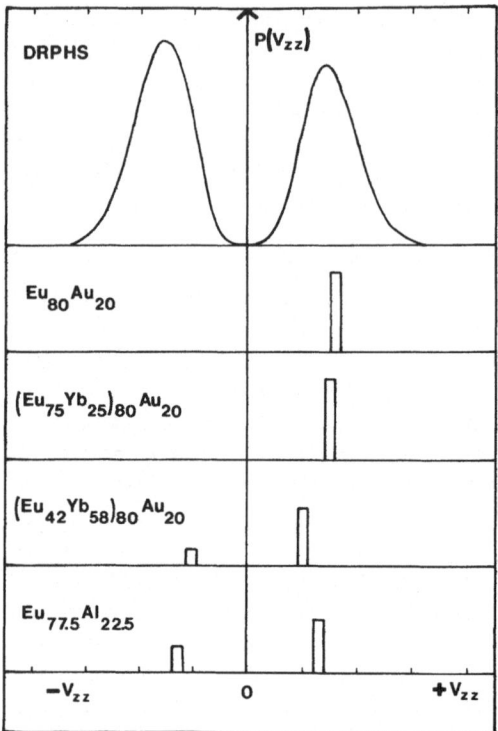

Fig. 11.1. Schematic distribution of V_{zz} (arbitrary scales) obtained by ^{151}Eu Mössbauer spectroscopy in various metallic glasses with $0 \leqq \eta \leqq 0.2$) [11.33]. The upper part of the figure sketches the $P(V_{zz})$ calculated by *Czjzek* et al. [11.31]

for the $Ni_{81.5}B_{18.5}$ metallic glass (whose composition is that of the eutectic) corresponds to a EFG distribution which is especially narrow ($\sigma/\bar{v}_Q \lesssim 0.15$) [11.34], with a mean value for η close to that in orthorhombic Ni_3B compound. Particularly informative with respect to the local structure of MG's is the progressive change of the EFG parameters observed by ^{11}B NMR spectroscopy when increasing the B content in the $Ni_{100-x}B_x$ MG's ($18.5 \leqq x \leqq 40$) [11.34]. Over this concentration range, one finds in the corresponding phase diagram the following compounds: Ni_3B (orthorhombic, strongly anisotropic, with $\eta = 0.6$ around B), Ni_2B (tetragonal, axial symmetry) and Ni_4B_3 (orthorhombic, three B sites with η equal to 0.2, 0.4, and 0.5, respectively). The local symmetry around boron in the amorphous NiB alloys remains nonuniaxial from 18.5 to 40 at. % B with a continuous decrease of the asymmetry parameter from its value in c-Ni_3B to that in c-Ni_4B_3. But the axial symmetry of c-Ni_2B is not preserved in the MG of same composition. It is of importance to note that a common motif in the c-Ni_3B and c-Ni_4B_3 (orthorhombic) compounds is made of trigonal prisms of Ni around the B atoms, with B substituting for Ni at the apex of the pyramids caping the prisms in the Ni_4B_3 compound. It is then tempting to agree with *Gaskell* [11.35] that the same motif is favoured in the amorphous modifications. The fact that ^{11}B in amorphous Mo_2B has the same axial symmetry as in the crystalline compound seems to be related to the fact that all the molybdenum borides are tetragonal. Thus, there is probably not a simple one-to-one correspondance between the local symmetry around a given constituent in a crystalline compound and its amorphous modification. But, in any case, the average symmetry around the constituents of an amorphous alloy has to be understood in relationship with the basic ingredients of the corresponding crystalline compounds in the phase diagram. This is well documented for MG's containing T and $s-p$ elements (i.e., amorphous alloys related to the presence of a deep-lying eutectic in the phase diagram). This is true also, but probably to a lesser degree, for MG's of the CuZr type [11.36].

In these circumstances, the DRPHS model represents an adequate approach to the atomic-scale structure only for MG's where the SRO is loosely defined. Computer calculations of $P(V_{zz})$ and $P(\eta)$ for DRPHS amorphous alloys [11.30, 31, 37] yield zero probability for both $V_{zz} = 0$ and $\eta = 0$. The probability is nearly the same of obtaining V_{zz} values of both signs (see the top of Fig. 11.1). On the other hand, the probability is high for having large values for η. Such a model is then unable to describe the large variety of local symmetries and of EFG distributions encountered in real MG's.

b) Crystal-Field Effects in Amorphous Alloys Containing RE

It was first emphasized by *Harris, Plischke*, and *Zuckermann* (HPZ) [11.38] that even in a DRPHS amorphous alloy, the CF effects on RE ions are not averaged out, but this electrostatic field strongly influences the bulk properties of the alloy, such as magnetic, calorimetric and transport properties. Calculations based on the DRPHS model showed that the quadratic terms in the CF Hamiltonian, absent by symmetry in most of the crystal structures, are predominant in amorphous alloys [11.13]. In addition, the CF in amorphous solids was assumed in the HPZ model to be uniaxial:

$$H^i_{\text{HPZ}} = -DJ^2_{z_i}$$

with $D > 0$, and with the easy axis z_i varying randomly in direction from site to site. This model was originally applied to magnetically-ordered alloys containing heaving heavy RE elements (large J values). Magnetic properties such as the approach to saturation were satisfactorily accounted for within the HPZ assumptions.

Nevertheless, these assumptions were questioned later on in view of several results. First, subsequent calculations based on the DRPHS model showed that the probability of having locally $\eta = 0$ is practically zero, which is contradictory to the uniaxial hypothesis [11.30, 31, 37]. Second, applying the uniaxial hypothesis to magnetic data in amorphous alloys containing heavy RE leads to a value of D which is always positive, while the sign of D should follow the change of sign of the Stevens coefficient α_J between Ho and Er [11.39]. Third, all the experimental results accumulated on amorphous alloys containing Pr show evidence of a rather large amount of Pr lying in a singlet ground state [11.40];

this is not compatible with the uniaxial model for which the ground state for all the RE ions is the doublet $|J_{z_i} = \pm J\rangle$. For all these reasons, *Fert* and *Campbell* suggested that the nonuniaxial terms in the quadratic CF Hamiltonian should be taken into account [11.41]:

$$H^i_{CEF} = C\left(\frac{3-\eta_F}{6} J^2_{x_i} + \frac{\eta_F}{3} J^2_{y_i} - \frac{3+\eta_F}{6} J^2_{z_i}\right),$$

where C is proportional to V_{zz} through the Stevens coefficient α_J, x_i, y_i, and z_i are the randomly-oriented local principal axes and η_F is an asymmetry parameter related to the asymmetry parameter η used in the EFG studies [$\eta_F = \pm 3(\eta-1)/(\eta+3)$ depending upon the sign of V_{zz}]. The parameter η_F is allowed to vary between -1 and $+1$. The limit $\eta_F = +1$ corresponds to a uniaxial CF with easy axis (HPZ model with $D=C$). The other limit $\eta_F = -1$ corresponds to a uniaxial CF with easy plane ($D=-C$). For values of η_F between -1 and $+1$, the CF is not uniaxial. The sign of η_F should follow the change of sign of α_J. Assuming the quadratic approximation for CF to be valid, the H_{CEF} Hamiltonian could be applied to significant experimental results in order to determine the mean values of C and η_F, together with the distribution of those parameters in an amorphous alloy. In principle, then, the CF effects can be used as a probe of the local symmetry around a RE ion in an amorphous matrix.

CF effects as evidenced by magnetic, transport and thermal properties in concentrated Ce base MG's have been analyzed by *Felsch* and co-workers [11.42] with respect to the average local symmetry. Experimental arguments are given for the existence of a local symmetry different in nature in $Ce_{80}Au_{20}$, $Ce_{89}Al_{11}$, and $Ce_{72}Cu_{28}$ alloys. But these concentrated alloys which are magnetically ordered at low temperature are probably not the ideal cases for an accurate investigation of the local symmetry in MG's. The dilute RE $La_{80-x}RE_xAu_{20}$ MG's constitute an amorphous system for which the CF effects have been extensively studied. Among dilute RE ions, Ce and Pr (small J values) are more sensitive to the surrounding local symmetry. The low-temperature susceptibility [11.43, 44] and specific heat (for Pr diluted alloys) [11.45] suggest a narrow distribution of the asymmetry parameter centered around a mean value close to $\eta_F = -1$. Analysis of the field dependence of the low-temperature magnetization in Pr diluted alloys was performed by following the method proposed by *Borchi* and *De Gennaro* [11.46]. A mean value of $\eta_F \gtrsim -0.80$ was then obtained [11.47]. The low-temperature field dependence of the anisotropic magnetoresistance in dilute RE alloys can be related to the degeneracy of the RE ground state [11.48]. Magnetoresistivity measurements carried out on $La_{80-x}RE_xAu_{20}$ MG's indicated for RE: Pr, Nd, Tb, Dy, Tm, respective ground states wich are compatible with a value of η_F close to -1 [11.49]. The concentration dependence of the superconducting temperature in amorphous $La_{80-x}Pr_xAu_{20}$ alloys is characteristic of that observed for superconductors containing Pr in a singlet ground state [11.50], which indicates that η_F is close to -1. Finally, let us note that all these observations are perfectly

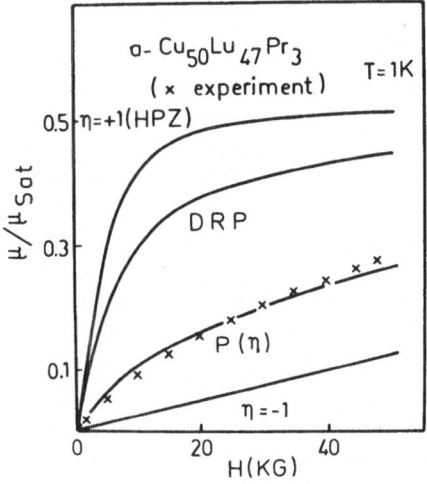

Fig. 11.2. Normalized magnetization versus applied field at 1 K for the amorphous sputtered $Cu_{50}Lu_{47}Pr_3$ alloy. Crosses indicate experimental results. Solid curves were calculated by using a value of $C/k_B = 70$ K and for several $P(\eta_F) : \eta_F = +1$ (HPZ model), $\eta_F = -1$ (planar model), $P(\eta_F)$ calculated by *Czjzek* et al. [11.31]. The fitted curve corresponds to a phenomenological $P(\eta_F)$ (see text) [11.52]

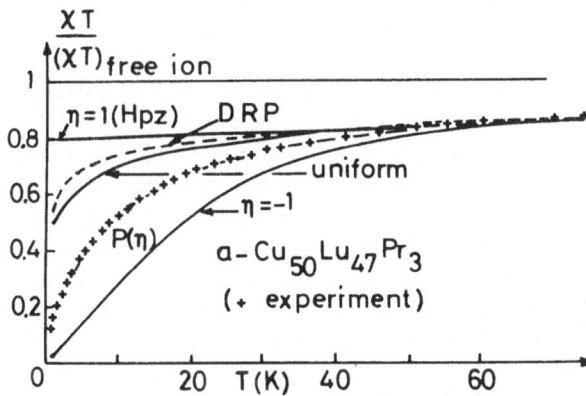

Fig. 11.3. Normalized values of χT versus temperature for an amorphous sputtered $Cu_{50}Lu_{47}Pr_3$ alloy. Crosses indicate experimental results. Solid curves were calculated for several $P(\eta_F)$, as for Fig. 11.2, [11.52]. Uniform distribution refers to [11.41]

consistent with the EFG studies of the $Eu_{80}Au_{20}$ alloy through ^{151}Eu Mössbauer spectroscopy [11.33]. A question can be raised about the use of the quadratic Hamiltonian for the CF in the amorphous $La_{80-x}RE_xAu_{20}$ alloys, where the short-range order is rigorously defined with a narrow distribution of the local symmetry parameters. Higher-order terms in CF should be significant for an average planar symmetry and then should be taken into account for an accurate determination of the energy level schemes for the RE ions.

The quadratic Hamiltonian of *Fert* and *Campbell* [11.41] was also used to determine the local symmetry around RE ions, such as Tb [11.51], Pr, Dy, Er, and Tm, diluted in $Cu_{60}Zr_{40}$, $Cu_{50}Lu_{50}$, and $Cu_{50}Y_{50}$ metallic glasses [11.52]. The field dependence of the magnetization measured at 1 K (see Fig. 11.2 for a-$Cu_{50}Lu_{47}Pr_3$) and the temperature dependence of the incremental susceptibility measured between 1 and 70 K (see Fig. 11.3 for a-$Cu_{50}Lu_{47}Pr_3$) were compared to model predictions and were found to differ markedly from both the limit cases of $\eta_F = 1$ (HPZ model) and $\eta_F = -1$ (planar symmetry).

Experimental curves also depart from calculated curves based on DRPHS distribution of the asymmetry parameter [approximated by $P(\eta) = 0.75\,(1 - \eta_F^2)$] or based on a uniform distribution of η_F between -1 and $+1$. The data for different ions in the same amorphous matrix were fitted by using a unique value of $C/|\alpha_J|$ (equal to $3350\,k_B$ for a-CuLu) and a unique distribution function for the asymmetry parameter. The phenomenological $P(\eta_F)$ thus obtained for a-CuLu does not differ significantly from those determined for CuY and CuZr metallic glasses [11.53]. These distributions have little in common with those calculated within DRPHS types of models.

Another approach to determine the distribution of the local symmetry in amorphous alloys containing Pr was suggested by *Bhattacharjee* and *Coqblin* [11.54]. The local symmetry at each Pr sites assumed to yield two low-lying singlet levels separated by an energy splitting Δ_i. This gap Δ_i varies randomly over the RE sites according to a Gaussian distribution. Experimental data, especially on $La_{80}Au_{20}$ MG's, were found to be compatible with a value of γ equal to about 0.5, γ being the ratio of the variance Δ_1 of the Gaussian function over the average gap Δ_0 between the two singlets ($\gamma = \Delta_1/\Delta_0$).

The knowledge of the local symmetry, rather well documented in non-metallic glasses, has remained poor in metallic glasses to date. There is a need in particular for direct measurements of $P(\eta)$ and for combined EFG and CF studies. Our knowledge of the local symmetry in MG's should also be improved by time-of-flight measurements in inelastic neutron scattering experiments on the MG's containing a dilute amount of RE [11.55]. It already seems clear that the HPZ model for CF in MG's, sufficient to describe the properties related to Kramers ions with large J values, is an inadequate approximation for the case of Ce ions or non-Kramers ions, which are more sensitive to the local symmetry. In most MG's studied so far, provided that the local symmetry was investigated by a sensitive probe, it appeared that the atomic coordinations were not distributed according to DRPHS types of predictions, but were governed by a chemical SRO whose origin has to be determined in the basic motifs of the compositionally related crystalline compounds.

11.2.2 Fluctuations in Metallic Glasses Over a Medium-Range Scale

The local symmetry parameters for an atomic species and their fluctuations in metallic glasses mainly concern the first surrounding atomic shells. But a characterization of the local structure of MG's is needed over a larger range for an understanding of the macroscopic magnetic properties, especially those properties which have a potential for technical applications such as coercive field, saturation magnetization, initial permeability, magnetocrystalline anisotropy, etc..... Macroscopic ($> 1\,\mu$m) and microstructural (10^2 to 10^3 Å) fluctuations in MG's are of particular significance for the domain structure and wall mobility, the approach to saturation, the temperature dependence of the magnetization, the magnetostrictive strains... [11.56]. Depending on their range, these fluctuations are observable by different techniques including microscopy or small-angle x-ray or neutron scattering. But bulk and local magnetic measurements can themselves yield valuable information on the medium-range fluctuations in MG's.

a) Bulk Magnetic Properties

The different aspects of the hysteresis loop in many ferromagnetic MG's were studied in detail by *Kronmüller* [11.57]. In particular, the field dependence of the magnetization as it approaches the saturation was analyzed with respect to the microstructural state of the sample. This approach to saturation is dominated by effects due to defect structures. These defect structures induce elastic stresses, producing extended spin inhomogeneities due to the magnetoelastic coupling energy. The contribution of defect structures to the departure from saturation, ΔM_{def}, is written as

$$\Delta M_{def} = a_{1/2} H^{-1/2} + a_1 H^{-1} + a_2 H^{-2},$$

where the three terms are attributed, respectively, to point-like defects $(a_{1/2})$, quasidislocation dipoles (a_1) and isolated dislocations (a_2). This interesting approach has to be applied in other ferromagnetic amorphous alloys where no doubt can be cast on possible fluctuations on the sign of the exchange. To that respect, amorphous Fe base alloys are probably not the best suited alloys to be analyzed with the Kronmüller method. Different contributions to the coercive field H_c in ferromagnetic MG's were also analyzed by *Kronmüller* and *Gröger* [11.58]. In magnetostrictive alloys, a relatively important contribution to H_c was concluded as arising from volume pinning of domain walls by defect structures. In conclusion of these studies of domain patterns, magnetization processes and defect structures, Kronmüller distinguished between short-range stresses which pin the domain walls and long-range stresses which determine the domain pattern. Short-range stresses are attributed to quasi-dislocation dipoles of spatial range 50–3000 Å. These quasi-dislocation dipoles would be a consequence of the free volume existing in the liquid state and partially trapped in the metallic glass. Long-range stresses would be due to inhomogeneities quenched in the lattice during the rapid solidification process. The liquid quenching would result in instantaneous coexistence of liquid and solid islands giving rise to regions of compressive and tensile stress. In positively magnetostrictive MG's, tensile stresses would generate wide and wavy laminae, while compressive stresses would produce narrow laminae with an easy direction perpendicular to the ribbon.

b) Hyperfine Field Distribution Studied by NMR in Ferromagnetic MG's

Spin-echo NMR studies of ^{59}Co hyperfine field (hf) distributions in Co based ferromagnetic MG's yield information on the microstructure of these MG's which is in the line with that obtained by Kronmüller on Fe based MG's. It has been widely recognized that hf distributions as measured in ferromagnetic MG's through various nuclear techniques can be related to the distribution of local environments [11.59]. In addition to the fact that the hf distribution is given directly in spin-echo NMR spectra without fitting procedures, the zero-external field NMR spectroscopy in ferromagnetic materials is especially well suited for studying the microstructure of soft magnetic MG's owing to some features that we recall briefly [11.60]. Nuclear levels in magnetic materials are not excited directly by the radio frequency (rf) field H_1, but they are excited through a local effective field H_1^* created by the rotation of the local electronic moment due to H_1. The NMR signal arises then from the rotating electronic magnetization excited by the rotating nuclear magnetization. This results in an enhancement factor of the local excitation field and then of the NMR signal intensity. This enhancement factor is not the same for all the atoms in a ferromagnetic material. In highly anisotropic materials, the NMR signal originates only from nuclei in domain walls parallel to H_1, at least for the low values (up to 10 Oe) of H_1 commonly used. In low-anisotropy materials, which is the case of most Co base MG's, the NMR signal is not restricted to nuclei in domain walls, but, depending on the strength of H_1, nuclei in other regions of the sample can contribute to the signal through domain rotation. The signal intensity will then be sensitive to the strength of the rf field.

A detailed NMR study was recently performed on ^{59}Co in $Co_{100-x}B_x$ MG's ($14 \leq x \leq 27$) [11.61]. The ^{59}Co zero-external field spin-echo NMR spectra exhibit structures which are not quadrupolar in origin. The experimental spectra were computer analyzed as the sum of three subspectra. For fixed excitation conditions, the centers of gravity of these subspectra were determined for the different alloy compositions. For a fixed alloy composition, the relative intensities of the subspectra were found to vary as a function of rf field strength, of sample orientation with respect to excitation field and a function of sub-T_g annealing treatments. But, in any case, neither the positions of the subspectra centroids for each alloy nor the position of the centroid for the whole spectrum were affected significantly by changing the experimental conditions. The concentration dependence of the ^{59}Co hf in CoB alloys being firmly established, it was then possible to associate a well-defined average composition to each subspectrum in an alloy. The centroid for the high-frequency subspectrum decreased rather linearly in frequency from its value of about 200 MHz for $x = 14$ down to about 170 MHz for $x = 27$. A linear extrapolation to zero B content yields a frequency close to that of fcc Co. This high-frequency subspectrum arises then from Co environments resembling a supersaturated CoB solid solution. The centroid of the central subspectrum does not vary in frequency significantly with the B content. It corresponds to Co atoms in zones where the metalloid content is 18–20 at. % in average, i.e., around the eutectic composition. The frequency of the centroid for the low-frequency subspectrum remains roughly constant at a value (about 110 MHz) close to that for the center of gravity of ^{59}Co hf in c-Co_2B. The concentration dependence of the relative intensities of these subspectra at fixed excitation

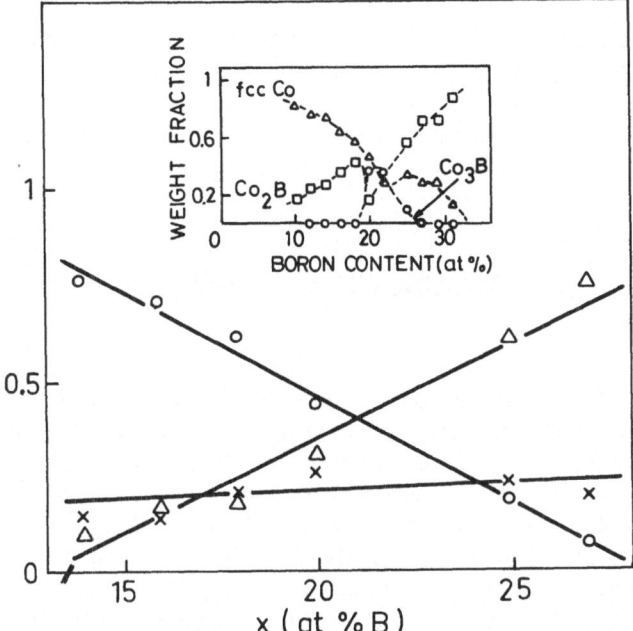

Fig. 11.4. Concentration dependence (with no correction for the enhancement factor from the walls) of the relative intensities of the Co^{59} NMR subspectra (see text) in $Co_{100-x}B_x$ metallic glasses (Inset: weight fractions of fcc Co, Co_2B, and Co_3B phases in crystallized CoB samples as a function of boron content [11.61])

conditions mimics somewhat the concentration dependence of the weight fractions of crystalline phases in crystallized CoB samples [11.61] (Fig. 11.4). These three types of Co environments in CoB MG's can be further characterized by analyzing the signal intensity of the subspectra as a function of the exciting conditions, namely, the H_1 strength, the sample orientation with respect to rf field and the annealing treatments. From these systematic studies it was concluded that the Co environment zones with 18–20% B (the B coordination number for Co, Z_B, being 2 in average) correspond to regions where the density of defects is high and where the magnetization is perpendicular to the plane of the ribbons. According to the Kronmüller scheme, perpendicular anisotropy would be attributed to in-plane tensile stresses resulting from inhomogeneous quenching rates. These regions would correspond to the last liquid islands to solidify, which is consistent with their concentration close in average to that of a deep eutectic. The high and low-frequency subspectra arise from Co nuclei located in regions where the density of defects is low and where the magnetization domains are in-plane. The NMR allows one to assign regions of well-defined environment to the various types of domains observed in these MG's by Kerr or Bitter techniques [11.62]. Structures in ^{59}Co NMR spectra were also attributed to regions of well-defined Co environment in CoPB [11.63] and CoBSi [11.64, 65] metallic glasses. Surprisingly enough, subspectra very similar to those analyzed in CoB MG's were also observed in amorphous electrodeposited CoP alloys [11.66]. The subspectrum in the medium frequency range might correspond to the columnar zones of perpendicular anisotropy observed by *Chi* and *Cargill* [11.67]. The question is raised as to whether these inhomogeneities are intrinsic to the amorphous state or dependent upon fabrication technique. Another question arises as to the role of the eutectic composition in the atomic structure and in the physical properties of these amorphous materials. Further experimental work along this line is expected in the near future.

11.3 Magnetic Properties of Metallic Glasses Based on Transition Metals

Assuming that SRO in MG is comprehensively determined together with the structural fluctuation on an atomic scale and over a medium range scale, it remains to elucidate among the magnetic properties of MG's which ones are primarily dependent on SRO and can be regarded as alloying effects and which ones basically reflect the fluctuations and the lack of periodicity. There is no a priori reason that the magnetic properties of MG's would be treated within models different from those commonly used in other metallic systems. But, the problem is how to take into account the structural disorder inherent to MG within the existing models when one has to describe the occurrence of a magnetic moment in a nonmagnetic host, or the interactions between magnetic impurities, or the properties of magnetically-ordered MG's at zero or finite temperature. We will review first this problem of the models. Then we will analyze the magnetic properties of MG's for cases ranging from the dilute impurity to the onset of ferromagnetic long-range homogeneous order. Finally, we will summarize the main features of the magnetic behaviour of amorphous ferromagnets.

11.3.1 The Models for the Magnetic Properties of Metallic Glasses

On a free atom, a magnetic moment exists whenever the atom has unpaired electrons. Atoms with partially filled f-shells generally assume a definite ionic valence state which is preserved when the atoms are embedded in metallic or nonmetallic solids. The value of the magnetic moment is then predicted by the Hund's rules. The crystal field interaction is a perturbation of order of 10^2 K as compared with spin-orbit coupling (about 10^4 K). Possible exceptions among the RE ions are those with almost emptied (Ce^{3+}) or almost filled (Yb^{3+}) f-shells, and more generally the RE ions with unstable valence (Sm^{3+}, Eu^{3+}, Tm^{3+}). It is clear that structural disorder would not then affect the moment of the normal RE ions. Possible effects of amorphicity will have to be searched along abnormal valent ions.

Atoms with partially filled d-shells, when diluted in nonmetallic hosts or when forming nonmetallic compounds, behave much the same as normal RE ions. But in the case of metallic hosts, the presence of broad conduction bands makes the situation more complex and the value of the magnetic moment no longer obeys the Hund's rule predictions. Let us note also that for the $3d$ series, the relative magnitude of spin-orbit and electrostatic interactions are inverted as compared with those for the $4f$ series. The problem of localized moments in metals can be treated from two different points of view. The first approach leads to a criterion for the occurrence of spin magnetism at zero temperature, which in the orbital (l) degenerate case reads

$$(U + 2lJ)n_d(E_F) > 1,$$

where $n_d(E_F)$ is the nonmagnetic d density of states at the Fermi level. U is an electrostatic Coulomb repulsion between two electrons with antiparallel spins on the same orbital or between electrons in different orbitals for both spin directions. J is the intra-atomic exchange energy arising from the difference between the energies of two d electrons with parallel and antiparallel spin. J and U take in metal effective values, owing to the screening by the conduction electrons. The second approach is commonly used to describe the behaviour of localized moments at finite temperature. In this model, referred to as the $s-d$ exchange model, the impurity spin S with a well-defined value interacts with the spin s of the conduction electrons through a Heisenberg-type interaction:

$$H = -\bar{J}_{sd} s \cdot S.$$

In the strong magnetic limit (large U) both approaches are equivalent. Let us note that the effect of local environment has to be taken into account in the Hartree-Fock treatment of the moment formation. The effect of structural disorder on the inputs of the above-mentioned models has been the subject of formal discussions. It is hard to predict how parameters such as U and J could be affected by disorder. More sensitive to disorder might be the d density of states $n_d(E_F)$ which depends on the average position of the d levels with respect to E_F, and upon the width Δ of the localized states. Disorder might result either in a smearing-out of singularities existing in the d band of corresponding crystalline materials or in a change of the width Δ. This latter effect was discussed on the basis of a Hubbard model by *Cyrot* [11.68], who treated the disorder by making some random sites in a lattice inaccessible to electrons. More phenomenologically, the influence of amorphicity on the problem of localized moments can be analyzed in terms of spatial range and fluctuations of J_{sd} or in terms of fluctuations in the local environment. These two latter approaches have been abundantly used by experimentalists, but mainly within the context of spin glasses and the onset of magnetic order, respectively, rather than about the appearance of localized moment in an amorphous matrix. This point will be discussed in following sections.

Concerning magnetically ordered systems there are two theoretical approaches, namely, the localized and the itinerant electron models [11.69]. In the localized model, each electron remains localized on an atom owing to large intra-atomic interactions. The interatomic exchange interactions are much smaller and compete with thermal disorder to define the magnetic order. This model is *a priori* well suited for magnetic insulators and for metallic systems containing normal RE ions. In the itinerant electron model, each carrier of moment moves in the average field of the other electrons and ions. The interatomic interactions (characterized by the band width W) are much larger than the intra-atomic electron-electron interactions ($U/W \ll 1$). This model is more appropriate for d metals and alloys.

Both models have been invoked to describe the properties of magnetic MG's. The localized approach has been the most commonly used. The effect of structural disorder was then introduced on the main conceptual inputs of the model, namely, the moment, the exchange coupling and, for MG's containing RE, the crystal field effect. Distributions on the value of the moment is reflected by the hf distribution through the hf coupling constant. The hf distribution for the RE based MG's is extremely narrow. For example, the relative width of the distribution is about 1 % for Dy [11.3], while it can reach 50 % in Fe or Co base amorphous alloys. In some MG's, nonmagnetic atoms can coexist with magnetic atoms, when some atoms have a coordination number smaller than the critical number of first neighbours required for an atom to carry a moment. Various exchange mechanisms exist in magnetic MG's depending on the nature of their constituents. The direct exchange mechanism is most important in transition metals. It originates from a direct overlap of d-orbitals between neighbor atoms. This overlap depends on the ratio of interatomic distance to orbital radius. The distribution of interatomic spacings in an amorphous solid can then lead to a distribution of exchange interactions. For Fe based MG's, interactions of both signs may coexist since the exchange integral changes sign for γ Fe with an interatomic separation of 2.55 Å. The distribution of the exchange can be related to the distribution of interatomic distances through the empirical Slater-Néel curve [11.70]. For all MG's made of T with metalloids, a superexchange mechanism was invoked for interactions between two next-nearest neighbor T atoms separated by a metalloid atom [11.71]. The experimental support for this mechanism is rather weak. Finally, for dilute $3d$ impurities or for RE ions, the magnetic exchange is of an indirect or RKKY type. The exchange between local moments is mediated by the spin polarized conduction electrons. A distribution on the parameters characteristic of these two latter mechanisms in MG's is likely but not easily determined from experiment.

If one adopts the itinerant electron model for magnetically ordered materials, the basic ingredients on which the disorder will have a possible influence are those discussed above for the onset of a localized moment, namely, the density of states and the effective electron interaction. We will not dwell here on formal discussions about the degree of fluctuation of the hopping integral which characterizes the band structure [11.6]. More directly connected to experimental data are several descriptions of the amorphous state in terms of a Landau-Ginzburg formalism. The expansion of the free energy F in terms of the magnetization M is written as follows:

$$F = \tfrac{1}{2}AM^2 + \tfrac{1}{4}BM^4 + \tfrac{1}{2}C|\nabla M|^2 - HM.$$

The spatial fluctuations of the magnetization in amorphous ferromagnets is expressed on one hand by the gradient term and on the other hand by letting the Laudau coefficients A and B be functions of position [11.72]. Within the same framework, paramagnetic and ferromagnetic Curie temperatures T_c, Curie Weiss constant and magnetization were calculated in terms of local density fluctuations [11.73], along with the pressure dependence of T_c and the Arrott plots for heterogeneous ferromagnets [11.74]. As an alternative approach, the spatial fluctuations of the magnetization were also discussed by *Wohlfarth* [11.75] within a tight-binding model. The problem of a ferromagnetic phase transition in amorphous materials with spatially fluctuating exchange interactions was studied recently by applying a nonlocal Landau-Ginzburg equation which is obtained in averaging the molecular field over a distance of the order of the spin correlation length [11.76]. More recently, a general formalism for the analysis of magnetization data in inhomogeneous materials was developed by *Brommer* [11.77]. All the aforementioned calculations were aimed at accounting for experimental observations such as flattened curves of reduced magnetization versus reduced temperature, nonlinear Arrott plots, curved temperature dependence of susceptibility just above T_c, etc. To what extent these singular behaviours are signatures of the amorphous state still remains in many cases an undecided matter.

11.3.2 From the Dilute Impurity Regime to the Onset of Long-Range Homogeneous Magnetic Order

The onset of a localized moment and the interactions between magnetic impurities have been studied in different amorphous matrices [11.78]. Most of the host MG's were based on transition metals. Thus, Co, Fe, Mn, Cr carry "good moments" in the Hartree-Fock sense in amorphous PdSi based alloys [11.79]. Fe and Mn bear good moments in amorphous $(Ni_{50}Pd_{50})_{80}P_{20}$ [11.80] and $Ni_{79}P_{13}B_8$ hosts [11.81], while Fe and Cr carry good moments in $Pd_{41}Ni_{41}B_{18}$ [11.82] and Cr and Mn in $(Ni_{50}Pt_{50})_{75}P_{25}$ [11.83]. Dilute Mn has a moment in $Cu_{60}Zr_{40}$ [11.84, 85], while for dilute Fe in the same host the experimental situation is controversial. As in crystalline hosts, Fe carries a small moment in MG's based on transition metals of the second series [11.84–86]. In other cases, a local moment appears on the impurity atom only when some local environment conditions are verified. An example of this is Co in amorphous $Ni_{78}P_{14}B_8$ [11.87]. Co atoms carry no moment in this matrix when they have zero or one Co first neighbor. When Co has 2 or 3 Co atoms as first neighbors, it carries a small moment of $0.50\ \mu_B$/at. When the number of Co first neighbors exceeds 3, the Co atoms in amorphous $Ni_{78}P_{14}B_8$ bear a moment of 1.15 μ_B/at, just as in amorphous $Co_{78}P_{14}B_8$. The environment conditions for Co to be magnetic in amorphous $Ni_{78}P_{14}B_8$ are very much the same as those deduced for Co in fcc Au [11.88]. Actually, due the lack of single-phase crystalline hosts of the same composition, an exact evaluation of the role of amorphicity in the onset of a localized moment in an amorphous host is not possible at present. Proper effects of atomic arrangements in amorphous metallic systems are more easily discussed in more concentrated magnetic MG's. We analyze in what follows three examples of possible manifestations of structural disorder in magnetic properties, namely, the inhomogeneous transition toward a magnetic regime, the RKKY interactions and the mictomagnetic alloys exhibiting a ferromagnetic-to-spin glass transition at low temperature and a ferro-to-paramagnetic transition at higher temperature.

a) The Inhomogeneous Character of the Appearance of Magnetism

The inhomogeneous character of the appearance of magnetism has been evidenced in many MG's. For example, in many Ni base alloys, a large moment was shown to appear on groups of Ni atoms, instead of individual Ni moments, around the critical concentration for magnetic order. These magnetic Ni clusters were evidenced in NiB [11.89] and NiPB [11.90] MG's, and also in electrodeposited amorphous NiP alloys [11.91]. Due to the distribution of the local environment in amorphous alloys, these clusters can survive at concentrations for which the crystalline counterparts exhibit pure paramagnetism. These giant moments in amorphous alloys can also interact to give rise to an ordering temperature. Thus, in crystalline NiP alloys, magnetization vanishes for 15%P, while the amorphous alloy of the same composition is still weakly ferromagnetic [11.92, 93]. Similarly, the amorphous $Y_{100-x}Co_x$ alloys remain magnetic for Co concentration down to $x = 40$, while the Co moment vanishes in crystalline YCo_2 which is a strongly exchange-enhanced Pauli paramagnet [11.23, 94]. A similar feature is observed in sputtered amorphous YNi alloys. The YNi_5 compound is a Pauli paramagnet and the amorphous alloy of the same composition has a T_c of 27 K [11.95]. The occurrence of a magnetic moment was accounted for in several Fe base amorphous alloys by a Jaccarino-Walker type of model [11.96]. The critical Fe concentration for the onset of a moment was found to be roughly the same (around 40 at. %) in many amorphous systems including evaporated [11.97–100] or sputtered [11.101] FeGe alloys, evaporated FeSi [11.102], FeSn [11.103], and FeSb [11.104] alloys, evaporated [11.100] and sputtered [11.25, 105] FeB alloys, sputtered FeLa, FeNb [11.106] and FeY [11.19] alloys, evaporated and liquid quenched FeZr alloys [11.107] and FeTh [11.108] metallic glasses. This was related to the fact that Fe needs to have an average number of 6 to 8 Fe atoms in the first atomic shell in order to carry a moment. Fluctuations in the Fe coordination number in amorphous alloys will then result in the coexistence of magnetic and nonmagnetic Fe sites, as deduced from Mössbauer spectroscopy. Owing to the great sensitivity of the exchange to the Fe interatomic distances, the structural disorder can alternatively be expressed in terms of fluctuations in Fe–Fe interatomic spacings, which is reflected by a broad distribution of the on-site moments in Mössbauer spectra on Fe base amorphous alloys over the critical concentration range for the onset of magnetism. This point was clearly illustrated by the dependence of the Fe hyperfine field on the size of the partner element in evaporated amorphous $LaFe_2$, YFe_2, and $LuFe_2$ alloys [11.21]. In summary, it has been abundantly proved that fluctuations in environmental conditions (coordination number and interatomic distances) favour the in-homogeneous nature of the onset of magnetism in amorphous alloys. This can result in marked differences between magnetic properties of amorphous alloys and corresponding crystalline compounds over the critical concentration range where magnetism appears.

b) The RKKY Interaction in Metallic Glasses

Interactions between magnetic impurities can induce within the dilute limit a random magnetic state, a spin glass, whose main experimental characteristics include (i) a cusp in the ac susceptibility whose shape and temperature are sensitive to measuring frequency and dc applied field, (ii) thermomagnetic history effects occurring at temperatures close to that of the cusp, displaced hysteresis loops, time relaxation effects on the remanence, (iii) for low enough concentration of magnetic impurities, all the above features scale with con-centration. Some metallic glasses were shown to exhibit at least some of the properties characterizing the magnetic behaviour of canonical spin glasses, namely, the PdSi alloys with dilute amount of Fe, Co, Mn, Cr [11.79, 109, 110], the NiPB alloys containing Fe [11.111] or Mn [11.81] and the CuZr metallic glasses containing Mn [11.85]. The spin glass regime in (Ni, Fe) MG's with various amounts of glass formers (B, P, Si, Al) has been the subject in recent years of extensive studies through bulk and local magnetic measurements along with specific heat and transport properties investigations [11.112–116]. In most of the above MG's, the concentration of magnetic impurities is too high for these alloys to be termed spin glasses in the canonical sense. The expression "cluster glasses" would be more appropriate. A detailed study of the con-centration dependence of the spin glass properties from the dilute case (< 1 at. %) was carried out in $La_{80-x}Gd_xAu_{20}$ metallic glasses [11.117]. It was shown that, provided one treats magnetic clusters as independent super-spins, some regularities of canonical spin glass properties can still be observed for Gd concentrations as high as 40 at. %. A spin-glass type of behaviour was also seen in concentrated sputtered amorphous alloys such as $Al_{63}Gd_{37}$ [11.118], MnSi [11.119], and YFe_2 [11.18, 120]. Analysis of bulk magnetic data and of neutron inelastic measurements demonstrated that these latter amorphous alloys behave in very much the same way as canonical CuMn crystalline spin glasses.

These strong similarities in the behaviour of amorphous and crystalline spin glasses raise the problem of the range of the RKKY interaction in an amorphous matrix. The oscillatory long-ranged RKKY function, which is the basic ingredient of simple theories for canonical spin glasses, can be expressed, in the case of an infinite electron mean free path (mfp), in its asymptotic form as

$$H_{RKKY} = - \frac{J^2 S_1 S_2}{r^3} \cos(2 k_F r),$$

k_F being the Fermi momentum and r the distance between two magnetic impurities with spins S_1 and S_2. The exchange integral J (assumed to be isotropic) should not be affected much by structural disorder, while larger effects are a priori expected on the spatial dependence of the RKKY interaction. Damping effects observed in crystalline spin glasses when the electron mfp is shortened are usually expressed by an exponential factor $\exp(-r/\Lambda)$, with Λ being mfp (for references see [11.117]). Such an effect should be particularly drastic in amorphous alloys where the high residual resistivity, interpreted within a free electron model, yields for mfp a value of 3 to 5 Å. In fact, amorphous spin glasses such as $La_{80-x}Gd_xAu_{20}$ were found to be truly RKKY spin glasses, where the scaling laws for magnetization with respect to concentration scaled field and temperature give support to a r^{-3} spatial dependence of the interaction [11.117]. Such regularities would not be observed in the case of an exponential damping. It seems thus that, at least in the extreme limit where mpf is of the order of interatomic spacing, the RKKY interaction is longer ranged than the electron mpf, so that a drastic attenuation takes place only for distances larger than about 3 atomic spacings [11.121]. This attenuation would seem to be even less drastic in simple metal amorphous hosts [11.122]. Indeed, calculations by *De Gennes* [11.123] showed that the charge (or spin) polarisation induced in its vicinity by an impurity cannot be damped, when the electrons mfp is finite, by an intuitive exponential factor. More recently, *De Châtel* [11.124] gave general arguments suggesting that the conduction-electron mediated indirect interaction between spins in amorphous metals is not exponentially damped. Thus, the electron mfp is not a measure of the range of the interaction. On the other hand, the strength of the RKKY interaction was estimated in dilute magnetic amorphous alloys, either from the approach to saturation magnetization in alloys containing Gd [11.117] or Fe [11.81, 110], or from the concentration dependence of the superconducting transition in Gd doped $La_{80}Au_{20}$ MG's [11.125].

Values thus obtained for the exchange integral J are quite comparable to those determined for equivalent crystalline systems containing Gd or Fe [11.117]. It seems then that despite some model predictions [11.126], the RKKY formalism for indirect exchange interaction in amorphous alloys needs not be drastically modified. The effect of structural disorder in amorphous spin glasses seems to be more clearly evidenced by measurements of the spin-lattice relaxation time for the conduction electrons, as can be achieved through EPR investigations [11.127]. Finally, note that, besides spin-glass and superconducting properties, there are other manifestations of the RKKY interaction in metallic glasses, namely, transferred hyperfine fields on metallic sites carrying no on-site moment within a ferromagnetic amorphous host [11.59], magnetic ordering temperatures and spin-disorder resistivity in RE based MG's. None of these extra probes showed evidence for a drastic alteration of conduction-electron polarisation mechanisms in an amorphous matrix.

c) The Ferromagnetic-to-Spin Glass Transitions

When continuously increasing the magnetic impurity concentration, a magnetic alloy undergoes a series of complex magnetic regimes spanning between standard RKKY spin glasses or cluster glasses and the onset of long-range homogeneous ferromagnetism. These concentrated intermediate systems have attracted great attention in the last few years. Near the critical concentration separating a ferromagnet from a spin glass, unusual magnetic states were discovered which are commonly termed "re-entrant ferromagnetism" (or, alternatively, "re-entrant spin glass") to mean that, by cooling down the sample, ferromagnetism appears and then disappears. The main controversy has been centered on the nature of this double transition, mainly on the ferro to spin-glass transition. Experimental features were found to be phenomenologically the same in various crystalline situations, namely, localized moments in noble metals (AuFe), positive or negative exchange interactions in an exchange-enhanced matrix (PdFe, PdMn, PdFeMn), disappearance of magnetism by virtue of competing interactions (FeCr, NiMn), off-stoichiometry compounds (Fe$_3$Al) and nonmetallic magnetic systems (EuSrS) [11.128]. Re-entrant ferromagnetism was indeed predicted by model calculations [11.129] to occur whenever the width of the distribution of exchange interactions is comparable to the mean value of the distribution. Such a situation is eminently favoured in magnetic MG's, which then offer an abundant reservoir of various alloys appropriate to the study of the magnetic re-entrance phenomena [11.130].

Magnetic phase diagrams of the type displayed in Fig. 11.5 for $(Ni_{100-x}Fe_x)_{79}P_{13}B_8$ metallic glasses were observed in many amorphous systems, including other (Ni, Fe) MG's with various combinations of glass formers such as P, B, Al, Si, and also (Fe, Mn), (Fe, Cr), (Co, Ni), and (Co, Mn) based metallic glasses. A similar diagram was also reported for binary amorphous. FeSn alloys produced by evaporation [11.103]. The magnetic phase diagrams were established from ac susceptibility measurements in zero as well as in small applied dc fields. The nature of the transitions along with the possible ground states for these different magnetic regimes were investigated through different approaches, including dc magnetization, thermoremanent magnetization, hysteresis loops [11.131], magnetic viscosity [11.132, 133], pressure dependence of the low-field ac susceptibility [11.134], temperature dependence of the ferromagnetic resonance linewidth [11.135], Mössbauer spectroscopy [11.136], and small-angle neutron scattering (SANS) [11.137]. Critical lines between the paramagnetic and the spin-glass regions (pg), the paramagnetic and the ferromagnetic regions (pf) and the ferromagnetic and the

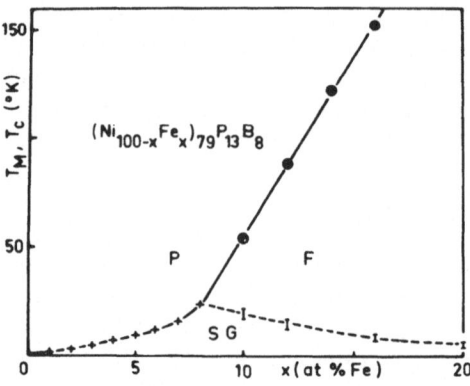

Fig. 11.5. Magnetic phase diagram of $(Ni_{100-x}Fe_x)_{79}P_{13}B_8$ metallic glasses (B. Loegel, J. Durand, quoted in [11.26])

spin-glass regions (fg) intersecting at a multicritical point (MCP) were defined by scaling analysis of *dc* magnetization data at finite field. Data for alloys over the double-transition concentration range can be fit to a magnetic equation of state and critical exponents and the transition temperatures can be determined under the scaling assumptions. Both pf and fg lines were found to be critical lines of second-order phase transitions, but with different critical exponents. It was deduced for both the $(Fe, Mn)_{75}P_{16}B_6Al_3$ and the $Pd_{82-x}Fe_xSi_{18}$ metallic glasses [11.110, 130], that the critical exponents for the fg line are closer to the mean-field values than those associated with the pf transition. More uncertainty remains about the pg line as a distinct critical line. Some criticism was recently formulated on the conclusions drawn from the above scaling analysis, arguing that no sharp feature is observed acting as a signature of a fg phase transition [11.137]. The fg transition temperature is indeed included very much as a parameter in the scaling analysis. In addition, large extrapolations to zero applied magnetic field are required. The picture which emerges from SANS experiments reported so far on alloys lying on the ferromagnetic side of the fg critical concentration is that of a gradual evolution with decreasing temperature into a spin-glass like state [11.137]. In turn it was argued that the values for the momentum transfer q used in the SANS experiments were still too large for the fg transition to be observed. Indeed, ESR linewidth measurements, which involve the long-wavelength limit ($q \to 0$), have proven sensitive to the fg transition [11.138]. Even more speculative than the transition itself appears to be the ground state of the low-temperature phase of those ferromagnetic MG's which are re-entrant. In particular, one wonders whether this low-temperature phase is continuous with the spin-glass region reached across the pg line. An attempt to approach this problem was to measure the spin-wave excitation spectra through inelastic neutron scattering in metallic glasses exhibiting re-entrant ferromagnetism [11.139]. In the ferromagnetic regime, well-defined spin-wave excitations were observed which obey the quadratic relation $E = Dq^2$, with D increasing with decreasing temperature. Around the temperature for re-entrance, D began to decrease when the temperature was lowered, while the spin-wave linewidth started to increase markedly. That the spin-wave excitations are still observed in the low-temperature phase seems to contrast this phase with the canonical spin-glass regime. The progressive smearing-out of the excitation spectrum at low temperature is attributed to a distribution of magnetic states and also to an increased scattering rate as the magnetic disorder increases.

Due to the frequent occurrence of re-entrant ferromagnetism in MG's, it would be of particular significance to determine at least qualitatively the influence of structural disorder on the parameters of the re-entrant phenomena. This was tried by comparing MCP's in $(Fe_xCr_{1-x})_{75}P_{16}B_6Al_3$ MG's (MCP at $x \simeq 0.65$ and $T \simeq 25$ K) and in Fe_xCr_{1-x} crystalline alloys (MCP at $x \simeq 0.20$ and $T \simeq 60$ K) [11.140]. It appeared that, even once the alloying effects were taken into account in the amorphous system compared with the crystalline one, it remained a difficult task to assess separately the effects of composition disorder and those of structural disorder on the distribution width of the exchange interaction. On the other hand, the situation is made more complex in alloys where exchange interactions of both signs are likely to occur such as in (Fe, Mn) based MG's. Along this line, the $Fe_{92}Zr_8$ metallic glass has to be regarded as a special case of magnetic re-entrance owing to its very high Fe content together with its intriguing Invar properties [11.141, 142]. The appearance of a low-temperature spin glass-like phase was first explained in terms of the coexistence of ferromagnetic and antiferromagnetic states as in other Invar alloys. But recent Mössbauer investigations of the Fe hyperfine field distribution at low temperature showed evidence for a low-field tail (down to $H = 0$) of the distribution, but with no antiferromagnetic component [11.143]. The low-temperature magnetic phase seems to be associated here with the coexistence of Fe with large and small moments. Finally, let us note that random anisotropy in metallic glasses can give rise to anomalous low-field critical behaviour of the type predicted by *Aharony* and *Pytte* [11.144]. Such deviations from normal ferromagnetic behaviour, but without any second, low-temperature transition to a spin-glass state, were pointed out in $(Fe, Mo)_{75}P_{16}B_6Al_3$ metallic glasses [11.145]. The origin of the random anisotropy in these MG's is attributed to the difference in the spin-orbit coupling constants of Fe and Mo. It is out of the scope of this review to discuss the implications on the temperature dependent electrical resistivity of the various manifestations of magnetic inhomogeneities encountered so far in MG's. Let us just mention that, owing to their very short mfp, electrons in MG's are expected to probe the small magnetic entities rather than the bulk of the magnetic alloys. This is a possible cause for the low-temperature resistivity anomalies observed in most of the apparently homogeneous ferromagnetic MG's.

11.3.3 Ferromagnetic Metallic Glasses Based on Transition Metals

Since the first report on a ferromagnetic FePC metallic glass [11.146] exhibiting high saturation magnetization and T_c, low coercive field, good stability at room temperature with respect to crystallization, and then offering a prospect for technical application, a very large variety of ferromagnetic MG's based on transition metals have been the subject of intense theoretical and experimental investigation. Among the basic magnetic properties, emphasis will be placed in the first section on those more directly characterizing the ground state, namely, the magnetic moment at zero temperature and the Curie temperature. In the second section we will analyze the magnetic properties at finite temperature, such as the magnetic excitations and critical phenomena. We will not review here the abundant literature on the linear saturation magnetostriction λ_s and its temperature and composition dependence in MG's. These data are also of fundamental significance [11.147], but they are more commonly reviewed along with the magnetic properties for technical applications.

a) Magnetic Moment and Curie Temperature

The effect of structural disorder on ground state properties of ferromagnetic amorphous alloys is not a problem that can easily be handled from a general point of view. Such an effect was found to depend considerably upon the family of amorphous alloys under consideration. We mentioned in the introduction the cases of amorphous alloys, such as those with compositions corresponding to Laves or Heusler compounds, where the effect of amorphicity was found to be drastic. More work is expected on these interesting amorphous materials in order to elucidate the mechanisms responsible for such dramatic effects (e.g., atomic arrangements, electronic band structure, exchange interactions). Unfortunately, metallic glasses of similar composition (if technically feasible) have not been studied so far, with few exceptions such as liquid-quenched YFe_2 [11.22]. Most of the ferromagnetic MG's investigated so far belong to the following families of amorphous alloys, namely, the alloys of T with sp elements (FeB or CoB type) and the alloys of early and late transition metals (FeZr or CoZr type). These MG's are produced over composition ranges which encompass stable or metastable crystalline compounds in the phase diagrams. The effect of amorphicity has to be discussed with reference to the magnetic properties of both the compositionally related crystalline compounds and ideally pure amorphous transition metals. As expected, magnetic moment and T_c are largely affected by structural disorder when Fe content increases in Fe based MG's, while the effects are tiny in Ni or Co based MG's. We analyse these points below.

The average Fe(Co) moment was found to be practically the same in crystalline and amorphous modifications of Fe(Co) compounds such as Fe_3B [11.148], Fe_3P [11.149], Co_3B [11.61], Co_2B [11.149], and Fe_3Si and Fe_5Si_3 [11.102]. In a compound such as Fe_3B, there are three inequivalent Fe sites with three values for the Fe moment. The Fe field distribution in amorphous Fe_3B as obtained by Mössbauer spectroscopy can be interpreted as resulting from a Gaussian distribution (half-width: 30 kOe) of the three crystalline lines occurring at 230, 275, and 300 kOe, respectively [11.150]. Although this analysis of the field distribution is not necessarily unique [11.151], it clearly demonstrates that the distribution of the Fe moments, as can be deduced from the hf coupling constant, is rather narrow around a mean value practically identical to that for Fe in the crystalline counterpart. The case of Fe_3B might be an exceptional one in terms of this moment distribution. The distribution is more asymmetrical in the FeP amorphous system [11.152]. On the other hand, the moment distribution is considerably broadened in amorphous alloys with smaller Fe content, so that the magnetic properties of the amorphous phase can be quite different from those of the corresponding crystalline compounds (e.g., FeSi, FeSn [11.102, 103]). The effect of topological disorder on the value of T_c seems to be more observable than on the average moment. For example, as a result of amorphicity, T_c is lowered by about 10% in Fe_3B [11.153, 154] and $Fe_{75}P_6B_{19}$ [11.155], and by about 20% in Fe_3P [11.149, 156]. Accurate comparison is made difficult in Co based MG's for they undergo crystallization before reaching T_c.

In fact, metallic glasses having a composition corresponding to a single-phase crystalline compound are exceptions more than the rule. Typical concentration ranges are 13 to 40 at.% B for the MG's of the CoB, FeB type (deep eutectic around 18 at.% B) and 9 to 12 at.% Zr for the ferromagnetic MG's of the CoZr, FeZr type. Evaporated and sputtered amorphous films can be produced over broader concentration ranges. The concentration dependence of the Co moment in CoB MG's, as deduced from bulk magnetic measurements and NMR spectroscopy, yields for amorphous Co a value close to that for crystalline Co. No significant difference is observed in the magnetization of crystalline CoSi and CoP alloys and of the corresponding amorphous alloys obtained by evaporation [11.157] and electrodeposition [11.158], respectively. The effect of amorphicity on the Co moment was found to be negligibly small in amorphous alloys of Co with various transition metals such as Ti [11.159], Zr [11.160], and Nb and Ta [11.161]. In all these latter alloys produced by sputtering or coevaporation, the gradient of the magnetization versus concentration of nonmagnetic element is about the same. Similarly, magnetization in amorphous evaporated CoSn [11.162] extrapolates rather well to that measured for crystallized CoSn solid solution. All these data are in good agreement with conclusions drawn from studies of amorphous pure Co films [11.12].

Measurements on amorphous Ni thin films [11.12], along with extrapolations from amorphous NiAg films [11.163], pointed to a small influence of structural disorder on the moment and the T_c of pure amorphous Ni. Values of $0.4 \mu_B$ and 540 K, respectively, are proposed for amorphous Ni, as compared with $0.6 \mu_B$ and 627 K for crystalline Ni.

The concentration dependence of magnetic properties in Fe base metallic glasses raises more complex problems. The average Fe moment decreases with the increasing content of a nonmagnetic element. Over the concentration range where amorphous alloys can be produced by liquid quenching, this decrease is

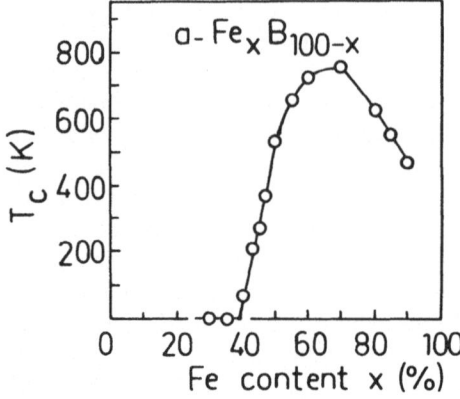

Fig. 11.6. Magnetic phase diagram of amorphous FeB alloys [11.25]

rather monotonous, except for a change of slope reported around the eutectic composition in FeB metallic glasses [11.148, 153]. But, in evaporated amorphous FeSi, FeAu, FeO films, a steep decrease of the Fe moment was observed for Fe concentrations of the order of 95 at. % [11.164], so as to suggest for amorphous Fe a very low value of the moment (values for pure amorphous Fe films range between 0 and $1 \mu_B$/Fe at. % [11.12]). Extrapolations made from magnetization in FeZr metallic glasses over the 9–12 at. % Zr range also yielded vanishingly small values for both the moment and the T_c of pure amorphous Fe [11.165]. In FeB metallic glasses, the T_c decreases with decreasing B content (down to 13 at. % B), while the Fe moment increases. For this family of amorphous alloys, the magnetic properties do not depend significantly upon the fabrication technique. One is thus allowed to use values obtained for sputtered FeB alloys in order to build up the magnetic phase diagram presented in Fig. 11.6. The T_c undergoes a maximum for concentration around 30 at. % B, then decreases to vanish at around 60 at. % B. The magnetic phase diagram for the amorphous FeZr system is basically the same, except for the position of the maximum at around 20 at. % Zr [11.141, 165]. Interestingly enough, both families of metallic glasses were found to exhibit Invar properties and, therefore, a series of anomalies which will be discussed below.

Both localized and itinerant electron models for ferromagnetism were used to account for the compositional dependence of T_c in amorphous FeB alloys. Among approaches of the first type, let us mention that of *Kaneyoshi* within a simple molecular field theory [11.166]. More directly connected to the Invar problem is the itinerant electron model approach [11.11, 167]. The FeB alloys are characterized as weak (both d subbands are partially filled at E_F) or strong (one group of d subbands is completely filled at E_F) ferromagnets depending on the B content. Use is made of the criterion that for weak itinerant ferromagnets, T_c and magnetization vary in a parallel manner on alloying, which is the case for B content larger than 30 at. % and, possibly, for very low ($< 10\%$) B concentrations. Over the intermediate concentration range where T_c and moment vary in an opposite way, the FeB alloys would be strong ferromagnets.

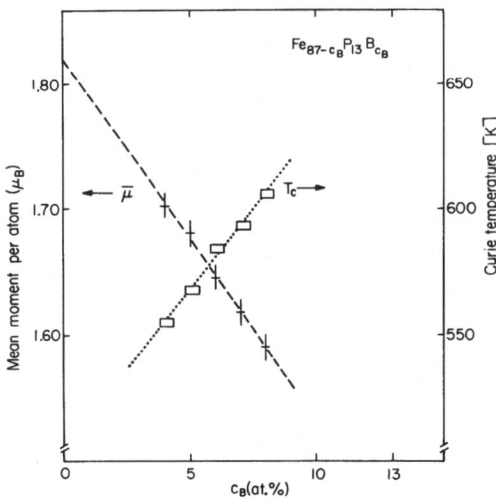

This interpretation is consistent with Wohlfarth's model for Invar properties in crystalline FeNi alloys.

A similar conclusion was drawn from analyses of the alloying effects on the magnetization of the $Fe_{79}P_{13}B_8$ metallic glass [11.111, 168]. The magnetization gradient for dilute transition-metal impurities substituted for Fe in amorphous $Fe_{79}P_{13}B_8$ was interpreted according to the virtual bound-state model of *Friedel* [11.169] for T impurities in crystalline Ni. The screening in the $d_\uparrow$ bands was found to be very small for T designating Ni and Co, while for Mn, Cr, and V, repulsive potentials are large enough to repell d bound states from the $d_\uparrow$ bands. This screening mechanism for T impurities in the $Fe_{79}P_{13}B_8$ metallic glass was concluded to be basically the same as in crystalline strong ferromagnets (Ni, Co) where, for small impurity excess charges, there is no displaced charge in the completely filled $d_\uparrow$ subbands. Note that in $Fe_{79}P_{13}B_8$, as in $Fe_{100-x}B_x$ ($13 \leqq x \leqq 25$) MG's, the moment and T_c vary in an opposite way, as shown in Fig. 11.7. This analysis of the screening effects on alloying illustrates the meaning of the displaced Slater-Pauling curves for (Fe, Co, Ni) PB alloys with respect to Slater-Pauling curves for crystalline Fe, Co, Ni alloys [11.170]. Such a rigid-band model approximation can hold only for impurities with small excess electronic charge with respect to the matrix. On the other hand, the displacement of the Slater-Pauling curves is not to be understood literally in terms of a charge transfer of sp electrons from P or B to the transition-metal d bands, but rather in the sense of $sp-d$ hybridization [11.171]. Finally, let us refer to an analysis of the magnetization of $Co_{80-x}T_xB_{20}$ glasses (T: Fe, Mn, Cr, V) in terms of electronic structure [11.172] by following the same method as for the $Fe_{79-x}T_xP_{13}B_8$ glasses (Fig. 11.8).

Systematic studies have also been performed on the influence of a small amount of transition metals on the saturation magnetization and T_c of

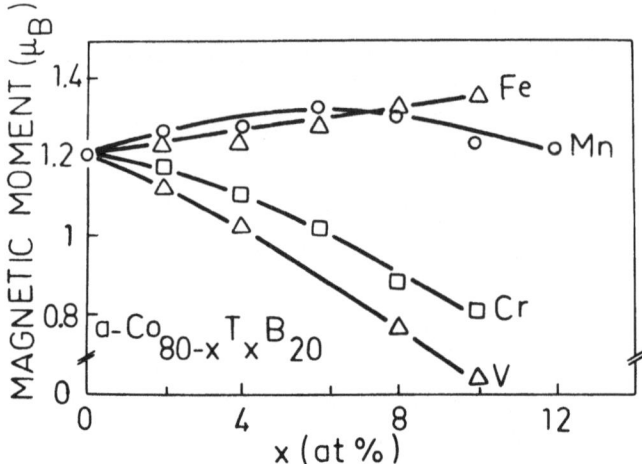

Fig. 11.8. Saturation moments per transition-metal atom versus T content at 4.2 K in $Co_{80-x}T_xB_{20}$ metallic glasses [11.172]

$Fe_{80}T_3B_{17}$ (T: transition metals of the second series) [11.173]. Substitution of a large amount of Mo, Cr, Mn for Fe in various metallic glasses was investigated by means of bulk and local magnetic measurements (see references in [11.59]). More easily accountable in terms of electronic structure are the concentration dependences of the moments and Curie temperatures in the cases of complete (Fe, Ni), (Fe, Co), (Fe, Mn), (Ni, Co), (Co, Mn) substitutions for different MG's containing various amounts of glass formers such as $(P, B)_{20}$, $(SiB)_{20}$, $P_{16}B_6Al_3, P_{20}, B_{20}, B_{14}, P_{13}C_7, P_{15}C_{10}, Zr_{10}$, etc. (references to original works can be found in [11.6–9, 59, 78]). Due to lack of space, it is not possible to review here all these data in detail. Two remarks seem to be of particular significance with respect to the atomic arrangements in these MG's. First, the concentration dependence of T_c (maxima in the Fe–Ni and Fe–Co substitutions) and of the individual moments per transition-metal atoms are basically the same in crystalline borides and phosphides and in corresponding MG's [11.59]. A second remark is about the concentration dependence of the saturation magnetic moment in $(Fe, Ni)_{86}B_{14}$ glasses, which is reminiscent of the Invar anomaly observed in crystalline (Fe, Ni) alloys [11.174].

A huge amount of experimental work has been devoted to studying the respective influence of different $s-p$ elements on the basic magnetic properties of ternary ferromagnetic MG's such as CoBC, CoBSi, CoBGe, CoBAs, FeBC, FeBAl, FeSi, FeBGa, FeBGe, FeBSiC, FePC, FePSi, FePGe, and FePAl (references to original works in [11.7–9]). Among the exotic amorphous alloys recently investigated, let us mention the Be-substituted $Fe_{82}B_{18}$ MG's [11.175] and the electron-beam deposited amorphous (Fe, Co)Bi alloys [11.176]. The whole of the experimental data is far from being completely understood. Charge transfer is invoked as the driving parameter for the concentration dependence of the moment, while metalloid size effect would be the predominant factor governing the variation of T_c [11.149]. Changes in the short-range order can be superimposed, as evidenced in FePB glasses [11.155]. An approximate picture of the metalloid dependence of both saturation magnetization and the T_c was proposed within the framework of the band magnetism model [11.177]. General trends were thus semi-quantitatively described. The concentration dependence of T_c in transition-metal substituted metallic glasses was fit to phenomenological expressions based on the molecular field theory to yield values for the nearest-neighbor interactions between different species of transition metals [11.178]. The metalloid size effect on T_c was expressed in terms of empirical Slater-Bethe curves (the equivalent for T_c of the Slater-Néel curve for the exchange interaction) [11.156].

The pressure derivative of T_c was measured on metallic glasses of both the FeB and FeZr types in relationship with Invar properties [11.179]. These data were discussed in detail within the formalism of weak itinerant ferromagnetism [11.74]. The relationship between the pressure derivative of T_c and spontaneous magnetization was generalized to the case of strong itinerant ferromagnets [11.180].

b) Magnetic Excitations, Invar Properties, and Critical Phenomena

Magnetic excitations, especially spin waves, in amorphous alloys have attracted a considerable amount of work among theoreticians and experimentalists [11.4]. The existence of spin-waves in amorphous ferromagnets was first ascertained by means of bulk magnetization [11.181] and inelastic neutron scattering experiments on the electro-deposited Co_4P and the FePC metallic glass [11.182]. In a simple spin-wave formalism, the spontaneous magnetization decreases with temperature T according to

$$M_s(T) = M_s(0)(1 - BT^{3/2} - CT^{5/2}...).$$

The coefficient B is related to the stiffness constant D, which in turn is proportional to the exchange integral, to the magnetic moment and the square of interatomic distance. D can be measured directly by low-angle inelastic neutron scattering experiments which allow one to verify the ferromagnetic dispersion relationship:

$$\hbar\omega = E_0(T) + D(T)q^2 + E(T)q^4.$$

D was also determined in MG's through Brillouin light scattering [11.183]. Ferromagnetic resonance was widely used to measure the temperature dependence of M_s [11.135]. When good agreement is observed between the values of the spin-wave stiffness constant obtained directly from neutron experiments and indirectly from magnetization measurements, then support is given to the idea that magnetic excitations are the dominant mechanism for the decrease in spontaneous magnetization. Such an agreement was ascertained for $(Fe_xNi_{1-x})_{75}P_{16}B_6Al_3$ MG's ($x \geq 0.5$) [11.184]. But in many cases, the coefficient B was found to be too large, yielding then for D a value significantly smaller than that measured by the neutron technique [11.184]. Extrinsic reasons for these discrepancies can be numerous [11.185]. First, both types of measurements have to be performed on the same sample. Second, magnetic inhomogeneities can be blamed. Third, in (NiFe) MG's with Fe content less or equal to 30 at. %, there are re-entrant ferromagnetism phenomena at low temperature which can make the spin-wave theory inadequate [11.186]. Fourth, in some cases, better agreement is obtained when the $T^{5/2}$ term in $M_s(T)$ is taken into account [11.187]. More intrinsically, it seems that such discrepancies are systematically related to the Invar effect in metallic glasses (FeB) as well as in crystalline alloys ($Fe_{65}Ni_{35}$, Fe_3Pt) [11.188, 189]. Note that in zero-magnetostrictive MG's, the same value for D is deduced from $M_s(T)$ and from spin-wave spectra measurements [11.190]. More experimental work is needed along this line, but there is no real evidence as yet for additional low-energy short-wavelength excitations which could contribute to $M(T)$ in MG's. Finally, let us point out that for a series of Fe and Co based MG's, D was found to scale with T_c as predicted by the model for weak itinerant ferromagnetism [11.191]. The ratio D/T_c gives an estimate of the range of the exchange interaction (about $0.2 \, \mathrm{meV \, \AA^2 K^{-1}}$).

An intriguing observation was reported on sharp magnetic excitations for values of q around the first diffraction maximum. The dispersion curve was found to exhibit a minimum at $3.05 \, \mathrm{\AA}^{-1}$ which was termed "roton-like" [11.185] and assigned to localized magnons [11.192]. These early conclusions were critically re-examined after recent neutron measurements of magnetic excitations on several Fe based amorphous ferromagnets [11.193].

Much attention has been paid in the literature to the overall temperature dependence of magnetization. In fact, the flattened curves of reduced magnetization versus reduced temperature are not an effect of structural disorder. These curves are practically indistinguishable for the crystalline and amorphous phases of $(Fe_{0.5}Ni_{0.5})_3B$ as well as for crystalline Fe_3Si and the $Fe_{78}B_{12}Si_{10}$ metallic glass [11.194]. On the other hand, similar features were observed in crystalline Invar alloys. It seems then again that the Invar problem is the key problem to understanding many apparent anomalies in the properties of ferromagnetic MG's.

Basic characteristics of the Invar behaviour include (i) very small or negative thermal expansion coefficients, (ii) a large positive volume magnetostriction, and (iii) strong ΔE effects [11.195]. FeB and FeZr based MG's exhibit these characteristics. Other aforementioned anomalies are also related to the Invar behaviour in both Invar metallic glasses and Invar crystalline FeNi alloys, namely [11.196], a large negative pressure derivative of T_c, an anomalous concentration dependence of the magnetic moment and of T_c [11.155, 197], flattened curves of reduced M_c versus reduced T, discrepancies between values of D obtained by neutron and magnetization measurements, large high-field magnetic susceptibility [11.198] and a large value for the γ coefficient in low-temperature specific heat [11.199]. The origin of this set of anomalies in crystalline FeNi alloys is still the subject of intense debate. Explanations span from an homogeneous approach based on band theory [11.200] to an inhomogeneous picture of finite clusters [11.201]. Between those two extremes, analogy is suggested between Invar alloys and mixed-valence systems, both having in common the near degeneracy of two possible ground states (ferro and paramagnetic for Invar alloys) that differ significantly in their volumes [11.202].

It has been demonstrated that the critical behaviour of homogeneous ferromagnetic MG's is not affected much by structural disorder. A well-defined magnetic transition is observed with values for critical exponents β, γ, δ (magnetization and susceptibility measurements) and α (resistivity and specific heat) which obey the standard scaling laws (references can be found in [11.6, 78]). However, several features widely observed in metallic glasses are noteworthy. First, the value of β in MG's tends to be slightly enhanced with respect to the model value for a three-dimensional Heisenberg ferromagnet. Secondly, due to superparamagnetic behaviour subsisting over a broad temperature range above T_c, the value of the exponent γ is very sensitive to the temperature range over which the susceptibility was fitted. In the temperature range $\varepsilon \leqq 0.06$ [$\varepsilon = (T - T_c)/T_c$], one obtains for γ a value which is close to that of the three-dimensional Heisenberg model ($S = \infty$). This was recently verified in $Fe_{40}Ni_{40}P_{14}B_6$, $Fe_{20}Ni_{60}P_{14}B_6$, $Fe_{10}Ni_{70}B_{19}Si_1$ and $Fe_{13}Ni_{67}B_{19}Si_1$ metallic glasses [11.203]. When γ is determined for temperatures higher than $\varepsilon = 0.06$, it is found to fit the susceptibility over a very broad temperature range (of the order 100 K) and its value is about 1.6 [11.204]. This strong temperature dependence of γ [11.205] is regarded as a direct consequence of the fluctuation in the magnetic interactions inherent to the amorphous nature of the MG's [11.56, 76]. Finally, it is remarkable that so far the exponent α has always been found to be negative [11.206] which is the necessary condition for a homogeneous phase transition to occur in disordered materials.

11.4 Rare-Earth Base Metallic Glasses

Studies performed on RE base MG's are considerably less abundant than investigations on RE base amorphous alloys produced by evaporation or sputtering. Magnetic properties of amorphous films have been studied mainly from an application-oriented viewpoint, while liquid-quenched ribbons or foils containing RE have remained so far a laboratory curiosity especially attractive for fundamentalists. These reasons, together with the scope of the book, speak for a rather restricted review of magnetic properties of RE base amorphous alloys. However, binary RE base MG's are with respect to concentration quite complementary to binary amorphous films because RE base MG's are easily produced on the RE rich side around the composition of a deep eutectic, which is a composition range hardly accessible to amorphous films.

The localized model for magnetism is quite appropriate for describing the properties of normal RE ions, with the following conceptual inputs: the moment predicted by Hund's rules, an exchange mechanism of the indirect type and, for non-S state RE ions, the local anisotropy. The picture is more complex for mixed-valence ions. We first review the magnetic properties of MG's containing RE in a S state (Gd^{3+}, Eu^{2+}), then the MG's with non-S state RE ions where anisotropy is expected to play a major role. In a third section, we will analyze the data on MG's with intermediate valence ions. Finally, we will present some preliminary studies of MG's containing uranium and of transition-metal MG's containing RE additives.

11.4.1 Metallic Glasses Containing Gd or Eu

A large variety of Gd base MG's has been studied, namely, Gd rich alloys with various amounts of sp elements (C, Al, Ga), or of noble metals (Cu, Au) or transition metals (Fe, Co, Ni, Mn, Pd, Rh, Ru, Pt) ranging between 18 and 30 at. %. The composition of these MG's is close to that of a deep-lying eutectic in the phase diagrams [11.23, 207–210]. Eu was found to be divalent in a series of $Eu_{100-x}Ag_x$ metallic glasses ($19 \leq x \leq 82$) [11.211] and in the $Eu_{80}Au_{20}$ glass [11.212]. The effect of disorder can be discussed on the moment, on the exchange and on the magnetic ordering temperatures.

The Gd atoms in crystalline Gd carry a moment of 7.55 μ_B, 7 μ_B arising from the f electrons and 0.55 μ_B being attributed to a polarisation of the conduction electrons. A possible effect of disorder could be a damping of the conduction electron polarisation so as to reduce the Gd moment to 7 μ_B in metallic glasses. Such an argument is frequently presented but it is not substantiated by experiments. Indeed, Gd carries a moment of 7.45 μ_B per atom in a $Gd_{65}Ni_{35}$ MG, where Ni is not likely to bear any moment [11.213]. Moreover, values of the Gd moment in crystalline intermetallic compounds scatter between 7 and 7.5 μ_B. In the $Eu_{80}Au_{20}$, the Eu moment was found to be about 7 μ_B per atom [11.33, 212]. Due to the electronic configuration of Eu^{2+}, the contribution to the moment arising from conduction electron polarisation is expected to be weaker than for Gd. However, it seems that the main effect of structural disorder would be to cause a slight misalignment of the Eu spins according to the schemes discussed by *Coey* [11.3]. In view of the above reasons, very small values (about 5.5 μ_B) reported for the Gd moment in the $Gd_{78}Al_{22}$, $Gd_{76}Pd_{24}$, and $Gd_{82}Rh_{18}$ MG's have to be attributed to an admixture of antiferromagnetic exchange or to some extrinsic reasons [11.209].

A remarkable feature in the Gd base MG's is that all of them so far have been found to be ferromagnetic, including those which have a composition close to an antiferromagnetic crystalline compound. This was also observed by *Boucher* in the sputtered GdAg alloy [11.214]. Such a prevailing ferromagnetism in Gd base MG's together with a conjectural damping of the RKKY interaction in MG's having a very short mfp led *Buschow* [11.209] to propose

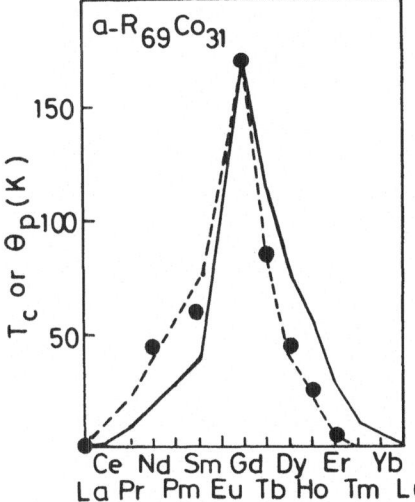

a-R$_{69}$Co$_{31}$

T$_c$ or Θ_p (K)

150

100

50

Ce Nd Sm Gd Dy Er Yb
La Pr Pm Eu Tb Ho Tm Lu

Fig. 11.9. Curie temperatures in various R$_{69}$Co$_{31}$ metallic glasses. The solid line represents the function $(g_J - 1)^2 J(J + 1)$ normalized to the Curie temperature of Gd$_{69}$Co$_{31}$. The broken line represents the correction indicated in the text [11.215]

an indirect exchange mechanism for RE in amorphous alloys, which is an alternative to the standard RKKY coupling. This second coupling scheme would be mediated only by the $5d$ electrons. As a result, this d polarisation would remain positive and it would be restricted to the nearest-neighbor atoms. The problem of the RKKY interaction in MG's was discussed above. This type of long-ranged indirect interaction is required to account for magnetic interactions between dilute impurities in MG's. There is no obvious reason why the same interaction should be inadequate in more concentrated systems. Fluctuations of the atomic environments in MG's can favour ferromagnetism, or an admixture of interactions of both signs with a prevalency of the positive ones, without requiring a new mechanism. Nevertheless, it is known that hybridized conduction electrons which interact with f electrons have a predominant d symmetry at the Fermi level. This holds for both crystalline and amorphous RE base compounds.

The indirect ($sp - d$ mediated or d mediated) interaction between RE ions in magnetic MG's is basically the same in nature and strength for a given series of RE base MG's of the same composition. This is illustrated in Fig. 11.9, where the ordering temperatures for RE$_2$Co MG's [11.215] follow qualitatively the variation of the de Gennes factor. However, systematic departures of experimental results with respect to model predictions (experimental results are above for light RE, and below for heavy RE) are observed, which Buschow suggested is accounted for by a correcting spin-orbit coupling term $[-0.25(g_J - 1)J(J + 1)]$. The same trends are observed for the ordering temperatures in RE$_{69}$Ni$_{31}$ and RE$_{60}$Fe$_{40}$ metallic glasses [11.216, 217]. In conclusion, it seems that fluctuations in MG's affect the sign of the average exchange interaction and their strengths, but these quantities remain the same for the RE series and are little altered by random anisotropy.

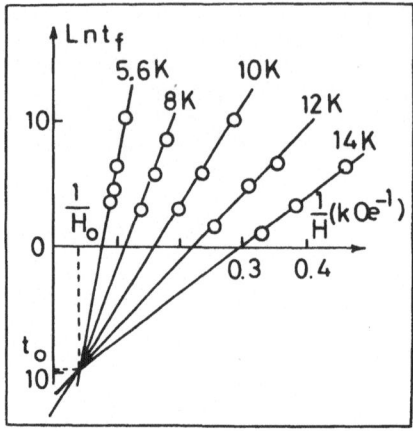

Fig. 11.10. Log Δt (with $\Delta t = t_f - t_i$) versus reverse applied field at various temperatures for the amorphous sputtered $Tb_{52}Ag_{48}$ alloy [11.222]

11.4.2 Metallic Glasses Containing Non-S State RE Ions

The major role of the magnetocrystalline anisotropy in magnetic MG's containing non-S RE ions has been stressed by many authors [11.13]. According to the HPZ model, the anisotropy in MG's was assumed to be completely random and of uniaxial type. The limitations of such a model have already been discussed for the nonmagnetic MG's containing RE ions. The HPZ approximation is likely to be more acceptable for concentrated magnetic alloys, although from polarised neutron experiments, evidence was shown for a correlation between the local easy-magnetisation axes in an amorphous sputtered $ErCo_2$ alloy [11.218]. Competition between single-ion anisotropy and exchange in amorphous alloys results in (more or less random) noncollinear magnetic structures whose ideal types were termed speromagnetic, asperomagnetic and sperimagnetic by *Coey* [11.3]. To what extent these magnetic ground states differ from concentrated cluster glasses is a difficult theoretical problem.

In fact, the phenomenology of the magnetic properties of these noncollinear magnetic MG's resembles very much that of canonical spin glasses. Very sharp peaks in the ac susceptibility were observed whose temperature depends upon the frequency [11.212, 219]. Thermomagnetic history effects were evidenced by many groups on various MG's containing large amounts of non-S RE ions [11.212, 215–217, 220, 221]. Both the remanence and the magnetization under applied fields exhibit magnetic viscosity phenomena, of the type shown in Fig. 11.10 [11.212, 219–223]. It is important to note that these phenomena are observed not only in spin glasses, but also in many low-symmetry crystalline compounds with noncollinear magnetic structures [11.209, 224]. However, it seems that the characteristic times involved in the magnetic aftereffects are in marked contrast to those measured in equivalent crystalline systems. The same conclusion was drawn from a Mössbauer study of a sputtered DyAg amorphous alloy [11.225]. Finally, anomalies occur in the critical behaviour of these magnetic amorphous alloys [11.223, 226] which were pointed out a long time ago [11.14] and which resemble those predicted by *Aharony* and *Pytte* [11.144]. But the existence and the nature of the predicted magnetic state remains an open question.

11.4.3 Metallic Glasses with Intermediate Valence RE Ions

The problem of intermediate valence RE ions in metallic glasses has gained little attention so far. However, it seems that such an investigation could shed some light on both the question of SRO in MG's and the problem of intermediate valence RE in itself. This will be illustrated by two examples, namely, the mixed-valence state of Sm at the surface of $La_{80-x}Sm_xAu_{20}$ MG's [11.227] and in the volume of a $Ce_{75.5}Co_{24.5}$ metallic glass [11.228]. One might expect, indeed, that in a glassy alloy, fluctuations in the coordination number and interatomic distances together with fluctuations in local symmetry parameters would stabilize a spatially inhomogeneous mixture of different valence configurations of instable RE ions. Such a reasoning is invoked, in particular, to account for the inhomogeneous admixture of Sm^{2+} and Sm^{3+} at the surface of pure crystalline Sm. If fluctuations were large enough within the volume of a Sm base MG, there should be no difference between the valence states at the surface and in the volume of the glassy alloy. In fact, combined x-ray absorption edge and ESCA measurements showed that it is not true. The same differences as in Sm metal were observed in $La_{80-x}Sm_xAu_{20}$ metallic glasses, which speaks for a rather well-defined SRO in this family of MG's, in agreement with information obtained from other different techniques. Similar conclusions can be drawn for the Ce ions in various MG's. The Ce ions were found to be purely trivalent and magnetic in $Ce_{80}Au_{20}$, $Ce_{72}Cu_{28}$ and $Ce_{89}Al_{11}$ MG's, while magnetic, thermal and transport properties of the $Ce_{75.5}Co_{24.5}$ metallic glass indicate a homogeneous mixed valence state of the Ce ions. This might also imply quite homogeneous atomic arrangement in the glassy matrix with rather minor local fluctuations.

11.4.4 Uranium Base MG's and Transition-Metal Base MG's with RE Additives

Metallic glasses of uranium with Fe, Co, Ni can be produced over a concentration range close to a deep-lying eutectic on the U rich side of the equilibrium diagram. Magnetization and susceptibility measurements were performed on $U_{66}Ni_{34}$, $U_{66}Co_{34}$, and $U_{66}Fe_{34}$ MG's [11.229]. The first alloy is not magnetic, while the alloys with Co and Fe are best described as cluster glasses with thermomagnetic history effects below about 100 K and a broad maximum of the dc susceptibility around 50 K. Application of an environmental model for the onset of a magnetic moment on Fe atoms yields an effective concentration of 0.3 at. % of Fe carrying a moment of $2\mu_B$ (elemental uranium is a Pauli paramagnet).

Recent investigations were undertaken on the influence of the addition of a few percent of RE in magnetic properties of the FeB and FeSiB type of alloys [11.230–234]. Magnetization decreases linearly with RE content, since the RE moments are antiparallel with respect to the Fe moments. The effect of RE on the T_c is more complex. The T_c can be slightly enhanced for small amounts of RE additives. Interest in this family of alloys has grown for technical reasons (magnetostrictive properties). But these novel alloys can also offer a new field of investigation for fundamentalists. Note that some MG's with high RE content such as $(Pr_{80}Ga_{20})_{80}Fe_{20}$ [11.235] amorphous alloys were found to be phase separated in the glassy state.

11.5 Concluding Remarks

Despite the considerable progress achieved in recent years concerning the knowledge of the atomic structure of MG's and concerning an understanding of their basic magnetic properties, several unsolved problems were encountered in this review. Questions are still raised as to the medium or long-range structure of those MG's having composition close to that of a deep-lying eutectic. These alloys are known to be prone to phase separation. The question is then about the relationship between their structure and their magnetic properties such as cluster glass behaviour, re-entrant ferromagnetism and Invar anomalies. Other problems remain as to the range of the indirect exchange interaction or the critical behaviour in these highly disordered materials. Also, the magnetic ground states of cluster glasses and noncollinear structures in amorphous alloys or crystalline compounds of low symmetry are far from being fully characterized. These problems and others make the study of the magnetic properties of MG's a fascinating field of research.

Acknowledgements. I would like to thank Drs. H. Beck and H. J. Güntherodt for their constant encouragement. I am grateful also to the authors who sent me their preprints. I appreciated very much the collaboration with Mrs. Brun and Mr. Kaul during the preparation of the manuscript.

References

11.1   P. Duwez: In *Glassy Metals I*, ed. by H. J. Güntherodt, H. Beck (Springer, Berlin, Heidelberg, New York 1981) pp. 19–23

11.2   For a bibliography on amorphous magnetism and magnetic materials over 1950–1976, see S. Kobe, A. R. Ferchmin: J. Mat. Sci. **12**, 1713–1749 (1977)

11.3   J. M. D. Coey: J. Appl. Phys. **49**, 1646–1652 (1978)

11.4   R. Alben, J. I. Budnick, G. S. Cargill III: In *Metallic Glasses*, ed. by J. J. Gilmann, H. J. Leamy (American Society for Metals, Metals Park, Ohio, USA 1978) pp. 304–339

11.5   S. Methfessel: In *Liquid and Amorphous Metals*, ed. by E. Lüscher, H. Coufal (Sijthoff and Noordhoff, Alphen 1980) pp. 501–521

11.6   K.Handrich, S.Kobe: *Amorphous Ferro- and Ferrimagnetika* (Akademie-Verlag, Berlin 1980)
       G.A.Petrakovskii: Sov. Phys. Usp. **24**, 511–525 (1981)
11.7   C.D.Graham, T.Egami: Ann. Rev. Mat. Sci. **8**, 423–457 (1978)
11.8   F.E.Luborsky: In *Ferromagnetic Materials*, Vol. 1, ed. by E.P.Wohlfarth (North-Holland, Amsterdam 1980) Chap. 6, pp. 451–529
11.9   T.Masumoto: In *Metallic Glasses: Science and Technology*, ed. by C.Hargitai, I.Bakonyi, T.Kemény (Central Research Institute for Physics, Budapest 1981) pp. 121–132
11.10 J.Schneider, K.Zaveta: In *Proc. of the Fourth Intern. Conf. on Rapidly Quenched Metals*, Sendai, ed. by T.Masumoto, K.Suzuki (The Japan Institute of Metals, Sendai 1982)pp. 1067–1072
11.11 E.P.Wohlfarth: IEEE Trans. MAG-**14**, 933–937 (1978); in *Amorphous Metallic Alloys*, ed. by F.E.Luborsky (Butterworths, London 1983) Chap. 15, pp. 283–299
11.12 P.J.Grundy: J. Mag. Mag. Mat. **21**, 1–23 (1980)
11.13 R.W.Cochrane, R.Harris, M.J.Zuckermann: Phys. Rep. **48**, 1–32 (1978)
11.14 J.J.Rhyne: In *Handbook on the Physics and Chemistry of Rare-Earths*, Vol. 2, ed. by K.A.Gschneidner, L.Eyring (North-Holland, Amsterdam 1979) Chap. 16, pp. 259–294
11.15 U.Gonser (ed): *Application of Nuclear Techniques to the Studies of Amorphous Metals*, Atom, En. Rev., Suppl. No. 1 (IAEA, Vienna 1981)
11.16 Zs. Kajcsos, I.Dézsi, D.Horváth, T.Kemény, L.Marczis, D.L.Nagy (eds.): *Proc. of the Intern. Conf. on Amorphous Systems Investigated by Nuclear Methods*, Balatonfüred (Central Research Institute for Physics, Budapest 1981) and Nucl. Instrum. Methods. **199** (1982)
11.17 A.E.Berkowitz, J.L.Walter, K.F.Wall: Phys. Rev. Lett. **46**, 1484–1487 (1981)
       S.Aur, T.Egami, A.E.Berkowitz, J.L.Walter: Phys. Rev. B**26**, 6355–6361 (1982)
11.18 D.W.Forester, N.C.Koon, J.H.Schelleng, J.J.Rhyne: J. Appl. Phys. **50**, 7336–7341 (1979)
11.19 J.M.D.Coey, D.Givord, A.Liénard, J.P.Rebouillat, J.Chappert: J. Phys. F **11**, 2707–2744 (1981)
11.20 Y.Nishihara, T.Katayama, S.Ogawa: J. Appl. Phys. **53**, 2285–2287 (1982)
11.21 N.Heiman, N.Kazama: Phys. Rev. B **19**, 1623–1632 (1979)
11.22 J.J.Croat, J.F.Herbst: J. Appl. Phys. **53**, 2294–2296 (1982)
11.23 K.H.J.Buschow, M.Brouha, J.W.M.Biesterbos, A.G.Dirks: Physica **91**B, 261–270 (1977)
       K.H.J.Buschow: J. Mag. Mag. Mat. **28**, 20–28 (1982), **29**, 91–99 (1982); J. Appl. Phys. **53**, 7713–7716 (1982)
11.24 R.C.Taylor, C.C.Tsuei: Solid State Commun. **41**, 503–506 (1982); L.Krusin-Elbaum, A.P.Malozemoff, R.C.Taylor: Phys. Rev. B**27**, 562–565 (1983)
11.25 C.L.Chien, K.M.Unruh: Phys. Rev. B**25**, 5790–5796 (1982)
11.26 For complementary reviews, see J.Durand: Rev. Phys. Appl. (Paris) **15**, 1036–1042 (1980)
       J.Durand: J. Phys. **41**, C8, 609–617 (1980)
       J.Durand, P.Panissod: IEEE Trans. MAG-**17**, 2595–2599 (1981)
11.27 See a recent review by G.Czjzek: In Ref. [11.16, pp. 93–115]
11.28 D.Sarkar, R.Segnan, E.K.Cornell, E.Callen, R.Harris, M.Plischke, M.J.Zuckermann: Phys. Rev. Lett. **32**, 542–544 (1974)
       R.W.Cochrane, R.Harris, M.Plischke, D.Zobin, M.J.Zuckermann: Phys. Rev. B**5**, 1969–1970 (1975)
11.29 P.Panissod, D.Aliaga Guerra, A.Amamou, J.Durand, W.L.Johnson, W.L.Carter, S.J.Poon: Phys. Rev. Lett. **44**, 1465–1468 (1980)
11.30 P.Heubes, D.Korn, G.Schatz, G.Zibold: Phys. Lett. **74**A, 267–270 (1979)
11.31 G.Czjzek, J.Fink, F.Götz, H.Schmidt, J.M.D.Coey, J.P.Rebouillat, A.Liénart: Phys. Rev. B**23**, 2513–2530 (1981)
11.32 D.AliagaGuerra: Thesis, Strasbourg (1980)
11.33 M.Maurer, J.M.Friedt, J.P.Sanchez: In [Ref. 11.16, Vol. 1, pp. 517–523]
       J.M.Friedt,M.Maurer, J.P.Sanchez, J.Durand: J. Phys. F**12**, 821–836 (1982)
11.34 P.Panissod,I.Bakonyi, R.Hasegawa: J. Mag. Mag. Mat. **31–34**, 1523–24 (1983)
11.35 P.H.Gaskell: J. Non-Cryst. Sol. **32**, 207–218 (1979)

11.36   H.J.Eifert, B.Elschner, K.H.J.Buschow: Phys. Rev. B25, 7441–7448 (1982)
        J.Abart, W.Socher,J.Voitländer: Z. Naturforsch. 37a, 1030–1034 (1982)
11.37   M.E.Lines: Phys. Rev. B20, 3729–3737 (1979)
11.38   R.Harris, M.Plischke, M.J.Zuckermann: Phys. Rev. Lett. 31, 160–162 (1973)
11.39   R.Asomoza, I.A.Campbell, A.Fert, A.Linéart, J.P.Rebouillat: J. Phys. F9, 349–371 (1979)
11.40   C.Pappa, B.Boucher: J. Mag. Mag. Mat. 15–18, 97–98 (1980);
        P.Garoche, A.Fert, J.J.Veyssié, B.Boucher: J. Mag. Mag. Mat. 15–18, 1397–1398 (1980)
11.41   A.Fert, I.A.Campbell: J. Phys. F8, L57–61 (1978)
        A.Fert, D.Spanjaard: J. Phys. 40, C5, 248–249 (1979)
11.42   W.Felsch, S.G.Kushnir, K.Samwer: J. Phys. 41, C8, 630–631 (1980)
        U.Ernst, W.Felsch, K.Samwer: J. Mag. Mag. Mat. 15–18, 1375–1376 (1980)
        H.v.Löhneysen, H.J.Schink, W.Felsch, K.Samwer, H.Schröder, Physica 107B, 631–632
        (1981)
11.43   J.Durand, S.J.Poon: J. Appl. Phys. 49, 1702 (1978)
11.44   N.Hassanain, A.Berrada, J.Durand, B.Loegel: J. Mag. Mag. Mat. 15–18, 1377–1378 (1980)
        A.Fert, P.Garoche, B.Boucher, J.Durand: In Crystalline Electric Field and Structural
        Effects in f-Electron Systems, ed. by J.E.Crow, R.P.Guertin, T.W.Mihalisin (Plenum Press,
        New York 1980) pp. 491–496
11.45   P.Garoche, J.J.Veyssié, J.Durand: J. Phys. 41, L357–360 (1980)
11.46   E.Borchi, S.DeGennaro: J. Phys. F11, L47–52 (1981)
11.47   A.Berrada: Thesis, Strasbourg (1981)
11.48   A.Fert, A.Friedrich: Phys. Rev. B13, 397–407 (1976)
11.49   D.Eichler, J.C.Ousset, S.Cantaloup, A.Berrada, J.Durand: J. Phys. 43, C9, 671–675 (1982)
11.50   D.Eichler: Thesis, Strasbourg (1982)
11.51   H.Hernandez, R.Ferrer, M.J.Zuckermann: Can. J. Phys. 58, 629–632 (1980)
11.52   J.B.Bieri, J.Sanchez, A.Fert, D.Bertrand, A.R.Fert: J. Appl. Phys. 53, 2347–2349 (1982)
11.53   J.B.Bieri: Thesis, Orsay (1981)
11.54   A.K.Bhattacharjee, B.Coqblin: J. Phys. F12, 2065–2077 (1982)
11.55   B.D.Rainford, V.Samadian, R.J.Begum, E.W.Lee, S.K.Burke: J. Appl. Phys. 53,
        7725–7727 (1982)
11.56   See the review of T.Egami: IEEE Trans. MAG-17, 2600–2605 (1981)
11.57   H.Kronmüller: J. Phys. 41, C8, 618–625 (1980); id. J. Appl. Phys. 52, 1859–1864 (1981)
11.58   H.Kronmüller: J. Mag. Mag. Mat. 24, 159–167 (1981)
        H.Kronmüller, B.Gröger: J. Phys. 42, 1285–1292 (1981)
        B.Gröger, H.Kronmüller: Appl. Phys. 24, 287–296 (1981)
11.59   J.Durand: In [Ref. 11.15, pp. 143–172]; see also a review on hf in MG's by P.Panissod,
        J.Durand, J.I.Budnick: Nucl. Instr. Meth. 199, 99–114 (1982)
11.60   C.P.Slichter: Principles of Magnetic Resonance, Springer Ser. Solid-State Sci., Vol. 1
        (Springer, Berlin, Heidelberg, New York 1980)
11.61   P.Panissod, A.Qachaou, J.Durand, R.Hasegawa: In [Ref. 11.16, Vol. 1, pp. 543–550]
        R.Hasegawa, R.Ray: J. Appl. Phys. 50, 1586–1588 (1979)
11.62   J.D.Livingston, W.G.Morris: IEEE Trans. MAG-17, (1981)
11.63   J.Durand, B.Lémius, R.Hasegawa, D.AliagaGuerra, P.Panissod: J. Mag. Mag. Mat.
        15–18, 1373–1374 (1980)
11.64   V.S.Pokatilov, Yu.A.Gratsianov, B.N.Kulagin: Sov. Phys. Dokl. 25, 206–208 (1980)
11.65   K.Inomata: In [Ref. 11.10, Vol. 1, pp. 547–550]
11.66   J.Durand, M.F.Lapierre: J. Phys. F6, 1185–1192 (1976)
11.67   G.C.Chi, G.S.Cargill: Mat. Sci. Eng. 23, 155–159 (1975)
11.68   M.Cyrot: In Amorphous Magnetism, ed. by H.O.Hooper, M.deGraaf (Plenum Press, New
        York 1973) pp. 161–167
11.69   F.Gautier: Itinerant Magnetism, in Magnetism of Metals and Alloys, ed. by M.Cyrot
        (North-Holland, Amsterdam 1982) Chap. 1, pp. 1–244
        M.Shimizu: Rep. Prog. Phys. 44, 329–409 (1981)
11.70   A.Herpin: Théorie du magnétisme (PUF, Paris 1968)
11.71   G.S.Grest, S.R.Nagel: Phys. Rev. B19, 3571–3580 (1979)

11.72 E.P.Wohlfarth: IEEE Trans. MAG-14, 933–938 (1978)
11.73 D.Wagner, E.P.Wohlfarth: J. Phys. F9, 717–725 (1979); J. Mag. Mag. Mat. 15–18, 1345–1346 (1980)
11.74 D.Wagner, E.P.Wohlfarth: J. Phys. F11, 2417–2428 (1981); Physica 112B, 1–5 (1982)
11.75 E.P.Wohlfarth: Phys. Lett. 69A, 222–224 (1978)
11.76 G.Herzer, M.Fähnle, T.Egami, H.Kronmüller: Phys. Stat. Sol. (b) 101, 713–721 (1980); J. Appl. Phys. 52, 1794–1796 (1981); J. Mag. Mag. Mat. 24, 175–178 (1981); J. Appl. Phys. 53, 2326–2328 (1982)
11.77 P.E.Brommer: Physica 113B, 391–399 (1982)
11.78 This aspect was reviewed in more detail by J.Durand: In *The Magnetic, Chemical, and Structural Properties of Glassy Metallic Alloys*, ed. by R.Hasegawa (CRC Press, Boca Raton, USA 1983) in press
11.79 R.Hasegawa, C.C.Tsuei: Phys. Rev. B2, 1631–1640 (1970); Phys. Rev. B3, 214–224 (1971)
11.80 R.Hasegawa: Mat. Sci. Eng. 23, 293–296 (1976)
11.81 J.Durand, S.J.Poon: J. Phys. 39 C6, 593–594 (1978)
       A.Amamou: In *Amorphous Magnetism II*, ed. by R.Levy, R.Hasegawa (Plenum Press, New York 1977) pp. 265–275
11.82 V.K.C.Liang: Solid State Commun. 9, 579–583 (1971)
11.83 A.K.Sinha: J. Appl. Phys. 42, 5184–5187 (1971)
11.84 G.R.Gruzalski, D.J.Sellmeyer: Phys. Rev. B20, 184–193 (1979)
11.85 A.Amamou: J. Phys. 41 C8, 670–674 (1980)
11.86 E.Domb, C.A.MacDonald, W.L.Johnson: Solid State Commun. 30, 775–779 (1979)
11.87 A.Amamou: IEEE Trans. MAG-12, 948–952 (1976)
11.88 J.Costa-Ribeiro, J.Souletie, D. Thoulouze, R.Tournier: J. Phys. 32, C1, 733–757 (1971)
11.89 S.N.Kaul, M.Rosenberg: Phys. Rev. B25, 5863–5874 (1982)
11.90 A.Amamou, J.Durand: Commun. Phys. 1, 191–197 (1976)
11.91 A.Berrada, M.F.Lapierre, B.Loegel, P.Panissod, C.Robert: J. Phys. F8, 845–857 (1978)
11.92 P.A.Albert, Z.Kovac, H.R.Lilienthal, T.R.McGuire, Y.Nakamura: J. Appl. Phys. 38, 1258–1259 (1967)
11.93 D.Pan, D.Turnbull: AIP Conf. Proc. 18, 646–650 (1974)
11.94 D.Gignoux, D.Givord, A.Liénard: J. Appl. Phys. 53, 2321–2323 (1982)
11.95 A.Liénard, J.P.Rebouillat: J. Appl. Phys. 49, 1680–1682 (1978)
       A.Liénard, J.P.Rebouillat, P.Garoche, J.J.Veyssié: J. Phys. 41, C8, 658–662 (1980)
11.96 V.Jaccarino, L.R.Walker: Phys. Rev. Lett. 15, 258–261 (1965)
11.97 G.Suran, H.Daver, J.C.Bruyère: AIP Conf. Proc. 29, 162–164 (1976)
       O.Massenet, H.Daver: Solid State Commun. 25, 917–920 (1978)
11.98 P.Terzieff, K.Lee, N.Heiman: J. Appl. Phys. 50, 1031–1034 (1979)
11.99 H.S.Randhawa, L.K.Malhotra, K.L.Chopra: J. Appl. Phys. 52, 1600–1602 (1981)
11.100 K.H.J.Buschow, P.G.vanEngen: J. Appl. Phys. 52, 3557–3561 (1981)
11.101 O.Massenet, H.Daver, V.D.Nguyen, J.P.Rebouillat: J. Phys. F9, 1687–1699 (1979)
11.102 G.Marchal, Ph.Mangin, M.Piécuch, C.Janot: J. Phys. 37, C6, 763–768 (1976); J. Phys. F7, L165–168 (1977)
       P.Mangin, M.Piécuch, G.Marchal, C.Janot: J. Phys. F8, 2085–2092 (1978)
11.103 B.Rodmacq, M.Piécuch, C.Janot, C.Marchal, P.Mangin: Phys. Rev. B21, 1911–1923 (1980)
       D.Teirlinck, M.Piécuch, J.F.Gény, G.Marchal, P.Mangin, C.Janot: IEEE Trans. MAG-17, 3079–3081 (1981)
11.104 T.Shigematsu, T.Shinjo, Y.Bando, T.Takada: J. Mag. Mag. Mat. 15–18, 1367–1368 (1980)
11.105 N.A.Blum, K.Moorjani, T.O.Poehler, F.G.Satkiewicz: J. Appl. Phys. 53, 2074–2076 (1982)
       N.A.Blum: J. Appl. Phys. 53, 7747–7749 (1982)
11.106 N.S.Kazama, H.Fujimori, H.Watanabe: J. Mag. Mag. Mat. 15–18, 1423–1424 (1980)
       C.L.Chien, K.M.Unruh, S.H.Liou: J. Appl. Phys. 53, 7756–7758 (1982)
11.107 K.H.J.Buschow, P.H.Smit: J. Mag. Mag. Mat. 23, 85–91 (1981)
11.108 K.H.J.Buschow, A.M.vanderKraan: Phys. Stat. Sol. (a) 53, 665–669 (1979)
11.109 A.Zentkova, P.Duhaj, A.Zentko, T.Tima: Physica 86–88B, 787–789 (1977)

11.110 G. Dublon: Sol. State Commun. **33**, 1195–1199 (1980); Phys. Stat. Sol. (a) **60**, 287–296 (1980)
G. Dublon, C. H. Lin, J. Bevk: Phys. Lett. **76** A, 92–94 (1980)
G. Dublon, Y. Yeshurun: Phys. Rev. B **25**, 4899–4902 (1982)

11.111 J. Durand: In *Amorphous Magnetism II*, ed. by R. A. Levy, R. Hasegawa (Plenum Press, New York 1977) pp. 305–318

11.112 J. Schneider, A. Handstein, R. Hesske: Phys. Stat. Sol. (a) **45**, K 47–50 (1978)

11.113 J. T. Prater, T. Egami: J. Appl. Phys. **50**, 1706–1709 (1979)
J. K. Krause, T. C. Long, T. Egami, D. G. Onn: Phys. Rev. B **21**, 2886–2896 (1980)

11.114 G. Hilscher, R. Haferl, H. Kirchmayr, M. Müller, H. J. Güntherodt: J. Phys. F **11**, 2429–2441 (1981)

11.115 S. N. Kaul: IEEE Trans. MAG-**17**, 1208–1215 (1981)

11.116 T. Shigematsu, W. Keune, V. Manns, J. Lauer, M. Naka: Physica **107**B, 629–630 (1981); id. in [Ref. 11.16, pp. 555–568]

11.117 S. J. Poon, J. Durand: Phys. Rev. B **18**, 6253–6264 (1978)
J. Durand, S. J. Poon: J. Phys. **40**, C9, 231–236 (1979)

11.118 B. Barbara, A. P. Malozemoff, Y. Imry: Phys. Rev. Lett. **47**, 1852–1855 (1981); id: Physica **108**B, 1289–1290 (1981) and references therein

11.119 J. J. Hauser: Solid State Commun. **30**, 201–205 (1979); id: J. Mag. Mag. Mat. **15–18**, 1387–1388 (1980)
R. W. Cochrane, J. O. Ström-Olsen, J. P. Rebouillat: J. Appl. Phys. **50**, 7348–7350 (1979)

11.120 A. P. Murani, J. P. Rebouillat: J. Phys. F **12**, 1427–1437 (1982)

11.121 S. J. Poon: Phys. Rev. B **21**, 343–346 (1980)

11.122 S. J. Poon, P. L. Dunn, L. M. Smith: J. Phys. F **12**, 1 101–106 (1982)
P. L. Dunn, S. J. Poon: J. Phys. F **12**, L 273–278 (1982)

11.123 P. G. deGennes: J. Phys. Rad. **23**, 630–636 (1962)

11.124 P. F. deChâtel: J. Mag. Mag. Mat. **23**, 28–34 (1981)

11.125 S. J. Poon, J. Durand: Solid State Commun. **21**, 999–1002 (1977)

11.126 See, for instance, T. Kaneyoshi, S. Iwabuchi: J. Mag. Mag. Mat. **15–18**, 1421–1422 (1980) and references therein

11.127 A. P. Malozemoff, G. Suran, R. C. Taylor: Phys. Rev. B **24**, 2731–2738 (1981)

11.128 For a review on re-entrant ferromanetism in crystalline systems, see G. J. Nieuwenhuys, B. H. Verbeeck, J. A. Mydosh: J. Appl. Phys. **50**, 1685–1690 (1979)

11.129 D. Sherrington, S. Kirkpatrick: Phys. Rev. Lett. **35**, 1792–1795 (1975)

11.130 For reviews on re-entrant ferromagnetism in metallic glasses, see M. B. Salamon, K. V. Rao, Y. Yeshurun: J. Appl. Phys. **52**, 1687–1691 (1981)
Y. Yeshurun, M. B. Salamon, K. V. Rao, H. S. Chen: Phys. Rev. B **24**, 1536–1549 (1981)
J. A. Geohegan, S. M. Bhagat: J. Mag. Mag. Mat. **25**, 17–32 (1981)
K. V. Rao, H. S. Chen: In [Ref. 11.10, Vol. 2, p. 1073–1078]
K. V. Rao: Phys. Scripta **25**, 742–748 (1982)

11.131 R. B. Goldfarb, K. V. Rao, H. S. Chen, C. E. Patton: J. Appl. Phys. **53**, 2217–2219 (1982)
M. A. Manheimer, S. M. Bhagat, H. S. Chen: Phys. Rev. B **26**, 456–458 (1982)

11.132 Y. Yeshurun, M. B. Salamon: J. Phys. C **14**, L 575–580 (1981)
Y. Yeshurun, L. J. P. Ketelsen, M. B. Salamon: Phys. Rev. B **26**, 1491–1494 (1982)

11.133 T. Kudo, T. Egami, K. V. Rao: J. Appl. Phys. **53**, 2214–2216 (1982)

11.134 C. W. Chu, M. K. Wu, B. J. Jin, W. Y. Lai, H. S. Chen: Phys. Rev. Lett. **46**, 1643–1647 (1981)

11.135 For a review, see M. L. Spano, S. M. Bhagat: J. Mag. Mag. Mat. **24**, 143–156 (1981)

11.136 H. Keller, K. V. Rao, P. G. Debrunner, H. S. Chen: J. Appl. Phys. **52**, 1853–1855 (1981)

11.137 G. Aeppli, S. M. Shapiro, R. J. Birgeneau, H. S. Chen: Phys. Rev. B **25**, 4882–4885 (1982)

11.138 T. F. Rosenbaum, L. W. Rupp, G. A. Thomas, W. M. Walsh, H. S. Chen, J. R. Banavar, P. B. Littlewood: Solid State Commun. **42**, 725–727 (1982)
D. J. Webb, S. M. Bhagat, K. Moorjani, T. O. Poehler, F. G. Satkiewitz: Solid State Commun. **43**, 239–242 (1982)

11.139 J. W. Lynn, R. W. Erwin, J. J. Rhyne, H. S. Chen: J. Appl. Phys. **52**, 1738–1740 (1981)
H. S. Chen, J. W. Lynn, R. W. Erwin, J. J. Rhyne: In [Ref. 11.10, Vol. 2, pp. 1153–1156]

11.140 Y. Yeshurun, K. V. Rao, M. B. Salamon, H. S. Chen: Solid State Commun. **38**, 371–376 (1981)
M. Olivier, J. O. Strom-Olsen, Z. Altounian, G. Williams: J. Appl. Phys. **53**, 7696–7699 (1982)
11.141 H. Hiroyoshi, K. Fukamichi: J. Appl. Phys. **53**, 2226–2228 (1982)
11.142 Y. Obi, L. C. Wang, R. Motsay, D. G. Onn, M. Nose: J. Appl. Phys. **53**, 2304–2306 (1982)
11.143 M. Ghafari, U. Gonser, H. G. Wagner, M. Naka: In [Ref. 11.16, pp. 463–477]
H. Yamamoto, M. Onodera, K. Hosoyama, T. Masumoto, H. Yamauchi: J. Mag. Mag. Mat. **31–34**, 1579–1580 (1983)
11.144 A. Aharony, E. Pytte: Phys. Rev. Lett. **45**, 1583–1586 (1980)
11.145 L. S. Barton, M. B. Salamon: Phys. Rev. B**25**, 2030–2033 (1982)
11.146 P. Duwez, S. C. H. Lin: J. Appl. Phys. **38**, 4096 (1967)
11.147 For phenomenological discussions of magnetostriction in MG's, see R. C. O'Handley: Phys. Rev. B**18**, 930–939 (1978)
M. Fähnle, T. Egami: J. Appl. Phys. **53**, 2319–2320 (1982)
11.148 R. Hasegawa, R. C. O'Handley, L. I. Mendelsohn: AIP Conf. Proc. **34**, 298–302 (1976)
11.149 N. S. Kazama, T. Masumoto, M. Mitera: J. Mag. Mag. Mat. **15–18**, 1331–1335 (1980)
11.150 I. Vincze, T. Kemény, S. Arajs: Phys. Rev. B**21**, 937–940 (1980)
11.151 J. M. Dubois, G. Le Caër: In [Ref. 11.16, pp. 729–748]
L. Takács, C. Hargitai: J. Phys. F**13**, 183–190 (1983)
M. E. Lines, M. Eibschütz: Solid State Commun. **45**, 435–439 (1983)
T. Kemény, F. J. Litterst, I. Vincze, R. Wäppling: J. Phys. F**13**, L37–41 (1983)
11.152 D. Musser, C. L. Chien, H. S. Chen: J. Appl. Phys. **50**, 7659–7661 (1979)
11.153 R. Hasegawa, R. Ray: J. Appl. Phys. **49**, 4174–4178 (1978); Phys. Rev. B**20**, 211–220 (1979)
11.154 C. L. Chien, D. Musser, E. M. Gyorgy, R. C. Sherwood, H. S. Chen, F. E. Luborsky, J. L. Walter: Phys. Rev. B**20**, 283–295 (1979)
11.155 J. Durand: IEEE Trans. MAG-**12**, 945–947 (1976)
11.156 H. S. Chen, R. C. Sherwood, E. M. Gyorgy: IEEE Trans. MAG-**13**, 1538–1540 (1977)
11.157 W. Felsch: Z. angew. Phys. **30**, 275–277 (1970)
11.158 D. Pan, D. Turnbull: J. Appl. Phys. **54**, 1406–1412 (1974)
11.159 G. Suran, J. Sztern, J. A. Aboaf, T. R. McGuire: IEEE Trans. MAG-**17**, 3065–3067 (1982)
11.160 R. Krishnan, M. Tarhouni, M. Tessier: J. Appl. Phys. **53**, 2243–2245 (1982)
11.161 M. Naka, N. S. Kazama, H. Fujimori, T. Masumoto: In [Ref. 11.10, Vol. 2, pp. 919–922]
11.162 G. Marchal, D. Teirlinck, P. Mangin, C. Janot, H. Hübsch: J. Phys. **41**, C8, 662–665 (1980)
11.163 J. J. Hauser: Phys. Rev. B**12**, 5160–5168 (1975)
11.164 W. Felsch: Z. Phys. 219, 280–299 (1969); Z. angew. Phys. **29**, 217–224 (1970)
11.165 T. Masumoto, S. Ohnuma, R. Shirakawa, M. Nose, K. Kobayashi: J. Phys. **41**, C8, 686–689 (1980)
11.166 T. Kaneyoshi: In [Ref. 11.10, Vol. 2, pp. 1087–1090]
11.167 A. P. Malozemoff, A. R. Williams, K. Terakura, V. L. Moruzzi, K. Fukamichi: J. Mag. Mag. Mat. **35**, 192–198 (1983)
11.168 J. Durand, C. Thompson, A. Amamou: In *Rapidly Quenched Metals III*, Brighton, ed. by B. Cantor (The Metals Society, London 1978) Vol. 2, pp. 109–116
11.169 J. Friedel: Nuovo Cim. Suppl. **7**, 287–311 (1958)
11.170 T. Mizoguchi, T. Yamauchi, H. Miyajima: In *Amorphous Magnetism I*, ed. by H. O. Hooper, M. de Graaf (Plenum Press, New York 1973) pp. 325–330
11.171 For recent band structure calculations of FeB metallic glasses, see T. Fujiwara: J. Phys. F**12**, 661–675 (1982)
11.172 R. C. O'Handley: Solid State Commun. **38**, 703–708 (1981)
B. W. Corb, R. C. O'Handley, N. J. Grant: Phys. Rev. B**27**, 636–641 (1983)
11.173 L. Granasy, A. Lovas, L. Kiss, T. Kemény, E. Kisdi-Koszo: J. Mag. Mag. Mat. **26**, 109–111 (1982)
11.174 R. Hasegawa: J. Phys. **41**, C8, 701–703 (1980)
11.175 R. Hasewaga: J. Appl. Phys. **52**, 1847–1849 (1981)
C. S. Severin, C. W. Chen, A. J. Bevols, M. C. Lin: J. Appl. Phys. **52**, 1850–1852 (1981)
11.176 D. W. Forester, J. H. Schelleng, P. Lubitz, P. D'Antonis, C. George: J. Appl. Phys. **53**, 2240–2242 (1982)

11.177 F.E.Luborsky, J.L.Walter, E.P.Wohlfarth: J. Phys. F **10**, 959–966 (1980)
        I.W.Donald, T.Kemény, H.A.Davies: J. Phys. F **11**, L 131–136 (1981)
11.178 F.E.Luborsky: J. Appl. Phys. **51**, 2808–2810 (1980)
11.179 K.Shirakawa, T.Kaneko, J.Kanehira, S.Ohnuma, H.Fujimori, T.Masumoto: In [Ref. 11.10, Vol. 2, pp. 1083–1086] and references therein
11.180 J.Inoue, M.Shimizu: Phys. Lett. **90** A, 85–88 (1982)
11.181 R.W.Cochrane, G.S.Cargill: Phys. Rev. Lett. **32**, 476–479 (1974)
        N.Kazama, T.Masumoto, H.Watanabe: J. Phys. Soc. Jpn. **37**, 1171–1175 (1974)
11.182 H.A.Mook, N.Wakabayashi, D.Pan: Phys. Rev. Lett. **34**, 1029–1032 (1975)
        J.D.Axe, L.Passell, C.C.Tsuei: AIP Conf. Proc. **24**, 119–123 (1975)
11.183 M.Grimsditch, A.Malozemoff, A.Brunsch: Phys. Rev. Lett. **43**, 711–714 (1979)
11.184 R.J.Birgeneau, J.A.Tarvin, G.Shirane, E.M.Gyorgy, R.C.Sherwood, H.S. Chen, C.L.Chien: Phys. Rev. B**18**, 2192–2195 (1978)
11.185 For a review on neutron magnetic scattering studies of amorphous solids, see A.C.Wright: J. Non-Cryst. Sol. **40**, 325–346 (1980)
11.186 S.M.Bhagat, M.L.Spano, H.S.Chen, K.V.Rao: Solid State Commun. **33**, 303–307 (1980)
11.187 S.M.Bhagat, M.L.Spano, K.V.Rao: J. Appl. Phys. **50**, 1580–1582 (1979)
11.188 Y.Ishikawa, Z.Xyanyu, S.Onodera, S.Ishio, M.Takahashi: In [Ref. 11.10, Vol. 2, pp. 1093–1096]
        Y.Ishikawa, K.Yamada, K.Tajima, K.Fukamichi: J. Phys. Soc. Jpn. **50**, 1958–1963 (1981)
11.189 J.J.Rhyne, G.E.Fish, J.W.Lynn: J. Appl. Phys. **53**, 2316–2318 (1982)
11.190 G.Suran, R.J.Gambino: J. Appl. Phys. **50**, 7671–7673 (1979)
11.191 J.J.Rhyne, J.W.Lynn, F.E.Luborsky, J.L.Walter: J. Appl. Phys. **50**, 1583–1585 (1979)
        F.E.Luborsky, J.L.Walter, H.H.Liebermann, E.P.Wohlfarth: J. Mag. Mag. Mat. **15–18**, 1351–1354 (1980)
        M.Maskiewicz: J. Appl. Phys. **53**, 7765–7767 (1982)
11.192 Y.Takahashi, M.Shimizu: Phys. Lett, **58** A, 419–420 (1976)
11.193 G.Shirane, J.D.Axe, C.F.Majkrzak, T.Mizoguchi: Phys. Rev. B**26**, 2575–2583 (1982)
        D.Mc.K.Paul, R.A.Cowley, W.G.Stirling, N.Cowlam, H.A.Davies: J. Phys. F**12**, 2687–2701 (1982)
11.194 I.Vincze, F.van der Woude, T.Kemény, A.S.Schaafsma: J. Mag. Mag. Mat. **15–18**, 1336–1338 (1980)
        P.J.Schurrer, A.H.Morrish: Solid State Commun. **30**, 21–25 (1979)
11.195 For a general review of magnetoelastic properties of MG's, see E.du Trémolet de Lacheisserie: J. Mag. Mag. Mat. **25**, 251–270 (1982)
11.196 For a review of Invar characteristics of FeB based MG's, see K.Fukamichi, M.Kikuchi, H.Hiroyoshi, T.Masumoto: IEEE Trans. MAG-**15**, 1404–1409 (1979), and for the FeZr based MG's, see: K.Shirakawa, S.Ohnuma, M.Nose, T.Masumoto: IEEE Trans. MAG-**16**, 910–912 (1980)
11.197 M.Takahashi, M.Koshimura: In [Ref. 11.10, Vol. 2, pp. 1061–1066]
11.198 C.J.Beers, H.W.Myron, C.J.Schinkel, I.Vincze: Solid State Commun. **41**, 631–636 (1982)
11.199 M.Matsuura, U.Mizutani: In [Ref. 11.10, Vol. 2, pp. 1291–1294]
11.200 M.Landholt, P.Niedermann, D.Mauri: Phys. Rev. Lett. **48**, 1632–1635 (1982)
11.201 J.C.Ododo: Phil. Mag. B**45**, 119–135 (1982)
        A.Z.Menshikov, A.Chamberod, M.Roth: Solid State Commun. **44**, 243–246 (1982)
11.202 A.R.Williams: Bull. Am. Phys. Soc. **27**, 171 (1982)
11.203 S.N.Kaul: Phys. Rev. B**24**, 6550–65665 (1981); **23**, 1205–1215 (1981)
        S.N.Kaul, M.Rosenberg: Phil. Mag. B**44**, 357–368 (1981); id.: Solid State Commun. **11**, 857–862 (1982)
11.204 R.Meyer, H.Kronmüller: Phys. Stat. Sol. (b) **109**, 693–703 (1982)
11.205 P.Gaunt, S.C.Ho, G.Williams, R.W.Cochrane: Phys. Rev. B**23**, 251–255 (1981)
11.206 O.Källbäck, H.Gudmundsson, K.V.Rao, H.U.Åström: Phys. Scripta **25**, 755–757 (1982)
11.207 S.J.Poon, J.Durand: Phys. Rev. B**16**, 316–330 (1977)
11.208 J.Durand, S.J.Poon: IEEE Trans. MAG-**13**, 1556–1558 (1977)
        J.Durand, K.Raj, S.J.Poon, J.I.Budnick: IEEE Trans. MAG-**14**, 722–724 (1978)

11.209 K.H.J.Buschow: Solid State Commun. **27**, 275–278 (1978)
K.H.J.Buschow, N.M.Beekmans: In *Rapidly Quenched Metals III*, Vol. 2, ed. by B.Cantor (The Metals Society, London 1978) pp. 133–136
K.H.J.Buschow, H.A.Algra, R.A.Henskens: J. Appl. Phys. **51**, 561–566 (1980)
11.210 J.A.Gerber, D.J.Miller, D.J.Sellmyer: J. Appl. Phys. **49**, 1699–1701 (1978)
11.211 K.H.J.Buschow, W.W.van den Hoogenhof: J. Mag. Mag. Mat. **12**, 123–126 (1979)
11.212 A.Berrada, J.Durand, N.Hassanain, B.Loegel: J. Appl. Phys. **50**, 7621–7623 (1979); id: In *The Rare Earths in Modern Science and Technology*, Vol. 2, ed. by G.J.MacCarthy, J.J.Rhyne, H.B.Silber (Plenum Press, New York 1980) pp. 307–312
11.213 T.Mizoguchi, J.I.Budnick, P.Panissod, J.Durand, H.J.Güntherodt: In [Ref. 11.10, Vol. 2, pp. 1149–1152]
11.214 B.Boucher: IEEE Trans. MAG-**13**, 1601–1603 (1977)
11.215 K.H.J.Buschow: J. Appl. Phys. **51**, 2795–2798 (1980)
11.216 K.H.J.Buschow: J. Mag. Mag. Mat. **21**, 97–100 (1980)
11.217 K.H.J.Buschow, A.A.van der Kraan: J. Mag. Mag. Mat. **22**, 220–226 (1981)
11.218 B.Boucher, A.Liénard, J.P.Rebouillat, J.Schweizer: J. Phys. F**9**, 1421–1431 (1979)
11.219 A.Berrada, J.Durand, T.Mizoguchi, J.I.Budnick, B.Loegel, J.C.Ousset, S.Askenazy, H.J.Güntherodt: In [Ref. 11.10, Vol. 2, pp. 829–834]
11.220 D.J.Sellmyer, G.Hadjipanayis, S.G.Cornelison: J. Non-Cryst. Sol. **40**, 437–445 (1980)
S.G.Cornelison, D.J.Sellmyer, G.Hadjipanayis: J. Appl. Phys. **52**, 1823–1825 (1981)
G.Hadjipanayis, D.J.Sellmyer, B.Brandt: Phys. Rev. B**23**, 3349–3359 (1981)
11.221 B.C.Giessen, W.A.Hines, L.T.Kabacoff: IEEE Trans. MAG-**16**, 1203–1205 (1980)
11.222 B.Boucher, B.Barbara: J. Phys. F**9**, 151–159 (1979)
11.223 J.M.D.Coey, T.R.McGuire, B.Tissier: Phys. Rev. B**24**, 1261–1273 (1981)
11.224 B.Barbara, C.Bécle, R.Lemaire, D.Paccard: J. Phys. **32**, C1, 299–304 (1971)
T.Egami: Phys. Stat. Sol. (a) **19**, 747–758 (1973); **20**, 157–165 (1973)
11.225 J.Chappert, M.Bogé, B.Boucher, B.Barbara: J. Phys. **41**, C8, 634–637 (1980)
J.Chappert, L.Asch, M.Bogé, G.M.Kalvius, B.Boucher: J. Mag. Mag. Mat. **28**, 124–136 (1982)
11.226 S. von Molnar, B.Barbara, T.R.McGuire, R.Gambino: J. Appl. Phys. **53**, 2350–2352 (1982)
11.227 G.Krill, J.Durand, A.Berrada, N.Hassanain, M.F.Ravet: Solid State Commun. **35**, 547–550 (1980)
N.Hassanain, B.Loegel, A.Berrada, J.Durand: J. Phys. **41**, C8, 770–773 (1980)
G.Krill, A.Amamou, A.Berrada, J.Durand, N.Hassanain: J. Phys. **41**, C8, 799–802 (1980)
11.228 W.Felsch, S.G.Kushnir, K.Samwer, H.Schröder, R.van den Berg, H.v.Löhneysen: Z. Phys. B**48**, 99–107 (1982)
11.229 S.G.Cornelison, G.Hadjipanayis, D.J.Sellmyer: J. Non-Cryst. Sol. **40**, 429–435 (1980)
11.230 N.C.Koon, B.N.Das: Appl. Phys. Lett. **39**, 840–842 (1981)
N.C.Koon, B.N.Das, J.A.Geohegan, D.W.Forester: J. Appl. Phys. **53**, 2333–2335 (1982)
J.A.Geohegan, N.C.Koon, B.N.Das: J. Appl. Phys. **53**, 7816–7818 (1982)
11.231 L.Potocky, L.Novak, A.Lovas, E.Kisdi-Koszo, J.Takacs: J. Mag. Mag. Mat. **26**, 112–114 (1982)
11.232 N.S.Kazama, H.Fujimori: In [Ref. 11.10, Vol. 2, pp. 799–802]
11.233 Y.Shimada, M.Yagi, H.Kojima: In [Ref. 11.10, Vol. 2, pp. 807–810]
11.234 L.Kabacoff, S.Dallek, C.Modzelewski, W.Krull: J. Appl. Phys. **53**, 2255–2257 (1982)
11.235 S.Cornelison, D.J.Sellmyer, J.G.Zhao, Z.D.Chen: J. Appl. Phys. **53**, 2330–2332 (1982)

Subject Index

Page numbers in *italics* refer to *Glassy Metals I*, Topics Appl. Phys., Vol. 46, ed. by H.-J. Güntherodt and H. Beck (Springer, Berlin, Heidelberg, New York 1981)

Springer-Verlag
Berlin Heidelberg
GmbH

Topics in Applied Physics

Founded by H. K. V. Lotsch

Volume 46

Glassy Metals I

Ionic Structure, Electronic Transport, and Crystallization

Editors: H. Beck, H.-J. Güntherodt
1981. 119 figures. XIV, 267 pages.
ISBN 3-540-10440-2

Contents:
H. Beck, H.-J. Güntherodt: Introduction. –
P. Duwez: Metallic Glasses – Historical Background. – *T. Egami:* Structural Study by Energy Dispersive X-Ray Diffraction. – *J. Wong:* EXAFS Studies of Metallic Glasses. – *A.P. Malozemoff:* Brillouin Light Scattering from Metallic Glasses. – *J. Hafner:* Theory of the Structure, Stability, and Dynamics of Simple-Metal Glasses. – *P.J. Cote, L.V. Meisel:* Electrical Transport in Glassy Metals. – *J.L. Black:* Low-Energy Excitations in Metallic Glasses. – *W.L. Johnson:* Superconductivity in Metallic Glasses. – *U. Köster, U. Herold:* Crystallization of Metallic Glasses.

Springer-Verlag
Berlin Heidelberg
GmbH

Glassy metals are a new type of material showing various exceptional and unexpected properties. The fascination of this field of solid state physics lies in the fact that it spans the whole range between basic problems of disordered matter and direct application in technology. In the present volume specialists present various aspects of the recent development in this domain of research, such as structural studies, transport properties and crystallization processes. This book is devoted uniquely to metallic glasses, summarizing the very rapid and still expanding development of the last decade.